GLOSSARY OF
CHEMICAL
TERMS

Periodic Table of the Elements

NOBLE GASES

IA	IIA	IIIB	IVB	VB	VIB	VIIB	VIII			IB	IIB	IIIA	IVA	VA	VIA	VIIA	NOBLE GASES
1 H 1.00797 ±0.00001																	2 He 4.0026 ±0.00005
3 Li 6.939 ±0.0005	4 Be 9.0122 ±0.00005											5 B 10.811 ±0.003	6 C 12.01115 ±0.00005	7 N 14.0067 ±0.00005	8 O 15.9994 ±0.0001	9 F 18.9984 ±0.00005	10 Ne 20.183 ±0.005
11 Na 22.9898 ±0.00005	12 Mg 24.312 ±0.005											13 Al 26.9815 ±0.0005	14 Si 28.086 ±0.001	15 P 30.9738 ±0.00005	16 S 32.064 ±0.003	17 Cl 35.453 ±0.001	18 Ar 39.948 ±0.005
19 K 39.102 ±0.0005	20 Ca 40.08 ±0.005	21 Sc 44.956 ±0.0005	22 Ti 47.90 ±0.005	23 V 50.942 ±0.0005	24 Cr 51.996 ±0.001	25 Mn 54.9380 ±0.00005	26 Fe 55.847 ±0.003	27 Co 58.9332 ±0.00005	28 Ni 58.71 ±0.005	29 Cu 63.54 ±0.005	30 Zn 65.37 ±0.005	31 Ga 69.72 ±0.005	32 Ge 72.59 ±0.005	33 As 74.9216 ±0.00005	34 Se 78.96 ±0.005	35 Br 79.909 ±0.002	36 Kr 83.80 ±0.005
37 Rb 85.47 ±0.05	38 Sr 87.62 ±0.005	39 Y 88.905 ±0.0005	40 Zr 91.22 ±0.005	41 Nb 92.906 ±0.0005	42 Mo 95.94 ±0.005	43 Tc (99)	44 Ru 101.07 ±0.005	45 Rh 102.905 ±0.0005	46 Pd 106.4 ±0.05	47 Ag 107.870 ±0.003	48 Cd 112.40 ±0.005	49 In 114.82 ±0.005	50 Sn 118.69 ±0.005	51 Sb 121.75 ±0.005	52 Te 127.60 ±0.005	53 I 126.9044 ±0.00005	54 Xe 131.30 ±0.005
55 Cs 132.905 ±0.0005	56 Ba 137.34 ±0.005	57 *La 138.91 ±0.005	72 Hf 178.49 ±0.005	73 Ta 180.948 ±0.0005	74 W 183.85 ±0.005	75 Re 186.2 ±0.05	76 Os 190.2 ±0.05	77 Ir 192.2 ±0.05	78 Pt 195.09 ±0.005	79 Au 196.967 ±0.0005	80 Hg 200.59 ±0.005	81 Tl 204.37 ±0.005	82 Pb 207.19 ±0.005	83 Bi 208.980 ±0.0005	84 Po (210)	85 At (210)	86 Rn (222)
87 Fr (223)	88 Ra (226)	89 †Ac (227)	104 Rf	105 Ha	106												

*Lanthanum Series

58 Ce 140.12 ±0.005	59 Pr 140.907 ±0.0005	60 Nd 144.24 ±0.005	61 Pm (147)	62 Sm 150.35 ±0.005	63 Eu 151.96 ±0.005	64 Gd 157.25 ±0.005	65 Tb 158.924 ±0.0005	66 Dy 162.50 ±0.005	67 Ho 164.930 ±0.0005	68 Er 167.26 ±0.005	69 Tm 168.934 ±0.0005	70 Yb 173.04 ±0.005	71 Lu 174.97 ±0.005

†Actinium Series

90 Th 232.038 ±0.0005	91 Pa (231)	92 U 238.03 ±0.005	93 Np (237)	94 Pu (242)	95 Am (243)	96 Cm (247)	97 Bk (247)	98 Cf (249)	99 Es (254)	100 Fm (253)	101 Md (256)	102 No (253)	103 Lw (257)

GLOSSARY OF
CHEMICAL
TERMS

SECOND EDITION

Clifford A. Hampel
Consulting Chemical Engineer

AND

Gessner G. Hawley
Editor, CONDENSED CHEMICAL DICTIONARY

 VAN NOSTRAND REINHOLD COMPANY
New York

Copyright © 1982 by Van Nostrand Reinhold Company Inc.

Library of Congress Catalog Card Number: 81-11482
ISBN: 0-442-23871-1

Manufactured in the United States of America

Published by Van Nostrand Reinhold Company Inc.
135 West 50th Street
New York, New York 10020

Van Nostrand Reinhold Company Limited
Molly Millars Lane
Wokingham, Berkshire RG11 2PY, England

Van Nostrand Reinhold
480 Latrobe Street
Melbourne, Victoria 3000, Australia

Macmillan of Canada
Division of Gage Publishing Limited
164 Commander Boulevard
Agincourt, Ontario M1S 3C7, Canada

15   14   13   12   11   10   9   8   7   6   5   4   3   2

Library of Congress Cataloging in Publication Data

Hampel, Clifford A.
  Glossary of chemical terms.

  1. Chemistry—Dictionaries. I. Hawley,
Gessner Goodrich, 1905-   II. Title.
[DNLM: 1. Dictionaries, Chemical. QD 5 H229g]
QD5.H34   1981   540'.3'21   81-11482
ISBN 0-442-23871-1   AACR2

PREFACE TO FIRST EDITION

Dr. Samuel Johnson, who compiled the first *Dictionary of the English Language,* once remarked that people need less to be informed than to be reminded. This generalization must have been a source of comfort and hope to all who have undertaken to present definitions in any area of human knowledge. It applies with particular force to the authors of this *Glossary,* whose purpose may best be explained by two additional definitions.

The first is that of the word *definition* itself. Primarily, it involves the setting of limits or boundaries to the meaning of terms and expressions. In chemistry, as in other fields, this is far more easily said than done, for there is no predetermined way in which such limits can be established. What may be quite satisfactory to one person may be only the beginning of an extended area of further knowledge to another. The inherently tricky nature of words is also a factor: many words have two or more quite different meanings even within the framework of a single major subject, and distinctions must be drawn carefully without obscuring their underlying relationship.

A useful definition should certainly tell *what* a substance or process or phenomenon is, with an appropriate example or two; but to explain *why* it is often leads one into ever more profound depths, the ultimate reason seeming to retreat in an endless succession of *why's.* Thus, it is necessary to set limits not only to the terms themselves but also to the informational background of those for whom the definitions are intended. Since definitions that a beginning chemistry student would find illuminating would be of little value to a professional chemist, it is essential that the definer have in mind the level of knowledge and experience of his expected audience.

The second definition concerns the word *glossary.* It is a group of definitions of *selected* terms in a field of knowledge, as opposed to *dictionary*—a much more pretentious and scholarly compendium, presenting intensive coverage of the terminology of a subject area.

This Glossary is intended for those who have had minor exposure to chemistry or who require a source of review information. Superficial though it may be by some criteria, it is the only volume of chemical definitions that serves this need. The several chemical dictionaries now existing are impressive and highly useful volumes which have established well-deserved reputations; they differ among themselves in respect to emphasis and treatment and are designed primarily for professional chemists, engineers, and industrial technologists. They are of little practical use to the introductory student or to those without considerable background in chemistry.

The emphasis in this Glossary has been placed on the following:

(a)   All major chemical classifications, e.g., aldehyde, alcohol, amine, sugar, protein, carbohydrate, gum, resin, wax, etc.

(b)   All important functional terms, e.g., catalyst, plasticizer, solvent, surface-active agent, antioxidant, etc.

(c)   Basic phenomena and processes, e.g., oxidation, photosynthesis, polymerization, optical rotation, distillation, filtration, vapor pressure, surface tension, etc.

(d)   All the chemical elements, both natural and man-made.

(e)   The most important compounds, e.g., ammonia, ethyl alcohol, acetone, carbon dioxide, acetic acid, etc. (the number of these has been purposely restricted).

(f)   General terms, e.g., acid, base, indicator, pH, bond, intermediate, etc.

(g)   Biographies of outstanding past contributors to the science of chemistry.

v

(h)   Common prefixes and suffixes, as well as a number of terms used in the argot of the chemical industries and miscellaneous terminology that is widely used but seldom explained.

(i)   Derivations and pronunciation where pertinent.

(j)   Abundant cross-references to call the user's attention to closely related topics.

Those who will find this a useful reference source include students in introductory or intermediate level courses in any of the numerous subdivisions of chemistry; professional scientists, engineers, and technologists in related fields, whose major interest is not chemistry; and management personnel in a wide range of industrial areas. It should be particularly valuable in the libraries of all high schools and colleges offering courses in general chemistry, as well as to public and industrial libraries.

CLIFFORD A. HAMPEL
GESSNER G. HAWLEY

PREFACE TO SECOND EDITION

The response to the authors' effort to compile an elementary dictionary of essential terms used in chemistry and the process industries has been sufficient to justify their belief in the need for such a book. This revised and enlarged edition includes recent developments in the energy field, such as thermonuclear research, synfuels, and solar energy; major achievements in recombinant DNA gene-splicing techniques; and expanded treatment of toxic materials and waste control. Entirely new features are chronological information on landmarks in the history of chemistry; background information on several important industries; a temperature conversion table; and a listing of miscellaneous equivalents (weights, volumes, lengths). A number of entries on topics of developing interest have also been added. The entire book has been thoroughly edited to ensure accuracy and ease of comprehension.

C.A.H.
G.G.H.

TEMPERATURE CONVERSION TABLE

$$°C = (°F - 32) \times ⅝ \text{ (or 0.55)} \qquad °F = °C \times ⅑ \text{ (or 1.8)} + 32$$

°F	°C	°F	°C	°F	°C	°F	°C
−40	−40	120	48.9	280	137.8	445	229.4
−35	−37.2	125	51.6	285	140.5	450	232.2
−30	−34.4	130	54.4	290	143.3	455	235.0
−25	−31.6	135	57.2	295	146.1	460	237.8
−20	−28.9	140	60.0	300	148.9	465	240.5
−15	−26.1	145	62.8	305	151.6	470	243.3
−10	−23.3	150	65.5	310	154.4	475	246.1
−5	−20.5	155	68.3	315	157.2	480	248.9
0	−17.78	160	71.1	320	160.0	485	251.6
+5	−15.0	165	73.9	325	162.8	490	254.4
10	−12.2	170	76.6	330	165.5	495	257.2
15	−9.4	175	79.4	335	168.3	500	260.0
20	−6.6	180	82.2	340	171.1	550	287.8
25	−3.9	185	85.0	345	173.9	600	315.5
30	−1.1	190	87.8	350	176.6	650	343.3
32	0	195	90.5	355	179.4	700	371.1
35	+1.6	200	93.3	360	182.2	750	398.9
40	4.4	205	96.1	365	185.0	800	426.6
45	7.2	210	98.9	370	187.8	850	454.4
50	10.0	212	100.0	375	190.5	900	482.2
55	12.8	215	101.6	380	193.3	950	510.0
60	15.5	220	104.4	385	196.1	1000	537.8
65	18.3	225	107.2	390	198.9	1100	593.3
70	21.1	230	110.0	395	201.6	1200	648.9
75	23.9	235	112.8	400	204.4	1300	704.4
80	26.6	240	115.5	405	207.2	1400	760.0
85	29.4	245	118.3	410	210.0	1500	815.5
90	32.2	250	121.1	415	212.8	1600	871.1
95	35.0	255	123.9	420	215.5	1700	926.6
100	37.8	260	126.6	425	218.3	1800	982.2
105	40.5	265	129.4	430	221.1	1900	1037.8
110	43.3	270	132.2	435	223.9	2000	1093.3
115	46.1	275	135.0	440	226.6		

MISCELLANEOUS EQUIVALENTS

1 kilogram (kg)	=	1000 grams = 2.20 lb
1 gram (g)	=	1/463.6 lb (i.e.., 1 lb = 463.6 g)
1 grain	=	0.064 gram (i.e., 1 g = 15.4 grains)
1 milligram (mg)	=	1 thousandth gram
1 microgram (μg)	=	1 millionth gram
1 nanogram (ng)	=	1 billionth gram
1 liter (l)	=	1.0567 quarts
1 milliliter (ml)*	=	1 thousandth liter
1 cubic centimeter (cc)*	=	1 thousandth liter
1 meter (m)	=	39.37 in.
1 centimeter (cm)	=	0.39 in. (i.e., 1 in. = 1.54 cm)
1 millimeter (mm)	=	0.039 in.
1 micron (micrometer) (μ)	=	1 millionth meter = 10,000 angstroms
1 millimicron (mμ)	=	1 thousandth micron = 10 angstroms
1 nanometer (nm)	=	1 billionth meter = 1 millimicron = 10 angstroms
1 angstrom (Å)	=	1 ten-thousandth micron
1 ml (1 cc) water at 4°C*	=	1 gram
1 gallon (gal) water	=	8.33 lb
1 cubic foot (ft^3) water	=	62.37 lb
1 atmosphere (atm) (at sea level and 0°C)	=	14.69 psi (760 mm Hg)

* A milliliter is only approximately equal to a cubic centimeter, but the difference is so miniscule that it is conventionally regarded as negligible; the two are used interchangeably except in the most precise calculations.

GLOSSARY OF
CHEMICAL
TERMS

A

A　(1) Abbreviation for absolute temperature scale.

(2) Abbreviation for angstrom, named after a Swedish physicist; Å and A.U. are also used.

(3) Unofficial symbol for the element argon; the approved symbol is Ar.

a-, an-.　A prefix meaning "not" or "without", as in asymmetric, acyclic, atactic, anhydrous, anaerobic.

abherent.　A substance used to prevent other materials from sticking together or from adhering to another substance. Dusting agents (talc, slate flour, starch) and liquid formulations (bentonite/water, soap/water) are widely used in the rubber, adhesives, and plastics industries; waxes and fats are used in the food industries. Coatings of fluorocarbon resins applied to cooking utensils are permanent abherents. *See also* dusting agent; antiblock agent.

abietic acid.　A carboxylic acid occurring in rosin, turpentine, and pine oil. *See also* oleoresin; rosin.

ablation.　This term, derived from the Latin "to carry away," denotes the act of absorbing and dissipating heat by a material at temperatures of 2800°C or more. Composites of epoxy or phenolic resins reinforced with glass or graphite fibers or metal whiskers have this unusual heat-transfer property. They are used to coat the surfaces of space capsules to protect them from the excessive heat due to atmospheric friction during reentry, and are often consumed in the process. The ablative coating effects cooling of the surface by convection and formation of gaseous decomposition products which prevent the heat from penetrating the interior.

abrasive.　Any substance used in particulate or bonded form to smooth, polish, or reduce the surface of another substance of lower hardness. Commonly used abrasives are alumina (emery), silica (sand), silicon carbide, rouge (iron oxide), and diamond dust. Abrasive wheels made of these materials compacted with heat-resistant binders such as water glass, frit, clay, or synthetic resins are widely used for grinding metals, sharpening tools, and buffing plastics. Abrasives vary in hardness from diamond (hardest), silicon carbide, and emery to feldspar and rouge (softest). *See also* Mohs scale.

absolute.　(1) The basic temperature scale used in scientific and engineering research, in which the theoretical zero point is −273.15°C (−459.72°F); also called the Kelvin scale. *See also* Kelvin.

(2) Relatively free from impurities, as absolute alcohol, which is 99% pure.

(3) Essential oils used in the perfume industry by first extracting flowers with solvents and then removing the solid (waxy) components.

absorption　(1) The action of a solid or liquid in taking up and retaining another substance uniformly throughout its internal structure. This action may be only mechanical, such as the absorption of a liquid by a solid (as by a sponge or a cellulosic material), the liquid being recovered by controlled pressure or heat. When the absorbed material is a gas, it may dissolve in a liquid to form a solution, or enter into a chemical reaction with it to form a new compound: i.e., carbon dioxide absorbed in caustic solution forms sodium carbonate. This process can be reversed and the absorbed material recovered by the opposite reaction, called desorption or stripping.

(2) In spectroscopy, this term refers to the ability of a substance to receive and retain certain wavelengths of radiant energy. Determination of the type and intensity of this radiation yields information about the nature of the absorbing substance and is the basis of modern methods of spectroscopic analysis. *See also* spectroscopy.

(3) In physical chemistry, this term denotes the ability of some elements to pick up or "capture" neutrons produced in nuclear reactors as a result of fission. This is due to the large capture cross section of their atoms, which is measured in units called barns; elements of particularly high neutron absorption capability are cadmium and boron.

ABS resin. A type of thermoplastic synthetic copolymer composed of various proportions of styrene, acrylonitrile, and butadiene; the term is an abbreviation of the three monomers involved (acrylonitrile, butadiene, and styrene). As these resins, and the plastics made from them, are extremely hard, their primary uses are for telephone handsets, shoe heels, automotive parts, and equipment housings. They do not burn readily and are reasonably resistant to chemicals, except for nitric and sulfuric acids, chlorinated hydrocarbons, and aldehydes. They can be metallized by either vacuum deposition or electroplating and are readily machined and fabricated. ABS resins are generally considered as engineering plastics because of their hardness and durability.

Ac **(1)** Symbol for the element actinium.

(2) Abbreviation for acetate group. *See* acetate.

accelerator. **(1)** Any compound that increases the rate of vulcanization of rubber, i.e., sulfur cross-linking. Up to about 1920, metallic oxides such as CaO and MgO were used. These had been known from the time of Goodyear and are still employed in low-quality products. The introduction of nitrogenous organic compounds (aniline derivatives and guanidines) greatly reduced vulcanization time and also gave a far better product. These were followed by still more active agents (thiazoles, dithiocarbamates, and thiuram sulfides), which permitted vulcanization times of five minutes or less; they also imparted antioxidant and heat-resistant properties. The dithiocarbamate group (called ultra-accelerators) is powerful enough to cause vulcanization at room temperature. *See also* vulcanization.

(2) In the photographic industry the term refers to any substance added to a developer to increase its rate of activity; among these are various alkaline agents that raise the pH; quaternary ammonium compounds are also used.

(3) An enclosed chamber, either circular or straight (linear), in which the speed of charged particles (protons, deuterons, ions, electrons) can be increased by application of an electric field; the circular type also requires a magnetic field. The circular devices are the cyclotron, the bevatron, the synchrotron, and the betatron. Linear accelerators utilize either a standing wave or a traveling wave to accelerate electrons, which can be made to approach the speed of light. The cyclotron was developed to provide a means of bombarding atomic nuclei with accelerated deuterons; it has been used to create synthetic elements (e.g., mendelevium). The other types are used for basic physical research. *See also* cyclotron; betatron.

acceptor. *See* donor.

acetaldehyde. A flammable, irritating, and poisonous liquid with the low boiling point of 20°C (68°F); its formula is CH_3CHO. It should be handled with the greatest care as it forms hazardous vapors at temperatures far below zero. Made by the oxidation of ethylene or by the dehydrogenation of ethyl alcohol, it is second only to formaldehyde in industrial importance. Its major use is in the production of acetic acid and its derivatives acetone and various acetates, and of numerous other organic chemicals. Minor uses are as a flavoring agent in food products and as a preservative and stiffening agent in proteinaceous materials such as leather. It polymerizes to form paraldehyde. Its IUPAC name is ethanal. *See also* aldehyde.

acetal resin. A group of stiff, crystalline, and extremely strong plastics made by polymerization of formaldehyde, giving straight-chain molecules containing more than 1500—CH_2O—units. Though classed as thermoplastic, they are useful up to temperatures of 93°C (200°F) for continuous service and 120°C (250°F) for intermittent service. Resistant to most solvents and dimensionally stable, acetal resins are more like metals than any other plastic, having excellent fatigue and abrasion resistance; thus, they belong to the group of engineering plastics. They are naturally white, but can be furnished in a variety of colors.

They are used as automotive components, oil pipelines, equipment housings, gears, bearings, and in similar mechanical applications. *See also* formaldehyde.

acetamide.   One of the most important amides, acetamide is a crystalline compound containing the $CONH_2$ group characteristic of organic amides, to which a methyl group is attached to give the formula CH_3CONH_2. It is made by reacting ammonium hydroxide with ethyl acetate. The crystals readily absorb water vapor and dissolve (i.e., are deliquescent); they melt at about 80°C (175°F). Acetamide has numerous uses in organic chemistry, e.g., in the stabilization of peroxides, in the manufacture of lacquers, as a solvent and surfactant, and in the synthesis of organic compounds. It is also a laboratory reagent. *See also* amide.

acetanilide.   *See* hydrogen peroxide; nitrile.

acetate.   A salt or ester of acetic acid (CH_3COOH) in which the terminal hydrogen atom of the acid is replaced either (1) by a metal or (2) by a hydrocarbon or other radical. The compounds so formed contain the acetate group CH_3COO. Reaction (2) is called esterification. In the case of cellulose, reaction with acetic anhydride forms cellulose acetate, the most important of all acetates, especially for synthetic fibers, films, and plastics; when the reaction is complete, three hydroxyl groups of the cellulose are esterified (cellulose triacetate). Other compounds of this type are vinyl acetate, from which polyvinyl acetate and polyvinyl alcohol are made, and ethyl acetate (a lacquer solvent). *See also* ester; cellulose.

acetic acid.   A relatively weak organic acid having the formula CH_3COOH; it is the first member of the fatty acid series and also belongs to the even larger group of carboxylic acids. It is made commercially by catalytic oxidation of butane (a petroleum gas) or of acetaldehyde; it also results from oxidation of alcohols formed by fermentation of carbohydrates and is the characteristic component of vinegar. Acetic acid is a tonnage chemical of broad application; among the industrial products made from it are acetic anhydride, numerous acetates, and many organic chemicals used in photography, drugs, and plastics; it is also used as a coagulant for rubber latex and as a solvent. The glacial grade, a common laboratory reagent, is 99.8% pure acetic acid, f.p. 16.6°C (62°F), b.p. 118°C (244°F).

Acetic anhydride ($(CH_3CO)_2O$, the most important derivative of acetic acid, is used in the manufacture of aspirin and of cellulose acetate and vinyl acetate plastics. Since both the acid and its anhydride are strong skin irritants, inflicting severe burns on contact, they should be handled with due caution. *See also* fatty acid; carboxylic acid; carboxyl.

acetic anhydride.   *See* acetic acid.

acetone.   The most important member of the ketone series, acetone (also called dimethylketone) is a liquid, b.p. 56.5°C (133°F), having a faint sweetish odor; its formula is CH_3COCH_3, comprised of the unsaturated carbonyl group ($C = O$) with a methyl group at both ends. It is quite a reactive compound, produced and used in tonnage quantities. Industrial production is by oxidation of either cumene, $C_6H_5CH(CH_3)_2$, or of a secondary alcohol (usually isopropyl), and by bacterial fermentation of carbohydrates. Miscible with water, as well as with organic liquids, its primary use is as a solvent in paints and lacquers, for cellulosic plastics, vinyl polymers, and most gums and waxes. A number of inorganic compounds are also soluble in acetone, for example, potassium iodide. Its high solvent power is applied in many laboratory procedures as a reagent and for extractive analyses in the rubber and plastics fields. The unsaturated compound ketene (CH_2CO) is obtained by heat treatment (pyrolysis) of acetone. Acetone is a fire risk, as it vaporizes readily at low temperature; it is relatively nontoxic. *See also* ketone.

acetophenone.   *See* ketone; Friedel-Crafts reaction.

acetyl.   The univalent group $CH_3C = O$, which occurs in acetic acid, acetic anhydride, and such compounds as acetyl chloride and acetylsalicylic acid. It contains one less oxygen atom than does the acetate group (CH_3COO). The process involving its introduction into other molecules is called acetylation. *See also* acetate.

acetylation.   *See* acetyl; acetylsalicylic acid.

acetylcholine.   A nitrogenous organic compound formed by the functioning of the nervous system in man and animals; it is a derivative of the important metabolic chemical choline, which is related to the vitamin B complex. When acetylcholine has completed its activity in conveying nerve impulses, it undergoes hydrolysis to acetic acid and choline; the enzyme that catalyzes this reaction is known as cholinesterase.

If left intact in the body, acetylcholine quickly overstimulates the nervous system so as to cause an imbalance in the muscle control of the breathing mechanism; this soon results in paralysis and death. Thus, it is an extremely toxic agent which has been used in an indirect sense in both military nerve gases and organophosphorus insecticides of the parathion type. These substances suppress the detoxifying action of cholinesterase, resulting in the accumulation of lethal concentrations of acetylcholine. *See also* cholinesterase inhibitor.

acetylene.   The first member of a series of aliphatic hydrocarbons (also called alkynes) having the generic formula C_nH_{2n-2}. Acetylene and its derivatives are unique with respect to the presence of a triple bond in the molecule, indicating a high degree of unsaturation. The parent compound is a gas with the formula $HC{\equiv}CH$. The suffix -yne indicates a triple-bonded structure, which is more stable to heat than are double or single bonds. Acetylene is made by the partial oxidation (combustion) of hydrocarbons at high temperature; one process occurs at about 1100°C (2000°F) and has a reaction time of 0.1 second. It can also be made by using an electric arc as a heat source. The original method involved reaction of calcium carbide with water: $CaC_2 + 2H_2O{\rightarrow}C_2H_2 + Ca(OH)_2$.

Acetylene is a reactive compound and requires careful handling, as it is flammable and explodes readily, especially under high pressure and when in contact with copper and silver. On heating it yields an electrically conductive type of carbon black; on vinylation it forms (with chlorine) the synthetic polymer neoprene and other vinyl polymers; and on reaction with hydrogen cyanide it gives acrylonitrile. Acetylene burns with an intense, hot flame and thus is used in welding and metal cutting. It can be polymerized to form polyacetylene and cycloöctatetraene, a ring compound having eight conjugated double bonds. The German chemist Reppe did much to develop the chemical technology of acetylene and its derivatives during World War II. *See also* polyacetylene.

acetylene black.   *See* acetylene; carbon black.

acetylide.   *See* explosive.

acetylsalicylic acid.   A white, water-soluble powder m.p. 143°C (289°F), made by an esterification reaction between salicylic acid and acetic anhydride. Generally known as aspirin, it is the most widely used pain-arresting drug. Its chemical formula is $CH_3COOC_6H_4COOH$. It should be protected from moisture, which will cause it to decompose by hydrolysis. It can bring about allergic reactions and may cause internal bleeding in susceptible individuals. While dosages of eight or ten standard 5-grain pills over a 24-hour period are not considered harmful, ingestion of 10 grams at once may cause death. It is classed as an analgesic and anti-inflammatory.

acicular.   Having the shape of a needle, one of the characteristic forms of crystals. Metal "whiskers" are of this shape, some being as long as 2 inches with a diameter of 10 microns.

acid.   A compound, either inorganic or organic, that (1) reacts with a metal to evolve hydrogen; (2) reacts with a base to form a salt; (3) dissociates in water solution to yield hydrogen (or hydronium) ions; (4) has a pH of less than 7.0; and (5) neutralizes bases or alkaline media by receiving a pair of electrons from the base so that a covalent bond is formed between the acid and the base (Lewis acid). A useful modification of the Lewis concept introduced in 1963 is the classification of acids (and bases) as either "hard" or "soft," depending on such properties as electronegativity, tendency to polarize, and oxidation state, these in turn relating to the availability of vacant orbitals. Hard acids are in a high oxidation state, and their outer electrons are relatively stable; they tend to associate with hard bases. Soft acids have the opposite properties.

All acids contain hydrogen. Inorganic acids (sometimes called mineral acids) are exemplified by sulfuric, nitric, hydrochloric, and phosphoric acids. Organic or carboxylic acids contain one or more carboxyl groups (COOH); simple types are formic and oxalic acids. Carboxylic acids include the broad subgroups of fatty acids and amino acids. Some acids, such as carbonic and acetic, are weak, forming few ions in solution, while others are very strong (hydrochloric, sulfuric, nitric) and ionize completely. Concentrated strong acids are extremely corrosive to tissue and should be handled with care. *See also* base; pH; carboxylic acid; fatty acid; amino acid; nucleophile.

acidulant.   An organic acid used as a multi-purpose food additive; among the more important functions of acidulants are prevention of rancidity, texture modification of doughs and baked goods, preservation of meats, enhancement and potentiation of flavors, and pH control. Citric, acetic, tartaric, and propionic acids are commonly used. Succinic anhydride is an excellent leavening and dehydrating agent.

acrylate. *See* acrylic resin.

acrylic acid. *See* monomer; acrylic resin.

acrylic resin. Any of a number of thermoplastic products resulting from the polymerization of acrylic or methacrylic acid esters or nitriles. The term acrylate is also used to identify these materials, e.g., methyl methacrylate. The monomers, such as acrylic acid, $H_2C = CHCOOH$, are quite unstable and tend to polymerize spontaneously, necessitating the use of an inhibitor during storage or shipment. Polyacrylic acid and polymethacrylic acid are water-soluble and may be components of soil conditioners, textile sizings, thickening agents, etc. Polymethylmethacrylate is a hard, transparent polymer with excellent light transmission properties; it may be used in the form of a water dispersion as a paint base or with organic solvents in the formulation of lacquers. Polybutylmethacrylate is used in finishes for textile and leather products. Acrylic resins can be cross-linked by copolymerization with low percentages of an anhydride. Polyacrylonitrile fibers are also a member of this group. *See also* acrylonitrile.

acrylonitrile. An unsaturated nitrile belonging to the family of monomers from which acrylic resins are prepared. It is derived from acetylene by a catalytic reaction with hydrogen cyanide, or from propylene, ammonia, and atmospheric oxygen passed over a fluid-bed catalyst. Its formula is $H_2C=CHCN$. Its major use is as a monomer in the production of nitrile rubber (a copolymer of butadiene and acrylonitrile) and of acrylic and modacrylic textile fibers; acrylic fibers contain 85% or more of acrylonitrile monomer units, whereas modacrylics have less than 85% but more than 35% acrylonitrile. The latter term has been adopted by the Federal Trade Commission. Acrylonitrile is also a component of ABS resins. As a fiber it has quite high tensile strength and can be blended with wool and other natural fibers. *See also* nitrile; acrylic resin; polyacrylonitrile.

actinide series. The sequence of radioactive elements immediately following radium in the Periodic Table; all are members of Group IIIB, starting with actinium (atomic number 89) and including all the remaining elements. Those be-

yond uranium do not occur in nature but are made synthetically; these are called transuranic. The officially approved name for this series is actinoid, but it is seldom used. All its members have properties similar to actinium; they are as follows:

actinium	berkelium
thorium	californium
protactinium	einsteinium
uranium	fermium
neptunium	mendelevium
plutonium	nobelium
americium	lawrencium
curium	elements 104, 105, 106

See also actinium.

actinium. An element.

Symbol	Ac	Atomic No.	89
State	solid	Atomic Wt.	227
Group	IIIB	Valence	3

Actinium, m.p.1050°C (1840°F), is a radioactive decay product of uranium-235; the 227 isotope occurs in nature, with a half-life of about 22 years. It has no stable form. It is obtained as a whitish metal, and forms compounds with the halogens and with oxygen. It has bone-seeking properties, and has limited use in tracer studies. Actinium is the first element of the actinide series. *See also* actinide series.

actinoid. *See* actinide series.

actinomycin. *See* antibiotic.

activated carbon. A highly porous form of carbon made by destructively distilling carbon-rich materials to eliminate the volatile portions, followed by high-temperature treatment of the carbon with steam or carbon dioxide. The product may have an internal surface area ranging from 420 to almost 1700 square meters per gram. This enables it to strongly adsorb molecules of gases and vapors or colloidally suspended particles from liquids. Nut shells are used as a carbon source for material of the smallest pore size, which is used for gas adsorption; coal, wood, and similar materials are the basis for carbons used for adsorption from liquids. Activated carbon (sometimes called activated charcoal) is used to deodorize gases, to reclaim solvents from the vapor phase, and to decolorize liquids; it also can effectively remove toxic contaminants from air, as in gas masks. *See also* surface area.

activated charcoal.   *See* activation (1); activated carbon.

activated sludge.   A biologically active sediment resulting from the repeated exposure of organic waste materials to atmospheric oxygen, permitting the growth of aerobic bacteria and other unicellular organisms that feed upon and destroy the pollutants in the waste. The activated sludge process is one of the major methods of treating municipal sewage and can also be applied to such industrial wastes as occur in canneries and other food-processing plants. It involves mechanically mixing air with the waste material for several hours to promote growth of bacteria, followed by a shorter settling period; this cycle is repeated until the purification is completed. About one pound of sludge is formed for every pound of organic waste digested. Phenolic wastes can also be purified in this way by the development of special types of bacteria which can utilize them as food. *See also* aerobic.

activation.   (1) Increasing the adsorptive capacity of a material, such as charcoal or alumina, by exposing it in a finely divided state to steam or carbon dioxide at high temperature; this treatment creates a high degree of porosity and thus increases the interior surface area.

(2) Supplying of air or oxygen continuously to sewage sludge to promote growth of aerobic bacteria which digest (decompose) the organic waste material.

(3) Increasing the activity of one substance by the addition of a second substance, usually in low concentration. Examples are the activating effect of fatty acids on thiazole accelerators, and of trace amounts of certain metals on phosphors. *See also* initiator.

activity.   This term has a number of specialized meanings in chemistry and physical chemistry, among which are the following:

(1) The tendency of a metal high in the electromotive series to replace another metal, lower in the series, from its compounds; for example, magnesium will replace copper in copper sulfate to form magnesium sulfate, releasing free metallic copper. The electromotive series is also called the activity series or the displacement series.

(2) The tendency of a metal, metallic oxide, or (rarely) an organic substance to accelerate the chemical combination of other substances; this

is known as catalytic activity. Some metals are active catalysts for given reactions and others are not. A fine state of subdivision usually increases the activity of a substance by exposing a maximum area of solid surface to the reacting gases.

(3) Surface-active agents (surfactants) reduce the surface tension of water and thus aid in emulsification, cleaning, stabilization, etc., at liquid-liquid or liquid-solid interfaces. The type of surface activity noted in (2) usually acts at solid-gas interfaces.

(4) In the physical chemistry of chemical systems, particularly solutions, the term activity relates to the apparent concentration of a substance in a reactive system. Activity involves the thermodynamics of equilibrium systems and the concentrations of their components. It indicates the deviation of a system from the ideal state. Best defined in mathematical terms, when the ratio of the activity, a, to the actual concentration, $[A]$, is designated as f, in the expression, $a = f[A]$, f is a factor denoting the extent to which the activity of a substance differs from the actual concentration. In this case, f is the *activity coefficient*.

(5) Other uses of the term are optical activity and radioactivity (which see).

activity coefficient.   The activity coefficient of a substance may be defined as the ratio of the effective contribution of the substance to a phenomenon to the actual contribution of the substance to the phenomenon. In the case of gases, the effective pressure of a gas is represented by the *fugacity, f,* and the actual pressure of the gas by P. The activity coefficient, γ, of the gas is then given by $\gamma = f/P$. In solutions the activity coefficient is equal to unity at infinite dilution of the solute and then deviates from unity as the concentration increases.

activity series.   *See* activity (1).

acyl.   A carbonyl group $(C = O)$ attached to an alkyl or aryl group. It may also be described as the group which remains after the removal of an OH group from an organic acid. Acyl groups occur in several series of organic compounds. Acetyl and benzoyl groups are specific types of acyl groups.

Adams, Roger   (1889-1971. American chemist, born in Boston; graduated from Harvard, where he taught chemistry for some years. After studying in Germany, he moved to the University of

Illinois in 1916, where he later became Chairman of the Dept. of Chemistry (1926-1954). His executive and creative ability made him an outstanding teacher, innovator and administrator. He strongly influenced the development of industrial chemical research in the U.S. Among his contributions were development of platinum hydrogenation catalysts, and structural determinations of chaulmoogric acid, gossypol, alkaloids, and marijuana. He held many important offices, including presidency of the ACS and AAAS, and was a recipient of the Priestley medal.

addition polymer. A macromolecule resulting from the catalytic combination of many identical monomer units; the combination is initiated by free radicals formed by decomposition of the catalyst in the presence of heat or light. For example, the ethylene monomer ($CH_2 = CH_2$) unites with others like it to give polyethylene, one of the most versatile synthetic resins; the same is true of such monomers as butadiene, styrene, and acrylonitrile, which are the basis of many synthetic elastomers. See also free radical; polymerization.

additive. A nonspecific term referring to any substance added to a food, a petroleum product, or a rubber or plastic mixture to impart a particular function or property. In most cases the proportion necessary to accomplish the desired results is 1 percent or less. Food additives include leavening agents, emulsifiers, antioxidants, flavors, and colorants; gasoline and lubricating-oil additives include gum inhibitors, knock suppressants, ignition improvers, and pour-point depressants; rubber and plastics additives include antioxidants, organic dyes, blowing agents, and accelerators. See also food additive; antioxidant.

adduct. See clathrate.

adenosine triphosphate. See phosphorus; Krebs cycle; nucleotide.

adhesive. A substance that has the property of attaching itself to the surfaces of other substances in such a way as to form a strong and relatively permanent bond between them. Many adhesives are natural substances of a proteinaceous or hydrocarbon nature, such as glues made from animal hides, blood, and tendons, fish skins, casein, soybeans, rubber, rosin, asphalt, etc. Others are semisynthetics (water glass, inorganic cements of various types). More important than these are synthetic polymers, both thermoplastic and thermosetting, which have a stronger bonding ability and are more resistant to deterioration than natural adhesives. Chief among them are epoxy and phenol-formaldehyde resins, widely used in plywood, reinforced plastics, and other composites; hot-melt adhesives (polyethylene, polyvinyl acetate, isobutylene); and various types of organic cements which are dispersions of an elastomer in a solvent. Another important group of synthetics is silicone resin cements. Adhesives have been introduced which give a heat-resistant bond; for example, an organic material (polybenzimidazole) is effective up to 260°C (500°F), and mixtures of silica and boric acid up to 1100°C (about 2000°F).

adiabatic. A term used in chemical engineering to characterize a process in which no heat is gained or lost as the conditions of operation are changed. For example, certain fixed-bed reactors used in petroleum refining operate adiabatically.

adipic acid. A carboxylic acid having the formula $(COOH)_2(CH_2)_4$, m.p. 152°C (306°F), made by catalytic oxidation of cyclohexane with air, followed by oxidation of the resulting mixture of cyclohexanol and cyclohexanone with nitric acid. It is also obtained by hydrolysis of adiponitrile, $NC(CH_2)_4CN$. Adipic acid occurs naturally in beets. It has some minor uses in the food industry (baking powder and as an acidulant in fruit essences) and in the manufacture of plasticizers. Its major use is in the synthesis of nylon 66 (the original nylon developed in 1933) by a condensation reaction with hexamethylenediamine, $NH_2(CH_2)_6NH_2$. See also nylon; Carothers.

adiponitrile. See adipic acid; nitrile.

adrenaline. A hormone produced by a gland located near the kidney; it has a benzenoid structure with the formula $C_9H_{13}O_3N$ (also called epinephrine). It is obtained by extraction from the adrenal glands of cattle, and is also made synthetically. Its effect on body metabolism is pronounced, causing an increase in blood pressure and rate of heart beat. Under normal conditions its rate of release into the system is constant, but emotional stresses such as fear or anger rapidly increase the output and result in temporarily heightened metabolic activity. Adrenaline is a pharmaceutical product which should be administered only by a physician. When spelled without the final *e*, the word is a registered trademark.

adsorption.   Attachment of the molecules of a gas or liquid to the surface of another substance (usually a solid); these molecules form a closely adherent film or layer held in place by electrostatic forces that are considerably weaker than chemical bonds. The finer the particle size of the solid, or the greater its porosity, the more efficient an adsorbent it will be because of the large surface area thus provided. Adsorption occurs on activated carbon, silica gel, and similar materials especially prepared for this purpose. It takes place only at the surface, which includes all openings, capillaries, cracks, depressions, and other types of irregularities. Adsorption is of primary importance in catalytic mechanisms, but should be distinguished from chemisorption, which involves much stronger attractive forces. *See also* chemisorption; catalyst; activated carbon.

aerobic.   Requiring the presence of air or oxygen to live, grow, and reproduce; the term is applied exclusively to micro-organisms of various types, especially those used in the activated sludge process for industrial and municipal waste treatment. Its opposite, anaerobic, refers to an organism that can live without air or oxygen, or to a digestion process from which air is excluded. *See also* activated sludge; bacteria.

aerosol.   Extremely small particles of a liquid or a solid which are suspended in air or other gas. If the particles are too small to settle out by gravity, the aerosol is permanent, at least until some force is introduced to bring about coalescence or precipitation. Colloidal suspensions of coal dust or other particles (smoke) and of fine water droplets (fog) are instances of aerosols that are undesirable. Aerosols may serve a useful purpose, as in the case of finely divided sprays of perfume, sanitizing agents, and pesticides. These are released by a propellant gas such as carbon dioxide, which expels the liquid through tiny orifices in special nozzles. A common method of eliminating aerosols in industrial effluent gases is precipitation by electrical means, as in the Cottrell process, invented by the American scientist Frederick G. Cottrell (1877-1948). *See also* dust; precipitate.

affinity.   The degree to which one substance is likely to react with another; the tendency of two substances to combine. An example is the read-

iness with which hemoglobin unites with carbon monoxide (over 200 times greater than with oxygen). The paraffin series of saturated hydrocarbons derives its name from the fact that these compounds have less affinity for other substances than do unsaturated hydrocarbons.

Ag   Symbol for the element silver, derived from the Latin word *argentum*. This form is sometimes used in naming silver compounds, i.e., argentic and argentous.

agar.   A starch-like natural polymer obtained from red algae, a form of seaweed. It will absorb as much as 20 times its weight of water, forming a strong gel at about 40°C (100°F) which melts at about 90°C (194°F). It is typical of the class of materials known as phycocolloids. Because of its strongly hydrophilic (water-holding) ability, it is used in the food and ice cream industries as a thickening and gelling agent, as a laxative, and in photographic emulsions. It is also used as a medium for bacteriological cultures. The gelling property results from the trapping of water within the framework formed by the long-chain polysaccharide (carbohydrate) macromolecules. As little as 1% of agar dissolved in hot water will set on cooling as the slowly diminishing rotary motion of the molecules causes them to become entangled and thus enclose water within the interstices. This is sometimes called the "brush-heap" theory of gel formation, which applies to other phycocolloids (carrageenan, algin) as well as to pectins. *See also* phycocolloid; polysaccharide.

agglomeration.   Combination or aggregation of small particles suspended in a liquid into clusters of approximately spherical shape. The food industry uses this term in the sense of increasing the particle size of powdered food products. Because such powders tend to be hydrophobic due to the high surface tension of water, agglomeration causes them to be more readily dispersible in water. This operation is known as "instantizing." The agglomerates have varying degrees of voids (open spaces), and are loosely bound, foamlike structures. They are formed by mechanical means in chamber spray dryers, tubes, or fluidized beds, usually in the presence of moisture. *See also* aggregation.

agglutination.   A term used by biochemists and immunologists to refer to a form of aggregation occurring in antigen-antibody reactions in which proteinaceous particles, such as bacteria, blood cells, etc., form clumps or clusters. The action is quite specific to the proteins involved. *See also* aggregation.

aggregate.   A miscellaneous assortment of gravel, pebbles, cinder, etc., used as a filler for concrete, paving asphalt, and in greenhouse benches for nutrient plant culture.

aggregation.   A general term describing the combination or coalescence of particles dispersed in a suspension into clusters or clots, which either precipitate or are removed by filtration. Such particles may be macromolecules of proteins, soaps, or other colloids (aggregates of which are called micelles), or larger particles, such as red cells in blood, rubber hydrocarbon in latex, and fat particles in milk. There are several types of aggregation: (1) agglutination, which is chiefly the cohesion of bacterial cells, as in antigen-antibody reactions; (2) coagulation, the irreversible clotting of proteins, etc., brought about by electrolytes, enzymes, or heat; (3) flocculation, the formation of clumps of dispersed colloidal solids such as carbon black in rubber mixes; and (4) agglomeration. *See also* coagulation; flocculation; agglomeration.

agitation.   The act of exerting force on a solid or liquid mixture in order to bring about intimate mixing and uniform dispersion of the components. This is usually effected by stirring, grinding, shaking, kneading, or "plowing" with any of a number of mechanical devices; it can also be done (in the case of an aerosol) by means of an air stream, as in the floatation of clays.

air.   The mixture of gases that envelops the earth's crust; the atmosphere. Its composition varies with the altitude at which a sample is taken. At sea level the composition is (in percent by volume): nitrogen 78, oxygen 20.9, argon 0.94, and carbon dioxide 0.033. There are also trace amounts of other gases as well as varying proportions of water vapor. A low percentage of dissolved air is present in ocean water. The density of dry air is 1.29 grams per liter at STP. Air can be liquefied by compression and cooling. Nitrogen, oxygen, and argon for industrial use are separated from air by distillation. Most of the oxygen is produced by photosynthesis, and the carbon dioxide by combustion and degradation of organic matter.

The nitrogen of air is not available to animals and man metabolically, but is used by plants to synthesize proteins. A method of using it to synthesize ammonia was discovered in Germany about 1912 by Haber. Air is used as such in the manufacture of blown asphalt and castor oil and in whipped food products. It is also used as a means of classifying particulate solids such as clays (air floatation). *See also* atmosphere; nitrogen fixation.

air pollution   (atmospheric pollution). Introduction into the atmosphere of substances that are not normally present therein and that have a harmful effect on man, animals, or plant life. Important among these are sulfur dioxide, which forms sulfuric acid on contact with water vapor; automotive exhaust emission products (carbon monoxide, lead compounds, polynuclear hydrocarbons, nitrogen oxides); toxic metal dusts from smelters, coal smoke, and other particulates; formaldehyde and acrolein; and radioactive emanations. Control of these is exercised by the Environmental Protection Agency. As conventionally used, the term does not apply to interior air spaces such as industrial workrooms. Tolerances for the latter are established by the American Conference of Governmental and Industrial Hygienists (ACGIH) and enforced by the Occupational Safety and Health Administration (OSHA).

Al   Symbol for aluminum. In erroneous typography it sometimes appears as AL.

alanine.   *See* pantothenic acid; amino acid.

albumin.   A mixture of water-soluble globular proteins occurring in the tissues and fluids of the body. It will crystallize on drying but can be reconstituted with water. It is heat-sensitive and coagulates irreversibly at about 80°C (180°F). Albumin is the chief component of egg white in the form of ovalbumin and conalbumin; it also occurs in lower percentages in egg yolk, milk, and blood serum. It is produced for commercial use in food products and for photographic emulsions. Its molecular weight is about 50,000. The term is derived from the Latin word for white. *See also* protein.

alcohol.   A broad class of organic compounds which contain one or more hydrocarbon groups and one or more hydroxyl groups. Alcohols may be aliphatic (straight- or branched-chain) or ring structures (alicyclic, aromatic, heterocyclic, and polycyclic). The presence of a hydroxyl group is indicated by the suffix -ol, used in IUPAC nomenclature (methanol, ethanol, etc.). Alcohols can be thought of as being derived from hydrocarbons in which one or more hydrogen atoms have been replaced by a hydroxyl group, or from water in which a hydrocarbon group has replaced a hydrogen atom.

Alcohols may be mono-, di-, tri-, or polyhydric, depending upon the number of hydroxyl groups per molecule, each attached to a different carbon atom. Examples are, respectively, ethyl alcohol, C_2H_5OH; glycol, CH_2OHCH_2OH; glycerol, $C_3H_5(OH)_3$; and pentaerythritol, $C(CH_2OH)_4$. Their physical properties vary with molecular weight; those having from one to four carbons are low-boiling, mobile liquids; from five to eleven carbons, the viscosity and boiling points increase; above eleven carbons, alcohols are waxy solids. Alcohols exhibit both acidic and basic behavior. Several are toxic, notably methyl and allyl, and all are combustible.

Though many alcohols occur widely in plants and animals, most are made synthetically by hydrolysis of esters, by oxidation of hydrocarbons, by condensation of hydrogen or of an olefin with carbon monoxide, or by fermentation. Among their industrially useful properties are their chemical reactivity, which makes them important intermediates; their solvent action on organic materials; and the polar nature of their molecules, which enables long-chain alcohols to act as detergents and emulsifiers. *See also* monohydric alcohol; glycol; glycerol; sterol.

alcoholysis.   A chemical reaction between an alcohol and another organic compound, analogous to hydrolysis. The alcohol molecule decomposes to form a new compound with the reacting substance, the other reaction product being water. Both hydrolysis and alcoholysis may be considered as forms of solvolysis. *See also* solvolysis.

aldehyde.   Any of a series of unsaturated organic compounds, both aliphatic and aromatic, containing a carbonyl group, in which one of the bonds of the carbon atom is attached to hydrogen, C—H; in conventional straight-line formulas this is usually written CHO, which is characteristic of aldehydes, e.g., formaldehyde is HCHO. Aldehydes are closely related to ketones, but are, in general, more chemically reactive. They are formed from alcohols by dehydrogenation, and the term aldehyde is derived from this reaction, i.e., *al*cohol + *dehyd*rogenation. For example, formaldehyde (HCHO) is

formed by removing two hydrogen atoms from a molecule of methyl alcohol (CH_3OH). The aldehydes are named for the acids they form when oxidized, e.g., acetaldehyde yields acetic acid, butyraldehyde yields butyric acid, etc.

Aldehydes are important synthesizing agents; they enter into condensation reactions, and are readily oxidized to organic acids. They also prevent deterioration of protein-rich substances, and thus are used as preservatives in the food and leather industries and for biological specimens. Among the more important aldehydes are formaldehyde, acetaldehyde, and benzaldehyde. *See also* ketone; aldol condensation; condensation.

aldol condensation.   A reaction between two aldehyde molecules or between one aldehyde and one ketone molecule in which the position of one of the hydrogen atoms is changed so as to form a single molecule having one hydroxyl and one carbonyl group. Since such a molecule is partly an alcohol (OH group) and partly an aldehyde (CO group) and represents a union of two smaller molecules, the reaction is called an ald-ol condensation. It can be repeated to form molecules of increasing molecular weight. The condensation of formaldehyde to sugars in plants, which on repetition builds up the more complex carbohydrate structures such as starch and cellulose, is thought to be a reaction of this type. It occurs most effectively in an alkaline medium. *See also* condensation.

aldose.   Any of a group of sugars whose molecule contains an aldehyde group and one or more alcohol groups. An example is glyceraldehyde ($HOCH_2 \cdot CHOH \cdot CHO$), specifically called an aldotriose because it contains three carbon atoms. *See also* glyceraldehyde.

algae.   A division of the plant kingdom, whose members usually reproduce by single cells, both sexually and asexually. They range from microscopic single-celled forms, mostly aquatic but a few terrestrial, to the wholly aquatic seaweeds, among which some of the giant kelps exceed 60 feet in length. The algae are divided into green, brown, and red, according to their visual appearance, though all contain chlorophyll. The so-called bluegreen "algae" are more closely related to bacteria than to algae. In some locations algae are harvested as a water crop and find use in foods, pharmaceuticals, soil conditioners, and as a minor source of iodine. The blue-green "algae" are often a serious water contaminant. The widespread distribution of algae in oceans accounts for the fact that well over half of the oxygen-producing photosynthesis in the world is carried out by these organisms, though much of the oxygen liberated remains in solution and is recycled by aquatic organisms. Their growth is so stimulated by excess nutrients that they over-reproduce and thus contaminate shallow areas with their decomposition products. *See also* eutrophication.

alginic acid. A carbohydrate polymer of the polysaccharide type; its formula may be represented as $(C_6H_8O_6)_n$. Derived from brown algae,

pounds whose carbon atoms are arranged in straight or branched chains (sometimes called open chains). They may be saturated (single bonds only) or unsaturated (one or more double or triple bonds). Those having only single bonds are alkanes or paraffins; those with one double bond are alkenes or olefins; those with two double bonds are alkadienes or diolefins; and those with triple bonds are alkynes or acetylenes. The aliphatic series includes hydrocarbons, aldehydes, ketones, alcohols, organic acids, and carbohydrates which have no ring structure. Examples are as follows:

		typical	Formula	
common name	IUPAC name	compound	molecular	generic
paraffin	alkane	methane	CH_4	C_nH_{2n+2}
olefin	alkene	ethylene	$H_2C{=}CH_2$	C_nH_{2n}
diolefin	alkadiene	isoprene	$\overset{\overset{\textstyle CH_3}{\mid}}{H_2C{=}C}{-}CH{=}CH_2$	C_nH_{2n-2}
acetylene	alkyne	acetylene	$HC{\equiv}CH$	C_nH_{2n-2}

in which it is a component of the cell walls, it has many applications in the food industry as a thickener and protective colloid, and has even been used as a textile fiber. Numerous derivatives (calcium, sodium, and potassium alginates) are well known for their hydrophilic and gel-forming properties. *See also* polysaccharide; algae.

alicyclic. A major series of cyclic organic compounds whose molecules have a closed ring structure, but whose properties are much more like those of aliphatics than those of aromatics. The term is a combination of *ali*phatic and *cyclic*. They may be either saturated (cycloparaffins) or unsaturated (cycloölefins). They include a large number of alcohols, e.g., cyclohexanol, menthol, terpineol, cholesterol. The rings vary greatly in size and may consist of from three to twelve or more constituents. Their chemical properties are strongly affected by the size of the ring, those with fewest constituents being the most reactive. Cycloparaffins were formerly called naphthenes, a name that is misleading and no longer in general use. *See also* cyclic compound.

aliphatic. Derived from the Greek word for "fat," this term refers to a major series of organic com-

See also aromatic; alicyclic.

aliquot. A portion of a sample which is a specific fraction of the total sample. For test purposes master lots are often prepared; these are divided into equal portions, to which various test substances are added; each portion may be any desired fraction of the master lot, e.g., one-half, one-quarter, one-tenth, etc. These are aliquot parts. *See also* moiety.

alizarin. *See* anthracene; lake; dye.

alkadiene. *See* diolefin.

alkali. This term is normally used to refer to hydroxides and carbonates of the metals of Group IA of the Periodic Table, as well as to ammonium hydroxide. All are bases, and their aqueous solutions range in pH from 7 to 14. Strong alkalis are also called caustics, their solutions being corrosive to the skin and other tissues. Alkali deposits occur widely in the American desert, and are either surfacemined or recovered from salt lakes. Very large tonnages of sodium carbonate in the form of trona, $Na_2CO_3 \cdot NaHCO_3 \cdot 2H_2O$, are mined from underground deposits in Wyoming. Most alkali hydroxides are made by electrolysis of alkali chloride solutions, e.g., sodium hydroxide (NaOH) from sodium chloride (NaCl) and potassium hydroxide (KOH) from potassium chloride (KCl). *See also* base; alkali metal.

alkali metal. The univalent metals lithium, sodium, potassium, rubidium, cesium, and francium, all of which are in Group IA of the Periodic Table. As they are electropositive, they evolve heat when in contact with water; they are all basic in chemical nature and (with the exception of francium) are comparatively soft. Care should be taken in handling them, as they are strongly corrosive in the presence of moisture. For further details, see individual element.

alkaline earth. One of three divalent metallic elements of Group IIA of the Periodic Table, namely barium, calcium, and strontium. They were originally called earths because of their noncombustibility and insolubility in water. All form strong bases. Their properties are given under specific headings. *See also* earth.

alkaloid. Any of a group of complex heterocyclic compounds containing nitrogen which have strong physiological activity, are often toxic, and have basic chemical properties. (There are a few exceptions to this definition.) Alkaloids are derived from plants, though they occur in only about 100 varieties. Of the 2000 or more compounds classified as alkaloids, perhaps the best-known are nicotine from tobacco, opium and its derivatives from poppies, quinine from cinchona bark, strychnine from nux vomica, and caffeine from coffee; less familiar examples are cocaine, reserpine, digitalis, atropine, and belladonna. Alkaloids can legally be used as drugs and stimulants when administered by a physician, as they have specific pharmacological actions. In the pure state most are extremely poisonous. The berries and leaves of some plants (poinsettia) may be lethal if ingested.

alkane. IUPAC name for straight-chain saturated compounds of the paraffin series which have the generic formula C_nH_{2n+2}. *See also* paraffin.

alkene. IUPAC name for straight-chain unsaturated compounds of the olefin series, which have the generic formula C_nH_{2n}. *See also* olefin.

alkyd resin. A type of polyester resin used principally in exterior paints, baked enamels, marine protective coatings, etc., made by reacting glycerol or ethylene glycol with phthalic anhydride; a third substance must be present to act as a modifier, usually a drying oil such as linseed,

which varies in amount depending upon the consistency desired. The resulting product is thermosetting. The name is a combination of <u>al</u>cohol and ac<u>id</u>. *See also* polyester, allyl resin.

alkyl. A univalent group derived from paraffinic hydrocarbons (alkanes), but containing one less hydrogen atom than the corresponding hydrocarbon; for example, the methyl group (CH_3) has one less hydrogen than methane (CH_4); similarly, the ethyl group (C_2H_5), and so on through the homologous series. Such groups are often represented in formulas by the letter R and have the generic formula C_nH_{2n+1}. *See also* aryl.

alkylate. A reaction product containing an alkyl group or radical characteristic of aliphatic hydrocarbons. (Here the suffix -ate means "product of", as in conden*sate*.) The term alkylate has two distinct and somewhat different applications in chemistry. (1) In the petroleum field it designates a product obtained by catalytic reaction of an alkane (paraffin) with an alkene (olefin) to give a branched-chain paraffinic molecule used in high-octane gasoline, e.g., neohexane. (2) In detergent chemistry it refers to a group of compounds known as alkylbenzene sulfonates, made by reacting a long-chain alkane or alkene with benzene. Detergent alkylates made from straight-chain compounds are more biodegradable than those derived from branched-chain compounds and are called linear alkylates. These are now the more generally used because they are more readily decomposed by bacteria and thus do not seriously contribute to water pollution. Since the term "alkyl" applies to both straight- and branched-chain structures the word "linear" is used to indicate that the alkylate is a straight, unbranched chain containing at least ten carbon atoms. *See also* alkylation; detergent.

alkylation. Addition of an alkyl group to an organic compound by means of catalytic or thermal processes; it involves the synthesis of higher molecular weight compounds rather than the breaking down or cracking of large molecules. The catalysts most widely used are aluminum chloride, hydrogen fluoride, and sulfuric acid, at low temperatures and pressures. Thermal processing is carried out at temperatures of about 480°C (900°F) and pressures of several thousand pounds per square inch. The petroleum industry utilizes alkylation to make isooctane, neohexane, etc., by reacting olefins such as ethylene or propylene with a branched-chain paraffinic hydrocarbon. Ethylbenzene and cumene are also prepared by alkylating benzene with ethylene or propylene. *See also* alkylate; neohexane.

alkylbenzene sulfonate. This term usually refers to a group of synthetic detergents made by combining a long-chain alkyl group (C_{10} or more) with benzene, followed by sulfonation. The alkyl portion may be either straight or branched, depending upon its chemical origin; the straight-chain or linear, type (called LAS) is preferred because of its greater biodegradability. A typical member of this group is dodecylbenzene sulfonate. *See also* detergent; alkylate (2); dodecylbenzene.

alkyne. The series of unsaturated aliphatic hydrocarbons characterized by the presence of a triple bond; their generic formula is C_nH_{2n-2}. The most important is acetylene, $HC\equiv CH$, the first member of the series. *See also* acetylene.

allergen. Any substance that induces the formation of an antibody in the bloodstream of a susceptible individual; the allergen and antibody combine to form a histamine-like substance, which in turn weakens the blood vessels sufficiently to release fluids. This causes the commonly experienced edema, sneezing, and swelling well-known to hay fever sufferers. Only those who are predisposed to certain allergens are capable of forming such antibodies. Severe allergens can cause extremely serious conditions (anaphylaxis), which have been known to be fatal. Protein-containing substances are the most common allergens (wool, hair, pollen, etc.), but allergic reactions may be experienced by a few individuals from exposure to almost any substance. *See also* antibody; antihistamine.

allo- A combining form generally meaning "other," as in allotropic and polyallomer. More specifically, it refers to the more stable of two geometrical isomers.

allotrope. One of a number of forms in which an element can exist. Both the chemical and physical properties of allotropes display marked differences for a given element. While most allotropes are those of crystalline metals, there may also be gaseous and liquid allotropes. For example, molecular oxygen has two allotropic forms, diatomic oxygen (O_2) and ozone (O_3); liquid helium also has two such forms (helium I and helium II). Probably best-known are the allotropes of carbon (diamond, graphite, carbon black), phosphorus (white and red), and tin (gray and white). Uranium, manganese, plutonium, and many other metals have two or more variations of crystal structure. Such modifications are often designated by the Greek letters alpha, beta, gamma, etc.

alloy. A mixture or solution of metals, either solid or liquid, which may or may not include a nonmetal. In certain types the components are not completely miscible as liquids or tend to separate on solidification; others, in which the components are miscible, can be considered as solid solutions. The most common nonmetal alloying agent is carbon, which is present in steel from 0.02 to 1.5%; other elements (chromium, molybdenum, etc.) are added to steel in amounts from 10 to 20% to give specific properties such as hardness, corrosion resistance, etc. The properties of an alloy are usually quite different from those of its components. The two major classes of alloys are ferrous and non-ferrous, depending upon whether or not they contain iron.

Fusible alloys, designed to melt at low temperatures, are represented by solders lead/tin and by Wood's metal (bismuth/cadmium); refractory alloys, which melt only above 1650°C (3000°F), are solutions of such metals as tungsten, cobalt, nickel, and molybdenum. Amalgams are alloys of mercury with another metal and may be prepared in liquid form. An alloying element is used for a constructive and beneficial purpose rather than as an adulterant; thus, copper is required in sterling silver and in yellow gold to effect a necessary increase in strength and hardness. *See also* amalgam; refractory.

allyl alcohol. A monohydric unsaturated primary alcohol having the formula $CH_2=CHCH_2OH$. It is an extremely toxic and irritant liquid which is easily absorbed by the skin. It gives off combustible vapor at 32°C (90°F) and is thus a fire hazard. It boils at 96°C (205°F). Adequate precautions should be observed in operations in which it is involved. Its major use is in the manufacture of synthetic organic chemicals, especially plasticizers, glycerol, and pharmaceutical products. It is prepared by hydrolyzing allyl chloride with aqueous sodium hydroxide, or by catalytic isomerization of propylene oxide.

allyl resin. A thermosetting polymer, and a comparatively recent addition to the family of polyesters. The monomers result from reaction between allyl alcohol and an acid anhydride, such as phthalic or maleic. Diallyl phthalate is the most important of these, and the name also applies to a partially polymerized stage called a prepolymer, which can readily be softened and molded by heat and presure. Allyl resins form particularly high-grade plastics having excellent resistance to electricity, chemical attack, heat, and distortion. They are used in composites and special laminated structures for rocket cases, insulating varnishes, specialized electrical components, and for decorative paneling. Their superior properties are also utilized in the form of molding compounds composed of blends of monomer and prepolymer, which cross-link during processing. Fillers such as pulverized acrylic and glass fibers are usually added. These molding compounds have a wide variety of applications. *See also* polyester; alkyd resin.

alpha   (or α). This term or symbol has several different meanings, as follows:

(1) As a prefix in the names of organic compounds, it indicates the position in the molecule of an atom or group which has replaced another atom; in acids this position is immediately to the left of the terminal COOH group (where, for example, chlorine can replace one of the hydrogens); in naphthalene compounds it indicates attachment of the substituting atom to the carbon at the number 1 position; in other cyclic structures it is used to show that the substituting atom is attached to a side-chain rather than to one of the ring carbons.

(2) In metallurgy it denotes the major allotropic form of a metal or alloy

(3) As a symbol it denotes specific optical rotation.

(4) In radiation chemistry it denotes a form of radiation consisting of helium nuclei. *See also* beta; gamma; alpha particle.

alpha particle.   A radioactive decay emanation of relatively low penetrating power. Though often called a ''ray,'' an alpha particle is actually a helium nucleus, which has a positive charge of 2, corresponding to the two protons present, and a mass number of 4. It is emitted by uranium and radium, as well as by artificial radioisotopes of many other metals. An atom of radium gives off one alpha particle in decaying to radon. The curie (unit of radioactive change) is defined as the number of alpha particles emitted per second by one gram of radium (3.7×10^{10}). Alpha particles are used to bombard other nuclei in creating artificial radioactivity and are the least

damaging to human tissue of all radioactive emanations. *See also* radiation.

alum.   *See* aluminum sulfate; hydration.

alumina.   *See* aluminum oxide.

aluminum.   An element.

Symbol Al	Atomic No. 12
State Solid	Atomic Wt. 26.9815
Group IIIA	Valence 3

Aluminum, m.p. 658°C (1216°F), is a strongly electropositive metal. It is obtained in relatively pure form from bauxite (a clay-like material) by refining with alkalis to remove silica (Bayer process), followed by electrolytic reduction in which cryolite (a fluoride) is the electrolyte (Hall process). Aluminum is amphoteric in nature and is notable for its ability to quickly form a surface coating of aluminum oxide (alumina) which adheres tightly and permanently. This prevents further oxidation and makes possible a high degree of corrosion resistance, one of the metal's most valuable properties. This, together with its light weight, makes it an ideal construction material for both the building and the automotive industries. The protection can be increased by an electrolytic process called anodizing.

Aluminum forms compounds with halogens and with both inorganic and organic groups. Its electrical conductivity is about two-thirds that of copper, making it economical for power transmission. It is almost completely nontoxic, but in powder form it is flammable, with strong evolution of heat. Its major uses other than construction are quite numerous: as an alloying element with copper, manganese, magnesium, silicon, and zinc; in special paints; as tubes and containers for ointments and toilet goods; as packing foil, etc. Some of its compounds are important catalysts, e.g., aluminum chloride and triethylaluminum. *See also* bauxite; anode; Hall process.

aluminum alkyl.   *See* propylene; Ziegler catalyst; triethylaluminum.

aluminum chloride.   An inorganic compound having the formula $AlCl_3$; it is usually in the form of solid crystals which vaporize at about 190°C (370°F). When dissolved in water, it reacts energetically, evolving hydrogen chloride. Precaution should be observed in handling aluminum chloride in the presence of moisture. It is made by reacting chlorine with molten aluminum. One of its most important chemical properties is its ability to act as an electron acceptor; thus, it catalyzes condensation reactions between aliphatic and aromatic compounds involved in the production of ethylbenzene (an intermediate in styrene production), alkyl benzenes (detergents), and such pharmaceuticals as phenolphthalein. These are known as Friedel-Crafts reactions, after the names of their discoverers. The use of aluminum chloride as a catalyst was important in the manufacture of synthetic rubbers in the United States during World War II. It also is a catalyst of reactions involved in petroleum refining such as cracking, isomerization, and polymerization of aliphatic compounds. *See also* condensation; Friedel-Crafts reaction.

aluminum chlorohydrate ($AlCl_3 • H_2O$). A topical antiperspirant or body deodorant which acts by constricting the pores, thus inhibiting water loss. This and other aluminum salts are produced by the cosmetics industry, principally for axillary application. *See also* deodorant.

aluminum fluoride.   An inorganic compound having the formula AlF_3, sublimes at about 1200°C (2300°F) without melting; also in the form of the hydrate, $AlF_3 • 3\frac{1}{2}H_2O$; each is slightly soluble in water. The former is made by the action of hydrogen fluoride gas on alumina trihydrate, or by the reaction of hydrofluoric acid on a suspension of aluminum trihydrate, followed by calcining the AlF_3 hydrate formed, or by the reaction of fluosilicic acid and aluminum hydrate. Anhydrous AlF_3 is used as an additive to the molten electrolyte of the aluminum production cell to lower the melting point and increase the electrical conductivity. Both forms are used in ceramics. The mineral cryolite found in Greenland ($3NaF • AlF_3$ or Na_3AlF_6) contains aluminum fluoride. Cryolite is made synthetically from fusion of sodium fluoride and aluminum fluoride; it is the major constituent of the aluminum cell electrolyte in which alumina dissolves for the electrolytic production of aluminum. *See also* aluminum; hydrofluoric acid.

aluminum oxide.   Generally called alumina, m.p. 2000°C (3632°F), this compound has the formula Al_2O_3. It spontaneously forms a protective coating on the surface of metallic aluminum, thus providing corrosion resistance. Commercially, it is made from bauxite by reaction with sodium hydroxide. In addition to its use in pro-duction of aluminum metal, it is prepared in an activated form of high surface area which is used to remove moisture from gases and vapors. It is also used as a catalyst. Aluminum oxide is also known as corundum and emery, used as abrasives. Such gemstones as ruby and sapphire are essentially aluminum oxide; after appropriate "doping" with other metals they are used as laser crystals and in other solid state devices.

aluminum silicate.   *See* clay; kaolin; bentonite.

aluminum sulfate.   Often loosely called alum, aluminum sulfate is a water-soluble powder obtained by digesting clay or bauxite with sulfuric acid. It is also available as a solution. The anhydrous compound has the formula $Al_2(SO_4)_3$; it also exists in the hydrated form $Al_2(SO_4)_3 • 18H_2O$. Its major uses are in sizing and coating papers, as a flocculating agent in the purification of water and sewage, and as mordant in textile dyeing. Closely related materials are the double salts aluminum potassium sulfate (potash alum) and aluminum sodium sulfate (soda alum). The latter is used in baking powders, in leather tanning, and as a medical astringent.

amalgam.   A liquid or solid alloy or mixture of mercury with another component, which is usually a metal but may be a nonmetal. The component metal must be soluble in mercury, for example, lithium, sodium, potassium, and cesium. There is a strong tendency for these components to combine to form compounds under certain conditions. A well-known use of amalgams is for the repair of dental caries; in this case, the amalgam contains silver and tin, forming a paste which quickly sets to a hard mass. Industrial applications of amalgams containing other components are in silvering mirrors, in special techniques in metallurgical analysis, and as catalysts. One method of recovering gold from ores is to form a gold amalgam from which the gold is separated by distilling off the mercury. Silver is similarly recovered from ores by the amalgam process. *See also* alloy; mercury.

amber.   *See* resin (1); fossil.

ambergris.   *See* fixative.

ambient temperature.   *See* temperature.

American Association for the Advancement of Science   (AAAS). A society founded in 1848 whose interests extend into all areas of natural science. Its major publication, *Science,* was established by Thomas Edison in 1880. It also publishes many symposium volumes of papers presented at its annual meetings. It is deeply concerned with chemical education at the secondary level. It also sponsors the Gordon Research Conferences, originated in 1931 by Dr. Neil E. Gordon. These conferences have been expanded to more than 30 technical meetings attended by chemists from many countries; the presentations and discussions are at the postdoctoral level. Society headquarters are in Washington, D.C.

American Chemical Society. (ACS). The nationally chartered professional society for chemists in the United States. One of the largest scientific organizations in the world, it was started in 1876, and now has over 110,000 members. Its publications include the world-famous *Chemical Abstracts, Journal of the American Chemical Society,* a number of high-level journals devoted to major subdivisions of chemistry, as well as technical books, including the ACS Monograph Series. It has 31 Divisions and many regional Sections. The Priestley Medal, considered the highest award in chemistry, was established by the ACS in 1923. Its offices are in Washington, D.C.

American Institute of Chemical Engineers (AIChE). Founded in 1908, the AIChE is the largest society in the world devoted exclusively to the advancement and development of chemical engineering. It has over 50 local sections and many committees working in a wide range of activities. Its offices are in New York City.

American Institute of Chemists (AIC). Founded in 1923, the AIC is primarily concerned with chemists and chemical engineers as professional people, rather than with chemistry as a science. Special emphasis is placed on the scientific integrity of the individual and on a code of ethics adhered to by all its members. Its offices are in New York City.

americium. An element.

Symbol	Am	Atomic No.	95
State	Solid	Atomic Wt.	241(?)
Group	Acti-	Valences	3,4,5,6
III B	nide		
	Series		

Americium, m.p. 1000°C (1832°F), is an artificial transuranic element, discovered in 1944, which exists only as radioisotopes. It is a whitish crystalline metal of the bone-seeking type, emitting gamma radiation. The isotope with longest half-life is 243 (7700 years); next longest is 241 (430 years). It forms a large number of compounds with fluorine, oxygen, potassium, etc. It is used in radiography because of its gamma emission and as a source of curium isotopes.

amide. An organic or inorganic compound closely related to or derived from ammonia (NH_3). Inorganic amides are made by reaction of an alkali metal with ammonia to form, for example, sodium amide ($NaNH_2$), from which sodium cyanide and cyanamide are formed by reaction with carbon monoxide and carbon dioxide, respectively. Organic amides contain an acyl group (CO attached to an organic radical), which replaces one of the hydrogen atoms of ammonia; thus, their characteristic molecule is $R = CONH_2$, except in formamide, where the R stands for hydrogen. Amides are versatile chemicals which enter into many useful reactions. Among the most important amides is urea, NH_2CONH_2, (carbamide), used in fertilizers, formaldehyde plastics, and animal feeds. It is formed in the body as an oxidation product of protein metabolism. *See also* polyamide.

amine. An organic compound of nitrogen in which one or more alkyl or aryl groups are attached to the nitrogen atom. Amines are ammonia derivatives and may be primary, secondary, or tertiary depending upon the number of the hydrogen atoms of ammonia (NH_3) replaced by the organic group. If one hydrogen is replaced, the amine is primary; if two, it is secondary; and if three, it is tertiary, i.e., it contains no hydrogen attached to the nitrogen. Thus methylamine (CH_3NH_2) is primary, dimethylamine ($CH_3)_2HN$, is secondary, and trimethylamine, $(CH_3)_3N$, is tertiary. Industrially important amines are aniline, the basis for an entire system of organic dyes; triethanolamine, a useful solvent, detergent and plasticizer; and hexamethylenetetramine, a catalyst used in the manufacture of formaldehyde resins. Many types of amines are also used as rubber accelerators and as chemical intermediates.

amino acid. An organic acid whose molecule contains both a carboxyl group (COOH) and an amino group (NH_2) coupled with an alkyl, aryl, or heterocyclic structure. Since amino groups are basic and carboxyl groups acidic, amino acids are amphoteric; in water solutions they form dipolar ions, i.e., ions having a positive electric charge at one end and a negative charge at the other. All except glycine are optically active, i.e., they contain one or more asymmetric carbon atoms; most have the left-handed (L) configuration. Amino acids can be obtained by the hydrolysis of proteins (which occurs in the digestive process), though many are now made synthetically. Amino acids have been created in the laboratory by passing an electric discharge through a mixture of ammonia, methane, and water vapor. It is believed that a similar reaction may have accounted for the origin of amino acids, and hence of life on earth. Eight amino acids are regarded as essential in the diet, since they cannot be synthesized within the body; twelve others are considered nonessential (see list below). Over 80 have been identified in the free or uncombined state; of these, 22 are known to be constituents of proteins. Amino acids combine in chains of great complexity in forming proteins, and molecular biologists have established the critical importance of their arrangement and sequence in these chains. Some of the most brilliant elucidative work in the recent history of science has been done in this field.

Nonessential		Essential
alanine	glycine	isoleucine
arginine	histidine	leucine
aspartic acid	hydroxyproline	lysine
cysteine	serine	methionine
cystine	tyrosine	phenylalanine
glutamic		threonine
acid		tryptophan
		valine

These names are conveniently abbreviated in chemical notation by using the first three letters, with initial capital and no period, e.g., Gly, His, Try, Lys, etc. *See also* protein; asymmetry; enzyme

aminobenzoic acid. *See* antagonism.

aminophenol. An aromatic compound derived from nitrophenol or nitrobenzene; it has the three isomeric forms characteristic of many benzenoid structures, with the formula $C_6H_4NH_2OH$. At room temperature it is a crystalline water-soluble powder, the color varying from white (meta form) to saffron (para form); these colors change slowly to brown or violet after long standing in light. Aminophenol (also called hydroxyaniline) is a primary source of organic dyes, which are especially suitable for furs and other forms of keratin; the para isomer is also used as an intermediate in the manufacture of pharmaceuticals and as a developing agent in photography. *See also* aniline.

aminoplast resin. *See* amino resin.

amino resin. A type of thermosetting high polymer (also called aminoplast); it results from a catalyzed condensation reaction between formaldehyde and either urea (an amide) or melamine (an amine) to form monomers containing either a triazine ring (melamine) or an amide unit (urea). These are then polymerized in carefully prescribed and closely controlled procedures. The properties of the resulting resins vary with the extent and degree of polymerization. On curing with heat and pressure, their internal structure becomes cross-linked. Amino resins were among the earliest types of plastics and were notable because they were the first to permit the manufacture of white and colored plastics.

Their range of applications includes molding compounds, textile and paper impregnants, protective coatings, electrical and automotive components, plywood adhesives, foundry molding sands, etc.; they can also be molded into attractively decorated and heat-resistant dinnerware, lampshades, and other household products.

ammine. A type of coordination compound in which the ligand, or electron-donating group, is ammonia; it usually contains either cobalt or platinum as the metal ion, which is connected directly to the nitrogen by a coordinate linkage. This term should not be confused with the more common "amine," in which nitrogen is bonded to carbon. *See also* coordination compound; ligand.

ammonia. A suffocating, combustible gas, b.p. $-33.35°C$ ($-28°F$); formula NH_3. It can be recovered in low yield by destructive distillation (coking) of coal (about 6 lb per ton). The principal source is commercial synthesis, using air as the source of nitrogen, first developed in 1912 by the German chemist, Fritz Haber (1868-1934), Nobel prize 1918; since then many variations of the original process have been used. The essential step is the catalytic combination of 3 parts hydrogen and 1 part nitrogen to form ammonia at high pressures (1000 atm) and temperatures. The basic material used is synthesis gas obtained from natural gas or coal. Ammonia is an end product of animal metabolism by the decomposition of uric acid.

It is a high-tonnage chemical used in the manufacture of sodium carbonate and bicarbonate by the Solvay process, of nitric acid (by oxidation), hydrogen cyanide, acrylonitrile, ammonium nitrate, etc. It is a component of fertilizers, a refrigerant in large-scale fish and food preservation, a catalyst, a detergent and household cleaning agent, and a stabilizer of rubber latex. It has some medicinal use in smelling salts and aromatic spirits. Dissolved in water in concentrations up to 30%, it forms ammonium hydroxide (NH_4OH).

Proper inhalation protection is necessary in handling ammonia and many of its compounds. It is a choking gas, causing intense irritation of eyes and lungs. It ignites at 650°C (1200°F) at concentrations of 16 to 25% in air and may explode if confined. *See also* Haber-Bosch process; synthesis gas.

ammonia-soda process. *See* Solvay process.

ammonium chloride. *See* battery.

ammonium cyanate. *See* rearrangement; urea.

ammonium hydroxide. *See* ammonia.

ammonium nitrate. An inorganic compound, usually a dry powder, having the formula NH_4NO_3, made by reacting ammonia with nitric acid. It evolves N_2O rapidly when heated and may decompose violently at 210°C (410°F). Ammonium nitrate is a detonating explosive of low sensitivity but extremely high intensity and has been the cause of several disastrous explosions. It explodes more readily when contaminated with combustible materials and should be kept out of contact with them. It should be carefully protected against fire, kept away from heat from any source, and preferably stored in unconfined areas. If accidentally ignited, it should be removed from confinement immediately and flooded with water. Though it has minor uses as an explosive and as an oxidizer in solid rocket fuels, its highest-volume application is in fertilizers.

ammonium sulfamate. *See* herbicide.

amorphous. Lacking a crystal structure (literally, without shape); having random atomic arrangement without a three-dimensional structure. This property is characteristic of liquids, in which the molecules are not densely ordered but, rather, are comparatively free to move. A few solids also may have amorphous forms (carbon, for example) when their crystal orientation is sufficiently disordered to permit random movement of the atoms. When a solid melts, it changes from the crystalline to the amorphous state, in which it remains until recrystallization takes place. An exception to this behavior is shown by silica (SiO_2), which does not recrystallize because of its exceptionally high viscosity on cooling. Thus, glass retains the amorphous nature of a liquid even though it is physically rigid. Silica gel, a porous adsorbent, is also amorphous. Some high-polymer substances, such as cellulose, are partly amorphous and partly crystalline; others, such as polyethylene, rubber, etc., can be converted from amorphous to crystalline by cross-linking or by synthesis with special (stereospecific) catalysts. *See also* glass; liquid; vitreous.

amphetamine. A drug which acts on the central nervous system. It is derived from benzyl methyl ketone and has the formula $C_6H_5CH_2CH$-(NH_2)CH_3. Amphetamine is a liquid, b.p. 200°C (392°F) (dec.), but the sulfate is a white powder. It is flammable, quite toxic, and habit-forming and should be administered only by prescription. *See also* hallucinogen.

amphibole. *See* asbestos.

amphoteric. Having the property of reacting with either an acid or a base. Many oxides and salts have this ability (aluminum hydroxide, for example); it is also exhibited by some ion-exchange compounds and by complex substances such as proteins (polypeptides). Keratins such as wool, hair, and feathers are amphoteric, which is a distinct advantage in dyeing since they can accept both acidic and basic types of dyes.

amyl. The hydrocarbon group containing five carbon and eleven hydrogen atoms (C_5H_{11}), used in naming members of the fifth order of the homologous series of hydrocarbons. This is also, more logically, called pentyl, but amyl continues to be used for certain isomers of amyl alcohol and for amyl acetate, amylamine, etc. Some amyl alcohols are also called pentanols, and since there are eight possible isomers of C_5H_{11} compounds, the names are a source of some confusion.

amyl acetate. Unless otherwise indicated, this term refers to a mixture of several isomers and is derived from either amyl alcohol or fusel oil by reaction with acetic acid. The formula is $CH_3COOC_5H_{11}$. Amyl acetate is a flammable liquid having a penetrating odor suggestive of bananas and is sometimes called banana oil. Its chief uses are as a solvent in leather polishes and for fingernail lacquers, as a food additive in flavoring fruit essences, and in textile finishing operations. It has also been used in mine rescue work, as the odor is detectable through fissures and pores in rocky soils. It has especially good solvency for cellulose derivatives and thus is often a component of nitrocellulose lacquers.

amyl alcohol.   See amyl; pentane; fusel oil.

amyl hydride.   See pentane.

amylose.   See dextrose; starch; malt.

anaerobic.   See aerobic.

analysis.   (1) Determination of the properties of elements, the structure and constitution of compounds, and the composition of mixtures comprises the field of analytical chemistry. Early methods were limited to determination either by weighing (gravimetric) or by titration (volumetric). These so-called "wet" methods served well for industrial qualitative and quantitative analyses for many decades and are still in general use. But more precise and sophisticated techniques were required by advances in theoretical science. In 1815 the composition of the sun was determined by spectroscopic techniques; a century later, the English physicist, Henry G. J. Moseley (1887-1915), applied x-ray methods to ascertain the exact location of elements in the Periodic System. Since 1920 numerous instrumental methods have been developed for rapid and accurate chemical analysis; these have largely replaced the older techniques for all but routine work. Among them are mass spectrometry, infrared and Raman spectroscopy, nuclear magnetic resonance, chromatography of various types, colorimetry, x-ray diffraction, polarography, and use of radioactive tracers. See also infrared spectroscopy; chromatography; mass spectrometry; colorimetry.

(2) Study of the three-dimensional arrangements possible within a molecule, that is, its conformations, by movement of its constituent atoms or groups into various positions. Cyclohexane is a compound that has been studied extensively in this way. Such research is called conformational analysis. See also conformation.

anaphylaxis.   See allergen; penicillin.

-ane.   A suffix indicating a saturated aliphatic or alicyclic compound, as in the names methane, octane, cyclopropane, etc.

anesthetic.   A chemical compound used in surgery to eliminate pain. Anesthetics may be broadly classified as general (affecting the entire body), local (affecting a small area at the site of the operation), and basal (spinal), used for operations on the lower portion of the body. Nitrous oxide and ethyl ether had been used in the early 1840's by dentists, but it was not until 1846 that the use of ether was demonstrated by Thomas Morton in a major operation at Massachusetts General Hospital in Boston. Chloroform, ethyl chloride, and nitrous oxide came into use as general anesthetics a few years later. Cyclopropane and ethylene were introduced in the 1920's, and trichloroethylene followed in 1934. The extreme flammability of ether and the hydrocarbon anesthetics, as well as the difficulty of administration during extended operations, resulted in the development of safer and more effective agents, especially barbituric acid derivatives. Nitrous oxide is still commonly used in dentistry. Local anesthetics include cocaine (applied topically) and the procaine group e.g., lidocaine, (administered by injection).

ANFO.   See detonation.

angstrom.   A linear dimensional unit, equal to one-ten thousandth micron or one-100 millionth centimeter (one-250 millionth inch). It is used to measure the spacing of atoms in crystals, the size of atoms and molecules, and the length of light waves. The diameter of a single atom is from 0.64 Å (hydrogen) and 1.54 Å (carbon) up to 5 Å for heavier atoms; the length of a carbon-to-hydrogen bond is 1.09 Å. Molecules range from about 3 Å for hydrogen to over 1000 Å for polymers. The shortest wavelength of visible light is about 4000 Å. Ten thousand angstroms equal one micron. The abbreviation is either Å or A.U. Named for Anders Jons Ångstrom (1814-1874), a Swedish physicist and authority on optics.

anhydride.   A compound formed by removing two atoms of hydrogen and one atom of oxygen from another compound, e.g., sulfur trioxide (SO_3) is the anhydride of sulfuric acid (H_2SO_4), and carbon dioxide is the anhydride of carbonic acid. In organic chemistry, phthalic anhydride $C_6H_4(CO)_2O$, is formed from phthalic acid, $C_6H_4(CO_2H)_2$, by the elimination of one water molecule. Amino acids can be considered as anhydrides of proteins, from which they are formed by hydrolysis. "Anhydride" should not be confused with "anhydrous." *See also* anhydrous.

anhydrous. A term literally meaning "without water"; it describes a substance in which no water is present in the form of a hydrate or water of crystallization. For example, anhydrous aluminum chloride is the pure compound $AlCl_3$, whereas the hydrated form contains six molecules of water ($AlCl_3 \cdot 6H_2O$). In such cases, the water molecule retains its identity rather than separating into its constituents, as in hydrolysis. "Anhydrous" is not synonymous with "anhydride" and should not be confused with it. *See also* anhydride; hydration.

aniline. A liquid aromatic amine b.p. 184.4°C (364°F), having the formula $C_6H_5NH_2$. It is made by reacting nitrobenzene vapor with hydrogen with a catalyst, removing two oxygen atoms from the nitrobenzene to form aniline and water; or by reacting chlorobenzene with ammonia. Diphenylamine and phenol are by-products of this process. Aniline has two major uses, both introduced early in the century; (1) as the starting point for a large and important family of organic dyes and (2) as a basic organic accelerator and antioxidant for rubber. It is also employed as a chemical intermediate and in photographic chemicals. Aniline is toxic and is absorbed by the skin; it should be handled with care. *See also* amine.

anion. A negatively charged ion. When a compound ionizes in water solution, the molecule divides into two ions of opposite electrical charge, those carrying a negative charge being anions and those with a positive charge cations. When electrodes are placed in the solution, one of them becomes the positive terminal, or anode, and the other the negative terminal, or cathode. When current is passed through the solution, the negatively charged anions migrate to the anode, and the positively charged ions (cations) move toward the cathode. This phenomenon is the basis of electrowinning and electrorefining of metals, electroplating, electrophoresis, and similar processes. The electrodes also emit positively and negatively charged particles; in vacuum tubes the electrons emitted by the cathode are called cathode rays. *See also* ionization; electroplating; anode.

anisotropic. A term used in crystallography to indicate a major group of crystalline structures in which the optical properties (e.g., refractive index) vary with the direction of the axes, that is, with the direction of the vibrations of the crystal. In isotropic crystals there is no such variation, as light passes through them at the same rate in all directions. Cubic crystals are isotropic, but other structures are generally anisotropic. The word literally means "not turning the same way." *See also* crystal; dichroic.

annealing. (1) An operation by which the strength, hardness, and durability of glass is greatly increased due to elimination of the internal stresses caused by rapid cooling. It is carried out by careful stepwise lowering of the temperature from a predetermined high point maintained for as long as three days (for plate glass) to successively lower temperatures maintained for shorter periods. This treatment, which is carried out in a specially designed oven called a lehr, brings the components into a more stable equilibrium state and prevents the distortions and weaknesses caused by premature crystallization.

(2) The term is also applied to metals, especially steels, for a process which involves heating and cooling, and is usually applied to induce softening. The term also refers to treatments intended to alter mechanical or physical properties, produce a definite microstructure, or remove gases. *See also* glass.

anode. The positive electrode in an electrolytic cell to which negatively charged ions, molecules, or colloidal particles migrate under the influence of an electric current. The anode may be in the form of a plate made of graphite, as in the manufacture of chlorine by electrolysis of sodium chloride solution, or it may be of aluminum (anodizing process) or other metal. Electrochemical changes occur at the point of contact between the anode and the electrolyte which involve electron transfer between them; the nature of the electrolyte determines whether the anode gives up electrons or receives them. All anodic reactions are of an oxidizing nature.

In a primary cell (battery) the anode is the negative terminal. In electrochemical corrosion, the material corroded, attacked, or dissolved is generally the anode in the galvanic circuit. *See also* electrode; cathode.

anodizing. An electrolytic method of applying a strongly adherent film or coating of oxygen on the surface of a metal, especially aluminum, in addition to the oxide film already formed by natural processes. This is accomplished by passing an electric current through a bath containing, for example, sulfuric or chromic acid as electrolyte; the negatively charged oxygen ions formed are deposited on the positively charged aluminum anode, building up a film of controllable thickness which increases the corrosion resistance of the metal. Anodizing is thus a type of electrodeposition. Such films can absorb particles of added colorant, so that the metal is in effect dyed by this method. *See also* anode; electrodeposition.

antacid. A mildly alkaline substance suitable for internal use to neutralize excess stomach acidity; commonly used are aqueous solutions of sodium bicarbonate, magnesium hydroxide, and potassium sodium tartrate (Rochelle salt). Strongly effervescent types should be avoided. Some have a laxative effect if taken in large doses.

antagonism. A term used in biochemistry and chemotherapy to describe the inhibiting or retarding effect of one biologically active substance on another of closely similar molecular structure. An example is the antagonism that was found to exist (1940) between sulfanilamide and *p*-aminobenzoic acid, a growth-promoting substance. The molecules of the latter have almost the same structure as those of sulfanilamide, the only difference being in the chemical nature of one substituent group on the benzene ring; yet this difference enables the sulfa drug to replace the *p*-aminobenzoic acid in an enzyme complex and to thus deactivate the enzyme. Structurally similar agents, therefore, compete with or ''antagonize'' each other, resulting in the impairment of enzyme growth functions. Substances that act as antagonists are also called antimetabolites or growth inhibitors. Inhibiting effects of this kind are characteristic of the behavior of antibodies, antihistamines, etc. *See also* antibody; antihistamine.

anthracene. An aromatic hydrocarbon, m.p. 217°C (423°F), derived from coal tar; its molecule consists of three fused (attached) benzene rings:

the formula being $C_6H_4(CH)_2C_6H_4$. It is chemically related to naphthalene and phenanthrene.

One of its unusual properties is the blue fluorescence exhibited by its crystals; another is its ability to act as a semiconductor, a characteristic possessed by only a few organic materials. Anthracene is the source of alizarin and anthraquinone dyes, as well as of phenanthrene and carbazole. The coal-tar fraction from which it is obtained (anthracene oil) has minor use as a wood preservative. *See also* phenanthrene.

anthracite. *See* coal.

anthraquinone. An unsaturated tricyclic compound, sublimes 286°C (547°F), belonging to the general class of quinones. It can be made by oxidation of anthracene with chromic anhydride or by condensation of phthalic anhydride with benzene, followed by dehydration. The semistructural formula is $C_6H_4(C=O)_2C_6H_4$. The two benzenoid rings and the two carbonyl groups function as chromophores (color bearers); thus, this compound is the parent of a large and important group of organic dyes. Many of the derivatives of anthraquinone have an added heterocyclic nucleus. Anthraquinone dyes include acid or mordant, disperse, and vat types. The disperse dyes, originally limited to acetate fibers, have grown in importance since the proliferation of synthetic fibers in recent years. Many of the vat (or reduced) dyes were developed in the early years of this century; the first of these was indanthrone, which dates back to 1901. Because of their alkaline nature, vat dyes are used chiefly for cellulosic fibers such as cotton and rayon. *See also* quinone; phthalocyanine; vat; fiber-reactive dye.

antibiotic. The name first used by Selman A. Waksman (1888-1973), an American microbiologist (Nobel Prize 1952), in 1942 to refer to organic compounds formed by specific strains of microorganisms which are able to stop the growth of and often destroy other microorganisms. Discovered by Sir Alexander Fleming in 1928, such compounds are produced by (1) bacteria, which yield polypeptide (protein-like) molecules; (2) fungi or molds, from which penicillin is obtained; and (3) actinomycetes, a bacillus occurring in the soil, from which actinomycin and streptomycin are derived. According to Waksman, the ''ability to produce antibiotics is characteristic not of the genus, nor even of the species, but of strains of particular organisms.'' The formulas of typical antibiotics are as follows: clavacin, $C_7H_6O_4$; streptomycin,

$C_{21}H_{39}O_{12}N_7$; penicillin, R—$C_9H_{11}N_2O_4S$; chloramphenicol, $C_{11}H_{12}N_2O_5Cl_2$. Many antibiotics are produced by fermentation processes.

The effectiveness of antibiotics varies widely; selection of preferred types for specific cases is determined on the basis of extensive laboratory research. About 75 antibiotics are approved for medical use and are manufactured on a commercial scale, penicillin being the most widely used. In a few instances restrictions have been imposed where damaging side effects have been demonstrated. In spite of these exceptions, antibiotics have constituted a tremendous advance in the science of chemotherapy since 1940 and have largely replaced sulfa drugs in the treatment of many infections. See also chemotherapy; penicillin; Fleming.

antiblock agent. In plastics terminology, an abherent material such as a silicate, wax, or finely divided mineral substance that is added to a mixture to prevent cohesion of films or layers of the plastic after processing. Waxes tend to "bloom" to the surface, forming a thin, dry coating which keeps the layers from sticking together; the mineral materials fulfill the same function by introducing small air pockets between the layers. The term "block" means the undesired cohesion of plastic surfaces. See also abherent.

antibody. A protein (globulin) formed or synthesized in the blood as a protective agency when a proteinaceous infective organism enters the body; the antibody is capable of destroying such an organism. An antigen is usually a protein, of lower molecular weight than the antibody, which is intentionally introduced into the blood for the purpose of causing antibody formation, thus immunizing the recipient to a particular infective organism for considerable periods of time. The ability of an antibody to act upon bacteria and viruses in this way has been found to be due to its ability to combine or couple with specific chemical groups present in the antigenic protein molecules, for example -COOH and -SO_3H. The forces holding the antibody-antigen combination together are of the relatively weak hydrogen bond or van der Waals type. The study of antibodies, their formation, and their manner of responding to antigens comprises the science of immunochemistry, the foundations of which were laid by Edward Jenner, an English physician (1749-1823), who in 1776 discovered the principle of inoculation for smallpox and later by Louis Pasteur, a French chemist (1822-1895).

anticaking agent. An additive used primarily in certain finely divided food products that tend to be hygroscopic to prevent agglomeration and thus maintain a free-flowing condition. Starch, calcium metasilicate, magnesium carbonate, and magnesium stearate are used for this purpose in table salt, flours, sugar, coffee whiteners, and similar products.

antichlor. A term used chiefly in the textile and paper industries to refer to any substance that will neutralize chlorine-containing bleaching agents (sodium hypochlorite) in order to improve the color of the product and to remove the undesirable odor and irritant effects of residual chlorine. The compounds most widely used for this purpose are sodium thiosulfate and sodium bisulfite. See also bleach.

anticoagulant. Any substance having the ability to extend the normal clotting time of blood. Many anticoagulants are polysaccharides or complex carbohydrate polymers; an example is heparin, one of the most widely used types, which inhibits or prevents formation of thrombin from its precursor prothrombin. The uses of such compounds are almost entirely medical, though there has been some application to the rodenticide field in which dosage leads to a lethal amount of internal bleeding.

antidote. A substance administered in cases of poisoning by any means (ingestion, inhalation, skin absorption, or injection) which has the property of counteracting the effects of the poison or of eliminating it from the body. Emptying the stomach by means of an emetic or by administering a substance that acts as a neutralizing or adsorptive agent is usually effective if the patient remains conscious. Prompt action is essential, and a physician should be called immediately. See also poison; toxicity.

antifertility agent. A natural or synthetic biologically active compound which is able to retard or prevent normal formation of ova in the female reproductive cycle. The steroid hormones progesterone and estrogen, which occur naturally in the female, were found to have this capability, but were not effective when taken orally. Many semisynthetic analogs of these hormones (including some non-steroid types) which inhibit conception when ingested as pills have been developed; some of these have adverse side effects, which have tended to reduce their use. Antifertility agents are regulated by the FDA and are available only by prescription.

antifouling paint. An organic protective coating especially designed for marine environments. Such paints are applied to ship bottoms, pilings, buoys, and other structures immersed in salt water for long periods to reduce attack by barnacles, teredos, and other sea-dwelling organisms. The active chemical compounds often present are naphthenates of copper and lead, organic tin compounds, mercuric oleate, and other mercury compounds having fungicidal properties.

antifreeze. See ethylene glycol; methyl alcohol.

antigen. See antibody; allergen.

antihistamine. One of about 25 synthetic organic compounds developed in recent years to inhibit the irritant and sometimes toxic effects of histamine. The latter substance is a degradation product of the amino acid histidine; it is a pharmacologically active compound formed in body tissues as a result of an external stimulus (allergen) in a manner similar to that of an antibody. The allergic effect is usually in the nasal tissues, and it is for this that antihistamines are generally used, though they are also effective in more severe disorders involving an allergic response to penicillin and other drugs. Chemically, they are complex aromatic amines. Since some have undesirable side effects, antihistamines must be administered by a physician.

antiknock. See knock.

antimetabolite. See antagonism; metabolite.

antimony. An element.

		Atomic No.	51
Symbol	Sb	Atomic Wt.	121.75
State	Solid	Valence	3,4,5
Group	VA	Isotopes	2 stable

Antimony is a metallic element, m.p. 630.5°C (1166°F), in the form of a crystalline solid, though one of its two allotropic forms is amorphous. It is not a good conductor of heat or electricity. The two major sources are stibnite ore (Sb_2S_3) found in Mexico, Peru, and Algeria and lead-containing scrap metal from batteries. Its chief uses are as an additive to harden lead and in metal bearings; it also has some application as a solder and in type metal. It is too soft for independent metallurgical uses. The highly purified metal has semiconducting properties. An-

timony forms many compounds which are quite toxic, as are fumes of the metal itself; the toxicity depends largely upon the chemical state of the metal or compound. One of the most poisonous compounds is antimony hydride (stibine).

antioxidant. An organic compound which has the power to greatly reduce the normal tendency of oxygen to combine chemically with hydrocarbons in petroleum products, animal fats, vegetable oils, rubber, and similar substances, with consequent degrading effects on the material. The protective effect is considered to be due to free radical mechanisms and is obtained with extremely low percentages. The maximum content of antioxidants in foodstuffs is 0.02% and some will function at a concentration of only 0.0025%. Antioxidants are added to gasoline to prevent oxidative gum formation; to natural and synthetic rubbers, in which they increase service life by a factor of from 5 to 10; and to foods to inhibit rancidity and spoilage. Chemicals having a strong antioxidant function are phenols, arylamines, and phosphites; sulfur compounds are also effective in lubricating oils and in preservation of some fruits. Some antioxidants are naturally present in rubber and foods (in the latter, tocopherols and ascorbic acid). The discovery of the effectiveness of these additives has been largely responsible for the extended shelf and service life of these essential products. See also additive; food additive.

antiperspirant. See aluminum chlorohydrate.

antiseptic. Any substance which will destroy or inhibit the growth of bacteria or other infective organisms when in contact with the tissue of humans or animals. The most commonly used antiseptics are organic mercury compounds (thiomersal), essential oils (thymol, menthol), silver compounds (Argyrol), and oxidizing materials (hydrogen peroxide, potassium permanganate). Alcohol also has antiseptic properties; iodine solutions (3%) have been widely used but are considered too drastic for most applications. Antiseptics should be distinguished from disinfectants, which are applied to things rather than to people because of their high toxicity (e.g., inorganic mercury compounds, cresol), and also from sterilants, which include agencies such as heat and ultraviolet radiation as well as chemical substances. Antiseptics were discovered and introduced into medical practice by Lister in the 1880's. See also bacteriostat.

antistatic agent. A compound incorporated in a thermoplastic polymer which will migrate to the surface to form a conductive film, thus reducing the tendency of the plastic to accumulate static electric charges. A number of polar materials have been used, including amines, phosphate esters, and quaternary ammonium compounds. Hygroscopic substances like polyethylene glycol are also effective, as are various water-repellent amides, which increase the ''slip'' properties of the thermoplastic. Such agents are especially desirable in articles made of nylon and other thermoplastic fibers, which attract unsightly dust particles because of their static surface charge.

aprotic. A term descriptive of a substance, usually a solvent, which neither gives nor accepts protons and thus behaves neither as an acid nor a base. The term protic has the opposite meaning.

aqua regia. An extremely strong oxidizing solvent used chiefly in metallurgical analysis; composed of concentrated nitric acid (one part) and concentrated hydrochloric acid (three parts), it is capable of dissolving such metals as palladium, platinum, and gold. The term literally means ''ruling water.''

aqueous. A solution or suspension in which the solvent is water. The term is often used to describe a gaseous compound dissolved in water, for example, aqueous formaldehyde, a form in which the gas is commercially handled. Aqueous solutions of hydrogen chloride, hydrogen fluoride, and hydrogen iodide are known as hydrochloric acid, hydrofluoric acid, and hydriodic acid, respectively. The standard abbreviation is aq.

Ar (or **A**) Symbol for the element argon. Ar is approved by IUPAC.

arabic gum. A water-soluble polymer composed of complex carbohydrate units (polysaccharide); the major constituents are galactose, glucuronic acid, arabinose, and similar sugars, the units being linked together to give a polymer structure having a molecular weight of about 250,000. Like other gums of this type (tragacanth, karaya, ghatti), arabic is formed on the bark of several varieties of tropical trees wherever they have been damaged, apparently as a natural wound-healing device. The dried material is removed in slabs or chunks and purified for commercial use. Its water-absorption property makes it a useful additive in such food products as ice cream and frozen desserts, where it acts as a protective colloid and thickener; it also has some use in cosmetic creams and pharmaceutical products. The water-soluble gums do not have the gel-forming properties of agar and pectins, though their chemical nature is quite similar. *See also* carbohydrate; polysaccharide.

arachidic acid. *See* vegetable oil; fatty acid.

aralkyl. *See* aryl.

aramid. Generic name for a distinctive class of highly aromatic polyamide fibers characterized by their flame-retardant properties. Some types are also suitable for protective clothing, dust-filter bags, and bullet-resistant structures. Their chemical nature and physical properties are different from those of nylon. *See also* polyamide.

arene. Class name for unsaturated cyclic (aromatic) compounds, of which benzene is typical; several rings may be present, either chemically bonded, as in diphenyl, or fused, as in anthracene. *See also* aromatic.

argentic,-ous. *See* silver.

argon. An element.

Symbol	Ar (or A)	Atomic Wt.	39.948
State	Gas	Valence	0
Atomic No.	18	Isotopes	3 stable

One of the few truly inert or ''noble'' elements, argon, b.p. $-186°C$ ($-303°F$), is a component of air to the extent of 0.94% by volume and is obtained as a by-product of liquid air distillation operations which produce oxygen and nitrogen. It does not form compounds with other elements in the usual sense, though ionic combinations with hydrogen have been observed in electric discharge tubes; quite stable hydrates and clathrates are also possible, but none of these involves actual chemical bonding. The most important industrial use of argon is in arc welding, where it serves as a shielding medium; it is also used as a filler gas in incandescent and fluorescent lamps and in electron tubes. Argon has a particularly stable electronic structure.

aromatic. A major series of unsaturated cyclic hydrocarbons (also called arenes) whose carbon atoms are arranged in closed rings typified by the benzene nucleus (C_6H_6); the rings may be either single or fused, that is, attached to each other in groups (naphthalene, anthracene, phenanthrene). Aromatics are characterized by their chemical reactivity and versatility as intermediates; the hydrogen atoms on the ring are readily replaced with other elements (chlorobenzene, C_6H_5Cl), with groups (phenol, C_6H_5OH), or with side chains of various types (dodecylbenzene, $C_6H_5C_{12}H_{25}$). Thus, the unsaturated ring structure occurs in aromatic alcohols, aldehydes, amines, acids, steroids, etc., collectively called benzenoid compounds. Aromatic compounds can be obtained from coal tar by fractional distillation, but the more important source is petroleum. They are derived by several processes, perhaps the most important of which is catalytic reforming, which involves a number of reactions defined under separate headings. *See also* benzene; reforming; dehydrocyclization; transalkylation; isomerization.

Arrhenius, Svante (1859-1927). Native of Sweden, Nobel Prize in chemistry, 1903. Best known for his fundamental investigations of electrolytic dissociation of compounds in water and other solvents and for his basic equation stating the increase in the rate of a chemical reaction with rise in temperature:

$$\frac{d \ln k}{dT} = \frac{A}{RT^2}$$

in which k is the specific reaction velocity, T is the absolute temperature, A is a constant usually referred to as the energy of activation of the reaction, and R is the gas law constant.

arsenic. An element.

Symbol As	Atomic No. 33	
State Solid	Atomic Wt. 74.9216	
Group VA	Valence 2,3,5	

Arsenic is a soft, solid substance classified as a nonmetal. It sublimes at 613°C (1135°F) at 760 mm pressure, and at its melting point of 817°C (1503°F), it has a vapor pressure of 28 atmospheres. The pure material has semiconducting properties and is used commercially in transistors and other semiconducting devices in both elemental and compound form (gallium arsenide). It has few other uses, though it is a metallurgical additive in some lead and copper alloys. It has several allotropic forms, one of which is amorphous. It readily forms compounds, some of which are used as pesticides; arsine (AsH_3) is an intensely toxic gas resulting from electrochemical reactions involving arsenic

compounds. Elemental arsenic is much less toxic than commonly supposed, but its dusts and vapors, as well as most of its compounds, are extremely poisonous (carcinogenic) and should be handled with great care. *See also* arsenic trioxide.

arsenic trioxide. The most important arsenic compound, arsenic trioxide is a white and extremely toxic powder; it is often called white arsenic, or loosely "arsenic," which of course is incorrect. The formula is As_2O_3; sublimation temperature is 195°C (383°F). It has a number of industrial uses, though some have been discontinued because of safety factors; it is a decolorizing agent 'in glass manufacture, a source of other arsenic compounds, a pesticide, and a preservative of skins (taxidermy). Not only ingestion, but also inhalation and skin contact must be avoided.

arsine. *See* arsenic; hydride.

arsphenamine. *See* chemotherapy.

aryl. A univalent group derived from aromatic hydrocarbons and having a ring structure of benzene, naphthylene, etc. Examples are phenyl (C_6H_5) and naphthyl ($C_{10}H_7$). Such groups are often represented in general formulas by the letter R. Compounds made up of both aryl and alkyl groups are sometimes called aralkyl (a shortened form of arylalkyl), depending on where the available bond is located. For example, the benzyl group ($C_6H_5CH_2$) is an arylalkyl group, since the open bond is on the alkyl portion of the molecule, and the tolyl group ($C_6H_4CH_3$, or methylphenyl) is an alkylaryl group, since the open bond is on the aryl portion of the molecule. *See also* alkyl.

asbestos. Fibrous magnesium silicate occurring in two forms: (1) chrysotile or serpentine asbestos, which is relatively pure and can be handled on textile machinery and (2) amphibole asbestos, which contains other silicates as impurities and has fibers that are too hard to weave or spin. Asbestos is mined from surface deposits in southern Quebec, northern Canada and the southwestern United States. Being noncombustible, its primary use is as a fireproofing additive to heavy textile products, brake linings, etc., as well as in rubber, plastic, and paint compositions, where heat resistance is a factor. Asbestos dust is strongly carcinogenic, and inhalation should be avoided.

ascorbic acid. An unsaturated aliphatic acid derived from sugars and occurring naturally in citrus fruits and some vegetables. Generally known as vitamin C, it is an important factor in a balanced diet; inadequate intake results in scurvy. Synthetic ascorbic acid is made from glucose. It protects against various types of infections, including the common cold. In addition to its metabolic value, it is used as a food additive, where it acts as an antioxidant and inhibitor of discoloration of peeled vegetables and also as a stabilizer of dough. Its molecular formula is $C_6H_8O_6$, and it is dextrorotatory. It oxidizes readily, thus losing its protective value; it also deteriorates on exposure to heat. *See also* vitamin; antioxidant.

-ase. A suffix characterizing the names of many enzymes, e.g., diastase, cholinesterase, etc. However, the names of some enzymes have the suffix -in (pepsin, rennin).

ash. The unburnable residue remaining after the complete combustion of a material; it consists of mineral matter, the amount usually being a specification requirement.

askarel. *See* dielectric; transformer oil; insulator.

asphalt. A thick, viscous mixture of hydrocarbons (bitumens) obtained chiefly as the residue of petroleum distillation. A small amount is also taken from natural sources, such as the asphalt "lake" in Trinidad; this contains about 33% of colloidal clays. Its ability to flow when heated and to solidify when cool enables it to be used effectively in road paving (alone or blended with other materials), roofing, waterproof sealants, and hot-melt coatings for pipelines, underground cables, etc. Besides hydrocarbons, asphalt also contains low percentages of heterocyclic compounds in which sulfur and nitrogen are present. Several unusual uses have been under experimentation in recent years: as a lining beneath a layer of sandy soil to prevent loss of water and as a nutrient medium for bacteria in the production of proteins from petroleum. A product known as blown asphalt or "mineral rubber" is made by permeating asphalt with air, giving a hard, brittle material used as a diluent in cheap rubber goods and as a roofing sealant. Liquid asphalt occurs naturally in California; it is used in road treatment and hot-melt adhesives. *See also* pitch.

asphyxiant. A gas which has little or no positive toxic effect but which can bring about unconsciousness or death by replacing air and thus depriving an organism of oxygen. Among the so-called asphyxiant gases are carbon dioxide, nitrogen, helium, methane, and other hydrocarbon gases.

aspirin. *See* acetylsalicylic acid.

assay. Determination of the content of a specific component of a mixture, with no evaluation of other components. Such determinations are made on ores of various metals (especially precious metals), on pharmaceutical products to validate the amount of drug present in a given unit, and on organisms (bacteria) to determine their reactions to an antibiotic or bactericide. The latter procedure is called bioassay. Modern assay methods utilize appropriate currently available physical, biological, and chemical analytical techniques.

assistant. *See* auxiliary; dye.

association. A loose chemical union between molecules of the same or different types, characterized by relatively weak bonding forces (hydrogen bonding of water molecules to each other or of acetic acid to water). In some cases, such as complex ion formation, coordinate covalent bonds are involved. Associated groups of molecules dissociate readily; thus, association is a reversible phenomenon, quite different from polymerization, which is irreversible. An example is the protein hemoglobin, made up of four chains of polypeptides which are associated, but which dissociate when the pH changes markedly in either direction; this dissociation is reversible. Another instance is that of water solutions of soaps or synthetic detergents, which are sometimes referred to as association colloids.

astatine. An element.

Symbol	At	Atomic No.	85
State	solid	Atomic Wt.	211
Group	VIIA	Valence	1,3,5

Astatine, m.p. 302°C (576°F), the heaviest of the halogen elements, is classed as a nonmetal. It occurs in nature in trace quantities, the total in the earth's crust not exceeding 50 grams; thus, its physical properties are not definitely known. All its isotopes are strongly radioactive, the longest-lived having a half-life of 8.5 hours. It has no stable form. It is capable of compound formation with other halogens, with oxygen, and with some organic groups. It is a thyroid-seeking element similar to iodine. Astatine was one of the first elements to be made synthetically (1940) by alpha particle bombardment of bismuth, but the largest quantity made at one time is less than 0.1 microgram.

Aston, Francis Williams (1877-1945). This noted English chemist and physicist carried out much of his work with J. J. Thomson at Cambridge. He was the pioneer investigator of isotopes and his method of separating the lighter from the heavier atomic nuclei provided the technique that later developed into the mass spectroscope, which utilizes a magnetic field for this purpose. Aston received the Nobel Prize for this discovery in 1922, just three years after Rutherford performed the first transmutation of elements. Aston also correctly estimated the energy content of a hydrogen atom and predicted the controlled release of this energy.

asymmetry. A molecular arrangement in which each bond of the central element (usually carbon) is attached to a different element or group. This results in three-dimensional optical isomers having mirror-image configurations called enantiomorphs (or enantiomers). Carbon is the most common asymmetric element, but the phenomenon also exists with sulfur, nitrogen, cobalt, and others. A compound may contain several asymmetric atoms; the number of possible isomers for any compound having 2 or more asymmetric carbons is 2 to the nth power, where n equals the number of asymmetric carbons. Each isomer displays optical rotation to the left or right under polarized light, a property of many sugars, amino acids, aldehydes, and terpenes. An asymmetric atom is conventionally denoted by an asterisk, for example, C^*. It is possible for a molecule to be asymmetric in the absence of an asymmetric atom, i.e., by having two different organic groups attached to one end of a carbon chain, as in allenes. *See also* enantiomorph; optical rotation; racemic; stereoisomer.

At Symbol for the radioactive element astatine.

atactic. A high-polymer molecule in which the substituent atoms or groups (R) are arranged in a random manner in reference to each other above and below the carbon chain, as indicated:

$$-\overset{\overset{\textstyle H}{|}}{\underset{\underset{\textstyle H}{|}}{C}}-\overset{\overset{\textstyle R}{|}}{\underset{\underset{\textstyle H}{|}}{C}}-\overset{\overset{\textstyle H}{|}}{\underset{\underset{\textstyle H}{|}}{C}}-\overset{\overset{\textstyle R}{|}}{\underset{\underset{\textstyle H}{|}}{C}}-\overset{\overset{\textstyle H}{|}}{\underset{\underset{\textstyle R}{|}}{C}}-\overset{\overset{\textstyle H}{|}}{\underset{\underset{\textstyle H}{|}}{C}}-\overset{\overset{\textstyle H}{|}}{\underset{\underset{\textstyle H}{|}}{C}}-\overset{\overset{\textstyle R}{|}}{\underset{\underset{\textstyle H}{|}}{C}}-\overset{\overset{\textstyle H}{|}}{\underset{\underset{\textstyle H}{|}}{C}}-\overset{\overset{\textstyle H}{|}}{\underset{\underset{\textstyle R}{|}}{C}}-\overset{\overset{\textstyle H}{|}}{\underset{\underset{\textstyle H}{|}}{C}}-\overset{\overset{\textstyle H}{|}}{\underset{\underset{\textstyle R}{|}}{C}}-\overset{\overset{\textstyle H}{|}}{\underset{\underset{\textstyle H}{|}}{C}}-$$

The term literally means "without specific order or arrangement." *See also* tacticity; stereospecific.

-ate. A suffix having two different meanings.

(1) In inorganic compounds, it indicates a salt whose metal or radical is in the highest oxidation state, as calcium sulfate, ammonium nitrate, etc.

(2) In engineering terminology, it means "result of", as in precipitate, condensate, alkylate, and the like.

atmosphere. (1) A pressure of 14.696 pounds per square inch; the pressure exerted by the air at sea level which will support a column of mercury 760 millimeters high (about 30 inches). This is considered to be the standard barometric pressure, though it varies slightly with local meteorological conditions. The term is sometimes used to indicate working pressures of steam, water, etc. The customary abbreviation is atm. *See also* air.

(2) An artificial gaseous environment (nitrogen, CO_2) in which a flammable solid is kept to prevent it from catching fire or in which harvested fruits and vegetables are stored to retard ripening (controlled-atmosphere storage).

(3) An ionic atmosphere (see entry).

atom. Derived (though erroneously) from the Greek, meaning "indivisible," an atom is the smallest unit of an element that can exist. Atoms have long been known to be made up of particles of matter called protons, neutrons, and electrons. The nucleus of an atom contains one or more protons, which are positively charged, and two or more neutrons (except in hydrogen), which have no electric charge. Each proton and neutron has a mass number of 1; they are bound together in the nucleus by exceedingly strong forces; in a few special cases, they can be separated by bombardment with high-energy particles from an external source. When this occurs, some of the binding energy is released (the fission process for nuclear energy). A few types of atoms are radioactive; these decay spontaneously, releasing their energy, in some cases over a period of centuries.

The electrons move around the nucleus in shells (groups of orbitals) which are determined by their energy levels or wave functions. An electron is a negatively charged unit which behaves as both a particle and a wave, i.e., it is partly mass and partly energy. Every atom has the same number of electrons as protons and is thus electrically neutral. It might be asked why the electrons, being negatively charged, do not combine with the positively charged nucleus in accordance with the law of attraction of oppositely charged bodies. The explanation lies in the dual properties of the electron; its wave function rather than its particulate nature is the controlling factor in the stability of the atom. Atoms which have gained or lost one or more electrons are called ions. *See also* element; electron; Periodic Table; orbital.

atomic energy. *See* uranium; fission.

atomic number. The number of protons, or positively charged mass units, in the nucleus of an atom, upon which its structure and properties depend. This number represents the location of an element in the Periodic Table; it is always the same as the number of negatively charged electrons in the shells. Thus an atom is electrically neutral, except in an ionized state, that is, when one or more electrons have been gained or lost. Atomic numbers range from 1, for hydrogen to 106, for the most recently discovered element. *See also* Periodic Table; atomic weight; mass number.

atomic weight. The average weight (or mass) of all the isotopes of an element, as determined from the proportions in which they are present in a given element, compared with the mass of the 12 isotope of carbon (taken as precisely 12.000), which is now the official international standard. The true atomic weight of carbon, when the masses of its isotopes are averaged, is 12.01115. The total mass of any atom is the sum of the masses of all its constituents (protons, neutrons, and electrons). The atomic weight of an element expressed in grams is called the gram atomic weight. *See also* atomic number; mole.

ATP. Abbreviation of adenosine triphosphate.

atropine. *See* alkaloid.

attapulgite. *See* clay.

Au Symbol for the element gold, derived from the Latin word *aurum*. This form is sometimes used in naming gold compounds, i.e., aurous, auric.

auric, aurous. Adjectival forms derived from the Latin word *aurum,* meaning gold, referring respectively to gold compounds in oxidation states + 3 and + 1. These terms are seldom used today, the newer terminology gold (III) chloride and gold (I) cyanide being preferred to auric chloride

and aurous cyanide. They persist, however, in names such as chloroauric acid.

austenitic steel. A steel in which the iron component is present in the gamma modification, namely, a face-centered lattice structure, which occurs at temperatures between 730 and 1100°C (1346 and 2012°F), in contrast to the alpha modification called ferrite, which has a body-centered lattice. These two modifications of iron are allotropic forms, the gamma (austenite) being nonmagnetic and the alpha, magnetic. In austenitic steels the carbon atoms are able to diffuse through the iron atoms, which is an important factor in making a steel susceptible to heat treatment. Some stainless steels are austenitic. *See also* steel; allotrope.

autocatalysis. *See* autoxidation; catalyst.

autoignition point. The lowest temperature at which a material will catch fire without the aid of a flame or spark; it is also called the ignition point. Some substances, such as white phosphorus, have such a low autoignition point that they will ignite spontaneously on exposure to air; materials like paper, thin wood shavings, and cotton fibers will burn when exposed to temperatures of from 200 to 260°C (392 to 500°F); nitrocellulose will ignite at about 135°C (275°F). The lowest ignition point for any solvent is 100°C (212°F) for carbon disulfide. This property should not be confused with flash point. *See also* flash point; combustible material.

autoxidation. An oxidation that takes place spontaneously in a substance exposed to air or oxygen at temperatures below 150°C (300°F). It is activated by a free radical chain mechanism (autocatalysis) and is characteristic of many hydrocarbons; thus it is applied to a number of industrial operations in the chemical industry, such as the production of acetone and phenol from cumene. Autoxidation also is responsible, at least in part, for development of rancidity in food products, the drying of certain types of vegetable oils, the formation of "gum" in petroleum products, and the deterioration of vulcanized rubber. *See also* free radical; oxidation.

auxiliary. A term used almost exclusively in the textile-processing industry to include a large number of functional materials which have no chemical relationship; among these are fats and oils used to lubricate fibers, natural and synthetic surface-active agents, film-forming materials (starches, gums, etc.) for coating and sizing, delustering agents, and waterproofing materials. The term "assistant" is used in almost the same sense in the industry, but it also applies to the fixing of dyes on fabrics, for example, a mordanting assistant. Dyestuffs are not usually regarded as auxiliaries as their function is essential rather than merely "helpful."

auxin. One of a group of hormones which regulate or control the growth of plants and vegetation; they affect a number of processes, including the cell wall structure of stems, enabling the plant to lean one way or another in response to sunlight, as well as the length and direction of roots, and the development of buds and fruit. The auxins include indoleacetic acid and 2,4-D; the latter is effective in very small amounts, as it occurs in the plant but can be lethal to some plants if applied in high concentrations. *See also* plant growth regulator; 2,4-D.

Avogadro number. In 1811 an Italian scientist, Amadeo Avogadro (1776-1856), advanced the hypothesis that the number of molecules of any gas, at 0°C (32°F) and 1 atmosphere, is the same for a given volume, regardless of the chemical composition and physical properties of the gas. It was later determined that this number, for a volume of 22.41 liters, is 6.023×10^{23}, which represents one mole (or gram molecular weight) of the gas. Thus, 32 grams of oxygen and 44 grams of carbon dioxide both contain 6.023×10^{23} molecules, and both occupy a volume of 22.41 liters. This number is a constant, which applies not only to gases but to all forms of matter and to all types of chemical units (atoms, ions, molecules, etc.). For many years oxygen was the standard element for establishing atomic weights; but because the presence of isotopic components introduced a slight variation, the 12 isotope of carbon was selected in 1961 to replace oxygen as a reference standard. Thus, the Avogadro number is now defined as the number of atoms in 12 grams (one mole) of carbon-12. It is one of the most important constants in chemistry, as it underlies the determination of molecular weights and stoichiometric relationships. It is conventionally represented in chemical calculations by the boldface letter **N**. *See also* mole; molecular weight.

azeotrope. A term used by chemical engineers to denote a mixture of two components in such proportions as to result in either the lowest or the highest constant boiling point attainable with any mixture of the same components. Thus, a blend of ethyl alcohol (b.p. 78.5°C (173.3°F)) and water (b.p. 100°C (212°F)), when mixed in the mole fraction of 89.4% alcohol and 10.6% water, has the lowest boiling point, 78.15°C (172.67°F), of any possible mixture of these substances. Similarly, HCl and water mixtures have a constant maximum boiling point. Hundreds of combinations of substances exhibit this behavior. The vapors of azeotropic mixtures have the same percentage composition as the original liquid mixture. In azeotropic distillation, complete separation of the components of the mixture being distilled is aided by adding another substance, e.g., benzene, which is known to form an azeotrope with one of these components; as the azeotropic mixture has a different boiling point, it readily distills off. *See also* eutectic.

azide. Any of a group of detonating explosive compounds whose molecules are comprised of a chain structure of double-bonded nitrogen atoms to which is attached a metal or metal complex, a halogen, hydrogen, an ammonium radical, or an organic radical. The most important industrial member of this class is lead azide, $Pb(N_3)_2$, which is used as a composition detonator or fuse cap in blasting devices and military explosives. Hydrogen azide and all the heavy-metal azides are highly sensitive and can be detonated by slight shock, heat or friction; less sensitive are the azides of barium, sodium, and calcium; organic azides lie midway between, in respect to sensitivity. Shipment and storage of azides are legally specified, as is the case with all high explosives.

azo dye. The most versatile and widely used group of dyestuffs; the term azo refers to the nitrogen content of the molecule. Two reactions are involved in their formation. The azo component is derived by reacting a primary aromatic amine with nitrous acid, plus a mineral acid. This reaction is called diazotization. It yields a compound containing an —N=N—, or diazo, group which performs the color-imparting function and thus is called a chromophore. The diazo compound then undergoes a second reaction, known as coupling, with other organic compounds, such as phenols or amines. Many azo dyes can be formed by means of diazotization and coupling, producing a wide variety of organic dyes and intermediates, suitable for both natural and synthetic fibers, as well as for non-textile materials. The dyes are bright, lightfast, and resistant to washing. *See also* dye; diazotization; coupling.

azote. The French word for nitrogen, derived from the prefix *a* = (not) and the base *zoo*, meaning "alive" (*zoon* = animal). This entry is included to facilitate recognition of nitrogen-containing compounds indicated by such chemical terms as azo, azide, thiazole, azobenzene, carbazole, etc. The derivation is due to the comparative chemical inactivity of nitrogen.

B

B Symbol for the element boron, the name apparently being a transliteration of an Arabic word, *baurach,* for borax or a similar compound.

Ba Symbol for the element barium, the name being derived from the Greek word for heavy.

babbitt. A group of bearing alloys used as coatings on steel; there are several modifications, including those in which lead, tin, or cadmium comprise the basis metal. They absorb heat developed by friction, retain lubricants, and thus protect the steel bearing from deterioration and seizing. They are named after their developer, Isaac Babbitt (1799-1862), an American inventor, and are sometimes capitalized.

bacteria. Microorganisms composed chiefly of proteins and nucleic acids. They have many important beneficial chemical functions, some of which are manufacture of ethyl alcohol, beer, and antibiotics by fermentation; baking industry (yeast); soil fixation of nitrogen; sewage treatment by the activated sludge process; formation of proteins from hydrocarbons; manufacture of sulfur from gypsum; organic oxidation processes; and reaction with cellulose for production of high-protein foods. Bacteria also have undesirable properties, as they are the cause of many types of food spoilage, toxic effects, and infectious diseases. They are large enough to be visible in an optical microscope and accept stains, (e.g., compounds of osmium, chromium, etc.) which make them readily distinguishable. One of the more dramatic results of recombinant DNA experimentation is the creation of synthetic bacteria, entirely new life forms made in the laboratory by gene-splicing techniques. A bacterium capable of digesting crude oil has been synthesized, and patents are pending on other types of synthetic microorganisms that have enzymic functions important to manufacture of many chemicals and agricultural products. *See also* fermentation; virus; recombinant DNA.

bacteriostat. A compound having the ability to inhibit or stop the proliferation and growth of infectious microorganisms without actually killing them. It thus acts in the manner of an antiseptic. *See* hexachlorophene; antiseptic.

Baekeland, L. H. (1863-1944). Born in Ghent, Belgium, Baekeland did early research in photographic chemistry and invented Velox paper (1893). After working for several years in electrolytic research he undertook fundamental study of the reaction products of phenol and formaldehyde, from which he developed a hard, infusible product, which he named Bakelite (1907). The reaction itself had been investigated by Bayer in 1872, but Baekeland was the first to learn how to control it to yield dependable results on a commercial scale. The Bakelite Co. (now a division of Union Carbide) was founded in 1910. *See also* phenolformaldehyde resin.

bagasse. The cellulosic residue of sugar cane after it has been disintegrated to extract the sugar. It is converted by high-temperature compression into a heavy board of high thermal insulating value which is useful for heat- and soundproofing in building construction. Paper can be made from bagasse on a commercial basis. Bagasse is also applied in the manufacture of furfural, as an extender in animal feeds. and is used as an on-site boiler fuel on sugar plantations (Louisiana, Hawaii). *See also* furfural.

baked enamel. A pigmented alkyd resin-based

coating, sometimes also containing nitrocellulose, urea, or melamine resin. After application by spraying, these so-called enamels are heat-cured at about 93°C (200°F), often by means of infrared radiation, to form a hard, glossy surface. They are used on such items as refrigerators, automotive parts, and various household appliances and are resistant to chipping and to chemical attack. *See also* paint; alkyd resin; protective coating.

"Bakelite".   *See* phenolformaldehyde resin; Baekeland.

baking powder.   A synthetic leavening agent used in the baking industry. There are several types, all of which contain a carbonate, a weak acid, and a filler. A typical composition is sodium bicarbonate, tartaric acid or monobasic calcium phosphate, and cornstarch. On exposure to water and heat, the acid reacts with the bicarbonate to evolve carbon dioxide, which produces a stable foam. The baked product is thus a dispersion of a gas in a solid. *See also* foam; blowing agent.

balance.   (1) By the law of Conservation of Mass, every atom present in the molecules taking part in a reaction must also be present in the reaction products; thus, every chemical reaction must balance, that is, the total number of atoms of each element entering the reaction must equal the total number of the same element which leave it. Balance can easily be determined by arithmetical count. For example, in the reaction $NaOH + HCl \rightarrow NaCl + H_2O$, the number of atoms on the input side is 2 hydrogen, 1 sodium, 1 oxygen, and 1 chlorine; exactly the same number of atoms of each element appears in the products, though in different combination. Catalysts are not counted, as they are not consumed in the reaction.

Ionic equations are balanced by use of the change in oxidation state of the atoms involved or by use of the electron transfer which occurs, especially when dealing with oxidation-reduction reactions. When an ionic equation is properly written, the number of atoms and the algebraic sum of the charges must be the same on one side of the equation as on the other. For example, if a solution of $FeCl_2$ is treated with chlorine, $FeCl_3$ is obtained: $2Fe^{++} + Cl_2 \rightarrow 2Fe^{+++} + 2Cl^-$; each ferrous ion loses one electron, and the chlorine molecule acquires two electrons; also, there are four positive charges

on the left side and six positive and two negative (four net positive charges) on the right side.

(2) By the law of Conservation of Energy, the amount of energy input to a process (or system) must balance the amount of energy output. Usually a heat balance is the only effect that must be considered for most processes or reactions.

(3) An extremely accurate beam-type scale used in analytical laboratories.

(4) In formulas showing the composition of alloys, the term means simply "remainder," e.g., 50Fe, 25Co, 20Be, bal.Cu. *See also* mass conservation (law); material balance.

balsam.   A sticky, resinous, fragrant substance obtained from several varieties of trees or shrubs throughout the world. It has a strong, pleasant aroma and is used as a base for perfumes, cough syrups, and pharmaceutical preparations. The chemical composition varies with the source; in general, it includes essential oils, terpenes, cinnamic acid, and benzoic acid. The best-known types of balsams are as follows: Peru, copaiba, and Tolu (South America); balm of Gilead (Near East); benzoin resin (Southeast Asia); and Canada balsam (North America).

banana oil.   *See* amyl acetate.

Banting, Sir Frederick   (1891-1941). A native of Ontario, Banting did his most important work in endocrinology. His brilliant research culminated in the preparation of the antidiabetic hormone which he called insulin (*q.v.*), derived from the "isles of Langerhans" in the pancreas. He received, in 1923, the Nobel Prize in medicine for this work, together with John J. R. MacLeod of the University of Toronto. In 1930 the Banting Institute was founded in Toronto. He was killed in an airplane crash during World War II.

barbiturate.   A class of hypnotic, habit-forming drugs, of which barbituric acid, $\underline{NHCONHCOCH_2CO}$, and barbital

$$\underline{NHCONHCOC(C_2H_5)_2CO},$$

are the best-known examples; they act by depressing the central nervous system. Chemically, they are pyrimidine derivatives, containing nitrogen in the molecule. About 25 types of barbiturates are now in approved medical use. They are highly toxic, and their administration is limited to prescriptions. Pheno- and mephobarbital function as anticonvulsants; other types are used as sedatives and anesthetics.

barium. An element.

Symbol	Ba	Atomic No.	56
State	Solid	Atomic Wt.	137.34
Group	IIA	Valence	2
		Isotopes	7 stable

Barium, m.p. 729°C (1344°F), is a heavy alkaline-earth metal derived from barytes (barium sulfate) and witherite (barium carbonate) found in Missouri, Georgia, Mexico, and Canada. It can be extruded and machined and is used in special alloys, as a "getter" in vacuum tubes, and as a thin-film lubricant in electronic instruments. It exhibits a green color when exposed to flame. Barium is quite active chemically, reacting with halogens, water, oxygen, and acids. In powder form it takes fire at room temperature and thus must be stored and handled in an atmosphere of inert gas. The most important compounds are the carbonate ($BaCO_3$), used to remove sulfates from brines before they are introduced into electrolytic cells, and as a component of ceramics and glass; the chloride ($BaCl_2$), from which precipitated barium sulfate (blanc fixe) is made; the nitrate ($Ba(NO_3)_2$), an ingredient of fireworks and flares for its green color emanation; and the sulfate ($BaSO_4$), used as a filler and weighting agent in oil-well drilling muds, etc. The metal and most of its compounds are toxic, with the exception of barium sulfate, which is harmless. *See also* barium sulfate.

barium carbonate. *see* barium.

barium chloride. *See* barium.

barium nitrate. *See* barium.

barium sulfate. A heavy white powder having a specific gravity of 4 to 4.5 and the formula $BaSO_4$. The natural product occurs as barite in the southwestern U.S., and is tan or yellowish before bleaching. Alternative names are barytes and basofor. It is made synthetically by reacting barium chloride with sodium sulfate, the white precipitate being called blanc fixe. It is one of the few barium compounds that is not toxic, as it is insoluble in water and acids. Most of its applications are due to its unusual density. Most important is its use in oil-well drilling muds, where weight is a significant factor; its opacity accounts for its value in suspensions used in x-ray examinations of the stomach and lower intestines. It is also a weighting filler in low-grade electrical tapes and other rubber and plastic products and in paper and textile coatings.

barn. The measurement unit used by physicists and nuclear scientists for the cross-section target area of atomic nuclei, equivalent to 10^{-24} square centimeter. For example, the absorption cross-scction of the beryllium nucleus for slow neutrons is 0.009 barn per atom; that of boron-10 is 3850 barns. *See also* absorption (3).

barytes. *See* barium sulfate.

base. During the first third of this century there was much theorizing about the nature of bases and their reactions with acids, by such chemists as Arrhenius, Franklin, Bronsted, Lowry, and Lewis. Several possible definitions have resulted, none of which is completely free from objection, as each has certain limitations. In approximate historical sequence, they are as follows. A base is a substance that (1) liberates hydroxyl ions when dissolved in water (Arrhenius); (2) that liberates negative ions of various kinds in any solvent (Franklin); (3) that receives a hydrogen ion (proton) from a strong acid to form a weaker acid (Bronsted-Lowry); (4) that gives up two electrons to an acid, forming a covalent bond with the acid (Lewis). Reactions (3) and (4) occur whether or not the substances are in solution. Lewis bases are considered "hard" if they have high electronegativity and are difficult to oxidize and "soft" if they have low electronegativity and are easily oxidized. These properties relate to the presence or absence of vacant orbitals. Bases react with acids to form salts and water; when the proper proportions are used, neutralization occurs. All bases give solutions having a pH of more than 7.0, the neutral point.

The term alkali is closely related in meaning; it refers to hydroxides and carbonates of the alkali metals (lithium, potassium, sodium, and others), which arc strongly electropositive; typical alkalies are sodium hydroxide, sodium carbonate, potassium hydroxide, and potassium carbonate, all widely used in industry. *See also* acid; pH; alkali.

basofor. *See* barium sulfate.

batch process. Any industrial procedure in which raw materials are mixed in individual lots or batches. Many basic materials are made in this way, e.g., soap, rubber, paper pulp, fermented liquors, and confectionery and food products. Batching is particularly necessary when special aging and enzymatic reactions are involved which require long holding periods for maturation purposes. Continuous processing is chiefly used in making finished products from the orignal batch-prepared material, as in the manufacture of sheet metals, paper, glass, etc.

bating.   A unit process in leather manufacture in which the dehaired skins or hides are treated with enzyme-containing materials to provide desirable softness and flexibility (plumpness) prior to tanning. The most effective enzymes for this purpose are pancreatin and trypsin.

battery.   An electrochemical source of electric power, which may be either reversible (secondary battery) or irreversible (primary battery). Early batteries were of the latter type, consisting of alternate plates (electrodes) of metals such as zinc and copper (which have different but opposite electrode potentials); between these plates were sandwiched sheets of heavy paper saturated with brine, which served as the electrolyte. Later, ammonium chloride was substituted to avoid the use of a liquid electrolyte. Such batteries, which became known as dry cells, are of limited service life, as one of the components is consumed during operation, and most cannot be recharged. The usual type of secondary or storage battery, such as is used in automobiles, consists of a series of lead plates arranged in cells and coated with a lead oxide paste; the cells contain sulfuric acid, which reacts reversibly with the lead and lead oxide to yield electric energy. The charge is maintained by a generator and can be completely renewed when necessary. There are many kinds of storage batteries for special types of service, e.g., nickel-cadmium, nickel-iron, silver-zinc. There is active current research, both private and governmental, on these and other types with a view to development of battery-powered automobiles. Zinc-chlorine batteries for this purpose are in production. Solar energy converters and fuel cells are often considered as batteries, but they operate quite differently. *See also* fuel cell.

Baumé   (pronounced bo-may). A scale introduced by the French chemist, Antoine Baumé (1728-1804), about 1800, for use in determining the specific gravity of liquids. A specially calibrated hydrometer is used, and from the reading obtained with this, the actual specific gravity can be calculated. The calculation is often unnecessary, since specifications usually state the degrees Baumé required. It is abbreviated Bé.

bauxite   (pronounced bok-site). A mineral mixture whose major component is hydrated aluminum oxide ($Al_2O_3 \cdot 3H_2O$). Bauxite is the chief ore of aluminum and is thus an important national resource. Discovered in France, it occurs in Arkansas and other states of the Midwest, as well as in Australia and the West Indies. Besides being the primary source of metallic aluminum and aluminum chemicals, it is used as a filler in plastics and rubber, in refractory brick, as an abrasive, and as a component of hydraulic cements. *See also* aluminum.

Bayer process.   *See* aluminum.

B complex.   A closely related but independently active group of growth-promoting substances (vitamins) produced by living cells; each member of the complex is essential, and no member can be substituted for any other member. Their physiological effect is additive, though each is a distinct compound which has specific functions in metabolism. The complex may be thought of as a rope, each strand of which is represented by an individual and unique substance. The complex occurs in all the cells of all organisms, both plant and animal, and among its components are thiamine, riboflavin, niacin, biotin, pyridoxine, folic acid, and cobalamin. *See also* vitamin.

Be   Symbol for the element beryllium (from beryl).

Becquerel, Henri   (1851-1908). French physicist who shared the 1903 Nobel Prize in physics with the Curies for the discovery of the radioactivity of uranium salts. He also discovered the deflection of electrons by a magnetic field, as well as the existence and properties of gamma radiation, which was originally named after him.

Beilstein, F. K.   (1838-1906). A German chemist noted for his compilation "Handbuch der Organischen Chemie," the first edition of which appeared in 1880. A multi-volume compendium of the properties and reactions of organic compounds, it has been revised several times and remains a unique and fundamental contribution to chemical literature.

beneficiation.   A term used in extractive metallurgy to refer to methods of concentrating metallic ores preparatory to the reduction process;

they include mechanical classification, fragmentation, magnetic separation, flotation, etc. *See also* ore dressing.

bentonite.   A unique type of clay, characterized by extremely fine particles of colloidal dimensions; originally found near Ft. Benton, Wyoming. It is notable for its ability to swell in water, a property which makes it especially useful for loading oil-well drilling muds (which must absorb much water), for lining dams and irrigation ditches, in road construction, and in making sand molding cores in foundries. It is also an adsorbent for colors and odors and is used as a component of ceramic products, and in the manufacture of catalysts. Chemically, it is aluminum silicate, with inclusions of magnesium and calcium silicates. *See also* clay.

benzaldehyde.   An aromatic aldehyde, b.p. 179.5°C (355°F), occurring naturally in oil of bitter almonds; it is a volatile liquid of pleasant odor and sharp taste, having the formula C_6H_5CHO. It is derived synthetically from toluene by oxidation or by chlorination followed by hydrolysis. Its chief uses are as a source of benzoic acid, as a dye intermediate, solvent, and in the manufacture of chemicals used in photography. It is combustible, but essentially nontoxic. *See also* aldehyde.

benzene.   The most important and versatile aromatic hydrocarbon; its molecule is made up of six carbon atoms arranged in a hexagonal ring, to each of which a hydrogen atom is attached (C_6H_6). This structure is called the benzene ring or nucleus. There are three double bonds, and this unsaturation contributes greatly to the chemical reactivity of benzene. It forms a vast number of derivatives by the replacement of one or more of its hydrogen atoms with other elements, groups, or side chains; these are used as dyes, plastics, detergents, elastomers, medicinal chemicals, insecticides, and solvents; among the chemicals made from it are nitrobenzene, ethylbenzene, dichlorobenzene, benzenesulfonic acid, styrene, phenol, aniline, and cyclohexane.

Benzene, b.p. 80°C (176°F), is a mobile liquid, lighter than water, with quite a strong odor; it blends well with other hydrocarbon liquids but is not miscible with water. It is made from toluene (a petroleum product) by hydrodealkylation, a catalytic process which removes the methyl group, or from coal tar by fractional distillation.

It is nonpolar and thus is a poor conductor of electricity. Its vapors are highly flammable and have a narcotic effect on inhalation. It is widely used as a solvent for oils, gums, fats, and waxes. *See also* benzene ring; aromatic.

benzene ring.   Also called the benzene nucleus, this carbocyclic structure is one of the most important in organic chemistry. The full representation, with the symbols included is shown in (1) below, and the once widely used outline in (2), though the double bonds are often omitted. The six carbon atoms may be numbered from the upper apex, as shown in (3), these numbers being used to designate the positions of the carbon atoms or of the elements or groups attached to them. Thus, 2 is the ortho- (or *o*-) position, 3 is the meta (or *m*-) position, and 4 the para (or *p*-) position. Numbers are utilized in some systems of chemical nomenclature in the name of the compound, e.g., 1,2-dichlorobenzene, to show the points of substitution of the chlorine atoms. The prefixes are also used; thus para-dichlorobenzene is another way of indicating 1,4-dichlorobenzene.

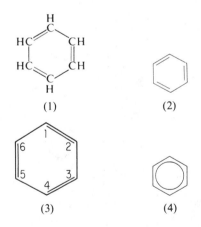

(1)   (2)

(3)   (4)

Benzene exhibits the phenomenon of resonance, which is a mathematical way of stating that the position of the double bonds cannot be precisely pictured by any one diagram but rather is intermediate among a number of representations; for this reason the structure shown in (4) has come into general use. *See also* ortho; meta; para; resonance.

benzenesulfonic acid.   An aromatic acid, m.p. 65°C (149°F) when anhydrous, made by reacting concentrated sulfuric acid with benzene; formula $C_6H_5SO_3H$. It is water-and alcohol-soluble and occurs in the form of long, slender crystals or flat plates, which are strongly deliquescent. Its major uses are in the manufacture of organic chemicals such as resorcinol and phenol, dyes, and similar synthetic products; it also has catalytic activity. Aromatic compounds are more readily sulfonated than are aliphatics.

benzenoid.   Any compound containing one or more benzene ring structures. This term was originally used to distinguish the entire range of aromatic compounds from the alicyclic ring compounds. It is still current in commercial and trade literature but is not much used otherwise. *See also* aromatic.

benzidine.   An organic intermediate and analytical reagent derived from nitrobenzene by reduction or electrolysis; its formula is $NH_2(C_6H_4)_2NH_2$. Besides its value in organic synthesis, it is used in police work to identify blood spots, as a stain for microscopic specimens, and as a source of orange, red, and yellow azo dyes. It has also been used to impart temporary stiffness to unvulcanized rubber mixtures to maintain their shape until cured. Benzidine and its derivatives are active cancer-producing agents; they should be handled with care, since they are absorbed by the skin.

benzoic acid.   An organic acid, m.p. 122°C (252°F), formed by attachment of a carboxyl group to a benzene ring, giving the formula C_6H_5COOH. It can be made from toluene either by oxidation or by chlorination to benzotrichloride, followed by hydrolysis. It can also be extracted from so-called benzoin gum (a balsamic resin) of which both benzoic acid and vanillin are components. Large amounts of benzoic acid were used to make phenol until recently, but this application has diminished in importance. It has been an antimicrobial preservative in food products since the early 1900's, in concentrations not exceeding 0.1%. It is also a mild antiseptic, vulcanization retarder for rubber, and an intermediate in organic synthesis. Various benzoate compounds are derived from it.

benzoin resin.   *See* balsam; benzoic acid. (It is *not* the same as the organic compound benzoin, obtained from benzaldehyde.)

benzopyrene.   An aromatic hydrocarbon, m.p. 179°C (354°F), composed of five fused rings, with formula $C_{20}H_{12}$. It is a strong carcinogen, and its presence in cigarette tars is one of the reasons for the broad concern over smoking as a cause of lung cancer, particularly when inhalation is involved. The concentration, though extremely low, is sufficient to be dangerous on a chronic basis.

benzotrichloride.   See benzoic acid.

benzoyl.   The univalent organic group $C_6H_5C\!\!=\!\!O$, which becomes benzaldehyde when a hydrogen atom is present on the aldehyde carbon and becomes benzoic acid when a hydroxyl group occurs on the aldehyde carbon.

benzoyl peroxide.   An organic peroxide with formula $(C_6H_5CO)_2O_2$, containing 6.5% atomic oxygen. It is in the form of white crystals and is made by reaction of sodium peroxide and benzoyl chloride. A strong oxidizing agent, it is used in bleaching organic materials such as flour, vegetable oils, etc. Its most notable property is its ability to form free radicals by decomposition; these initiate the molecular chain reactions involved in polymerization. Thus, the most important use of benzoyl peroxide is as a source of free radicals for addition polymerization; it can be used to vulcanize rubber in the absence of sulfur. Benzoyl peroxide may explode when heated and is a fire hazard when in contact with combustible materials. *See also* peroxide; free radical.

benzyl.   The univalent organic group $C_6H_5CH_2$; it is derived from toluene and occurs in many related benzenoid compounds. It is conventionally abbreviated Bz.

benzyl alcohol.   A monohydric aromatic alcohol, b.p. 206°C (403°F), having the formula $C_6H_5CH_2OH$; it is obtained by hydrolyzing other benzyl compounds, such as the chloride or the acetate. It is a good solvent for cellulose esters, waxes, resins, and similar materials. It is not as flammable as many other alcohols and has little toxic effect. It is used in suntan preparations and other cosmetics and to some extent as a local anesthetic; also in soaps, inks, and coating formulations.

berkelium.   An element.

Symbol	Bk	Atomic No.	97
State	Solid	Atomic Wt.	243(?)
Group	Actinide Series	Valence	3,4

Berkelium was discovered in 1949, at Berkeley, California, in the form of Bk^{243} as a product of the cyclotron bombardment of americium isotope 241 with helium ions. The eighth member of the actinide transition series, it is a radioactive transuranium element. Berkelium metal, m.p. 986°C (1807°F), has been prepared and several compounds, chiefly oxides and halides of oxidation states 3 and 4, have been identified. Larger amounts of the 249·isotope have been made so that its properties can be investigated. This isotope is made by intense neutron bombardment of curium 244. There are as yet no known uses of berkelium.

Berthelot, M. P. E. (1827-1917). French chemist who did fundamental work on the organic synthesis of hydrocarbons, fats, and carbohydrates. Opposed the then current idea that a "vital force" is responsible for synthesis. Did important work on explosives for French government. He was one of the first to prove that all chemical phenomena depend on physical forces that can be measured.

Berthollet, Claude Louis (1748-1822). French chemist. Followed Lavoisier, but did not accept the latter's contention that oxygen is the characteristic constituent of acids. He was the first to propose chlorine as a bleaching agent. His essay on chemical physics (1803) was first attempt to explain this subject. His speculations on stoichiometry, especially as regards relative masses of reacting atoms, profoundly affected later theories of chemical affinity.

beryl. *See* beryllium.

beryllium. An element.

Symbol	Be	Atomic No.	4
State	Solid	Atomic Wt.	9.01218
Group	IIA	Valence	2

Beryllium, m.p. 1280°C (2336°F), is a unique metal in many respects: (1) it is lighter (sp. gr. 1.85) than aluminum, with a high ratio of strength to weight; (2) it is extremely penetrable by x-rays; (3) it is an excellent neutron reflector (better than graphite for slow neutrons); (4) it is a good source of neutrons, especially when mixed with radium. Hotpressed blocks of beryllium are used for aerospace instruments and inertial guidance equipment, as well as in nuclear reactors. The metal is obtained from the ore called beryl, $Be_3Al_2(SiO_3)_6$, by oxidation and chemical processing. In metallurgical applications, it is an alloying agent (with copper) in the manufacture of springs and diaphragms. Both the metal and its compounds are quite toxic, especially in the form of powder, dust, or fume. Beryllium is classified as a rare metal.

Berzelius, J. J. (1779-1848). A native of Sweden, Berzelius was one of the foremost chemists of the 19th century. He made many contributions to both fundamental and applied chemistry: he coined the words *isomer* and *catalyst*; classified minerals by chemical composition; recognized organic radicals which maintain their identity in a series of reactions; discovered selenium and thorium and isolated silicon, titanium and zirconium; did pioneer work with solutions of proteinaceous materials which he recognized as being different from "true" solutions.

beta (or β) (pronounced bay-ta). A term or symbol having several meanings analogous to those of alpha. Its most common chemical designations are (1) the position of a substituent atom or radical in a compound, (2) the second position in a naphthalene ring, or (3) the fact that a chemical unit is attached to the side-chain of an aromatic compound. Another meaning used by metallurgists, refers to an allotropic modification of a metallic or crystalline structure. Physicists use the term to denote a type of radioactive decay. *See also* alpha particle; beta particle; alpha, gamma.

beta particle. An electrically charged particle emitted by an atomic nucleus (for example, radium) during radioactive decay. If the particles are negatively charged, they are identical with electrons; they may, however, also be positively charged (positrons). They have moderate energy (up to 4 million electron volts) and can be damaging to human tissue; effective shielding can be obtained with thin metal sheets. Collectively, such particles are often called beta rays or beta radiation.

betatron. An electromagnetic device for accelerating electrons (beta particles). Its action is similar in principle to that of an electric transformer, in which the secondary windings are replaced with focusing magnets. The electrons travel around the core in a vacuum tube placed between the magnets. At each revolution around the core the electrons pick up the same energy as the voltage that would have been induced in one turn of wire at that point. The betatron can generate electron beams up to 320 MeV. Invented by D. W. Kerst in 1940, it is used chiefly for basic research.

bi-. A prefix either meaning "two," as in the words binary, bidentate, etc., or indicating the presence of hydrogen in the names of certain inorganic compounds such as sodium bicarbonate ($NaHCO_3$), sodium bisulfate ($NaHSO_4$), etc. In a few compound names, it is interchangeable with di-, but the latter is preferred, e.g., carbon disulfide, diphenyl, etc. *See also* bis-.

bidentate. The literal meaning of this term is "having two teeth;" it refers to a ligand, in a coordination compound, which is bound to two metal atoms. *See also* ligand; coordination compound.

bile. A complex organic liquid secreted by the liver of vertebrate animals and comprised chiefly of cholic acid and its derivatives. It functions in the body as an emulsifying agent of ingested fats and is stored in the gall bladder. The bile acids are steroids having 24 or more atoms and one or more hydroxyl groups. Organic chemists extract important steroid complexes for research purposes from the bile of sharks, oxen, and reptiles. *See also* cholesterol.

binary. In reference to chemical compounds, this term indicates a compound composed of two elements, of an element and a group (hydroxyl, methyl, etc.), or of two groups, e.g., oxalic acid (two carboxyl groups). In reference to solutions, it refers to the presence of two phases or components, as in alloys, glass, etc.

binder. In general, any soft, tacky substance used to hold together such dry solids as, for example, propellants, nonwoven fabrics, and other composites. Specifically, in paint technology, the resin or drying oil that acts as the film-forming component of the paint; acrylic latexes and linseed oil are examples of commonly used paint binders.

binding energy. The energy that holds the protons together in an atomic nucleus. Since protons are positively charged, they exert strong mutually repellent forces, and tremendous energy is required to keep them from flying apart. This energy is so great that it results in a slightly lower value for the mass of nucleus taken as a whole than for the sum of its constituents taken individually. This phenomenon is of vast significance, for it means that a small fraction of mass has been converted into energy within the nucleus, as shown by Einstein's equivalence equation $E = mc^2$. Thus, when a uranium nucleus (92 protons) is split in the fission process, a portion of its binding energy (equivalent to the mass difference) is released; it amounts to about 200 million electron volts per nucleus. *See also* mass defect; fission; energy.

bio-. A prefix indicating life or the presence and activity of living organisms, as in biochemistry, biodegradable, bioluminescence, etc. It also appears in words in other sciences as well as in literature (biology, biography).

bioassay. *See* assay.

biochemistry. The science devoted to the chemistry of living organisms, including plants, bacteria, animals, and man. Within its scope are all the elements and compounds necessary for nutrition and metabolic functions in the organism; activating substances, such as enzymes, vitamins, hormones, etc.; infectious bacteria and viruses and the medicines used to combat them; the chemistry of immunity; the nature of the cell and its nucleus, including genes and chromosomes; and the complex reaction cycles which maintain the chemical balance of the organism. It includes certain aspects of both inorganic and organic chemistry, as well as of colloid and physical chemistry. *See also* molecular biology.

biocide. General name for any substance that kills or inhibits the growth of microorganisms, such as bacteria, molds, slimes, fungi, etc. Many of them are also toxic to humans. Biocidal chemicals include chlorinated hydrocarbons, organometallics, halogen-releasing compounds, metallic salts, organic sulfur compounds, quaternary ammonium compounds, and phenolics. *See also* antiseptic, disinfectant, fungicide.

biocolloid. A general term including all biochemically active aggregates in the colloidal size range, such as protein macromolecules, nucleic acid complexes, and various coenzymes, bacteria, and viruses.

biodegradability. The susceptibility of an organic material to decomposition as a result of attack by microorganisms. The term is applied especially to detergents and insecticides, which vary widely in this respect. For ecological reasons, a high degree of biodegradation is desirable to minimize the adverse environmental effect of such materials. Sewage is highly biodegradable, and present methods of treatment are based on this property. Phosphate compounds and chlorinated hydrocarbons such as DDT are not biodegradable, and it is largely for this reason that their use is restricted. Biodegradable plastic containers have been developed to ameliorate the solid waste problem.

bioengineering. See recombinant DNA.

biogas. Methane derived from animal manures. Large-scale production is under way in several states, as well as in India and China. *See also* biomass.

biological oxygen demand. *See* BOD.

biomass. Plant growth in all its forms, from algae and terrestrial vegetation to agricultural products and their residues, garbage, and metabolic wastes (manures). Biomass is becoming a renewable energy source of increasing importance. Its most important component is wood, both by direct combustion and by chemical or bacterial conversion to alcohols and other liquid fuels. The sap of the copaiba tree of Brazil has been found to be chemically similar to diesel fuel. Intensive cultivation of this and other types of trees as a long-range energy source has been proposed. Biomass may provide 3% or more of U.S. energy requirements by 1985. *See also* gasohol; wood.

biomaterial. Any material used either within the body or as a prosthetic device in contact with external tissue under the supervision of a surgeon or dentist. Such materials must meet standards of tissue compatibility, resistance to degradation, toxicity, and functional dependability. Among the materials used are titanium, tantalum, cellulose, graphite, and reprocessed animal bone; alloys (amalgams, stainless steel); many kinds of plastics, including nylon; ceramics, composites, and elastomers of various kinds. Tremendous progress has been made in recent years in this field. *See also* compatibility.

biotin. A member of the vitamin B complex, biotin is a product of human metabolism; it is an essential growth factor, occurring in milk, egg yolk, and liver and is also made synthetically. The formula is $C_{10}H_{16}N_2O_3S$. Its only use is in medicine as a dietary supplement. *See also* B complex.

birefringent. A term used in crystal optics to describe a type of crystal which has the property of resolving the divergent components of a light ray into two beams that are at right angles to each other. These perpendicular beams separate as they emerge from the crystal, so that two images appear, each of which is the result of a light ray that vibrates in only one direction. This is called plane-polarized light; crystals having this property, such as Iceland spar (a form of calcite), are used in Nicol prisms. *See also* Nicol.

bis-. A prefix used in the names of certain organic compounds to indicate that a particular group occurs twice. For example, in bisphenol, $(CH_3)_2C(C_6H_4OH)_2$, the ''bis'' refers to the phenolic group C_6H_4OH, as the name indicates, though two methyl groups are also present. The prefixes di- and bi- have a similar meaning, as in dibutylamine $(C_4H_9)_2NH$ and biphenyl $(C_6H_5)_2$, but the three are not interchangeable. In the element bismuth, the first syllable is not a prefix, but an integral part of the word.

bismuth. An element.

Symbol	Bi	Atomic No.	83
State	Solid	Atomic Wt.	208.9806
Group	VA	Valence	2,3,4,5

Bismuth, m.p. 271°C (518°F), is a hard, brittle metal whose crystals have a pronounced rhombohedral structure; it has only one stable form. It has two unusual properties: it expands 3.3% when it solidifies, and it is strongly repelled by magnetic fields, assuming a position at right angles to the lines of magnetic force. It forms compounds with halogens, oxygen, nitrogen, and sulfur; it also occurs in a few organic compounds (tartrates) especially prepared for therapeutic purposes. The nitrate and subnitrate are derived from concentrated solutions of bismuth in nitric acid. The most important uses of bismuth are (1) as a component of a number of low-melting (fusible) alloys used in fire protection equipment, metal joining, heat-transfer media, etc.; (2) in pharmaceutical and medicinal products for stomach trouble and treatment of syphilis; (3) in electrical equipment (thermoelectric alloys and permanent magnets); and (4) as a catalyst, especially in the manufacture of acrylonitrile.

An unusual aspect of bismuth is the existence of five naturally radioactive isotopes, at least one of which is involved in the actinium decay series. Though it is not normally considered a radioactive element, it is next above lead in the Periodic Table and thus is not completely stable.

bisphenol-A. An aromatic compound made by reacting phenol and acetone, this substance is a white solid with formula $(CH_3)_2C(C_6H_5OH)_2$. It has become increasingly important in recent years as a result of the development of polycarbonate, epoxy, and polysulfone resins, of which it is a primary ingredient. It is combustible, but not flammable, and is relatively nontoxic, unless directly ingested. *See also* bis-.

bitumen. *See* asphalt.

Bk   Symbol for the element berkelium, named from Berkeley, California, where it was first identified in 1949.

black.   As a noun, this term is used to denote any of the following products: (a) carbon black, a finely divided form of carbon obtained by thermal decomposition of natural gas; (b) furnace black, a type of carbon black made by incomplete combustion, in a closed chamber, of either natural gas or a hydrocarbon oil; the latter is also called oil black; (c) acetylene black, resulting from thermal treatment of acetylene; (d) bone black (about 10% carbon), made by calcining animal bones; (e) vegetable black, a form of carbon derived by destructive distillation of wood; (f) platinum black, a fine-ground platinum powder; (g) ivory black (similar to bone black), made by calcination of elephant tusks; (h) aniline black, a dye used on cotton fabrics.

black liquor.   *See* liquor (a); lignosulfonate; tall oil.

black powder.   *See* deflagration; explosive.

blackstrap.   A term used in sugar technology to refer to the molasses which remains after all the economically recoverable sugar has been removed from cane or beet syrup. It can be used as a source of alcohol and other fermentation products and as an animal feed base. *See also* molasses.

blanc fixe.   *See* barium sulfate.

blanching.   A term used in the food industry to describe the brief heating of fruits and vegetables at 93-100°C (200-212°F), primarily to inactivate enzymes that would cause deterioration or loss of flavor. This process usually precedes canning, dehydration, freezing, and other preservation methods. Notwithstanding the name, it has little effect on color.

blank.   *See* control.

bleach.   To remove or decolorize impurities occurring in numerous materials, including textiles, paper pulp, vegetable and mineral oils, as well as such earth products as barytes and clays, in order to achieve whiteness. Bleaching usually involves a chemical reaction such as oxidation or reduction. Agents commonly used for this purpose are chlorine, hydrogen peroxide, and sodium or calcium hypochlorite; chlorine dioxide and peracetic acid are used in bleaching of fine papers. Oils may be treated with special clays of high adsorptive power to achieve the same purpose. Simple exposure to air and sunlight has a similar effect over a long time period.

block.   (1) Undesirable cohesion of the surfaces of plastics. *See* antiblock agent.

(2) A type of combination polymer. *See* block polymer.

block polymer.   A combination long-chain macromolecule comprised of any given polymer (A) and either (1) another polymer of different chemical nature (B) or (2) a so-called coupling group (C) of relatively low molecular weight. It may be roughly diagrammed as follows:

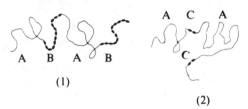

(1)

(2)

A stereoblock polymer is a chain comprised of only one chemical entity (e.g., polypropylene) in which orderly and random molecular configurations occur in alternate sequence. The properties of block and stereoblock polymers are notably modified by their structure and composition; indeed, the segments may have very different properties, resulting in a material which can have more than one form (that is, hard and soft) depending on the solubility properties of its components. *See also* graft polymer; polyallomer.

blood (human).   Considered to be a liquid tissue, blood is a heterogeneous suspension containing the following major components: (1) Red cells, or erythrocytes, which contain hemoglobin, an iron-protein complex; they are not actually living cells, but rather are aggregates of hemoglobin, lipids, and water. (2) White cells, or leucocytes, which are true biological cells; they contain nucleoproteins and lipids and function as a protective mechanism against infections. (3) Platelets, made up of cephalin and other proteins, which play a part in blood clotting. Besides these major components, blood contains many proteins, sugars, cholesterol, fatty acids, and inorganic substances in ionized form. Blood acts in the body as a transport agent for nutrients and for oxygen and carbon dioxide. It is a strongly buffered system and maintains a pH of from 7.3 to 7.5. The coagulation mechanism involves the protein fibrinogen; serum is the fraction remaining after coagulation. Plasma is blood from which the red and white cells have been removed; it is used clinically for reestablishment of the fluid balance in the body and for its protein, carbohydrate, lipid, and mineral content. The human body contains about six quarts of whole blood. *See also* fibrinogen; coagulation.

bloom. (1) Migration of one or more ingredients to the surface of a finished product. This term is used particularly in the rubber industry to refer to uncombined sulfur which appears on the surface of a vulcanized article after long standing, or to waxes added to protect the product from degradation due to sunlight.

(2) Name of a scale used to determine the viscosity of gelatin.

(3) Any sudden increase in the population of marine plants such as algae.

blowing agent. A compound, suitable for incorporation in plastics, rubber, or food products, which will evolve an inert gas (carbon dioxide or nitrogen), usually at high temperature; those used in foamed-in-place plastics react at room temperature. Sodium bicarbonate is a typical blowing agent, as it may be used to make sponge rubber, cellular plastics, and bread and other baked products, all of which are dispersions of a gas in a solid. Sodium bicarbonate and similar compounds give off carbon dioxide at a temperature just below that at which the blown material ''sets up,'' and the globules of gas are thus permanently entrapped within the solid, forming a stable foam. *See also* foam; baking powder.

blown oil. A vegetable or mineral oil through which air is passed at moderately high temperature. Such vegetable oils as linseed, soybean, and castor are thickened by this treatment, as a result of oxidation and partial polymerization, to become suitable for the modification of lubricants, paints, plasticizers, and the like. The most widely used blown mineral oil is an airtreated bituminous residue, the product being known as blown asphalt, or mineral rubber; it is a hard, brittle solid used as a diluent in low-quality rubber goods, as an impregnant for burlap and other coarse fabrics, and in hot-melt coatings. *See also* castor oil; linseed oil.

blue. (1) Any of a number of compounds that have a pigmenting effect in the blue color range, many of which are derived from cyanogen (CN-containing) compounds, often combined with iron (ferricyanides). Examples are Turnbull's blue, which occurs in blueprints, the several types of iron blue, and the organic copper phthalocyanine blues. Ultramarine blue is made by heating sulfur, clay, and an alkaline reducing agent. Iron blue and Ultramarine blue are used as white intensifiers in laundry ''bluing.'' Cobalt blue, a compound of cobalt and alumina, is a stable blue colorant used in glass.

(2) The term is also used in cases where the color blue is apparent, though the materials described are not primarily pigments; for example, blue gas for water gas, blue vitriol for copper sulfate, blue lead for lead sulfate, and the indicator methylene blue. *See also* cyanogen; iron blue; methylene blue.

blush. Condensation of atmospheric moisture on a freshly painted surface as a result of too rapid evaporation of solvent from the paint; the rapid evaporation causes a lowering of the temperature of the air near the surface, which precipitates water vapor and gives an unsightly appearance. It can be avoided by using a less volatile solvent and by applying the paint under conditions of minimum humidity.

BOD. Abbreviation for biological oxygen demand or its alternative name, biochemical oxygen demand. It is the amount, in milligrams per liter, of dissolved oxygen required by aerobic bacteria to decompose organic matter in water solution. The BOD is determined by adding oxygen-saturated water to a sample of the contaminated water and then measuring the dissolved oxygen in the mixture at once and again after several days. The BOD value is thus an index of the contamination of the water. It is an important consideration in sewage treatment, in which proper aeration of the organic material is essential. *See also* activated sludge.

bodied oil. A vegetable oil, such as linseed, whose viscosity has been increased by heating.

boiled oil. *See* linseed oil.

boiler scale. An extremely hard incrustation of calcium and/or magnesium compounds formed on the inner surfaces of boilers and hot-water lines after so-called "hard" water has been in contact with them for a long period of time. The deposition of these insoluble carbonates and sulfates results in heat loss and increasingly impairs the efficiency of the boiler; it also clogs feed lines and ultimately may block them almost completely. The best prevention is to test the water for hardness and to treat it prior to use; it can be demineralized or "softened" by ion-exchange methods. Scale may be prevented in small, low-pressure boilers by use of soda ash, tannins, or phosphates which precipitate the contaminating materials as a sludge. *See also* ion-exchange.

boiling point. For an unenclosed liquid or mixture of liquids, the temperature at which the upward pressure of molecules escaping from the surface (vapor pressure) equals the downward pressure of the atmosphere (14.7 pounds per square inch at sea level). When the liquid is in a closed container, the boiling point may be defined as the temperature at which the liquid and its vapor are in equilibrium, that is, when evaporation and condensation are occurring at the same rate at constant atmospheric pressure. The temperature is the same in both cases. The boiling point is increased somewhat by the presence of molecules of dissolved substances and substantially by the presence of ions. It decreases with decreasing atmospheric pressure. *See also* equilibrium; vapor pressure.

Boltzmann, Ludwig (1844-1906). Born in Vienna, Boltzmann was interested primarily in physical chemistry and thermodynamics. His work has importance for chemistry because of his development of the kinetic theory of gases and the rules governing their viscosity and diffusion. His mathematical expression of his most important generalizations is known as Boltzmann's Law, still regarded as one of the cornerstones of physical science.

bond. An electrostatic attraction acting between atoms of the same or different elements enabling them to unite to form molecules or metallic crystals. The number of bonds that characterize each element, that is, the number of combining forces it can exert, is called its valence and is determined by its electronic configuration. The energetic value of a carbon bond is about 5 electron volts. The several types of bonds (covalent, ionic, metallic, and hydrogen) are described in separate entries. In conventional structural formulas a single bond is indicated by a dash(—), a double bond by two dashes (=), and a triple bond by three (≡); dots are used in electronic formulas. *See also* valence; orbital.

bone. Industrial source of several products (bone char, bone oil, bone meal). Bone char, or bone black, results from calcination of animal bones and is used as an adsorbent in sugar refining to remove color and as a pigment in shoe polishes, etc. Bone oil is obtained by destructive distillation of bones; it has been used as a denaturant for alcohol because of its repellent odor. Bone meal is made by steaming and pulverizing animal bones. Its chief use is as a fertilizer and soil conditioner for its high phosphorus content, as well as an ingredient of fine china.

bone black. *See* bone; black.

bone char. *See* bone.

bone meal. *See* bone.

bone seeker. Any element that tends to concentrate in the bones and teeth of animals and man. Most important are calcium, an essential component of bony structures, and fluorine, which in very low concentrations (1 ppm) can inhibit tooth decay. Strontium is also in this class; the radioactive isotope strontium-90 is extremely dangerous, as it replaces calcium and causes radiation damage in blood- and marrow-forming structures. It occurs in the fallout from atomic bombs, is picked up by vegetation, and finds its way into the body via milk. *See also* fluoridation.

borane. *See* hydride; hydroboration.

borax. A hydrated form of sodium borate ($Na_2B_4O_7 \cdot 10H_2O$) found in the salt beds of the western United States. The anhydrous grade is made from the natural hydrated material. Its most notable industrial use is in glass, where it greatly increases heat resistance; such glasses are often referred to as borosilicate glasses. Borax also has value as a mild detergent and as a flux in the manufacture of ceramics.

Bordeaux mixture. An insecticide composed of copper sulfate and lime, used primarily on potato bugs and other common garden pests. It is applied as a suspension in water and should be kept out of reach of pets and livestock.

boric acid. A comparatively weak acid (H_3BO_3) derived from borax by adding a mineral acid and crystallizing. It is used as an eyewash in water solution and as a mild antiseptic; it also has neutron-absorbing capacity which is utilized in the so-called swimmingpool nuclear reactors. As a component of glass it imparts heat resistance and is also useful in baths for electroplating of nickel. Its esters are used as solvents, fire-retarding agents, and plasticizers.

boric oxide. *See* borosilicate.

borneol. *See* terpene.

borofluoride. *See* fluoride.

boron. An element.

Symbol B	Atomic No. 5
State Solid	Atomic Wt. 10.81
Group IIIA	Valence 3
	Isotopes 2 stable

Boron is a hard, nonmetallic solid, m.p. 2300°C (4172°F), obtained from the ores colemanite and kernite. It may be either a crystalline solid or an amorphous powder, depending on the method of production. It can be produced in several ways: (a) by reduction of boron oxide with magnesium or aluminum; (b) by electrolysis of fused potassium halide salts in which boron-containing compounds are dissolved; (c) by passing gaseous boron compounds and hydrogen over heated metal wires, upon which pure boron deposits. This method is unique, but is slow and expensive.

Boron has a number of unusual properties. The 10 isotope has the highest neutron-absorbing capacity of any known substance. In addition to forming conventional chemical compounds with hydrogen, nitrogen, oxygen, and the halogens, boron combines with carbon and silicon to form carbides, carborane, and various borosilicates. In combination with metals, it also forms a large array of refractory borides, which are crystalline structures in which the boron atoms are uniquely bonded to produce extremely hard, stable materials of high thermal conductivity and very high melting points.

The elemental uses of boron include high-strength filaments used in composites, neutron-absorbing devices for nuclear reactor control, alloying agent with iron, manganese and other metals, and deoxidizing additive in steels and copper. It is an essential plant micronutrient and is nontoxic in low concentrations. *See also* borax; carborane; diborane; decaborane.

boron carbide. An extremely hard, high-melting solid, m.p. 2450°C (4442°F); its formula is usually given as B_4C, but it may be a solid solution resulting from electric furnace heating of boron oxide and carbon. It is used as control rods in nuclear reactors, as a polishing agent and abrasive, as nozzles for sandblast equipment, and as high-strength fibers for composite structures for aerospace applications.

boron hydride. A compound of boron and hydrogen, e.g., diborane (B_2H_6, gas), pentaborane (B_5H_9, liquid), and decaborane ($B_{10}I_{14}$, solid). These highly energetic materials have been explored exhaustively for possible use as rocket fuels, though they have not been widely adopted. However, they are used as polymerization catalysts, in synthesizing organic boron compounds, and as oxygen scavengers in metals. Decaborane reacts with acetylenic compounds to form carborane ($B_{10}C_2H_{12}$). Boron hydrides are highly flammable and are toxic when inhaled; they explode in contact with moisture in any form and must be stored and handled with extreme caution.

boron nitride. An unusual compound (BN) having a graphite-like plate structure; it is a powder of extremely small particle size (1 micron), with a very high melting point 2980°C (5400°F), and superior electrical resistance. Under a compression load of a million pounds per square inch, it can be made as hard as diamond. It has the following uses: furnace-lining refractory, laboratory ware (crucibles), aerospace shielding, plasma technology, dielectric materials, high-strength filaments, and solid lubricant. It is made in an electric furnace from a blend of boric acid and tricalcium phosphate in an ammonia atmosphere.

borosilicate. A heat-resistant glass in which the principal ingredient in addition to silica, soda, and lime, is boric oxide (or boric acid), at a concentration of about 5%. The inclusion of boron greatly reduces the heat expansion property and raises the softening point of such glasses to approximately 650°C (1200°F). It also increases the transmission of ultraviolet radiation compared with that of regular glass. Borosilicate glasses have become commonplace for cooking utensils, laboratory ware, and other high-temperature uses, as well as in sun lamps. Their use for high-level nuclear waste disposal is under consideration. *See also* glass.

bound water. A thin film of water adsorbed on a molecule or other particle, as a result of electrical affinity. It is often found on protein molecules and is extremely difficult to remove. *See also* hydration (2).

Boyle, Robert   (1627-1691). A native of Ireland, Boyle devoted his life to experiments in what was then called "natural philosophy," i.e., physical science. He was early influenced by Galileo. His interest aroused by a pump which had just been invented, Boyle studied the properties of air, on which he wrote a treatise (1660). Soon thereafter he stated the famous law that bears his name (see next entry). Boyle's group of scientific enthusiasts was known as the "invisible college," and in 1663 it became the Royal Society of London. Boyle was one of the first to apply the principle which Francis Bacon had described as "the new method"—namely inductive experimentation (as opposed to the deductive method of Aristotle)—and this became and has remained the cornerstone of scientific research. Boyle also investigated hydrostatics, desalination of sea water, crystals, electricity, etc. He approached, but never quite stated, the atomic theory of matter; however, he did distinguish between compounds and mixtures and conceived the idea of "particles" becoming associated to form molecules.

Boyle's Law.   One of the most fundamental gas laws, discovered in about 1660 by Robert Boyle, an English chemist (1627-91), namely, the volume of any confined gas at constant temperature varies inversely as the pressure applied to it. This rule strictly applies only to an ideal gas, since all actual gases deviate from it slightly, especially carbon dioxide. The principle may be visualized by a piston-in-cylinder arrangement; when the pressure on the piston doubles, the piston moves into the cylinder to a point where the volume of the gas is halved. *See also* gas.

b.p.   Abbreviation for boiling point.

Br   Symbol for the element bromine, the name being derived from a Greek word meaning odor or stink.

brake fluid.   A type of hydraulic fluid composed of lubricants, solvents, and various stabilizers. The lubricant is either a processed castor oil or a polymer of propylene glycol; the solvents used are mixtures containing alkylene glycol ethers, polyglycols, and other aliphatic alcohols; the stabilizers are antioxidants, corrosion inhibitors, and buffers. Specifications govern such properties as boiling and freezing points, degrading

effect on rubber system components, water content, and chemical stability. *See also* hydraulic.

branched chain.   A molecular structure induced in C_4 to C_8 paraffinic hydrocarbons and the corresponding alcohols by a catalytic refining process called isomerization. Such molecules are characterized by a subordinate chain of one or more carbon atoms attached to the principal chain, as indicated below.

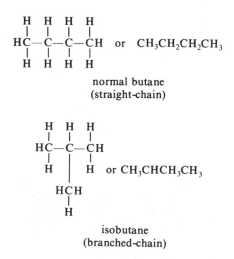

normal butane
(straight-chain)

isobutane
(branched-chain)

Branched-chain compounds are isomers of the normal or straight-chain compounds from which they are derived, and are designated by the prefix iso-. They are much more reactive than the normal compounds and are particularly effective in motor fuels, for which they were originally developed; for example, isooctane has a much higher antiknock rating than its straight-chain isomer octane. *See also* isomerization; octane number.

brass.   An alloy of copper and zinc, often containing some tin. There are many special types of brass, each designed for a specific use, for example, red brass (85% copper, 15% zinc), admiralty brass (73% copper, 26% zinc, 1% tin), and cartridge brass (70% copper, 30% zinc). Brass is corrosion resistant and is widely used for marine equipment, fire-hose nozzles, couplings, etc. *See also* bronze.

brazing.   A type of welding involving the use of nonferrous alloys as filler materials between the pieces of metal to be joined and temperatures above 540°C (1000°F). The "joins" formed by brazing are stronger than those made by soldering, as the filler alloy cannot be remelted once the union is complete. Numerous ternary and quaternary alloys are used.

breeder. A type of nuclear reactor, under development for power generation, which is far more efficient than conventional reactors. It permits conversion of about 70% of the energy in uranium-238 compared to the less than 2% in current practice. In fast-breeder reactions the heavy uranium isotope 238 is irradiated by neutrons from fission of U-235 and is thus converted to plutonium; any unconverted uranium-238 can be reprocessed and returned to the breeder until its energy content has been fully exploited. Thus, the reaction yields more nuclear fuel than it uses. Liquid sodium is used as a coolant, enabling the breeder to operate at a higher temperature, with a corresponding gain in efficiency. No full-scale installation is yet operating in the U.S., but there are several in Europe.

brewing. The production of beer and ale by enzymatic conversion of the starches contained in barley malt to wort, or malt extract. This is then hydrolyzed to sugars and proteins by a process known as mashing, followed by boiling with hops and fermentation with yeast. These processes (all of which are of the batch type) involve complex ractions induced by enzymes and require extremely close control and careful selection of ingredients. Barley is the grain chiefly used as a source of malt; its enzymes are activated by controlled cycles of water absorption and aeration, which induce partial germination of the barley. Beer and ale also contain colloidally suspended proteins which cause foaming; they are clarified by filtration through adsorptive clays or diatomite. *See also* fermentation; yeast.

bridge bond. *See* hydrogen bond.

brightener. (1) In the electroplating field, this term refers to an additive used in certain nickel-plating formulations which imparts increased brightness and reflectivity to deposits on smooth surfaces; they may be sulfonic acid derivatives, metallic ions of zinc or cadmium, or such compounds as thiourea.

(2) An organic fluorescent compound (an ultraviolet absorber), which emits enough blue radiation to conceal off-colors of paper, textiles, and the like. These products are often called optical brighteners and have much the same effect as bleaching agents.

brine. A water solution of sodium chloride of a higher concentration than that of sea water (2.6%). Natural brines of high concentration occur in central New York, Michigan, and in the western deserts. The brines of many lakes in the latter area are virtually saturated salt solutions. Brines are used in the chemical industry as a source of chlorine and caustic, sodium bicarbonate, soda ash, and sodium chloride; they are also used for curing and pickling hides and as supercooling media. *See also* sodium chloride.

brisance. *See* high explosive.

Brix. A density scale used chiefly in the sugar industry to indicate the sucrose concentration of a solution. The results are given in degrees Brix.

bromine. An element

Symbol Br	Atomic Wt.	79.904
State Liquid	Valence	1,3,5,7
Group VIIA	Isotopes	2 stable
Atomic No. 35		

A highly reactive element, bromine is one of the very few that exist as a liquid at room temperature, though it is highly volatile. It does not occur naturally in elemental form, but always in chemical combination. It is obtained commercially from brines, and was the first element to be recovered from sea water (1933). The isotopic composition of bromine is unusual, the percentage of its two stable isotopes (79 and 81) being almost equal. Both liquid and vapor are diatomic (Br_2). The vapor is over fives times as dense as air, and the liquid form boils at 58.8°C (138°F). Bromine reacts violently with aluminum and potassium but does not attack most other metals unless moisture is present. An extremely strong oxidizing agent, it is widely used in bleaches for textiles and paper, and in water purification. Its major industrial use has been in antiknock gasoline (as ethylene bromide), where it acts to keep cylinders free from lead oxide deposits; this use may be expected to decline in the future. Bromine is also important in organic synthesis, as a laboratory reagent, and general sanitizer; it forms a large number of inorganic and organic compounds, e.g., hydrobromic acid, bromates, and bromides. It is toxic by inhalation and strongly irritant to eyes and skin and should be handled with proper protective clothing.

bronze. (1) A mixture or alloy, normally comprised of 90 to 95% copper and 5 to 10% tin, which gives a dark reddish-brown appearance which does not take a high polish. Special types of bronze also contain small percentages of aluminum, silicon, or phosphorus, the last being called phosphor bronze. These alloys are used for special electrical applications, cutting tools, springs in precision equipment, and for the special wire used in papermaking on the fourdrinier machine. In powder form, suspensions of bronze particles in a solvent are used as a pigment for metal coating and decoration.

(2) Descriptive of various inorganic pigments, e.g., iron blue and Red Lake C, as well as of suspensions of bronze powder in organic solvents (bronzing liquids). *See also* brass.

Brownian motion. The discontinuous movement of colloidal particles in a gaseous or liquid suspension, giving the appearance in a microscope of a "dance" or zigzag effect. This was discovered and named after Robert Brown, a Scottish botanist (1773-1858), who noted it in his observation of botanical phenomena. It was later shown by Einstein that the motion is caused by the impact of the molecules of the suspending medium, and it thus provides visible evidence of molecular kinetics. The phenomenon can be observed in rubber latex using an optical microscope.

browning reaction. A condensation reaction occurring in many food products between amino acids and sugars, without the aid of enzymes. The hydroxyl groups of the sugar, e.g., glucose, react with the amide groups of protein to form undesirable brownish end products, which impair appearance, quality, and storage life. In the case of egg white, the glucose can be removed by bacterial fermentation or enzyme action. It is also called the Maillard reaction.

Bu Symbol often used in chemical formulas for the butyl group, C_4H_9.

buffer. An acid-base balancing or control reaction by which the pH of a solution is protected from major change when acids or bases are added to it. The protection is afforded by the presence in the solution of a weak acid and related salt (for example, acetic acid and sodium acetate), which maintain the equilibrium by means of ion transfer and neutralization. The same effect can be obtained by use of a blend of two acid salts; phosphates, carbonates, and ammonium salts are common buffering agents. An example of biochemical buffering in nature is the extremely close maintenance of the acid-base balance of the blood, which contains a weak acid (carbonic) and its salt (bicarbonate). Many commercial products are appropriately buffered to retain their original strength. *See also* acid; base; pH.

builder. A detergent additive which promotes the effectiveness of a natural or synthetic surface-active agent. Phosphates have been used for many years for this purpose but have been considered undesirable because they, along with other nutrients, promote the growth of algae; as a result, detergent wastes are ecologically destructive, especially when discharged into lakes and rivers. Nitrilotriacetic acid is also an effective builder, but its use has been restricted due to possible toxicity. Other compounds that have shown promise in this field are carboxymethylcellulose and ethylenediaminetetraacetic acid (EDTA). *See also* detergent.

bulk density. *See* density.

bunker oil. *See* fuel oil; residual.

Bunsen, Robert Wilhelm (1811-1899). Born in Germany, Bunsen is remembered chiefly for his invention of the laboratory burner named after him. He engaged in a wide range of industrial and chemical research, including blast furnace firing, electrolytic cells, separation of metals by electric current, and production of light metals by electric decomposition of their molten chlorides. He also discovered two new elements—rubidium and cesium.

burette. A liquid-measuring device used extensively in chemical laboratories. It is a vertical glass tube, open at the top and supported on a bracket; it is equipped with scale graduation marks and a hand-operated stopcock at or near the bottom. The liquid to be dispensed is flowed in at the open end and can then be withdrawn dropwise by manual adjustment of the stopcock.

burnt sienna. *See* iron oxide.

butadiene-1,3. An aliphatic conjugated alkadiene (diolefin) having the formula $CH_2{=}CH—CH{=}CH_2$; it is derived by dehydrogenation of petroleum gases (butane, butene), as well as from ethylene. It is a highly reactive hydrocarbon; because of its structural similarity to isoprene, it formed the basis of research leading to the development of synthetic rubbers and related elastomers. It is a flammable gas, b.p. $-3°C$ (27°F), which has a strong tendency to polymerize in the presence of oxygen. Its chief use is in SBR rubbers, where it is copolymerized with styrene, as well as in nitriles and polybutadiene resins. Butadiene is dangerous to handle, as it may explode when exposed to air; shipment is in steel cylinders and specially equipped tanks. There is also a 1,2-isomer which is not conjugated. *See also* elastomer; conjugated compound.

butane. A saturated hydrocarbon gas, b.p. $-0.5°C$ (31°F), obtained from petroleum refining; formula C_4H_{10}. Highly flammable, it is shipped as a liquid in steel cylinders and is used as a household and industrial fuel. It is a basic source of elastomers and is a valuable organic intermediate. It also has some application as a refrigerant and aerosol propellant. It is not soluble in water and has no corrosive or toxic properties. It should be handled with caution, as it may explode when heated in a container. When compressed, it is a liquefied petroleum gas (LPG).

butanediol. *See* butylene glycol.

butanol. *See* butyl alcohol.

butene (also called butylene). Two unsaturated hydrocarbon gases are included under this name: butene-1 (formula $CH_2{=}CHCH_2CH_3$) and butene-2 (formula $CH_3CH{=}CHCH_3$); the latter has both *cis-* and *trans-* isomers. The butenes are derived from petroleum refinery gases and are used in making butadiene, high-grade gasolines, and as sources of synthetic chemicals containing four and five carbon atoms in the chain. They are highly flammable and explosive and are shipped in steel cylinders. Butene-1 can be polymerized to polybutene to form a useful elastomeric plastic. Butenes are widely used in liquefied form as household and industrial fuels (LPG).

butter. (1) The glycerides of the fatty acids present in milk, of which about half is butyric acid, from which the product is named. The fat particles coalesce to a semisolid as a result of mechanical agitation, which ruptures the protective protein coating and allows them to cohere. The product is colorless, but permissible colorants are added for acceptability.

(2) Obsolete term for various metallic chlorides whose hygroscopic nature gives them a viscosity suggestive of butter, e.g., butter of tin.

butyl. The univalent group C_4H_9, the fourth member of the homologous series of paraffin hydrocarbons; it is derived from butane by dropping one hydrogen atom. There are four isomeric forms. When a second hydrogen atom is dropped from butane, the divalent butylene group results, $—C_4H_8—$, which represents the corresponding olefin. The abbreviation Bu is often used in formulas.

butyl alcohol. A monohydric alcohol (butanol) having four isomeric forms:

normal	$CH_3CH_2CH_2CH_2OH$
iso-	$(CH_3)_2CHCH_2OH$
secondary	$CH_3CH_2CHOHCH_3$
tertiary	$(CH_3)_3COH$

All these, except tertiary, are liquids having limited solubility in water, of varying degrees of flammability, and of moderate toxicity. Their chief industrial value lies in their solvent power. The normal compound was formerly derived by fermentation of carbohydrates but is now made synthetically by aldol condensation of acetaldehyde or by oxidation of petroleum gases; its major uses are in nitrocellulose lacquers and other resin coatings, as a selective solvent for pharmaceutical products, in plasticizers, and in manufacturing chemically related organics. Isobutyl alcohol has quite similar properties and uses. Secondary butyl alcohol, made from butylene derived from petroleum cracking, is mainly used to manufacture methyl ethyl ketone; other uses include solvent blends for alkyd resin coatings, and in synthetic perfumes, dyes, etc. The tertiary form, also petroleum-derived, is a crystalline solid at room temperature; it is also completely water-soluble and the most flammable of the group. Besides being an effective solvent, it has specialized uses as a dehydrating agent, in preparing artificial musk, and as a denaturant for ethyl alcohol.

butylene. *See* butene.

butylene glycol. A dihydric alcohol (also called butanediol) existing in several isomeric forms and having the general formula $C_4H_8(OH)_2$. The two hydroxyl groups are attached to different carbon atoms in the chain. The 1,3- and 1,4-isomers are liquids; the 2,3-isomer is a solid. All forms readily absorb atmospheric moisture. Butylene glycols are used as solvents and plasticizers, as humectants, in the manufacture of various resins (polyesters, polyurethanes), in flavoring extracts (1,3-), and as ingredients of printing inks. They are neither toxic nor flammable and are water-soluble. *See also* glycol.

butyl rubber.   A synthetic elastomer produced by catalytic copolymerization of isobutene (isobutylene) with no more than 3% of isoprene; the latter is necessary to permit vulcanization with sulfur-accelerator systems. An early form of synthetic rubber, butyl can be used as a component of tire carcases, though it is not especially suitable for treads. It is superior to natural rubber in resistance to attack by oxygen and ozone, in dielectric strength, and particularly in ability to retain gases, a property which makes it valuable as an inner lining for tubeless tires. In latex form it is an excellent insulating material for wires and cables, as well as for coating fabrics and composites. *See also* elastomer.

butyric acid.   A straight-chain fatty acid, having the formula $CH_3(CH_2)_2COOH$, which occurs in the milk fat of mammals in the form of a glyceride. It is not found in any other type of fat; its concentration in milk fat is about half the total fatty acid content. Butyric acid is a liquid characterized by a rather powerful unpleasant odor suggestive of sour milk. It is nonetheless used in synthetic perfumes and flavors, as well as in emulsifying and shortening food additives. It is essentially nontoxic and nonflammable, but its uses are limited by its objectionable odor.

butyric anhydride.   *See* cellulose acetate.

butyrolactone.   *See* lactone.

C

C (1) Symbol for the element carbon, the name being derived from the Latin word for charcoal.

(2) Abbreviation of centigrade (or Celsius) temperature scale.

Ca Symbol for the element calcium, the name being derived from the Latin word for limestone.

cadaverine. *See* ptomaine.

cadmium. An element.

Symbol Cd	Atomic Wt.	112.4
State Solid	Valence	2
Group IIB	Isotopes	8 stable
Atomic No. 48		

Cadmium is a soft metal, m.p. 320.9°C (610°F), which is subject to corrosion and tarnishing. Two notable properties are its extremely high capture cross-section for neutrons (2500 barns), which makes it unique in its ability to control the rate of nuclear reactions in a reactor, and its toxicity, which is severe enough to cause poisoning by inhalation of fume or dust and also to prevent its use in any material in contact with foods. It should be handled with adequate respiratory protection. It is used in brazing alloys, electrical contacts, and in electroplated coatings; nickel-cadmium storage batteries are another important application. It is also the basis for a series of pigments used chiefly in ceramics and baking enamels. Cadmium-based alloys find use in heavy-duty bearings. Control rods in nuclear reactors are made of cadmium. All its compounds share in its toxicity.

caffeine. An alkaloid occurring in plants from which coffee and tea are derived. Though not as toxic as most other alkaloids, it stimulates heartbeat and thus causes insomnia, nervousness, and irritability. Most of the caffeine naturally present in coffee can be removed without taste impairment. The permissible soft drink content is 0.02%. It has a few specific medical uses. The formula is $C_8H_{10}N_4O_2 \cdot H_2O$. *See also* alkaloid.

cage compounds. *See* clathrate; gas hydrate; molecular sieve.

calciferol. *See* ergosterol.

calcination. Heating an inorganic material such as limestone, gypsum, magnesium carbonate, or a metallic ore to a temperature below its melting point but high enough to effect evolution of carbon dioxide, and removal of bound water and of other impurities by oxidation, reduction, etc. Calcination also induces phase changes in some metal ores and in glass. In metallurgy it is called roasting when oxidation is involved and smelting when a reduction reaction occurs. *See also* pyrolysis; smelt; roast.

calcite. *See* calcium carbonate.

calcium. An element.

Symbol Ca	Atomic Wt.	40.08
State Solid	Valence	2
Group IIA	Isotopes	6 stable
Atomic No. 20		

Calcium, m.p. 845°C (1553°F), is a soft metal of the alkaline earth series; it can be derived in the pure state but does not occur naturally. It oxidizes on exposure to air and causes water to decompose and evolve hydrogen; for this reason metallic calcium should be kept from contact with air. It has no toxic properties. Calcium is a metabolically essential mineral which is deposited in bones and teeth, a process which is catalyzed by vitamin D; the best nutrient source is milk. Though it is best known in the form of compounds, the metal has a few specialized uses as a reducing and alloying material for other metals and as a getter (scavenger) in vacuum tubes.

calcium carbide. *See* acetylene; carbide.

calcium carbonate. One of the more abundant inorganic substances in nature, calcium carbonate ($CaCO_3$) is the calcium salt of carbonic acid (H_2CO_3). It is a major component of sedimentary rocks (limestone) and occurs in a relatively pure state as calcite (Iceland spar), marble, and chalk, as well as in oyster shells. Its very slight solubility in water is the chief cause of "hardness." Calcium carbonate reacts when heated to evolve carbon dioxide, lime (CaO) being the industrial product. When $CaCO_3$ is treated with an acid, CO_2 is evolved and the calcium salt of the acid is formed. Important chemical uses are in the manufacture of sodium carbonate by the Solvay process and in the smelting (reduction) of iron ore in steel manufacture. It is also a source of calcium carbide when reacted with coke and of calcium choride on reaction with hydrochloric acid. As a bulk solid it is widely used as a building stone, especially for decorative purposes (marble). In pulverized form it is known as whiting. *See also* calcium oxide; calcium chloride; whiting; Solvay process.

calcium chloride. A deliquescent solid made by reacting calcium carbonate with hydrochloric acid; its has the formula $CaCl_2$. It is also obtained from brines and from the manufacture of sodium compounds by various methods. It is corrosive to metals unless promptly removed by washing. It often occurs in the form of a hydrate with one, two, or six molecules of water. Calcium chloride has a number of uses including pulp and paper treatment, deicing of highways, as a desiccating agent, and in food products as a sequestering agent. It is essentially nontoxic.

calcium hydroxide. *See* calcium oxide.

calcium hypochlorite. *See* bleach.

calcium oxide. A grayish-white powder (CaO), often called lime or quicklime. It is obtained by heating (calcining) limestone, as a result of which carbon dioxide is evolved from the calcium carbonate. Calcium oxide is used as such in glass and hydraulic cement mixes, in the treatment of

organic wastes, in food processing, and in poultry feeds. The process called slaking (slacking) involves reaction of calcium oxide with water to form calcium hydroxide [$Ca(OH)_2$] or hydrated lime; a slurry of calcium hydroxide is commonly called milk of lime. $Ca(OH)_2$ is irritating to tissue in the presence of moisture and will decompose organic materials. It is used in plasters and cements, as an accelerator in low-grade rubber products, as a disinfectant and unhairing agent in the leather industry, and as a sanitizing agent. *See also* calcination.

calcium phosphate *See* baking power; phosphor; phosphorus.

calcium sulfate. A grayish-white heavy powder, which occurs in both anhydrous ($CaSO_4$) and hydrated ($CaSO_4 \cdot 2H_2O$) forms, the latter being known as gypsum. It is found in nature in many localities and is also a byproduct of many chemical reactions. The anhydrous product is used to control the setting rate of Portland cement, as a weighting agent in paper sizes, and for coating textile products. It is also a neutralizing agent in soil conditioning. The hydrated form (gypsum) has the property of quickly setting to a hard solid in the presence of water; it is the chief component of plasters and cements, such as plaster of paris, which is the hemihydrate, $CaSO_4 \cdot \frac{1}{2}H_2O$. It has also recently been developed as a commercial source of sulfur.

californium. An element.

Symbol Cf	Atomic Wt.	245
State Solid	Valence	2,3,4
Group IIIB (Actinide Series)		
Atomic No. 98		

Californium is an artificial radioactive element of the transuranic portion of the actinide series; it was made in 1950 by helium bombardment of curium. The 249 isotope has a half-life of 360 years. It forms compounds with halogens and oxygen. The 252 and 249 isotopes have been made in minute quantities, Californium is a bone-seeking element for which the body has a very low tolerance; it has a few specialized medical applications.

calomel. *See* electrode.

calorie. *See* kcal.

calorimetry. Measurement of the heat gained or lost by a substance when its physical state is altered or when it takes part in a chemical reaction. Under these circumstances the gain in the internal energy is equal to the difference between the heat absorbed and the work done by the system. Calorimetric measurements are useful in determining important thermodynamic properties of substances, for example, heat capacity, entropy, and free energy.

calorizing. *See* cementation.

camphene. *See* camphor.

camphor. An unusual vegetable product, camphor is derived by distilling the wood of a species of tree found chiefly in Taiwan. Chemically it is a ketone, with the formula $C_{10}H_{16}O$ and is characterized by a strong, penetrating odor. It is made synthetically from pinene by heating with sodium acetate to form camphene ($C_{10}H_{16}$), from which camphor is then derived by reaction with nitrobenzene and acetic acid. Camphor is flammable and evolves explosive vapors on heating. It is a plasticizer for nitrocellulose and was so used in the first synthetic plastic (Hyatt, 1869). It is also used as an insecticide and mothproofing agent and in medicine (camphor oil).

Canada balsam. *See* balsam; Nicol.

Cannizzaro, Stanislao (1826-1910). Born in Italy, Cannizzaro extended the research of Avogadro on the molecular concentrations of gases and thus was able to prove the distinction between atoms and molecules. His investigations of atomic weights helped to make possible the discovery of the Periodic Law by Mendeleev. His research in organic chemistry led to the establishment of the Cannizzaro reaction involving the oxidation-reduction of an aldehyde in the presence of concentrated alkali.

caprolactam. The monomer used in the manufacture of nylon 6, originally developed in Germany; it is derived by several methods from cyclohexane or from phenol, catalytic synthesis being involved, together with complex molecular rearrangements. In addition to its use in synthetic fibers, it is an ingredient of paints and plasticizers and also serves as a cross-linking agent. Its formula is $CH_2(CH_2)_4NHCO$. It is furnished in both solid and liquid forms and is water-soluble. It can be used to synthesize the amino acid lysine. *See also* nylon.

capryl alcohol. *See* octyl alcohol

capture. *See* absorption **(3)**.

carbamide. *See* urea; amide.

carbaryl. *See* insecticide.

carbazole. *See* anthracene.

carbene. A short-lived and strongly reactive group containing a carbon atom having only six instead of eight valence electrons. The composition of this group, which is derived by the action of ultraviolet radiation on ketone, is $:CH_2$. Because of its lack of electrons, the carbon atom has only two valences, which accounts for the exceptional activity of carbene and its value as an initiator of organic reactions. Its behavior is similar to that of carbonium ions, carbanions, and free radicals, all of which contain unstable carbon. The term methylene is used synonymously. *See also* carbonium ion; free radical.

carbide. A compound composed of two constituents, namely, carbon and a metal or nonmetal; examples are calcium carbide (CaC_2), from which acetylene is made, aluminum carbide (Al_4C_3), silicon carbide (SiC), boron carbide (B_4C), and tungsten carbide (WC). Carbides are usually made by heating the reacting substances in an electric furnace. Some are characterized by extreme hardness and resistance to high temperature; these are used as abrasives and refractory furnace linings (silicon carbide) and for metal-cutting drills and similar tools (tungsten carbide). Carbides may be pulverized, mixed with a similarly powdered metal such as cobalt or iron, and compressed at high temperature to form so-called cemented carbides.

carbocyclic. Any cyclic compound in which only carbon atoms appear in the ring, e.g., cyclohexane (alicyclic) and benzene (aromatic). *See also* heterocyclic.

carbohydrate. A class of compounds which are basic components of plants of all types; they collectively constitute the most abundant group of natural organic substances. They include both simple compounds (sugars) and polymers of these (starches and cellulose), the latter being called polysaccharides. Carbohydrate molecules are composed of carbon, hydrogen, and oxygen, the ratio of the last two being the same as exists in water. The atoms are so arranged that each molecule contains a hydroxyl group and a carbonyl group—a structure known as a saccharose unit.

The simple sugars include fructose (levulose), glucose (dextrose), and sucrose; the polymers comprise all varieties of starches, seaweeds, natural gums, and cellulose. *See also* sugar; starch; cellulose; polysaccharide; seaweed; gum.

carbolic acid.   *See* phenol.

carbon.   An element.

Symbol	C	Atomic No.	6
State	Solid	Atomic Wt.	12.0111
Group	IVA	Valence	(2),4
		Isotopes	2 stable
			4 radio-
			active
			(natural)

Classed as a nonmetal, carbon characterizes all organic compounds and is essential to the photosynthetic reaction on which all living organisms depend. It also occurs in a few inorganic compounds (carbon oxides and metallic carbonates). Though inactive at room temperature, it is highly reactive above 540°C (1000°F). Carbon occurs in compounds that number in the hundreds of thousands, e.g., hydrocarbons (petroleum) and carbohydrates (plants). It is the only element that can form four covalent bonds, resulting in stable chains in which carbon combines both with itself and with other nonmetals (hydrogen, oxygen, sulfur, nitrogen). It forms binary compounds called carbides with many metals and some nonmetals. Its electronegative nature makes it a reducing agent for metallic oxides; and its high electrical conductivity accounts for its use as electrodes, brushes, contacts, and other electrical equipment. It is also a constituent of organometallic compounds such as Grignard reagents, tetraethyllead, and triethylaluminum.

Carbon exists in several allotropic forms, two of which (diamond and graphite) are crystalline, the others being amorphous (coal, carbon black, activated carbon, and charcoal). The latter (except coal) are generally known as industrial carbon. A unique feature of carbon is its ability to notably affect the properties of metals in which it is present in very low concentrations; this is especially true of steel, in which carbon content varies from 0.02 to 1.5%. Since the 12 isotope comprises over 99% of carbon, its mass is taken as the international standard for determination of atomic weights. The natural radioisotope 14 (radiocarbon), with a half-life of over 5000 years, is used in dating geological eras and archeological specimens and in biochemical tracer research. A few compounds (carbenes) are known which contain divalent carbon. *See also* nonmetal; industrial carbon; diamond; graphite; carbon black; carbene; carbide; organometallic; Grignard.

carbonate.   A compound formed by the reaction of carbonic acid (H_2CO_3) with either a metal or an organic compound; thus, carbonates are either salts (inorganic) or esters (organic). The inorganic carbonates are among the more common substances; all of them decompose on exposure to high temperature or to a strong acid, evolving carbon dioxide. Calcium carbonate ($CaCO_3$), also called limestone or calcite, is a component of sedimentary rocks; other important materials of this class are the magnesium, sodium, and potassium carbonates and sodium bicarbonate. The organic carbonate esters (diethyl and diphenyl carbonates) are used as solvents and plasticizers and in the synthesis of polycarbonate resins. *See also* polycarbonate.

carbon black.   A type of amorphous carbon notable for its extremely small particle size, which results in a surface area as high as 18 acres a pound (for channel black). There are four grades: (1) channel black, made by high-temperature decomposition of natural gas, the particles being deposited on a metal bar or channel immediately above the flame (sometimes called impingement black for this reason); (2) furnace black, made by partial combustion of a mixture of hydrocarbon oils, gases, or vapors in an enclosed chamber; (3) thermal decomposition black, made by heating the gas in the chamber without air being present; and (4) acetylene black, made by decomposition of acetylene in a special refractory unit. Furnace black is now the most widely used. By far the largest volume of carbon black is consumed by the rubber industry, as its high surface area makes it an excellent reinforcing and abrasion-resistant agent, especially in tire treads. It is also widely used as a pigment in paints and printing inks, in carbon paper, and in plastics. Acetylene black imparts electrical conductivity to products in which it is used. *See also* reinforcing agent.

carbon cycle.   This term refers to two quite different phenomena, one occurring in the sun and other stars, the other in the earth's biosphere. As to the first, carbon acts as a catalyst of the thermonuclear reactions responsible for the sun's energy, namely, the formation of helium from four hydrogen nuclei (protons). The carbon is not transformed in the process. The second meaning involves the complex reaction known as photosynthesis, in which carbon is absorbed from the air by plants as carbon dioxide, then moves through photochemical reactions catalyzed by chlorophyll, and finally through decay or combustion sequences wherein it reappears as carbon dioxide. *See also* photosynthesis; chlorophyll; fusion.

carbon dioxide.   A gas at room temperature, carbon dioxide (CO_2) is the product of combustion reactions involving carbon-containing substances and of organic degradation processes. It is present in the air to the extent of 0.033% by volume, as a result of the decay of vegetation, the respiration of organisms, and the burning of fuels. A small amount comes from natural sources in the ground. Carbon dioxide is 1.5 times as heavy as air but diffuses uniformly at sea level. It is an asphyxiant at high concentrations but exerts an important regulating effect in the body in normal concentrations. It is relatively unreactive at room temperature and is often used as an ''inert atmosphere'' for storing chemicals which react with the oxygen or water vapor in the air. It is the source of carbon in the photosynthetic process.

Carbon dioxide is readily generated by the action of an acid on a carbonate or by exposing a carbonate to heat; it is also a product of the fermentation of carbohydrates and yeast. In gaseous form it is used in fire extinguishment, in carbonated beverages, as a propellant for aerosol foams, and in special types of lasers. It is also the blowing agent in bakery products, cellular rubber, etc.

The gas can be condensed by pressure to a snow-like solid called dry ice, which has wide use as a refrigerant; it vaporizes (sublimes) in direct contact with air at $-79°C$ ($-110°F$). The extremely low temperature of dry ice can cause skin burns. When the gas is dissolved in water, carbonic acid (H_2CO_3) is formed, of which carbon dioxide is the anhydride. *See also* carbonic acid; carbonate; greenhouse effect.

carbon disulfide.   A yellowish, mobile, heavy liquid (CS_2), b.p. 46.3°C (115°F), having a distinctive and unpleasant odor. Its most noteworthy property is its sensitivity to heat; it has the lowest ignition point of any industrially used liquid, 100°C (212°F), and a flash point of $-30°C$ ($-22°F$). Even frictional heat is sufficient to ignite it, and it is likely to explode if heating occurs when the liquid is partially confined, as in a beaker or flask. The fumes of carbon disulfide are poisonous, and the greatest care should be exercised in all aspects of handling and storage.

Manufactured by heating carbon, or such carbon-containing materials as charcoal, with sulfur in a furnace, its major uses are as a solvent for many organic materials (cellulose, rubber, synthetic resins, waxes, etc.), in the manufacture of regenerated cellulose products (cellophane and viscose rayon), and in the production of carbon tetrachloride. It also finds use as a solvent extraction medium and as a fumigant. It is classed as an inorganic compound. *See also* flash point; carbon tetrachloride.

carbonic acid.   A weakly acidic compound formed by the reaction $H_2O + CO_2 \rightarrow H_2CO_3$; it has never been obtained as such as it is very unstable. Since it is an acid, it will react with metallic bases to form salts, which in this case are called carbonates. The CO_3 group is divalent, and the water-soluble compounds in which it occurs tend to ionize, yielding positive metal ions and carbonate ions $(CO_3)^=$. Carbonic acid can also react with alcohols and other organic compounds to form esters, such as diethyl carbonate, which can be polymerized to form polycarbonate resins. *See also* carbonate; polycarbonate.

carbonium ion.   An unstable form of carbon in which the carbon atom has only six valence electrons instead of the normal eight. This results in a positively charged carbon ion which is an extremely active reaction intermediate in, for example, catalytic cracking of petroleum and many aromatic reactions. The carbonium ion is one of the most important of such unstable initiators, others being free radicals and carbenes.

carbon monoxide.   An almost colorless and odorless gas (CO) produced for industrial purposes by several processes: reaction of carbon with metallic oxides, carbonates, or steam at high temperatures (synthesis gas), and partial combustion of hydrocarbons. Carbon monoxide is evolved whenever a carbon-rich fuel is incompletely burned, as in automobile cylinders. It is a substantial component of producer, blast-furnace, and coke-oven gases. At high temperature it is a powerful reducing agent. It is both highly flammable and extremely poisonous; the hemoglobin of the blood adsorbs it more than 200 times as readily as it does oxygen. It often occurs in the shafts of coal mines, where it is an occupational hazard. It is classed as an inorganic compound.

Carbon monoxide is used industrially as a reducing agent, removal of sulfur dioxide from stack gases, preparation of iron ore for steelmaking, and other metallurgical processes. When combined with hydrogen in a high pressure-catalyzed synthesis, it is the chief industrial source of methanol (methyl alcohol). It is also a versatile intermediate in a number of organic reactions, among which are catalytic hydrogenation yielding a wide variety of compounds, Oxo reactions for conversion of unsaturated hydrocarbons to aldehydes and alcohols; reaction with acetylene to form acrylic acid; and phosgene synthesis. A unique feature of carbon monoxide is its ability to form metal carbonyls, which are then converted to pure metals by heat treatment, as in the Mond process for nickel production. *See also* Oxo process; synthesis gas; methyl alcohol; phosgene.

carbon tetrachloride. A heavy liquid (CCl₄), spr. gr. 1.585, b.p. 76.74°C (170°F), with a sweetish, distinctive odor. It is a member of the large group of chlorinated hydrocarbons, in which the hydrogen atoms of methane are replaced by chlorine. It is widely used as a solvent for heavy oils, crude rubber, and greases, as a cleaning agent for metal surfaces, as a fumigant, and in the manufacture of fluorocarbon refrigerants. Though noncombustible, it should not be used to extinguish fires since one of its decomposition products is phosgene, COCl₂. Carbon tetrachloride is quite poisonous and narcotic; skin contact and prolonged inhalation of its fumes should be avoided. *See also* chlorinated hydrocarbon.

carbonyl. The unsaturated group C=O which occurs in a wide range of organic compounds, including aldehydes, ketones, sugars, and organic acids. Its reactivity is strongly affected by the element or group to which it is attached (hydrogen, oxygen, hydrocarbon, amine, etc.). For example, it is much more active in aldehydes and ketones than in organic acids. In industrial processes, it results from catalytic oxidation of alcohols at high temperature, as well as from oxidation of hydrocarbons. A carbonyl group may form a coordination compound with either one or two atoms of a heavy metal, such as

nickel, iron, or cobalt, by reaction with carbon monoxide. Of these metal carbonyls, nickel carbonyl, Ni(CO)₄, is noted for its toxicity, as it evolves carbon monoxide above 38°C (100°F). Together with several other metal carbonyls, it has knock-suppressing properties in gasoline. *See also* aldehyde; ketone.

carbonyl chloride. *See* polycarbonate; phosgene.

carborane. The parent substance of a group of unique organic compounds containing boron, carbon, and hydrogen (B₁₀C₂H₁₂); some members also contain unsaturated organic groups (isopropenylcarborane). Carborane is made by reaction of acetylene with decaborane (B₁₀H₁₄) and diethyl sulfide. The unusual crystal structure is that of a 20-sided polyhedron. The nucleus common to these compounds (B₁₀C₂) often has a closed, cagelike form. Carborane is a solid, melting at about 260°C (500°F); it burns with intense heat and is highly toxic. Because of their value as an energy source, the carboranes have been researched for possible use as aerospace propellants and for initiation of high-polymer synthesis. *See also* boron hydride.

carboxyl. A univalent group or radical which is characteristic of organic acids (generally called carboxylic acids for this reason). It is shown in formulas as either —COOH or —CO₂H and is made up of a carbonyl group and a hydroxyl group; its structure is

The carbon-to-oxygen unsaturation within the carboxyl group, indicated by the double bond, is of a different order from the carbon-to-carbon unsaturation in the alkyl chains of unsaturated fatty acids. For this reason, fatty acids in which no double bond is present except that in the carboxyl group are called saturated. A carboxyl group normally occupies the terminal position in fatty acid molecules, of which it is the hydrophilic (negatively charged) portion. The hydrogen atom can be replaced by other elements in reactions involving the formation of metal salts and soaps. Two carboxyl groups occur in some acids, in which case they are called dicarboxylic. *See also* carbonyl; carboxylic acid; fatty acid.

carboxylic acid. Organic acids containing one

or more carboxyl groups (COOH); the simplest members of this series are formic acid (HCOOH), which occurs in the venom of bees and ants, and oxalic acid (HOOCCOOH), formed by oxidation of carbohydrates. A well-known carboxylic acid is acetic (CH_3COOH). These three acids have toxic properties, though this is not true of most organic acids. All the amino acids are carboxylic. *See also* fatty acid; amino acid; carboxyl.

carboxymethylcellulose. A water-soluble cellulose derivative, sometimes called cellulose gum, made by impregnating cellulose with caustic soda (NaOH) and reacting the mixture with sodium monochloroacetate, to give a product which is a sodium salt of the carboxylic acid $ROCH_2COONa$ (where R stands for cellulose). The molecular weight may run as high as half a million. This material has many useful properties; its solubility in water makes it uniquely applicable in food products where it acts as a thickener and stabilizer of emulsions; it is also used in detergents and as a coating agent for paper and textiles. It is one of the most widely used of the many types of modified cellulose. *See also* methylcellulose; cellulose.

carburetion. Addition of carbon, usually in the form of hydrocarbon gases or vapors, to another gas (air) as in the carburetor of an internal combustion engine. Carbureted water gas is a mixture of water gas (derived from steam and coke) and so-called oil gas (hydrocarbon gases from thermal dissociation of petroleum).

carcinogen. Any substance which has been found to induce the formation of cancerous tissue in experimental animals. Such findings, as in the case of cyclamates, have usually been determined by either feeding abnormal quantities of the suspected agent or by applying concentrated solutions directly on the skin of the animal; in many cases, these tests are far from conclusive with regard to normal ingestion of the substance by humans. A number of benzenoid compounds are carcinogenic, e.g., benzopyrene (in cigarette tars), benzidine, and phenanthrene, as well as other types such as the nitrogen mustards. The most dangerous is benzopyrene, since it involves chronic inhalation affecting the large surface area of the lungs. The number of suspected carcinogens is constantly increasing; some (such as asbestos fibers) are well documented, but others are not.

carnauba wax. An extremely hard, brittle wax, m.p. 84-86°C (183-187°F), obtained from the leaves of a Brazilian palm tree. Its high cost limits its uses, one of the most important being in carbon paper; it also finds application in floor polishes, leather dressing, and electrical insulation. It is excellent for preventing deterioration of rubber products due to sunlight. Being nontoxic, it is an acceptable ingredient of chewy confectionery and other food products. *See also* wax.

carnotite. *See* radium; uranium.

carotene. *See* carotenoid.

carotenoid. One of a group of natural pigments found in plants, the best-known of which is carotene ($C_{40}H_{56}$), a provitamin from which vitamin A is formed in the liver and which is especially abundant in carrots. It is extracted and used for coloring butter, margarine, etc., and in specialized pharmaceutical products. Xanthophyll is an oxygenated carotenoid present in the red and yellow colors in autumn foliage and other flower pigments, as well as in egg yolk. *See also* vitamin; xanthophyll.

Carothers, Wallace H. (1896-1937). Born in Iowa, Carothers obtained his doctorate in chemistry at University of Illinois. He joined the research staff of DuPont in 1928 and made several notable contributions to synthetic organic chemistry. One of these was polychloroprene, the first successful synthetic rubber, later called neoprene. Further research in the detailed chemistry of polymerization culminated in the synthesis of nylon, the reaction product of hexamethylenetetramine and adipic acid. Carothers' work in the polymerization mechanisms of cyclic organic structures was brilliant and productive, and he is regarded as one of the most original and creative chemists of the early 20th century.

carrageenan. A seaweed component which is a carbohydrate polymer of the polysaccharide group. It is a colloidal polyelectrolyte which has marked water-absorbing properties; some forms also have gel-forming ability. Biologically it is from one of the red algae technically named *Chondrus crispus* and popularly known as Irish moss. It is harvested chiefly off the New England coast. It is an ideal emulsifier in dairy food products, where it acts as a protective colloid; it is also used in cosmetics and pharmaceutical preparations. *See also* polysaccharide; polyelectrolyte.

carrier. A term which has several meanings in chemistry. It may refer to a catalyst carrier, which is a layer or bed of an inert material on which a solid catalyst is placed; to a stable element, containing radioactive isotopes of the same element, which "carries" the latter through a tracer sequence; or to a transport gas used in vapor-phase chromatography.

casein. A protein occurring in milk, cow's milk containing about 3%. It is closely associated with calcium, forming a complex molecule called calcium caseinate. By chemical classification it is a polyamide, characterized by the presence of $CONH_2$ groups, as well as of phosphorus. It is separated from skim milk either by addition of an acid or of the enzyme rennet; coagulation results, and the casein is extracted mechanically. The acid-coagulated type is used in cheese making, as well as for paper coating and textile sizing. Rennet casein (also called paracasein) is used for plastic items such as buttons. The industrial use of casein for adhesives, fibers, plastics, and water-based paints has decreased considerably in recent years because of the development of the more efficient synthetic polymer types.

casinghead gasoline. *See* natural gasoline.

casting. *See* mold (2).

castor oil. A nondrying oil obtained from the castor bean and consisting chiefly of glycerides of ricinoleic acid. A liquid of moderate viscosity, it is well-known for its repulsive taste, which restricts its use as a laxative. Castor oil and its derivatives have industrial application as plasticizers, as a source of nylon monomers (sebacic and aminoundecanoic acids), in Turkey red oil, and as lubricants and hydraulic brake fluids. When dehydrated by heating with catalysts, castor oil has the properties of a drying oil; when "blown" with air and heated, it can be used as a modifier in paints and protective coatings. It can also be hydrogenated to a waxy solid and polymerized. *See also* vegetable oil.

catalysis. One of the most important chemical phenomena in nature, catalysis is the "loosening" of the bonds of two or more reactants by another substance in such a way that an extremely small percentage of the latter can greatly increase the rate of a reaction while itself remaining unconsumed. *See also* catalyst.

catalyst. An element or compound that accelerates the rate of a chemical reaction but is nei-

ther changed nor consumed by it. The catalysts most used in industrial processing are inorganic substances, chiefly metals or metallic oxides, usually in the form of powders or porous masses. They function by either physical or chemical adsorption (chemisorption). Some of the best-known are vanadium pentoxide (for sulfuric acid), aluminum chloride (Friedel-Crafts syntheses), manganese dioxide, and pulverized or sponged metals such as cobalt, iron, platinum, and nickel. Only a few parts per million of catalyst are required to activate a large-scale reaction, and since it is not consumed, the same catalyst charge can be used for 1000 hours or more. To be effective, catalysts must be in a finely divided state to furnish maximum surface area for adsorption as the reacting substances pass over or through them; their effect is due to submicroscopic projections on the surfaces of their particles (active points), where a bond is formed with the reactants. If the adsorption is physical, these bonds are relatively weak; if it is chemical they are about as strong as chemical bonds.

Catalysts are particlarly useful in petroleum refining, where among other changes, they permit conversion of straight-chain to branched-chain paraffins, and of alicyclic compounds to aromatic compounds. Platinum is particularly effective for high-octane gasoline. The hydrogenation of vegetable oils to solid fats is carried out with pulverized nickel as catalyst. A major part of the synthetic organic chemical industry depends upon the action of catalysts of this type. The polymerization reactions by which many synthetic resins are made require catalysts that are highly specific in their action, giving molecular structures that can be closely controlled (Ziegler and Natta catalysts). Water has catalytic effects, as in the corrosion of metals.

Organic catalysts called enzymes are essential in the functioning of animal metabolism; the part played by chlorophyll in photosynthesis is an example of catalysis in plants, promoting formation of sugars from carbon dioxide and water.

Following is a partial list of catalysts and the types of reaction they promote; the asterisk indicates a destructive effect.

aluminum chloride	condensation (Friedel-Crafts)
aluminum alkyl + titanium chloride	Ziegler catalyst for stereospecific polymers

aluminum oxide	hydration; dehydration
ammonia	condensation (polymers)
chromic oxide	methanol synthesis; aromatization; polymerization
cobalt	hydrocarbon synthesis; Oxo process
copper salts	oxidation (of rubber)*
hydrogen fluoride	alkylation; condensation; dehydration; isomerization
iron	ammonia synthesis; hydrocarbon synthesis
iron oxide	dehydrogenation (oxidation)
manganese dioxide	oxidation
molybdenum oxide	dehydrogenation; polymerization; aromatization; partial oxidation
nickel	hydrogenation (oils to fats)
phosphoric acid	polymerization; isomerization
platinum metals	hydrogenation; aromatization; oxidation
silica-alumina	cracking hydrocarbons
silver	hydration; oxidation
sulfuric acid	isomerization; corrosion*
vanadium pentoxide	oxidation (sulfuric acid)
water (esp. + NaCl)	oxidation (corrosion)*

See also adsorption; chemisorption; enzyme; stereospecific; Ziegler catalyst.

catenate. A term used by chemists to indicate the tendency or ability of an element to unite with itself to form chains; carbon is notable in this respect, and several other elements, such as oxygen, phosphorus and silicon, have it to a limited extent.

cathode. The negative electrode of an electrolytic cell through which current is being passed. It is the electrode at which reduction takes place and at which electrons enter (current leaves) an electrolyte. Among the reducing reactions occurring at the cathode are discharge of positive ions, formation of negative ions, and reduction of chemical elements from higher to lower valence state. In a battery the cathode is the positive terminal. In electroplating, cations (positively charged ions) migrate to the cathode, upon which they are deposited to form a coating. In an electron or vacuum tube, the negative electrode is the cathode. It emits electrons which are collectively called cathode rays.

cation. An ion, having one or more positive charges, which migrates to the cathode in a solution when an electric current is applied. *See also* anion.

caustic. *See* alkali; sodium hydroxide.

Cb Symbol for the element columbium, which is identical with niobium, the latter name being preferred by chemists.

Cd Symbol for the element cadmium, being named from cadmia, a term for calamine (zinc carbonate).

Ce Symbol for the element cerium, the name being derived from the mythological goddess of agriculture, Ceres; the asteroid of the same name had just been discovered, which led to adoption of this name for the element also.

cell. (1) In electrochemistry, this term refers to a device for producing an electric current by chemical means (battery) or for performing electrolysis, the essential difference being that the former generates electricity whereas the latter consumes it.

(2) In biochemistry, a cell is the basic unit of tissue structure, composed of an outer wall or membrane, an interior fluid called cytoplasm, and a nucleus which contains chromosomes and genes (nucleic acids).

(3) The word cell may also mean a hollow structural unit of a plastic foam (cellular plastic). *See also* electrolytic cell; battery.

cellophane. A form of regenerated cellulose, similar to rayon, produced by the viscose process and extruded in the form of thin sheets. It has adequate strength and a high degree of transparency, which have made it an excellent packaging material for general use. It is not as good a moisture-vapor barrier as more recently developed plastic films; it is also quite flammable but may be made less so by treatment with flame-resisting chemicals. Cellophane tends to lose its strength when exposed to heat and becomes useless above 155°C (300°F). *See also* viscose process; rayon.

cellular. A term used in the plastics and rubber industries to describe a foam-like structure, either flexible or rigid, comprised of closed pores formed by the action of a blowing agent; this gives a soft, spongy feel to flexible products, such as "sponge" rubber. Rigid cellular structures can be made in a similar way. Such products are of light weight, are excellent heat insulators, and are impenetrable to moisture. The word "cellular" is not synonymous with "porous," since the latter includes both closed and open pores. *See also* poromeric.

"Celluloid." *See* plasticizer; Hyatt.

cellulose. The most abundant organic compound in nature, cellulose is a carbohydrate polymer belonging to the general class of polysaccharides. Its formula can be represented as $(C_6H_{10}O_5)_n$, in which the n stands for an undetermined number of carbohydrate units. It is a straight-chain molecule comprised of glucose units connected by oxygen atoms and can be converted to simpler forms (starches, etc.) by hydrolysis. The molecular weight ranges from 150,000 to over 500,000. The molecule has been shown to have both crystalline and amorphous segments. Cellulose will not dissolve in any solvent, though it swells in dilute caustic solution (NaOH) to form alkali cellulose, which is an intermediate in the manufacture of viscose. For chemical structure, see identity period.

Cellulose can be chemically modified to form nitrocellulose (used in lacquers and explosives), rayon, and cellophane (regenerated cellulose). Other types of reactions yield carboxymethylcellulose and related ethers and esters, e.g., cellulose acetate. Cellulose is the basis of the cotton textile industry, the paper industry, and the wood product field. It is entirely nontoxic and is a component of all common vegetables; synthetic food items have been made from it. It can be decomposed by composting (digestion with the aid of oxygen and/or microorganisms) to give animal feed supplements, fertilizers, etc. It is combustible, but not flammable, except in the form of dust or fine fibers. *See also* carbohydrate; nitrocellulose; viscose process.

cellulose acetate. A reaction product (ester) of cellulose and acetic acid (or anhydride), in which sulfuric acid acts as a catalyst and also tends to break down the cellulose polymer chain; the process is completed by controlled hydrolysis to form a thermoplastic solid which can be extruded in the form of fibers. It can be used in solutions for lacquers and similar coatings, for photographic film, magnetic tape, cigarette filters, and the like; also as woven fibers for textile products and for dialysis membranes. In some types, acetylation of the cellulose is only partial; in others it is complete (three acetate groups attached to each polymer unit) to form cellulose triacetate. It is soluble in acetone and a few organic solvents, but is insoluble in water. It is also quite flammable. Butyric anhydride is sometimes used in conjunction with acetic anhydride in making cellulose acetate, in which case the product is called cellulose acetate butyrate, with properties and uses similar to those of the normal acetate, but with somewhat better resistance to weathering and water absorption.

cellulose gum. *See* carboxymethylcellulose.

cellulose nitrate. *See* nitrocellulose.

cellulose xanthate. *See* viscose process.

cellulosic. A collective term referring to any of the numerous compounds and products made by reacting cellulose with various chemicals, usually involving substitution of groups of one kind or another for the hydroxyl groups of the cellulose. The products of these reactions are either plastics or gum-like in nature; they include rayon, nitrocellulose, cellulose acetate and its derivatives, and various types of methylcellulose.

Celsius, Anders (1701-1744). A Swedish scientist whose interests were devoted chiefly to astronomy; he invented and established the use of the centigrade temperature scale, now often called the Celsius scale, on which zero degrees is the freezing point of water and 100 degrees the boiling point of water.

cementation. A term used in metallurgy to refer to a coating (usually on steel) formed by heating the steel in contact with another metal or nonmetal which then diffuses into the base metal to give an alloy or intermetallic compound. The coating agent may be a powdered metal, a gas, or a liquid; high temperatures are required. The purpose of cementation is to protect the base metal from corrosion and to give it exceptional surface hardness, as in case-hardening (nitriding). Other cementation methods are known as carburizing (carbon), sherardizing (zinc alloy), chromizing (chromium), and calorizing (aluminum). Silicon is also used. *See also* corrosion; protective coating; intermetallic compound.

cemented carbide. *See* carbide.

cement, hydraulic. Classified as a type of ceramic, a hydraulic cement is a blend of varying percentages of calcium oxide (CaO), silica (SiO_2), alumina (Al_2O_3), and sometimes also iron oxide (Fe_2O_3), which is converted from a powder to a soft, plastic mass and then to a rigid solid by addition of water. Hydrates of these compounds are formed by chemical reaction with the water. There are two basic types: those containing about one-third alumina (aluminous cement) and those containing little or no alumina (Portland cement, named for Portland, England). The mixture is often diluted with sand before addition of water; plastics can also be added in liquid or latex form to impart a degree of flexibility. Hydraulic cements are manufactured as dry powders, the water and modifying ingredients being admixed shortly before pouring. Besides their use in the construction industry, such cements are good absorbers of neutrons and high-energy radiation and thus are used for shielding purposes. *See also* concrete (1); hydraulic.

cement, organic. (1) A mixture of a sticky, adhesive substance such as glue, casein, rubber hydrocarbon, etc. with a solvent, used to secure a temporary bond between wood veneers, paper, fabrics, and the like.

(2) A mixture of a high-polymer resin such as epoxy, silicone, or highly accelerated rubber with a solvent; strong permanent bonds can be obtained when these mixtures cross-link on exposure to air at room temperature. They are used chiefly in the furniture and woodworking fields.

centigrade. The temperature scale universally used by scientists in which the freezing point of water is represented by 0 and its boiling point by 100; it is also called Celsius, after its inventor, Anders Celsius, a Swedish astronomer (1701-1744). When no indication of a scale is given in a scientific article, it may be assumed that centigrade is meant; 20 to 25 degrees centigrade is considered to be "room temperature." To convert centigrade to Fahrenheit, multiply °C by $\%$ and add 32 to the product; for example, $250°C = (250 \times \%) + 32 = 450 + 32 = 482°F$. The temperature value for both centigrade and Fahrenheit scales is the same at $-40°$. *See also* Fahrenheit; Réaumur; Temperature Conversion Table.

centipoise. *See* poise.

centrifugation. Separation of the components of a liquid colloidal system by means of centrifugal force. This operation is usually confined to dispersions whose components have closely similar specific gravity, such as rubber latex and blood. In such fluids, water is the suspending medium with specific gravity of 1, while that of the dispersed phase is about 0.95. To effect separation, the liquid is placed in a container called a centrifuge, which rotates with extreme rapidity (10,000 rpm or more); the resulting centrifugal force tends to concentrate the lighter fraction near the center while the heavier water moves to the circumference. The separation obtained with latex is not complete; the smaller particles of rubber remain in the serum. Separation of blood fractions and similar medical operations is quite effective. This principle has also been used for separation of uranium isotopes, determination of molecular weights and other highly critical separations. The term ultracentrifuge is applied to small units, some of which exert a force of 250,000 times gravity.

ceramic. Any product made from earth-derived materials such as clays, silicates, sand, and the like, usually requring the application of high temperature in a kiln or oven at some stage in the process. They include the following: all types of household china, pottery, etc.; brick, tile, and similar structural products; porcelain and porcelain enamels; glass in all its many forms; refractories; hydraulic cements; silicon carbide and fused alumina, used as abrasives. Hydraulic cements set by the action of water and do not require high-temperature treatment.

cereal. Any plant of the grass family that yields edible grains or kernels rich in carbohydrates (starches) and proteins (gluten); most important are wheat, corn, oats, barley, rice, and rye. All are used in food products, either directly or indirectly, e.g., by fermentation. The protein content ranges from 6.5% for rice to 13% for wheat. The mineral content of rice is relatively high (6%). Carbohydrate content ranges from 60 to 70%, and there is also a low percentage of fats. Cereal chemistry includes the study of milk, soybeans, minerals, enzymes, and vitamins, as well as evaluation of such processing methods as grinding, milling, and refining.

cerium. An element.

Symbol	Ce	Atomic Wt.	140.12
State	Solid	Valence	3,4
Group	IIIB (Lan-		
	thanide		
	Series)		
		Isotopes	4 stable
Atomic No. 58			

A reactive metal, m.p. 798°C (1468°F), and a member of the rare earth family. It occurs naturally in monazite sands and as the ore bastnasite in California; allanite and cerite are alternative sources. Cerium oxidizes so strongly that it is a dangerous fire hazard at termperatures above about 300°C (572°F); it ignites at temperatures in the range of 150-180°C (302-356°F). It is used as an oxidizing agent when in the $+4$ valence state, and the metal is used as a reducing agent. *See also* misch metal.

cermet. A term derived from the combined first syllables of *cer*amic and *met*al. It refers to a finely divided composite of such metals as aluminum, tungsten, chromium, or titanium and a ceramic component, i.e., metallic carbides, oxides, borides, and nitrides. Of these, only the oxide- and carbide-based types are in active commercial use. Cermets display physical properties superior to those of either component alone. They are manufactured by powder metallurgy methods: after the powders are blended they are compressed and subjected to high temperature in a process called sintering, as a result of which the particles interlock and bond together, forming a ceramic-in-metal matrix. Cermets are used for refractory purposes, in aerospace equipment, in high-speed tools, and in nuclear engineering devices. *See also* composite; sinter.

cesium. An element.

Symbol	Cs	Atomic No.	55
State	Liquid	Atomic Wt.	132.9055
Group	IA	Valence	1

Cesium has a notably low melting point, 28.5°C (83°F), and is one of the three metals that are liquid at room temperature; it is classified as an alkali metal and is in the same group as sodium and potassium. It is intensely reactive with oxygen and members of the halogen group, causing ignition, and with sulfur and phosphorus it reacts explosively. Thus, it must be protected from contact with air by storage in sealed bottles or steel cylinders. Early uses of cesium were in photoelectric cells and as a getter in vacuumtubes; more recently its power-generating ability has been studied for possible thermoelectric applications, especially for spacecraft. Cesium has only one stable form.

cetane number. A series of values obtained by blending the hydrocarbon cetane ($C_{16}H_{34}$) with its isomer heptamethylnonane in various proportions, to establish a power rating scale for diesel fuels. It is analogous to the octane rating scale for gasoline.

Cf Symbol for the element californium, named for the state of its discovery.

cgs. Abbreviation of centimeter-gram-second, referring to the units internationally accepted and used by physicists and chemists. *See* table of Miscellaneous Equivalents.

Chadwick, Sir James (1891-1974). A British physicist who was awarded the Nobel Prize in 1935 for his discovery of the neutron (1932), the existence of which had been predicted by Rutherford. This was an immensely important advance in the knowledge of subatomic particles. *See also* neutron.

chain, molecular. The sequence of carbon atoms in an organic molecule. A straight or open chain is one in which the carbons extend in a linear sequence; such chains may have one or more branches. A closed chain is a cyclic compound in which the carbon atoms form a ring (benzene) or other geometric figure (cyclohexane); a ring in which one or more of the atoms is not carbon is called heterocyclic. A side chain is a straight chain attached to one of the atoms of a closed chain. *See also* aliphatic; aromatic; branched chain.

chain reaction. (1) A chemical reaction initiated by a free atom, ion, or radical of transitory existence as a result of rupture of a chemical bond. Under certain conditions these set up a limited sequence of reactions ending in the formation of a stable compound. For example, free chlorine atoms formed in a mixture of hydrogen and chlorine ($Cl_2 \rightarrow 2Cl^{\cdot}$) can induce the sequence: $Cl^{\cdot} + H_2 \rightarrow HCl + H^{\cdot}$; $H^{\cdot} + Cl_2 \rightarrow HCl + Cl^{\cdot}$, and so on, terminating in $H^{\cdot} + Cl^{\cdot} \rightarrow HCl$ (stable). Such reaction chains often occur in photochemistry and also in reactions involving polymerization, combustion, and oxidation of fats, oils, rubber, etc.

(2) A nuclear reaction in which each of the two or three neutrons released by the fission of one atomic nucleus splits another atomic nucleus, thus creating a continuous and self-sustaining energy release. *See also* (1) free radical; (2) fission.

change, chemical. An alteration in the atomic constitution or molecular structure of a substance; this always results in the formation of one or more new and different substances. Such changes usually occur by chemical reactions of various types (combination, decomposition, replacement, etc.), involving the union and rupture of chemical bonds. For example, water results from the combination of two gases under the influence of high temperature, the product being a liquid between 0 and 100°C (32 and 212°F): $(2H_2 + O_2 \rightarrow 2H_2O)$. A chemical change may also occur by rearrangement of atoms within a molecule, as in the formation of urea from ammonium cyanate: $NH_4OCN \rightarrow (NH_2)_2CO$. Chemical changes should be distinguished from (1) physical changes, in which only the form of a substance is changed, e.g., the melting, freezing, and evaporation of liquids; and (2) nuclear changes, which involve alteration of an atomic nucleus rather than of the bonds between atoms. *See also* reaction.

channel black. *See* carbon black; impingement.

charcoal. A highly porous form of carbon obtained from wood by destructive distillation or from animal bones by calcination; it is also called char. It is used to remove odors and other impurities from gases by adsorption, as a component of low explosives, and for recovery of solvents from their vapors. Its efficiency is increased by steam activation. *See also* activated carbon.

Chardonnet, Hilaire (1839-1924). A native of France, he has been called the father of rayon because of his successful research in producing what was then called artificial silk, from nitrocellulose. He was able to extrude fine threads of this semisynthetic material through a spinneret-like nozzle and the textile product was made on a commercial scale in several European countries. He was awarded the Perkin medal for this work in 1914, only a few years before the discovery of rayon.

charging stock. *See* feedstock.

Charles' Law. A gas law stated in 1787 by a French scientist, Jacques Alexandre Cesar Charles (1746-1823), as a result of his own experiments, namely, that the pressure of a confined gas is proportional to its absolute temperature if the volume remains unchanged. This was later modified by Gay Lussac (1802) to the more familiar form: at a constant pressure the volume of a given gas varies directly with the absolute temperature. *See also* Gay-Lussac's Law.

chelate. An organic coordination (complexing) compound in which the metal ion is bound to atoms of nonmetals, e.g., nitrogen, carbon, or oxygen, to form a heterocyclic ring having coordinate covalent bonds. The nonmetal atoms are called ligands; these may be attached to the metal ions by from one to six linkages. They are called uni-, bi-, tridentate, etc., meaning one-, two-, or three-toothed. Cobalt, copper, nickel, zinc, and platinum are metal ions that are commonly involved in chelate structures, many producing extremely stable compounds, e.g., copper phthalocyanine. *See also* coordination compound; ligand; ethylenediaminetetraacetic acid.

chemical engineering. That aspect of the science of engineering which applies the fundamental laws of physics, chemistry, and thermodynamics to the problems involved in expanding laboratory-scale experimental procedures first to pilot plant and then to full-scale production. Basic to the calculations are knowledge of material and energy balances and the rates at which chemical reactions occur under given conditions of concentration, temperature, and pressure, as well as of the operations to be performed, e.g., adsorption, diffusion, evaporation, heat transfer, fluid flow, mixing, etc. Chemical engineering also involves familiarity with materials of construction, safety factors, and cost evaluation.

chemical flooding. *See* enhanced oil recovery.

chemiluminescence. *See* luminescence.

chemisorption. A phenomenon related to adsorption in which atoms or molecules of reacting substances are held to the surface atoms of a catalyst by electrostatic forces having about the same strength as chemical bonds. Chemisorption differs from physical adsorption chiefly in the strength of the bonding, which is much less in adsorption than in chemisorption. The surface at which chemisorption takes place is usually a metal or metal oxide; the chemisorbed molecules are always changed in the process, and often the molecules of the surface are changed as well. Hydrogen and hydrocarbons are readily chemisorbed on metal surfaces, the hydrocarbons being so modified that they yield active initiating groups (carbonium ions, etc.). Thus, chemisorption is an essential feature of catalytic reactions and accounts in large measure for the specialized activity of catalysts. *See also* catalyst; adsorption.

chemistry. A basic physical science which is primarily concerned with the nature, properties, and composition of substances, with the reactions that occur between them in any of their various forms, and with the development of laws and theories which interpret chemical phenomena in a logical manner. It overlaps physics on one hand (physical chemistry) and biology on the other (biochemistry). The origin of the term is obscure. Chemistry evolved from the medieval practice of alchemy, and at least partially as a result of the empirical philosophy of Francis Bacon (1561-1626), which provided the rationale of modern scientific method, namely, the principle of inductive reasoning from objective evidence. Its foundations were laid by such men as Boyle, Lavoisier, Priestley, Berzelius, Avogadro, and Dalton, over a period of 150 years. *See also* experiment.

chemosterilant. A chemical compound which adversely affects the reproductive ability of an insect exposed to it. Such compounds, developed in the decade 1960-70, have been successfully used on both male and female insects of several species, including the screw-worm fly. This method of control may eventually considerably reduce the need for insecticides. The chemicals used are antimetabolites, organotin, and boron compounds. *See also* insecticide.

chemotherapy. The development and use of chemical compounds which are specific for the treatment of diseases. Since early in this century, the notable accomplishments in chemotherapy have been arsphenamine, developed in 1910 by Paul Ehrlich, German bacteriologist (1854-1915) (Nobel Prize 1908), who may be regarded as the founder of this science; synthetic antimalarials, developed during World War II to replace quinine; the spectrum of sulfa drugs; mold-derived antibiotics; and steroid hormones such as cor-

tisone. The chemical approach to cancer has been studied exhaustively without positive results. *See also* immunochemistry; cortisone.

chemurgy. A term introduced in 1935 to describe development of new, nonfood uses for certain crops, as well as conversion of farm, agricultural, and forest wastes to useful products. Many such products have resulted, at least in part, from this endeavor, e.g., soybean protein for plastics, paper from bagasse, furfural from oat hulls, etc. More recently conversion of manures and garbage to hydrocarbon gases for fuel use has been practiced, mostly on a small scale. Expanding efforts along this line may be expected in view of growing demands for conservation and fuel sources. *See also* biomass.

china clay. *See* clay; kaolin.

chiral, chirality. Derived from the Greek word meaning "hand," this term is applied by chemists to molecules that display the type of optical activity characteristic of enantiomorphs, i.e., having structures which are mirror images of each other, analogous to the left and right hands. *See also* enantiomorph; asymmetry.

chitin. A polysaccharide having a basic structure similar to that of cellulose but also containing amine (nitrogen) groups; it forms hard, whitish-to-gray layers or accretions and is found in the integuments of shellfish and some insects. It is insoluble in water and most solvents but is attacked by concentrated acids. It is of interest to biochemists but has no significant industrial use.

chloral. *See* DDT.

chloramphenicol. *See* antibiotic.

chlorate. A salt of chloric acid, $HClO_3$, in which chlorine has a valence of $+5$. The most common and typical chlorate is sodium chlorate, $NaClO_3$, a colorless crystalline solid salt, m.p. 255°C (491°F), soluble in water and a powerful oxidizing agent. When heated above the melting point, it decomposes to form NaCl and oxygen or sodium perchlorate, $NaClO_4$; when treated with an acid, it may explode due to the formation of unstable chlorine dioxide. It can be formed by the action of chlorine on a hot solution of NaOH, but the principal process is the electrolysis of a cooled solution of sodium chloride in a cell with graphite anodes and steel cathodes. The resultant solution of $NaClO_3$ and residual NaCl is fractionally crystallized to separate pure sodium chlorate. It is used as a source of chlorine dioxide for bleaching paper pulp and other materials; as a source of sodium perchlorate by further electrolysis; as a herbicide and defoliant; in matches, explosives, flares and pyrotechnics; and in the manufacture of other chlorates, such as $KClO_3$. It must be handled carefully to avoid reaction with reducing or organic material, which can lead to fire and explosion. *See also* chlorine dioxide; perchlorate.

chloride. The salt of hydrochloric acid (HCl), in which chlorine has a valence of -1. All metallic and most nonmetallic elements form chlorides. The most common chlorine compounds, which occur chiefly as sodium chloride (NaCl) and to a lesser extent as potassium chloride (KCl), obtained from sedimentary deposits formed by the evaporation of ancient seas and from inland salt lakes, as well as by solar evaporation of sea-water. The oceans contain about 2% of chloride ion or about 2.6% NaCl. The chloride ion is vital in the diet of mammals and is the source of hydrochloric acid secreted in the stomach during digestion. Chlorides are used industrially as sources of chlorine, alkalis and numerous chlorine compounds, and are thus basic raw materials for the chemical industry. *See also* sodium chloride; potassium; brine.

chlorinated hydrocarbon. Any of a large group of synthetic organic chemicals made by replacing one or more of the hydrogen atoms of a hydrocarbon with chlorine. Well-known and widely used examples are carbon tetrachloride, chloroform, trichloroethylene, *p*-dichlorobenzene, and chloronaphthalene. The hydrocarbons may be either aliphatic or aromatic and the materials either solid or liquid; they are used as cleaning fluids, fumigants, and insecticides. Many are quite toxic (carbon tetrachloride) and often have a sharp and distinctive odor. Some also have a narcotic and anesthetic effect (chloroform); their handling and use in any concentrated form should be performed with caution. *See also* insecticide; narcotic.

chlorine (pronounced klo-reen). An element.

Symbol	Cl	Atomic Wt.	35.453
State	Gas	Valence	1,3,4,5,7
Group	VIIA	Isotopes	2 stable
Atomic No.	17		

Discovered in 1774 by Carl Wilhelm Scheele, a Swedish chemist (1742-1786), chlorine at room temperature is a heavy, yellowish gas but can

easily be liquefied, b.p. $-34.5°C$ ($-30°F$). A highly reactive chemical, it is a member of the halogen family; it is a strong electron acceptor (electronegative) and thus is a powerful oxidizing agent. In addition to its elemental form, chlorine exists in the valence states of -1, $+1$, $+3$, $+4$, $+5$ and $+7$, indicative of its great reactivity. It is made industrially by subjecting sodium chloride to electrolysis in solution (electrolytic cell) or in molten form. The isotopic composition of chlorine is unique, as one of its two isotopes comprises about 25% of the element. Chlorine is unusually capable of forming ionic bonds and free radicals. It is toxic when inhaled as a gas and is a strong skin irritant as a liquid.

A versatile element, chlorine is used at the rate of millions of tons per year in the manufacture of a broad range of chemicals, both organic and inorganic; among these are insecticides, plastics, and hydrochloric acid. It is an intermediate in making ethylene oxide and tetraethyllead and a constituent of many chlorinated hydrocarbons used as solvents (carbon tetrachloride). It is a basic ingredient of bleaches and an important bactericide, for which purpose it is used in water purification. *See also* oxidation; electronegative.

chlorine dioxide. A yellow gas, ClO_2, b.p. $10°C$ ($50°F$), is a strong oxidizing agent which is very reactive and unstable; it reacts with alkalis to form equal moles of chlorite and chlorate. Because of its unpredictably explosive nature, it must be diluted with air to a maximum concentration of about 10% as generated and used. It is prepared at the site of use by the reaction of sulfuric acid on a chlorate in the presence of a reducing agent such as sulfur dioxide or methanol. Chlorine dioxide is widely used in the bleaching of paper pulp and cellulose textiles to a high whiteness without loss of strength of the cellulose; this use constitutes the major consumption of sodium chlorate. Fats, oils, and flour are also bleached by it, and it is used to purify and remove tastes from potable water. *See also* bleach; chlorate; chlorite.

chlorite. A salt of chlorous acid, $HClO_2$, in which chlorine has a valence of $+3$. The most common chlorite is sodium chlorite, a white or yellowish-white solid salt, $NaClO_2$, m.p. about 200°C (392°F) (dec.), soluble in water, and a powerful oxidizing agent. It is made (a) by the reduction of chlorine dioxide, ClO_2, in a solution of NaOH, and (b) by the hydrolysis of ClO_2 in a solution of NaOH and KOH to form equimoles of sodium chlorite and potassium chlorate, which are separated by fractional crystallization. Sodium chlorite is used in the bleaching of paper pulp, textiles, soap, etc.; in water purification; and in the preparation of ClO_2 by reaction with acid or chlorine. The salt is sensitive to decomposition by heat and flame and is a fire and explosive hazard when in the presence of reducing materials; its solution is irritant to skin and tissue. *See also* chlorine dioxide; sodium chlorate.

chloroacetic acid.   *See* 2,4-D.

chlorobenzene.   *See* aniline; dichlorobenzene; aromatic.

chlorodifluoromethane.   *See* fluorocarbon.

chlorofluorocarbon.   *See* fluorocarbon.

chloroform.   A heavy, volatile narcotic liquid made by reacting acetone with chlorinated lime or by chlorination of methane. Its formula is $CHCl_3$, its specific gravity 1.485, and b.p. 61.2°C (142°F). It has a sweetish odor and is nonflammable. Chloroform is well-known for its anesthetic effect, but it is toxic, and over-inhalation of its fumes may cause death. It has minor use as an extractive solvent and as a fumigant. *See also* chlorinated hydrocarbon.

chlorohydrin.   Any of a number of aliphatic compounds containing one or more hydroxyl groups and at least one chlorine atom; thus, they are both alcohols and organic chlorides and comprise quite a reactive series of compounds. A general method of preparation is the addition of hypochlorous acid to alkenes, e.g., $RCH{=}CH_2$ + $HOCl{\rightarrow}RCHOHCH_2Cl$. Chlorohydrin itself is made by reacting glycerol with hydrogen chloride, giving a liquid that is a good solvent for cellulose derivatives. Its formula is $CH_2OHCHOHCH_2Cl$, b.p. 213°C (397°F) (decomposes).

chloronaphthalene.   *See* chlorinated hydrocarbon.

chlorophyll.   The naturally occurring organic green pigment in vegetation which catalyzes the synthesis of carbohydrates from carbon dioxide and water (photosynthesis). It is able to effect conversion of light (radiant energy) into chemical energy and is unique in this respect. The mechanism of its action has been researched for many years, for example, by Melvin Calvin, American chemist, Nobel Prize, 1961. It belongs to the porphyrin group of substances and contains a centrally located atom of magnesium; the formula of chlorophyll a, the most important type, is $C_{55}H_{72}MgN_4O_5$. There are two other forms, b and c, with similar formulas and structures. Its physical nature is that of a microcrystalline wax. Chlorophyll a has been synthesized; it can also be obtained by extracting green plant tissue with alcohol. It has limited use in medicine, toothpaste, and as a colorant in various products. *See also* photosynthesis; porphyrin.

chocolate.   *See* theobroma oil.

cholesterol.   An unsaturated polycyclic alcohol (sterol) having one hydroxyl group and one double bond. It is closely related to phenanthrene in structure and falls into the general classification of lipids. Cholesterol is synthesized by all higher animals; its oxidation products in the body are bile acids and steroid hormones. In esterified form it is a component of the saturated fatty acids occurring in animal fats such as butter; it is thought by some to have an adverse effect on the circulatory system, but according to recent authorities, this has not been proved. The complex biochemical mechanisms in which it plays a part have been investigated with radioactive carbon tracers. *See also* sterol; lipid.

cholinesterase.   An enzyme in the body of animals and man which reacts with the acetylcholine formed by nerve functioning; as a result of this, the extremely toxic acetylcholine splits into acetic acid and choline and is thus detoxified. Some compounds, especially esters of phosphoric acid, are able to deactivate cholinesterase so that it can no longer decompose acetylcholine; when this occurs, death is certain to follow within a short time. Such compounds are called cholinesterase inhibitors and are present in a number of insecticides of the parathion type, as well as in so-called nerve gases. *See also* acetylcholine; parathion.

chromatin.   *See* chromosome.

chromatography.   A method of fractionating, or separating, the components of gaseous or liquid mixtures, originated by Tswett in 1906, which has been developed intensively during the last 40 years. The best-known types of chromatography are gas, liquid, paper, ion-exchange, and thin-layer, the name depending on the characteristic aspect of the method. Basically, chromatography involves passing a gaseous or liquid mixture such as refinery gases or gasoline, called the mobile phase, through a column of porous material sometimes coated with a nonvolatile liquid, called the stationary phase. The components of the mixture are adsorbed selectively and pass through the porous material at different rates, emerging as distinct separation zones at the bottom the column. A carrier gas or liquid is necessary to provide movement of the material through the adsorbent. When the components have been thus separated, the identity of each can be determined by an appropriate analytical method. In thin-layer chromatography, the adsorbent is placed on a glass plate in a layer less than 0.5 millimeter thick, and a solution of the mixture to be separated is placed near one edge of the thin layer; separation occurs by adsorption as the components move through the layer at different rates. The adsorbents used are silica gel, cellulose (paper), alumina, and diatomaceous earth. *See also* analysis; fraction.

chrome green.   *See* lead chromate; green.
chrome tanning.   *See* tanning.
chrome yellow.   *See* lead chromate.
chromic acid.   A water-soluble inorganic acid which, in the solid state, consists of dark red granules derived from sodium chromate by reaction with sulfuric acid, or from chromite ore. It is an extremely strong oxidizing agent and is corrosive to tissue; safe handling requires that it be kept away from organic materials and reducing agents, as it may ignite or even explode on contact. It has a variety of uses, among which are as a chemical intermediate, as a colorant in ceramic products, in metal cleaning, and in plating. Its chemical formula is CrO_3 (anhydrous), whereas in solution it is H_2CrO_4. *See also* chromium.
chromium.   An element.

Symbol Cr	Atomic Wt. 51.996
State Solid	Valence 2,3,6
Group VIB	Isotopes 4 stable
Atomic No. 24	

Named from a Greek word meaning color, chromium is a grayish metal, m.p. 1875°C (3407°F), though many of its compounds display color. It is derived by reduction of chromite ore found in South Africa, U.S.S.R., and the Philippines. It is an excellent coordination metal and owes much of its commercial use to its corrosion-resistant properties. The metal itself is not toxic

except in the form of fume, but some of its compounds are irritant and damaging to the skin, especially those involving hexavalent chromium. Its major uses are as an alloying agent in stainless steel and other alloys and for electroplated coatings on exposed metal parts. It may also be plated on plastics. One of its major important compounds is chromic acid. Chromic oxide (Cr_2O_3) has major use as a ceramic colorant where a stable heat-resistant green is required. In combination with lead ($PbCrO_4$), chromium produces a yellow pigment (chrome yellow). Chromium sulfate is a widely used tanning agent for leather. *See also* chromic acid; lead chromate.

chromium sulfate.   *See* chromium.
chromizing.   *See* cementation.
chromophore.   Any molecular grouping in an organic compound which causes a characteristic color to appear in its spectrum; the word itself means "color-bearing." Typical chromophore groups in dyes are N≡N, −NO, and −NO₂. *See also* anthraquinone; azo dye.
chromosome.   A protein-nucleic acid complex formed from chromatin in the nucleus of the biological cell; it is made up of DNA, the genetic code carrier, and various other nucleo-proteins. Chromosomes are unit structures formed during cell division from strands of the parent complex, chromatin, the number of chromosomes in a given cell being characteristic of a particular plant or animal species. They contain the genes, which are the specific agents of hereditary transmission. The word is derived from "chrome," or color, because of the susceptibility of the material to biological staining techniques. *See also* cell; DNA; gene.
chrysotile.   *See* asbestos.
ci.   Abbreviation for curie.
cinchona.   *See* quinine.
cis-.   A chemical prefix taken from the Latin, meaning "on this side;" its opposite, *trans-*, means "on the other side" or "beyond." These prefixes indicate the position of substituent atoms or groups in relation to the double-bonded carbons in so called geometric isomers, positions which are determined by stereochemical factors. An example is natural rubber (*cis*-1,4-polyisoprene), in which the methyl group lies above the hydrocarbon chain, as opposed to the *trans* form (gutta percha) in which the methyl group is below the chain. A further example is

that of fumaric and maleic acids, which are geometrical isomers with the following spatial configurations:

$$\begin{array}{cc} HOOC-CH & HC-COOH \\ \parallel & \parallel \\ HC-COOH & HC-COOH \\ \text{fumaric acid} & \text{maleic acid} \\ \textit{trans} & \textit{cis} \end{array}$$

See also geometric isomer.

citric acid. An organic acid, m.p. 153°C (307°F), in the anhydrous form, containing three carboxyl groups; its formula can be written HOOC—CH_2—C(OH)(COOH)—CH_2—COOH. It is formed in nature by a complex series of enzymic reactions known as the Krebs cycle; in plant and animal metabolism, it catalyzes the oxidation of tissue structures and thus is an important factor in respiration. In solid form it is a water-soluble crystalline hydrate. Citric acid occurs in many fruits and vegetables, especially lemons, oranges, tomatoes, and limes. It is obtained commercially by fermentation of beet and sugarcane molasses. Its major chemical uses are in the food industry as an acidulant, soft-drink flavoring, antioxidant activator, and color-fixing aid; it also acts as a dispersing and sequestering agent and as a mordant in dyeing and textile printing. *See also* Krebs cycle.

citronella oil. *See* geraniol; odor.

Cl Symbol for the element chlorine (pronounced klo-reen), the name being derived from the Greek word for green. Although discovered by Scheele (see chlorine), Humphry Davy gave chlorine its name and delineated it as an element.

cladding. A term used by metallurgists to designate a type of protective coating consisting of a relatively thick layer of a metal such as nickel, mechanically applied to a substrate metal (steel or copper). The coating is not less than 5% of the total thickness and usually is much more (up to 25%). A permanent intermetallic bond is formed between the coating and the substrate. Clad metals are used widely in the electrical field, for semiconductors, corrosion-resistant equipment, and nuclear fuel elements. *See also* protective coating.

clarification. Removal of colloidally dispersed solids from a liquid either by coalescence and sedimentation or by adsorption on active material surfaces. For example, aluminum sulfate causes flocculation and settling of solid particles in purification of potable water; finely divided clays, carbon and certain proteins effect clarification of vegetable oils, sugar solutions, beverages, etc. by adsorbing colloidal impurities. The operation is sometimes called decolorizing.

classification. (1) Separation or grading of fragmented solids (ores, coal, crushed stone) or products of varying unit size (peas, beans, cherries, etc.) into groups having the same or nearly the same dimensions, by passing them through screens, sifters, or a similar selective device, or by air floatation.

(2) A grouping of substances having closely similar chemical and physical properties, for example, alcohols, aromatic compounds, solvents.

clathrate. Derived from the Latin word for "grating," the term clathrate describes a unique type of solid mixture in which an atom or molecule of one substance (for example, argon) is entirely enclosed within the crystal lattice of another (e.g., hydroquinone). The two substances do not form a chemical compound, the only attraction between them being relatively weak, such as hydrogen bonds or van der Waals forces. Clathrate structures known as gas hydrates can be formed with water and several gases, e.g., bromine, propane, and hydrogen sulfide. Urea can form clathrates with aliphatic compounds. Clathrate structures are often called "cage" complexes, adducts, or inclusion complexes. They are useful in chemical separation processes such as desalination of seawater, petroleum refining, etc. *See also* molecular sieve; gas hydrate.

clavacin. *See* antibiotic.

clay. A group of crystalline, finely divided earthy materials having similar physical and chemical properties; some types have a particle size within or slightly above the colloidal range. The shape of the particles varies widely from one type to another. With some exceptions, clays can be considered as hydrates of alumina and silica, with iron oxide and magnesia as common minor components; a generalized chemical formula might be $Al_2O_3SiO_2 \cdot nH_2O$. The best-known types are kaolin (china clay), attapulgite, montmorillonite, and bentonite. Clays readily adsorb water to form a plastic matrix that can be shaped at room temperature and permanently hardened by baking or firing in a kiln; in the terminology of plastics, they are thermosetting.

The highest-volume use of clays is in the ceramics industry for fine china, whitewear, pottery, etc. They are also used as reinforcing agents in rubber and plastics products, as weighting materials for oil-well drilling muds, as coatings for paper (kaolin), as refractories (fireclay), and as decolorizing agents (adsorption). Clays are components of many types of soil and were derived originally from the weathering of rocks; they occur in high concentrations in Georgia, Texas, Wyoming, as well as in England, France, and U.S.S.R. Clay particles are size-classified by passing a stream of air through them, an operation called floatation. *See also* bentonite; bauxite; fuller's earth.

cleaning.   Removal of soil or other foreign matter from a material by the use of (1) detergents in aqueous solution, for light-weight textiles, plastics, linoleum, etc.; (2) solvents for dry-cleaning (Stoddard solvent), for suitings and other heavy fabrics; (3) solvents for metal degreasing, chiefly such chlorinated hydrocarbons as carbon tetrachloride, trichloroethylene, etc.; (4) abrasives (sandblast), for building stone, granite, etc.; (5) pressurized steam, for cement and concrete.

cleave.   To separate, or divide, into two or more pieces. The term is applied both to the chemical division of an organic molecule e.g., a protein, by hydrolysis, and to the separation of units of a crystalline material, which takes place along lines (or planes) determined by the structure of the crystal. *See also* hydrolysis.

clinical chemistry.   A subdiscipline of chemical science which deals with the behavior and composition of all types of body fluids, including the blood, urine, perspiration, glandular secretions, etc. It involves the analysis and testing of these fluids for content of numerous metabolic constituents as well as foreign materials; thus, it also includes toxicological factors. *See also* medicinal chemistry.

clotting.   *See* coagulation; precursor; fibrinogen.

Cm   Symbol for curium, named in honor of the Curies.

Co   Symbol for the element cobalt. The name is derived from the German *kobold,* meaning goblin or demon, because of the extreme difficulty in separating it from other metals in the ore,

which were at that time considered more valuable.

coagulation.   A type of aggregation in which semisolid particles or macromolecules of a suspension combine irreversibly into a cohesive mass or clot. Agents which induce coagulation are usually electrolytes (acids or alkalies) which form strongly charged ions in the suspension, thus neutralizing the oppositely charged particles. As a result, the particles coalesce and precipitate. Enzymes also have effective coagulant properties. Familiar examples of coagulation are: (1) blood, initiated by the enzyme thrombin; (2) skim milk, caused by an acid or an enzyme (rennet), resulting in precipitation of casein; (3) rubber latex, where coagulation is brought about by addition of acetic or formic acid; (4) the proteins of egg white, coagulation being caused by heat; (5) coalescence of fat particles of whole milk or cream by agitation or beating, which mechanically ruptures the protein coating of the particles, allowing them to cohere; (6) coalescence of colloidal impurities in water by addition of aluminum sulfate. *See also* aggregation; protective colloid.

coal.   A carbonaceous, combustible solid material formed throughout several geological eras from trees and other vegetation by pressure, heat, and bacterial action out of contact with air. The first stage of development of coal is peat, which is still used as fuel in low-lying areas of some countries. The next grade upward is lignite, or brown coal, and the progression goes on through subbituminous to bituminous (soft coal) and anthracite (hard coal). On destructive distillation, bituminous coal yields coal gas, coke, and coal tar; the coke is used largely as blast-furnace fuel and the coal tar for synthetic chemicals, dyes, etc. Most of the bituminous type is used for electric power generation and coking; virtually all the anthracite is used for fuel. *See also* coke; coal tar; peat.

coal conversion.   *See* gasification.

coalescence.   *See* coagulation.

coal tar.   A thick, malodorous liquid obtained as a by-product of destructive distillation of coal; about 9 gallons of coal tar are obtained from one ton of coal. It is one of the most prolific sources of organic compounds, especially those used in dyes, medicinals, and solvents. When distilled it yields fractions called light oil, heavy oil, anthracene oil, and pitch. Among the organic chemicals derived from coal tar are benzene, cumene, naphthalene, phenol, cresol, and anthracene. The fumes of heated coal tar are quite toxic. *See also* coke.

cobalamin. An organic cobalt compound, also called vitamin B_{12} which is a member of the B complex; it has several forms, one of which (nitrocobalamin) has the formula $C_{62}H_{90}N_{14}O_{16}PCo$. It is an essential metabolic factor, affecting the formation of nucleic acids and the interconversion of fats and carbohydrates by its complexing activity. It is effective in counteracting anemia and certain forms of acute neuritis. It occurs in low concentrations in meat, milk, and eggs and is stored in the liver; it is recovered from sugar beet and skim milk residues. *See also* B complex; cobalt.

cobalt.   An element.

Symbol	Co	Atomic No.	27
State	Solid	Atomic Wt.	58.9332
Group	VIII	Valence	2,3

Cobalt ores occur in Canada and in various parts of Africa. The metal is obtained by calcining and subsequent reduction. Its m.p. is 1493°C (2720°F). Cobalt has exceptional properties as a coordinating element and is also unusual for its magnetic properties which make it useful in alloys used for magnets. Some 80% of the cobalt consumed is used in the manufacture of several important nonferrous alloys. It is one of the trace elements required for plant nutrition. Some cobalt compounds are used as blue pigments in glass and cosmetics. Other uses of cobalt are as a drier (oxidation accelerator) in paints and printing inks and as a catalyst. Cobalt itself is essentially nontoxic; however, the highly radioactive artificial isotope cobalt-60 is intensely dangerous unless used under medical supervision, as for cancer therapy. The natural element has only one stable form.

cobalt blue. *See* blue (1).

cobalt naphthenate. *See* drier; naphthenate; paint.

cocaine (pronounced co-ca-in). *See* alkaloid.

cocoa butter. *See* theobroma oil.

coconut oil. *See* lauric acid.

codeine (pronounced co-de-in). *See* narcotic.

coenzyme. A complex organic substance, formed in animal metabolism, including yeast bacteria, which unites with and activates enzymes and is thus essential for cellular oxidation and other chemical transformations. A number of the coenzymes are nucleotides; all play an important part in vitamin activity and related functions. For example, Coenzyme I catalyzes oxidations, Coenzyme A activates acetylation (acetylcholine formation), and Coenzyme Q is involved in the citric acid cycle. The members of the B complex are components of coenzymes; their mechanism of action is that of a prosthetic group. *See also* prosthetic; B complex; enzyme.

cohere. Of a material or substance, to stick tightly together (as opposed to adhere, meaning to stick to another substance, as an adhesive). Cohesion is a property of most resins, gums, mucilages, and other materials that are sticky in the raw state; many of them also display good adhesive properties. The term "coherent," as applied to light rays emanating from lasers, is an extension of this meaning; it refers to a high-energy beam of light having only one wavelength and one frequency, which can be used as a penetrating or cutting device of tremendous efficiency and accuracy.

coke. The end-product of the destructive distillation of bituminous coal or of the cracking of petroleum hydrocarbons. Coal-derived coke is used chiefly for blast-furnace fuel, the coal producing about 1400 pounds of coke per ton; it is about 90% carbon. Petroleum-derived coke (also called metallurgical coke) is used to make electrolytic and electrothermic electrodes and also as refractory electrothermal furnace linings. It is somewhat purer, running about 95% carbon. Coke is generally regarded as a form of industrial carbon. An important chemical use of coke is in the manufacture of synthesis gas and water gas. *See also* carbon; coal tar; synthesis gas.

collagen. A protein occurring in the connective tissue of the body, especially skin, hide, tendon, etc. It is a fibrous polypeptide chain comprised of many amino acids. It has the unique property of shrinking in hot water within a specific temperature range (from 63-65°C (145-150°F) for cowhide). This behavior is a critical factor in leather tannage, for in leather, the shrinkage temperature increases with the extent of tannage. In addition to leather (which is actually tanned collagen), it is used to some extent in the manufacture of animal glues (the term means "glueformer") and in such medical applications as dialysis membranes, sutures, etc. Collagen is converted into gelatin by hydrolysis; after purification it is used as a protective colloid and gel-former in food products, for sausage casings, and in coating photographic film. *See also* protein; protective colloid.

collodion. *See* nitrocellulose; filtration; membrane; osmosis.

colloid chemistry. The study of phenomena occurring when any form of matter has at least one dimension that is less than 1 micron (the limit of resolution of the optical microscope) and more than 1 millimicron (10 angstrom), which is the approximate size of an average molecule. Thus, it is concerned with the size range that lies between molecular dimensions and particles that are just visible in a compound microscope. (Colloidal particles can be resolved in the electron microscope). It is important to note that in colloid chemistry it is the dimension that is significant, rather than the nature of the material; thus it involves not only particulate matter but films, foams, fibers, interfaces, and surface irregularities such as occur, for example, in catalysts. Materials having emulsifying action (chiefly proteins) are called protective colloids.

The term was first used in 1860 by Thomas Graham, a Scottish chemist (1805-1869), to distinguish materials that would not pass through a parchment membrane; as many of these were sticky and glue-like, he named them "colloids," a term derived from the Greek word for "glue." Materials that passed through the membrane he called crystalloids. The physical chemistry of colloids is highly complex, involving electrical charges, adsorption, solvation, surface and interfacial tension and other physicochemical phenomena. *See also* aerosol; foam; carbon black; suspension; emulsion; protective colloid; surface; microscopy.

colloid mill. *See* milk; homogenization.

colorant. A general term for any substance that imparts a distinctive and relatively permanent color. There are both natural and synthetic colorants which are classified as dyes and pigments. Most dyes are synthetic organic compounds; pigments may be either natural or synthetic and either organic or inorganic. Some metals have colorant properties, e.g., iron, cobalt, and chromium, especially in the form of compounds. Natural organic colorants are chlorophyll, xanthophyll, carotene, and other plant pigments. *See also* dye; pigment.

colorimetry. An analytical method by which the amount of a compound in solution can be determined by measuring the strength of its color by either visual or photometric methods. Such determinations depend on the light absorption and transmission properties of the substance measured. Colorimetric techniques can be used to measure extremely low concentrations, for example, of chromium in steel. They are also used in medical analyses and in food and pharmacological laboratories. *See also* analysis.

columbium. The original and alternative name for the element niobium, now used chiefly by metallurgists. See niobium for details.

combination. A type of chemical reaction in which one compound is formed, either from two elements: $2H_2 + O_2 \rightarrow 2H_2O$, from an element and a compound: $C_2H_2 + 2Br_2 \rightarrow C_2H_2Br_4$, or from two simpler compounds: $2NH_3 + H_2SO_4 \rightarrow (NH_4)_2SO_4$. This term is also loosely used in the sense of any union of molecules, either weakly bonded or cross-linked. *See also* association; polymerization.

combining weight. *See* equivalent weight.

combustible material. Though technically defined as any material that will burn when exposed to a sufficient degree of heat in an atmosphere of air or oxygen, the term is usually applied to materials which ignite above 65°C (150°F) and burn relatively slowly. In this sense, "combustible" is distinguished from "flammable." Combustible materials are typified by paper, wood, and other cellulosics in their gross or bulk state, as distinct from a finely divided state. Combustible materials are nonflammable, that is, they do not ignite instantaneously, as do flammable materials. "Nonflammable" should be distinguished from "noncombustible," which means literally "incapable of combustion" (carbon dioxide, water, oxygen, etc.). *See also* flammable material; combustion; autoignition point.

combustion. Most commonly, the chemical combination of oxygen with another element or compound, usually induced by high temperature, resulting in the formation of one or more new compounds (oxides). It is often called burning. In the case of organic materials, where decomposition of the original compound occurs, combustion is the reverse of photosynthesis; the reaction is usually rapid and is accompanied by release of energy as heat, and usually light as well. Combustion of hydrogen results in the formation of water, the oxide of hydrogen; in the case of phosphorus, combustion occurs at room temperature. Combustion will take place in air when an organic material is heated to its autoignition point, either by an external flame or spark or by exposure to a sufficiently high ambient temperature. An organic combustion reaction yields two oxides (carbon dioxide and water) as end products; for example, a unit volume of gasoline, when completely burned, will produce an equivalent volume of water, plus some carbon dioxide. Slow combustion of some organic substances such as cellulosics and high-protein compounds can be induced by the action of bacteria. Combustion, with evolution of light and heat, can also occur in the absence of oxygen. Examples are the burning of a mixture of hydrogen and chlorine to form HC1, of a mixture of natural gas and chlorine, of thin copper ribbons in sulfur vapor, and of antimony in a stream of chlorine. All of these represent oxidation reactions, as does combustion involving oxygen. *See also* oxidation; autoignition point.

comminution.   Reduction of the particle size of solids by mechanical action; for very large sizes this is done in a crusher, and further size reduction is carried out in grinding machines of one kind or another, e.g., ball mill or hammer mill. Under special conditions, particle sizes approaching the colloidal range (1 micron diameter) can be achieved. Industrial comminution occurs in the manufacture of cement, paints, pigments, ceramics, etc. *See also* particle size.

compatibility.   The ability of two (or more) materials to exist together in the form of mixtures, laminates, or composites without exerting a damaging or harmful effect on one another over long periods of time. The term is usually applied to solvents which can be blended in all proportions and do not tend to separate, for example, water and alcohol. It is also applied to mixtures of polymers of different molecular structure. Oils, fats, and waxes are not compatible with water. An incompatible adhesive would be one that causes warping, wrinkling, or other damage to the substrate materials.

complex, complexing agent.   *See* coordination compound; chelate.

component.   One of the elements or compounds present in a system such as a phase, a mixture, a solution, or a suspension, in which it may or may not be uniformly dispersed. For example,

chromium is a component of stainless steel, and water is a component of milk. *See also* constituent; mixture.

composite.   Any combination or blend of solid materials whose particles or fibers are large enough to be visible to the unaided eye and which are chemically different from each other. There are a number of different kinds of composites. Notable among them are combinations of extremely high-strength fibers (graphite, silicon carbide) with thermosetting plastics; these are used in specialized applications in aerospace technology, rocket parts, and the like. Another important type includes reinforced plastics, which usually involve glass fibers. Many laminated structures such as plywood, tire carcases, and electric insulating panels are composites made from resins, wood, rubbers, and numerous other materials. Even blends of natural and synthetic fibers in both woven and nonwoven fabrics can be considered as composites. The definition also includes cermets. *See also* laminate; reinforced plastic.

composting.   *See* digestion (**2**); cellulose.

compound.   A chemical combination of two or more elements, called constituents. Compounds are represented by formulas which specifically indicate their atomic or ionic content and the proportion by weight of their elements; some types of formulas also show the spatial arrangement of the elements in the compound. Chemical bonds or valences hold the compound together. The molecular weight of a compound is the sum of the atomic weights of its atoms. Its properties are almost invariably different from those of the individual constituents. A significant feature of compounds is their homogeneity, that is, the fact that any given quantity of a compound has the same properties throughout. The constituents can be separated only by a chemical change induced by a reagent, by heat, or by electrical energy. For example, calcium carbonate ($CaCO_3$) is decomposed by heat to form calcium oxide (CaO) and carbon dioxide (CO_2); it can also be decomposed by reacting it with an acid. Water is decomposed by passing an electric current through it. *See also* mixture; homogeneous; formula, chemical.

compressed gas.   Any gaseous element or compound that is subjected to specific pressure and introduced into metal cylinders or special tanks for shipment in either liquid or non-liquid form.

Common inorganic gases handled in this way are nitrogen, oxygen, ammonia, chlorine, carbon dioxide, helium, fluorine, hydrogen, and nitrous oxide. Liquefied petroleum gases (LPG) comprise a subgroup of compressed organic gases; they include butane, butene, propane, and propylene and are chiefly used as household and industrial fuels. Other organic gases in compressed forms are vinyl chloride, acetylene, and butadiene. The reasons for compression are economy and convenience in shipping, storage, and handling, and, in the case of flammable gases, the safety factor.

concentration. The quantity of any substance, either solid, liquid, or gaseous, present in a specified weight or volume of a mixture. The units used for mixtures of solids are usually percent by weight or by volume. These may also be used for liquids, e.g., seawater is a 2.6% sodium chloride solution containing 2.6 grams of NaCl per 100 grams of solution. In solution chemistry, concentrations are more accurately designated by the terms molal, molar, and normal (separately defined). Square brackets are often used in formulas to indicate concentration, i.e., $[CO_2]$. A concentration is said to be critical when even the smallest addition to the component in question would result in a notable change in the properties of the system. *See also* solution.

conchoidal. A term taken over from mineralogy by chemists to describe a type of surface formed by fracturing a hard solid by impact. Certain materials present involutely curved fracture surfaces suggestive of the shape of the shells of bivalves (conch), from which the term is derived. Examples are glass, blown asphalt, and numerous minerals.

concrete. (1) In the construction industry, a Portland cement mixed with sand, small stones, gravel, and the like; in reinforced concrete steel rods are embedded in the concrete for strengthening purposes.

(2) In perfume technology, a waxy product obtained from roses by extraction with benzene; the wax is then dissolved out with alcohol to yield the perfume.

condensation. (1) In organic chemistry this term denotes a reaction, one of whose products is water, a simple alcohol, or ammonia. Such reactions often involve aldehydes and ketones and characterize certain types of polymerization. There are a number of special types designated by such names as aldol, Claisen, and Friedel-Crafts. A condensation polymer is a linear or three-dimensional macromolecule resulting from a catalyzed reaction between two organic molecules, often of the aldehyde or ketone type, usually with the formation of water or an alcohol as a by-product. Several kinds of condensation polymers are made by such repetitive or multistage reactions (called polycondensation), e.g., phenolformaldehyde, polyamides, polyesters, and polyurethanes. *See also* aldol condensation; Friedel-Crafts reaction; polymerization.

(2) The change of water or other substances from vapor to liquid phase on contact with a cool surface, as in distillation.

conductance. A term usually used in reference to the conductivity of electrolytic solutions. See conductivity (2).

conductivity. (1) The property of a substance or mixture of transmitting heat uniformly throughout its mass, the energy being passed from one atom to another with little if any loss. Crystalline solids (especially metals and alloys) are good thermal conductors because of their high density; liquids (water, glass) and high polymers (rubber, cellulose) usually are not.

(2) The property of a substance of transmitting an electric current by flow of electrons through a dense solid (metal) or by movement of the ions of a dissolved electrolyte to the electrodes when a potential difference exists between them. In the case of solids, no chemical change occurs when the current is passed, but in electrolytic solutions, decomposition of the electrolyte takes place. The term "conductance" is often used in preference to "conductivity" when referring to electrolytic solutions. *See also* electrolysis; ionization.

configurational formula. A graphic device used to emphasize or portray the fact that a given molecule has a steric configuration, that is, it exists in three dimensions. This is especially true of high-polymer substances, and configurational formulas are essential to a proper understanding of their structure. Though all molecules have three dimensions, configurational formulas are seldom used for the simpler structures. Representations of the molecular structure of cellulose, sugars, alcohols, and high polymers often include heavy lines or shaded areas to suggest their three-dimensional nature. *See also* ster-.

conformation. A term used by organic chemists

to refer to any possible three-dimensional or steric pattern of substituent atoms in a molecule as they rotate around single carbon-carbon bonds. Since the substituent atoms or groups (for example, hydrogen) tend to restrict rotation by repelling each other, the preferred conformation of a given organic molecule is one in which these atoms are most widely separated. Study of molecular structure along these lines is called conformational analysis and has been carried out extensively on alicyclic compounds such as cyclohexane. *See also* configurational formula; analysis **(2)**.

Congo Red.   *See* indicator.

conjugated compound.   An organic compound which contains two (or more) double bonds with a single bond positioned between them. Such compounds are unsaturated and because of their structure are chemically reactive An example is butadiene (H_2C=CH—CH=CH_2). In some cases three double bonds are present, two of which are in sequence, followed by the single bond and finally by the third double bond.

conservation.   *See* energy; mass conservation (law); balance **(1)**.

constant composition (law).   Also called the law of definite composition, this principle states that any given chemical compound always has a specific elemental composition by weight, the formula weight of the compound being the sum of the atomic weights of its constituent atoms.

constituent.   One of the elements or groups present in a chemical compound, or one of the molecules in a macromolecule or high-polymer substance. For example, hydrogen is a constituent of water; amino acids are constituents of proteins. *See also* component; compound.

contact process.   *See* sulfuric acid; vanadium.

contaminant.   Any foreign and often deleterious matter present in an otherwise relatively pure substance or mixture. Though generally similar in meaning to "pollutant," it applies to any extraneous agent that has found its way into a normal material or organism. For example, sulfur dioxide is an air contaminant (or pollutant); insecticidal residues are water and food contaminants, though they are often called "unintentional additives." The term "contaminant" is specifically used in referring to ionizing radiation and to infectious bacteria. *See also* impurity; decontamination.

control.   A neutral reference unit (also called a "blank"), which serves as a basis for comparison of experimental results. For example, in the evaluation of a rubber antioxidant, a control would be a portion containing no antioxidant whatever, taken from the same mixture or batch as the portions to which the test antioxidant is added. A second control would be a portion containing an antioxidant of known properties. It is essential that a control be subjected to the same conditions of processing and environment as are the treated portions. In biological and medical research, the control may be one or more laboratory animals of known health and behavior characteristics, maintained under the same conditions as the test animals. *See also* experiment.

cooking.   **(1)** Changing a foodstuff from the raw state to a softer and more palatable condition by heat treatment of various types from 90 to 260°C (200 to 500°F). A number of chemical and physical changes occur, for example, hydrolysis of the collagen in the connective tissue of meats, causing tenderness; softening and partial hydrolysis of cellulose and starches to sugars; denaturation and cleavage of proteins in meat, resulting in a less tightly ordered structure; coagulation of egg albumin and wheat gluten (cooked eggs and bakery products); modification or "shortening" of wheat flour by added fats, which coat the particles to form a laminate; evolution of carbon dioxide by the carbonates in baking powder, which causes bread and cake to "rise;" browning of bakery products due to reaction between sugars and amino acids (nonenzymatic); deactivation of enzymes. One adverse effect is the large loss or deactivation of water-soluble vitamins. Cooking also contributes to flavor by physicochemical reactions still not completely understood.

(2) A term used in the paper industry to refer to digestion of wood pulp with such chemicals as sodium bisulfite, sodium hydroxide, and sodium sulfide, to separate the cellulosic content from the lignin.

coolant.   *See* heat transfer.

coordination compound.   A molecule (often called a complex), either charged or neutral, which is formed by the attachment of a transition-metal ion to another molecule or ion by means of a coordinate covalent bond, i.e., a covalent bond in which both the shared electrons are furnished by the same atom. Such compounds were first studied early in the century by Alfred Werner (1866-1919), a Swiss chemist, (Nobel Prize 1913), and were formerly known as Werner complexes. Most are inorganic in nature; however, organic complexes also exist which are called chelates, and these contain a heterocyclic ring structure. The ionic or molecular group to which the metal ion of a coordination compound is bound is known as a ligand. Ligands have electron pairs available for donation to metal ions; the number of locations at which ligands are joined with the metal ion is the coordination number; this is usually 2, 4, or 6, representing the oxidation state of the metal. Typical metal ions involved in coordination are cobalt, nickel, iron, and the platinum group; the most common ligands are ammonia or other nitrogen compounds and chlorine, though many others are possible. When the ligand is ammonia, the compound is called an ammine. The charge on the complex is the algebraic sum of the charges on the ligands and the metal ion and is often zero. Examples of coordination complexes are: $[PdCl_6]^{-2}$;$[Ni(NH_3)_6]^{+2}$;$[Pt(NH_3)_2Cl_2]^0$.

A number of geometric isomers are possible, usually in the form of parallelograms, with the metal ion in the central location and the ligands at the vertices. Metal complexes are important in both inorganic and organic catalysis (enzymes). Their discovery led to the development of so-called "sandwich" molecules (metallocenes) in the early 1950's; these react like aromatic hydrocarbons and have a broad range of applications. *See also* ligand; chelate; sequestering agent; metallocene.

copolymer.   A high-polymer substance, usually an elastomer, made up of two or more different kinds of monomer, for example, styrene and butadiene. Copolymers are made by simultaneous polymerization of the monomers in the same operation, usually in emulsion form; as a result, the constituent monomers are combined into a common macromolecule, in contrast to the blending of two separately polymerized monomers (called homopolymers). When the latter method is used, the product is called a polyblend, which is a mixture of homopolymers; the components of polyblends are likely to be incompatible. Copolymerization techniques are helpful in manufacturing plastic products of

predetermined properties, which are often quite different from those of the respective homopolymers. *See also* homopolymer; emulsion polymerization.

copper.   An element.

Symbol Cu	Atomic Wt. 63.546
State Solid	Valence 1,2
Group IB	
Atomic No. 29	Isotopes 2 stable

Copper is a reddish metal, m.p. 1083°C (1981°F), occurring in the form of sulfide and oxide ores in the western states, as well as in Chile, Canada, Mexico, and Peru. It is a relatively soft metal but is distinctive in its high electrical conductivity and in its resistance to corrosion; both these properties contribute to its premier use in telephone and other communication wiring and in many other electrical applications. It is also widely used as an alloying metal (brass and bronze). Copper is a trace element required for the proper nutrition of plants and animals. It is essentially non-toxic. Copper compounds are oxidation catalysts for many organic materials; and a number of them are used as fungicides and insecticides.

copperas.   An old and misleading name for ferrous sulfate, a hydrate having the formula $FeSO_4 \cdot 7H_2O$, probably so called from the reddish or "copperish" color of this compound.

copper naphthenate.   *See* antifouling paint.

copper sulfate.   *See* insecticide; Bordeaux mixture.

cork.   A spongy, compressible form of cellulose peculiar to a unique variety of oak tree indigenous to Spain, but now cultivated in California and Florida. It is an excellent sound deadener, vibration damper, and heat insulator and is thus used in acoustic paneling and flooring components. Its extremely light weight makes it useful in life preservers, marker buoys, etc.; its other applications include bottle seals and stoppers for chemical glassware.

corona.   An electrical discharge effect which causes ionization of oxygen and the formation of ozone. It is particularly evident in the air close to high-tension wires and spark-ignited automotive engines. The ozone formed has a drastic oxidizing effect on wire insulation, cable covers, and hose connections; for this reason, such accessories are made of such oxidation-resistant materials as nylon, neoprene, and other synthetics. *See also* ozone.

corrosion. An electrochemical change in a metal surface caused by reaction of the metal with one or more substances with which it is in contact for long periods; the effect is usually deleterious. As air is usually the environment to which metals are exposed, oxidation is one of the commonest forms of corrosion, as in the rusting of iron. This reaction (catalyzed by moisture) forms ferric oxide, identical with the ore from which the iron was obtained. Thus, corrosion involves a return of a pure metal to its original state. Oxidation may be beneficial to metals; aluminum derives its immunity to corrosion from the tenacious film of aluminum oxide, formed on its surface, which inhibits degradation. Another form of resistance called passivity is due to much thinner oxide films on chromium and nickel. This is not true of copper, which resists corrosion because of its lower position in the electromotive series. Corrosion is accelerated by such impurities in the environment as ionizing acids and alkalies, which act as electrolytes, e.g., sulfur compounds, salt, etc. Metals subject to corrosion and exposed to weathering can best be protected by painting; in confined systems, as a boiler or heat exchanger, inhibitors or neutralizing agents can be added to the corroding environment.

Corrosion is basically an electrochemical reaction in which the metal is ionized and assumes positive charges and is accompanied by evolution of hydrogen at the cathode and generation of a detectable electric current; thus, it requires the presence of an anode, a cathode, and an electrolyte. *See also* activity (**1**); electrochemistry.

cortisone. A steroid hormone either extracted from the adrenal cortex of animals or synthesized from steroid-containing sources such as the acids found in animal bile. It is a specific curative for acute arthritis and other degenerative diseases; caution is required because of its possible adverse side effects. Its chemical composition is $C_{21}H_{28}O_5$ (17-hydroxy-11-dehydrocorticosterone). *See also* steroid; chemotherapy.

corundum. *See* aluminum oxide; sapphire; ruby.

co-solvent. *See* latent solvent.

cotton. The most important of the natural fibers containing 90% or more of cellulose. It is readily dyed by a number of standard methods and can be blended with synthetic fibers in any propor-

tion. In addition to textile end-products (clothing, bedding, upholstery, etc.), it is used in fabricated rubber items such as belting, hose, and footwear, as well as for medical purposes. Cotton has been replaced by synthetic fibers for many applications but is still a fundamental natural resource. The seeds are used for production of cottonseed oil and the "linters" for manufacture of rayon and nitrocellulose. *See also* cellulose; linters.

cottonseed oil. *See* cotton; vegetable oil; hydrogenation.

Cottrell, Frederick G. (1877-1948). An American chemist and a native of California, Cottrell obtained his doctorate from Liebig in 1902. His major contribution to industrial chemistry was his discovery of a practical method of dust elimination by electrical precipitation. Used in factory stacks and other large units, this process has contributed greatly to purifying the atmosphere of industrial areas. The principle involves charging a suspended wire with electricity. This creates a field which ionizes the surrounding air, the particles assuming the charge on contact and then moving to the wall of the stack where they are electrically discharged and precipitated.

coumarin. *See* salicylic acid.

coumarone-indene resin. *See* resin (2).

countercurrent flow. A term in chemical engineering denoting the movement relative to one another of two or more streams of materials (gas, liquid, or solid) to effect chemical, physical, or thermal changes between the respective streams. This type of flow is most frequently applied to multistage unit operations such as drying, evaporation, distillation, leaching, washing of solids, solvent extraction, etc., to achieve maximum efficiency and economy. Heat exchangers utilize countercurrent flow to obtain maximum heat transfer between a material being cooled and a material being heated. In the laboratory, a good example is the common water-cooled condenser, wherein the coldest water entering at the bottom is across the condenser wall from the cooled condensate exit, while the warmer water leaving at the top is near the hot distillate vapor inlet. *See also* extraction; fraction; reflux.

coupling. (**1**) A chemical reaction which takes place in the union of amino acids to form proteins; the reaction is essentially a condensation polymerization, which can be achieved synthetically only by suppressing the active sites on the amino acid molecules; when this is done, a peptide linkage can be formed, i.e., coupling can occur.

(2) Polymerization of certain phenols (especially those containing two substituted methyl groups) by means of oxygen and a nitrogenous catalyst to form a thermoplastic polymer; the process is called oxidative coupling.

(3) A low molecular weight group which forms a linkage in block polymers (*q.v.*).

(4) In the chemistry of dyeing, a reaction between an electron donor, such as a phenol or arylamine, and an electron-accepting diazonium compound, to form an azo dye.

covalent bond. A type of chemical bond in which atoms of the same or different elements combine to form a molecule (or in some cases a crystal) by sharing pairs of electrons; the molecule so formed is stable and does not ionize. An example of the formation of a molecule from two atoms of the same element is hydrogen, where H • + H • becomes H:H (the dots represent electrons). Thus, hydrogen is a covalently bonded diatomic molecule. Atoms of different elements also form covalent bonds, as in water, where the six outer electrons of oxygen plus the two hydrogen electrons arrange themselves in this way, H:Ö:H so that the oxygen is surrounded by eight electrons, which is the ideal number. The covalent bond is of particular importance in organic chemistry, since carbon readily shares its valence electrons with other elements, as in methane, where C• + 4H • combine to form

$$H$$
$$H:\overset{..}{C}:H$$
$$H$$

In the case of diamond, the carbon atoms are covalently bonded to one another in a tetrahedral structure, forming a nonmetallic crystal rather than a molecule. Coordination bonds are special types of covalent bonds in which both the bonding electrons are supplied by the same atom. *See also* carbon.

covering power. *See* hiding power.

cp. Abbreviation of centipoise.

Cr Symbol for the element chromium, the name being derived from the Greek word *chromos*, color.

cracking. A refining process involving decomposition and molecular recombination of organic compounds, especially hydrocarbons obtained by distillation of petroleum, by means of heat, to form simpler molecules suitable for fuels, monomers, and other industrial uses. A series of condensation reactions takes place accompanied by transfer of hydrogen atoms between molecules which brings about fundamental changes in their structure. Thermal cracking, the older method, exposes the distillate to temperatures of about 540 to 650°C (1000 to 1200°F) for varying periods of time; it is no longer used for gasoline but is still of value in producing hydrocarbon gases for plastics monomers. The development of premium fuels for airplanes and automobiles resulted from the use of catalysts in cracking; in this process, hydrocarbon vapors are passed at about 400°C (750°F) over a metallic catalyst (e.g., platinum); the complex recombinations (alkylation, polymerization, isomerization, etc.) occur within seconds to yield high-octane gasoline. Among the chemical changes induced are conversion of alicyclic compounds (cyclohexane) to aromatic compounds (benzenoid) and of straight-chain to branched-chain structures. Cracking reactions are exothermic. Free radicals, carbonium ions, and other chain-initiating agents are involved in these rearrangements. The fluidized bed technique is often used. Modified aluminosilicate compounds (zeolites or molecular sieves) are largely replacing many cracking catalysts formerly used. *See also* gasoline; catalyst; fluidized bed; reforming; zeolite.

cream. This common term has several meanings:

(1) In the dairy industry, it is defined as a fat-in-water emulsion in which the fat content varies from 20% (light) to 40% (whipping).

(2) In the beverage industry, its general meaning is that of sweetness obtained by addition of sugar; examples are cream soda, cream sherry, and creme de menthe.

(3) In the cosmetics industry, it refers to an emollient mixture of vegetable oils and glycerol with a powder base, known as cold cream.

(4) In the baking industry, the term cream of tartar (so called for no apparent reason) designates a derivative of tartaric acid used as a leavening agent.

creep. *See* plasticity.

creosote. *See* cresol.

cresol. An organic liquid derived from coal tar or by reacting methyl alcohol with phenol. It is closely related to phenol, of which it usually contains low percentages as impurity. The basic formula for cresol is $CH_3C_6H_4OH$; it is available either as a mixture of three isomers (ortho-, meta- and para-) or as any one of these. It has a strong odor suggestive of a disinfectant; the pure material is a skin irritant, and its fumes are also corrosive to the lungs. Its major uses are as a disinfectant, as a chemical intermediate, and in phenolic resins. A similar product, creosote, is also derived from coal tar and is used chiefly as a wood preservative. Both cresol and creosote can also be obtained from distillation of wood tar.

critical mass.   The minimum amount of fissionable material required to achieve an uncontrolled chain reaction, as in an atomic bomb. It is about 33 pounds of uranium-235.

criticality.   A term used by nuclear technologists to denote the condition of a reactor when the flow of fission-generated neutrons is just sufficient to produce electric power. Below this level, the reactor is said to be subcritical.

critical point.   The point at which a significant and notable change occurs in the properties of a substance or in a state of matter, for example, the glass transition temperature. The point may be that of temperature, pressure, concentration, electrical charging, rate of nuclear fission, etc. *See also* critical mass; concentration; isoelectric point.

cross-linking.   The union or binding together of two or more polymer chains either (1) by heating in the presence of a substance capable of forming a chemical bond between the two chains, or (2) by exposing the polymer to ionizing radiation. Both can be illustrated by the vulcanization of rubber in which (1) sulfur or an organic peroxide forms a bond with one or more carbon atoms in the hydrocarbon chains, a schematic example of which would be:

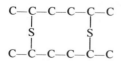

or (2) radiation effects the union by removing two or more hydrogen atoms, thus permitting direct carbon-to-carbon linkage. Free radical mechanisms play a part in both methods. Conversion of a thermoplastic to a thermosetting material, with accompanying increase in strength and durability, is an important result of cross-linking in rubber, polyethylene, and other high-polymers; wool, for example, is a naturally cross-linked protein. The polypeptide chains of hide and skin proteins are cross-linked by treatment with various chemicals in the tanning process. Cross-linking produces a more highly ordered molecular structure in which partially amorphous materials tend to become crystalline. *See also* polymerization; vulcanization; stereospecific.

crown glass.   A type of optical glass designed to have low index of refraction and low light dispersion; both these interrelated properties are critical factors not only in prescription lenses but in scientific instruments as well. Crown glass is of the soda-lime type. *See also* optical glass.

crucible.   A small, cup-like vessel, made of a refractory material, used for laboratory calcination and combustion work. Some types are equipped with a cover. The term is also used in the metal industry for a special type of furnace provided with a cavity in which molten metal collects. Crucible is derived from the Latin, *crux,* (cross), from the practice of alchemists who placed a cross on or near the container to exorcise evil spirits who might interfere with the results of their experiments.

cryogenics.   A branch of physics dealing with the properties of matter at extremely low temperatures. Chemistry is involved to the extent that such temperatures, which approach absolute zero, can be attained only by use of liquefied gases (oxygen, nitrogen, helium, hydrogen). Several practical operations have resulted from this research, such as the quick-freezing of foods by liquid nitrogen and the use of liquid oxygen with fuel for blast furnaces.

cryolite.   *See* electrolysis; aluminum fluoride; Hall process.

crystal.   The fundamental unit of a solid substance. Crystals characterize the solid state of matter, from the molecular to the visible size ranges. They represent a highly ordered state of molecular structure in three dimensions, as opposed to the disordered amorphous or liquid state. It is possible for both conditions to exist in certain high-polymer molecules such as cellulose. Crystals have a limited degree of vibration, and when enough energy is supplied in the form of heat to overcome the restraining forces, the crystals melt and the substance changes to the liquid phase. The shape of a crystal is an indication of its orderly internal structure, called a lattice; the atoms comprising the lattice act as a diffraction grating which permits determination of the structure by means of x-rays. Solids are made up of crystals of various forms, or "habits," united in aggregates which can be split apart along their cleavage planes (adjacent crystal surfaces); mica furnishes a good example of a substance whose crystals separate easily. The shapes of crystals always involve flat surfaces and straight lines and may be cubes, rhomboids, lozenges, needles, plates, etc. (but never spheres). Crystals can be made synthetically (grown) by various methods.

Crystals have both magnetic and optical properties which make them useful in lasers, computers, and similar devices. A substance can often be identified by the refractive index of its crystals and may be isotropic or anisotropic, depending on whether or not the transmitted light travels at the same velocity in all directions as it passes through the crystal. The behavior of semiconductors is based on the movement of electrons through crystal structures containing imperfections (called "holes") and atomic impurities. The science of crystals is called crystallography, which has an extensive vocabulary of its own. *See also* refractive index; liquid; amorphous; liquid crystal; habit; growth(2); hole.

crystal growth. *See* nucleation.

Cs Symbol for the element cesium, the name being derived from the Latin for blue, the frequency of its spectral lines.

Cu Symbol for copper, derived from the Latin, *cuprum;* this form also appears in the names of copper compounds, i.e., cupric, cuprous.

cumene. A light organic liquid, b.p. 152°C (306°F), having the formula $C_6H_5C(CH_3)_2$. It can be made by a catalytic reaction of propylene with benzene or as a distillation product of petroleum or coal tar. It is quite toxic and is absorbed by the skin. It also catches fire readily. The chief use of cumene is as a source of phenol and its by-product acetone which result from its oxidation. It has some application as a solvent.

cupric, cuprous. *See* copper.

curie. The accepted unit of radioactive decay, which has been found to be 3.7×10^{10} disintegrations per second. The term is abbreviated ci. *See also* alpha particle.

Curie, Marie S. (1867-1934). Born in Warsaw, Poland, she and her husband Pierre made an intensive study of the radioactive properties of uranium. They isolated polonium in 1898 from pitchblende ore. By devising a tedious and painstaking separation method they obtained a salt of radium, receiving the Nobel Prize in physics for this achievement in 1903, jointly with Becquerel. In 1911, Mme. Curie alone received the Nobel Prize in chemistry. Her work laid the foundation of radiochemistry which culminated in control of nuclear fission.

curing. A process, applied to many types of organic materials, which often involves application of heat and/or chemicals to bring about physicochemical changes that make the product suitable for use. Tobacco is cured by exposure to a temperature ranging from 27 to 83C (80 to 180F) for several days; this reduces moisture content, converts starches to sugars, and eliminates the green coloration. Meat is cured by addition of salt, sodium nitrite, sugar, etc.; some types are smoked after curing. Leather is cured by admixture of tannins, which change the soft collagenic material into a hard and serviceable one. The curing or vulcanization of rubber involves addition of sulfur and accelerator and short exposure to heat (121 to 204C (250 to 400F)); the change here is from an unstable thermoplastic to a thermosetting material as a result of cross-linking.

curium. An element.

Symbol Cm	Atomic No. 96
State Solid	Atomic Wt. 242(?)
Group IIIB	Valence 3,4
(Actinide Series)	

A radioactive and electropositive metal, m.p. 1340°C (2444°F), which is classed as a bone-seeker. The 242 and 244 isotopes produced in small quantities are useful for power generation in spacecraft. The chief compounds are the hydroxide, fluoride, and oxalate. The element was discovered in 1944 and named for the Curies.

current density. *See* density (2).

cyanamide. *See* amide.

cyanogen. The univalent group CN; also the compound C_2N_2, also called oxalonitrile. The term is derived from the Greek, meaning "blue-maker," because of the color produced when this group is introduced into aromatic compounds, as in phthalocyanine pigments. The compound cyanogen is a gas from which hydrocyanic acid is derived by hydrogenation; it is used as a chemical intermediate and as a fumigant. Like many CN-containing compounds, it is exceedingly toxic. *See also* nitrile; blue.

cyclamate.   *See* nonnutritive sweetener.

cyclic compound.   An organic compound containing one or more closed rings. If the ring contains only carbon atoms, it is called carbocyclic, represented by alicyclic compounds (cyclohexane) and aromatics (benzene). If one or more atoms in the ring is an element rather than carbon, e.g., nitrogen, sulfur, etc., the compound is called heterocyclic. Alicyclic rings are saturated and have various geometrical shapes; aromatic rings are unsaturated, with six sides; heterocyclics may be either hexagonal or pentagonal in form and are unsaturated. *See also* ring.

cyclohexane.   A saturated alicyclic compound, b.p. 80.7°C (177°F), having the formula C_6H_{12}; it occurs in petroleum and can also be made by hydrogenating benzene with the aid of a catalyst. Its chemical properties are similar to those of alkanes (paraffins), but its molecule (like those of other alicyclics) is in the form of a closed ring, often having the configuration of a "boat" or "chair." It is flammable but not very toxic. Cyclohexane is used commercially as a solvent and in the manufacture of nylon and has also been studied extensively on a theoretical basis in a branch of advanced chemistry called conformational analysis.

cyclohexanone.   *See* ketone.

cyclohexene.   *See* Diels-Alder reaction.

cyclonite.   *See* detonation.

cyclooctatetraene.   *See* acetylene.

cycloolefin.   *See* alicyclic.

cycloparaffin.   *See* alicyclic; cyclohexane; gasoline.

cyclopentadiene.   *See* diolefin.

cyclotron.   A circular electromagnetic device for accelerating positively charged particles (protons, deuterons, alpha particles). It was invented in 1929 by E. O. Lawrence (1901-1958). The acceleration is achieved by successive applications of small accelerations at low-voltage synchronized with the rotational period of the particles in a magnetic field. Energies of over 700 MeV for protons and over 900 MeV for alpha particles have been attained. The energized particles emerging from the cyclotron impinge upon a target nucleus, resulting in formation of radioactive isotopes, neutrons, and ionizing radiation. The first plutonium was made in a cyclotron in 1939. Powerful cyclotrons with huge electromagnets are in use in research laboratories throughout the world. *See also* accelerator (3); lawrencium; mendelevium; neptunium; plutonium.

cytochemistry.   A subdivision of biochemistry which deals with the chemical nature of the living cell and the behavior of its components, such as membranes, nucleus, protoplasm, etc., including nutrient and metabolic factors. Both plant and animal cells are included in this science.

D

D Symbol for the hydrogen isotope deuterium (heavy hydrogen), the name being derived from the Latin, meaning "second of two."

D- A prefix conventionally appearing as a small capital indicating the molecular structure or configuration of an optically active substance; it stands for "dextro," the "right-handed" enantiomorph of an optical isomer. It does not indicate that the substance is dextrorotatory. *See also* glyceraldehyde; *d-*.

d-. A prefix indicating that a compound is dextrorotatory, that is, it turns the plane of polarized light to the right. A plus sign (+) is preferably used instead of the letter. Do not confuse with D-, which refers to molecular configuration only.

2,4-D. An herbicide in the form of a white to yellow powder, m.p. 138°C (280°F), having the formula $Cl_2C_6H_3OCH_2COOH$; it is made by reacting chloroacetic acid with 2,4-dichlorophenol. It is usually applied as a spray, in the form of a long-chain ester. It is an effective weed-killer for specific types of unwanted vegetation and is also used as a defoliant. Like most other chlorinated organics, it is quite toxic and persistent, and its use has come under considerable criticism in recent years, resulting in restrictions in some areas. 2,4-D is especially effective on broad-leaf weeds and plants and may kill small trees and woody undergrowth of this type. *See also* herbicide; silvicide.

Dalton, John (1766-1844). The first theorist since the Greek philosopher Democritus to state that matter is composed of small particles, i.e., the atomic theory on which all succeeding chemical investigation has been based (1807). His essential concept of the indivisibility of the atom was not qualified until 1910, when radioactive decay was established by Rutherford.

Dalton's theories relating to pressure of gases and atomic combinations led to the basic generalizations stated in the law of multiple proportions, the law of constant composition, and the law of conservation of matter. Dalton's law of partial pressures states that the partial pressure of a gas in any mixture of gases is the pressure that any one of the component gases would exert if that component occupied the same volume as the mixture at the same temperature. Thus, if a mixture of oxygen, carbon dioxide, and nitrogen occupied a given volume, the partial pressure of any one of the three is the same as the pressure it would exert if the other two were not present and the one gas occupied the same volume as the original mixture. The law also states that the total pressure of the mixture is equal to the sum of the partial pressures of the gases. *See also* vapor pressure.

dammar. A natural resin of the group known as "recent" in the geological sense; it is obtained from various types of East Indian trees. Like other such resins, it is soluble in hydrocarbon solvents but insoluble in water. It is used to some extent in baked enamels, lacquers, and varnishes, as well as being a component of coatings for fabrics, paper, leather, etc. *See also* resin **(1).**

dating, radiocarbon. *See* radiocarbon dating.

daughter. Any member of a series of nuclides formed by radioactive disintegration of an unstable element, e.g., uranium, which is called the parent. The daughter nuclides in turn emit energy in the form of alpha or beta particles, so

79

that a sequence results, terminating in a stable nucleus. For example, radium and radon are daughters of uranium. *See also* decay; radio-activity.

Davy, Sir Humphry (1778-1829). Born in Cornwall, Davy was the first to isolate the alkali metals and recognize the identity of chemical and electrical energy. A pioneer in the science of electrochemistry, he carried out basic studies of electrolysis of salts and water, and his application of electricity to the decomposition of molten caustic potash led to the isolation of metallic potassium.

DDT.   A persistent (i.e., nonbiodegradable) insecticide which is a mixture of isomers of dichlorodiphenyltrichloroethane (a chlorinated hydrocarbon) having the formula $(ClC_6H_4)_2CHCCl_3$. It is derived from chloral and chlorobenzene by a condensation reaction. In the United States, it was once used on cotton and tobacco, as well as for mosquito control; it is still widely used as a general insecticide in India. DDT has been reported to be harmful to man only when ingested as such, but there is no evidence of any fatalities or direct harm to man from its legitimate use. Some studies have found that DDT is harmful to certain species of fish and bird life, but other studies have not confirmed this. However, despite the lack of agreement among scientists about its ecological effects, its use in the United States was prohibited in 1972. This order was later modified to permit its use to control the destructive tussock moth, against which it is the only effective insecticide.

Deacon process.   A method of converting hydrogen chloride (HCl) to chlorine by oxidation of HCl with oxygen at 400 to 500°C (752 to 932°F) over a copper salt catalyst; $2HCl + O_2 \rightarrow Cl_2 + H_2O$. Developed over a century ago and then largely abandoned, it has been reactivated recently as a means of producing chlorine without caustic and of utilizing the large amounts of by-product HCl from the chlorination of organic compounds. When conducted in the presence of an organic compound which reacts with the chlorine formed it is known as oxychlorination, e.g., $CH_2 = CH_2 + 2HCl + \frac{1}{2}O_2 \rightarrow CH_2ClCH_2Cl + H_2O$.

Debye, Peter J. M.   (1884-1966). A Dutch, and later Canadian, chemist and physicist who received the Nobel Prize in 1936 for his pioneer studies of molecular structure by x-ray diffraction methods. The interference patterns are still called Debye-Sherrer rings. He also made outstanding contributions to knowledge of polar molecules and to fundamental electrochemical theory.

deca-.   A prefix meaning ten, as in decaborane, decahydrate, etc. In the names of such organic compounds as decane, decanol, and the like, it indicates a chain of ten carbon atoms; in hydrates, it denotes the presence of ten molecules of water.

decaborane.   *See* carborane.

decane.   A moderately flammable liquid hydrocarbon, b.p. 174°C (345°F), obtained from petroleum refining; it has the formula $CH_3(CH_2)_8CH_3$, a straight-chain structure containing ten carbon atoms. It belongs to the alkane, or paraffin, group of hydrocarbons. Its major uses are as a solvent and in the synthesis of other organic compounds. It has the comparatively low ignition point of 249°C (480°F) and a flash point of about 43°C (110°F).

decay.   Spontaneous conversion of a portion of the mass of a natural or artificial radioactive element or nuclide into energy in the form of alpha, beta, and gamma radiation. This results from an unstable nuclear structure. Uranium decays through protactinium and thorium to actinium, radium, and radon and eventually to the stable 206 isotope of lead; it thus loses (or changes into energy) 32 mass units over the series (from 238 for the original uranium to 206 for lead). *See also* radioactive.

decolorizing.   *See* clarification; clay; diatomaceous earth.

decomposition.   A type of chemical reaction in which one compound divides or splits into two or more simpler substances, which may be either elements or compounds. The decomposition of an inorganic compound into two elements is exemplified by the reaction $2H_2O \rightarrow 2H_2 + O_2$; decomposition of an inorganic compound into two compounds is shown by the reaction $CaCO_3 \rightarrow CaO + CO_2$. Energy must be supplied for both these reactions; the first is accomplished by an electric current and the second by heat. Another example is the decomposition of an organic compound, induced by bacteria, which serve as catalysts in fermentation reactions:

$C_6H_{12}O_6 \rightarrow 2C_2H_5OH + 2CO_2$. *See also* degradation.

decontamination. In general, the removal of dirt, stains, infectious organisms, etc., from an area or a material; specifically, this term refers to treatment of clothing, equipment, buildings, and the like to remove radioactive substances to which they have been exposed. Standard methods of decontamination include thorough washing with a soap/water solution, or with potassium permanganate solution with subsequent application of 5% sodium bisulfite. Sequestering agents such as ethylenediaminetetraacetic acid are also effective. Steam blast can be used on buildings and metals. The effectiveness of the decontamination can be checked with a radiation counter. Clothing contaminated with radiation products should be treated as described, rather than by commercial cleaning methods. *See also* antiseptic; disinfectant.

defecation. A term used chiefly in sugar technology to refer to the clarification or removal of impurities from sugarcane or beet juice by heating in the presence of lime to precipitate undesirable solids. The resulting mother liquor is then drained off and concentrated by evaporation and boiling. *See also* clarification.

deflagration. Rapid and self-propagating combustion of a flammable solid, usually containing potassium or sodium nitrate (black powder) or nitrocellulose (smokeless powder). It is induced by heat, friction, or impact. Deflagration occurs at the particle surfaces of materials classified as low explosives, in which the combustion products flow away from the unreacted material. It differs from detonation, characteristic of high explosives, where the decomposition products flow toward the unreacted mass under high pressure. When a low explosive deflagrates in a confined area, the combustion rate increases rapidly to the point of detonation (nitrocellulose). Some high explosives will deflagrate if unconfined. *See also* detonation; explosive.

defoliant. A type of herbicide which is especially effective in dropping leaves from deciduous trees, shrubbery, etc., usually for military purposes. Most compounds used for this purpose are ecologically damaging, as they tend to deposit residues of varying degrees of toxicity in the soil, which may contaminate water supplies. Chemically, defoliants comprise a number of groups, including organic nitrogen compounds (thiocyanates, trichloropicolinic acid), phenoxyacetic derivatives, arsenates, etc. *See also* herbicide; silvicide; 2,4-D.

degradation. The breakdown of complex organic structures to simpler compounds by the influence of bacteria, usually accelerated by oxygen and sunlight. High-polymer substances are especially subject to this type of decomposition, which is used industrially in the treatment of municipal wastes. Infectious organisms can bring about degradation of proteins in the nervous system of the body. *See also* biodegradability; decomposition.

degras. *See* grease; lanolin.

degree of polymerization (D.P.) A number indicating the extent to which the molecules of a monomer have combined to form a polymer; it is determined by calculating the number of such molecules present in an average molecule of polymer in a sample. The degree of polymerization of viscose rayon, for example, is 700, while for pure natural cellulose it is 3000; for synthetic polymers it is usually much higher. In plastics manufacture, the properties of the product can be modified and controlled to a considerable extent by keeping the degree of polymerization within specified limits. *See also* polymerization.

dehydration. (1) Removal of 95% or more of the water from a food product by intensive oven drying, spraying, or other means, for the purpose of saving space, weight, and transportation cost. Solid materials are usually oven-dried in tunnel or vacuum dryers, while liquid or semiliquid products are sprayed through small orifices into hot air (milk, egg white, soaps, etc.). A number of common foods are available in dehydrated form (potatoes, bouillon cubes, fruit juices), requiring only addition of water to restore them to edible condition (reconstitution). There has been extensive use of dehydrated foods by the military, explorers, etc. A process called dehydrofreezing removes only about half the water content, after which the vegetable is frozen.

(2) Removal of chemically combined water from castor oil by heating it in the presence of a catalyst, thus converting it from a nondrying to a drying oil suitable for use in protective coatings.

(3) Excessive or abnormal loss of water from the body as a result of dysentery, diarrhea, or

vomiting, which results in severe chemical imbalance.

(4) Removal of a molecule of water from an alcohol by heating in the presence of a catalyst, an operation which is useful in synthetic processes: $CH_3CH_2OH \rightarrow CH_2{=}CH_2 + H_2O$.

dehydrochlorination. A process whereby a molecule of hydrogen chloride is removed from an organic chloride leaving a double or triple bond in the organic compound. It is the principal process for manufacturing vinyl chloride from ethylene dichloride (1,2-dichloroethane), by the reaction $CH_2ClCH_2Cl \rightarrow CH_2{=}CHCl + HCl$. It can be conducted thermally at an elevated temperature or at lower temperatures in the presence of an alkali which reacts with the liberated HCl.

dehydrocyclization. A catalytic petroleum reforming reaction by which straight-chain paraffin hydrocarbons containing up to five carbon atoms are dehydrogenated to unsaturated structures, which are then converted into aromatic (ring) compounds at 400°C (about 750°F). This greatly increases the octane number and is one method of making high-quality motor fuels. *See also* gasoline; reforming.

dehydrogenation. A chemical reaction, normally requiring a catalyst, in which hydrogen atoms are removed from a saturated compound (an alcohol, paraffin, or alicyclic compound) to form unsaturated aldehydes, olefins, or aromatic compounds. This is a form of oxidation (electron transfer), since each hydrogen atom contains one electron; for example, acetaldehyde is formed by removing two hydrogen atoms from ethyl alcohol: $CH_3CH_2OH \rightarrow CH_3CH{=}O + H_2$. The saturated cyclic hydrocarbon cyclohexane is dehydrogenated to benzene in the catalytic reforming process. *See also* oxidation; reforming.

deliquescent. A term used to characterize water-soluble salts usually in finely divided form (small plates, crystals), which not only absorb moisture from the air, but tend to soften and even dissolve as a result of this absorption. Examples of common deliquescent materials are calcium and magnesium chlorides, sodium hydroxide, and calcium nitrate. All such substances should be kept in well-stoppered bottles or in tightly closed containers. *See also* hygroscopic; hydrophilic.

demineralization. *See* ion exchange; electrodialysis; desalination.

Democritus (460-362 B.C.). A Greek philosopher, the first thinker of record to conceive of matter as existing in the form of small, indivisible particles, which he called atoms. However, this concept was overshadowed by Aristotle's theories and it was not until some 2000 years later that it was developed by John Dalton in England—an astonishing length of dormancy for one of the most creative ideas in the history of science. *See also* Dalton.

demulsification. The intentional destruction of the undesirable emulsions which often occur in petroleum extraction, where subterranean pressure has created a highly stable emulsion of saline water in the petroleum. A wide variety of materials and methods has been used to "break" these emulsions, including alternating current, neutralization by multivalent ions, centrifugation, etc. *See also* emulsion.

denaturant. A substance added to ethyl alcohol to prevent its being used for internal consumption. This applies chiefly to ethyl alcohol intended for industrial use, the contaminating material being added to insure that it is not diverted to beverage use (for taxation purposes). Methyl alcohol is used in Completely Denatured Alcohol because of its toxicity; Specially Denatured Alcohol may contain a number of specified materials, among which are tert-butyl alcohol, brucine, quassin, and sucrose octaacetate. Many other additives have been used unofficially in the past with the idea of making the alcohol obnoxious as to odor, taste, or internal effect.

denaturation. (1) A reordering of the molecular structure of some proteins (globulins) induced by a number of environmental factors, such as heating to just below the boiling point, change of pH, or exposure to various detergents. These cause rupture of the hydrogen bonds in the molecule, but not of the stronger peptide bonds. As a result, the structure of the protein becomes somewhat more random, and its enzymatic activity is greatly reduced. The process is reversible, restoring the protein to its original state. Denatured proteins are less soluble than normal proteins and lose their ability to crystallize.

(2) *See* denaturant.

denier. A term used in the textile industry to designate the weight per unit length of a filament, i.e., its diameter or "fineness;" a filament is called 1 denier if 9000 meters of it weigh 1 gram. Tenacity is the strength per unit weight, expressed as grams per denier; it is converted to tensile strength in pounds per square inch by the formula: T.S. = tenacity (grams per denier) × 12,800 × specific gravity of the material. *See also* grex.

density. (1) The ratio of weight (mass) to volume of any substance, usually expressed as grams per cubic centimeter. If 1 cc of a substance weighs 2 grams, its density is 2 grams per cc. Density is closely related to specific gravity, which is the ratio of the weight (or mass) to the weight of the same volume of a standard substance, usually water. Since 1 cc of water weighs almost exactly 1 gram, water is taken as the reference material for liquids and solids; air is used for gases. The specific gravity of a solid having a density of 5 grams per cc thus equals 5/1, or 5; a liquid with a density of 0.8 gram per cc has a specific gravity of 0.8/1, or 0.8. Apparent or bulk density, used for very light materials (cork, balsa wood), often differs greatly from the true density of the same material in the form of a compact solid; an example is magnesium carbonate, whose natural form (magnesite) has a density of 3 grams per cc, whereas the synthetic product is a fluffy powder having a bulk density of about 4 pounds per cubic foot, or 0.064 gram per cc.

(2) In electroplating technology, current density is the amperes per square decimeter of current being delivered to the surface being plated; other units of area are also commonly used, e.g., amperes per square foot.

deodorant. A substance, either with or without an independent odor, which has the effect of reducing, removing, or replacing the undesirable odors of other substances. It acts in several ways. For example, activated carbon adsorbs molecules of malodorous materials; aluminum chlorohydrate neutralizes acidic components of perspiration; materials having a pungent but pleasant smell replace or mask the unwanted odor, e.g., peppermint, pine oil, menthol, camphor, etc. *See also* odor.

deoxyribonucleic acid. This term is commonly abbreviated DNA; it is also frequently spelled desoxyribonucleic acid. It is an extremely complex nucleoprotein (a macromolecule made up of a nucleic acid bound to a protein) whose constituent groups are arranged in a double helix configuration composed of two interlocked chains, each containing several thousand chemical units. These chains are able to reproduce themselves as they separate by a process known as replication, thus forming new DNA molecules. The sequence of amino acids in the chains, which is controlled by DNA, determines the so-called genetic code, which dictates or programs the formation of body cells into tissues, organs, etc., as well as the transmission of hereditary characteristics. DNA is closely related to ribonucleic acid (RNA), the two working together in assembling the amino acid sequences which control cellular differentiation. The essential constituents of the DNA structure, first elucidated by two English biochemists Crick and Watson in 1953, are nitrogenous bases, phosphoric acid, and deoxyribose. The DNA molecule was synthesized in 1957. *See also* recombinant DNA; genetic code; replication; gene.

Derris. *See* retenone.

desalination. Removal of the mineral salts from seawater or saline waters (brines) to produce potable fresh water conforming to standard purity requirements. This can be done in a number of ways, not all of which are economically possible. The three methods that have proved most satisfactory on a large scale are flash distillation, reverse osmosis, and electrodialysis. Ordinary solar evaporation has been utilized on a small scale in several localities. Desalination techniques can also be applied to concentrated brines and acid mine waste water. An alternative term with virtually the same meaning is demineralization. *See also* osmosis; distillation; electrodialysis.

desiccation. Removal of water vapor from a material by a hygroscopic substance, such as calcium chloride or silica gel, placed in an airtight container with the material to be dried (often under vacuum). The term is conventionally restricted to laboratory control of chemical reagents, test samples, and the like and is not used for large-scale drying or dehydrating operations. Misspelling can be avoided by consideration of the derivation from the Latin *de* + *siccus* (dry). *See also* drying; dehydration; hygroscopic.

desorption. *See* absorption.

destructive distillation. *See* distillation.

detergent. A surface-active compound, either natural or synthetic, which acts as a cleansing and suspending agent by emulsifying the oils and greases occurring in soils on fabrics, ceramics, etc., as a result of its ability to reduce interfacial tension between dissimilar liquids.

Oil-soluble detergents are used in gasoline and lubricating oils to hold particulate impurities in suspension. Among the water-soluble detergents, alcohols and common soaps are effective. Still more so are the synthetic types (syndets) such as sulfonated dodecylbenzene and similar alkylates. The linear alkyl sulfonates (LAS) are more biodegradable than are the branched-chain type; they are made from petrochemicals by various cracking and polymerization techniques. Detergent action is promoted by phosphatic compounds called builders. Detergents may be nonionic, cationic, or anionic, depending on the nature and behavior of their molecules in emulsification. *See also* surface-active agent; emulsion; wet; builder.

detonation. The almost instantaneous decomposition of a high explosive as a result of ignition, shock, or impact, resulting in the propagation of an intense high-pressure wave traveling from 1 to 5 miles a second at the point of initiation. Well-known detonating explosives are trinitrotoluene (TNT), nitroglycerin, dynamite, tetryl, cyclonite, pentaerythritol tetranitrate (PETN), mercury fulminate, lead azide, and ammonium nitrate-fuel oil mixtures (ANFO). Detonation is also possible with some low explosives (nitrocellulose) when confined and exposed to friction or impact. *See also* deflagration; explosive.

deuterium. The rare natural isotope of hydrogen having atomic weight of 2.014 instead of the normal 1.008; the extra mass is due to the presence of a neutron in the nucleus in addition to the proton. Deuterium is often called heavy hydrogen, and water containing it, heavy water (deuterium oxide). The natural occurrence of this isotope is one in every 6500 parts of normal hydrogen. All the other properties of deuterium, including high flammability, are identical with those of hydrogen. The nucleus of the deuterium isotope is called a deuteron. Deuterium was discovered in 1931 by Harold C. Urey, who received the Nobel prize for this achievement in 1934. *See also* hydrogen; heavy water; Urey.

deuteron. The nucleus of a deuterium atom; it contains one proton and one neutron and thus has a mass of 2 and a positive charge of 1. Because of their greater mass, deuterons are used to bombard other nuclei to produce radioactive isotopes, as in a cyclotron. *See also* deuterium.

developing agent. (1) In photographic chemistry, a reducing compound which reacts with silver bromide crystals to form metallic silver under the influence of light, which activates the reaction. Hydroxyl- or amino-containing benzene compounds are among the most effective developers, e.g., hydroquinone. The action of developing agents is accelerated by an alkaline environment.

(2) In dyeing technology, an organic compound which will unite or combine with another compound in or on the fiber to form or "develop" a new color, often having improved properties.

devitrification. *See* vitreous.

dewatering. *See* drying.

dextran. *See* gel; plasma.

dextrorotatory. An organic compound containing an asymmetric carbon atom, which has the property of turning the plane of polarized light to the right; such compounds usually exist in the form of so-called optical isomers, one of which is dextrorotatory and the other levorotatory, the latter turning the plane of polarized light to the left. Many sugars, amino acids, and some alcohols exhibit these optical properties, e.g., dextrose and levulose derive their names from this phenomenon. Dextrorotatory is indicated either by the italic letter *d* prefixed to the name, or by a plus sign (+); levoratory isomers are indicated by italic *l*, or by a minus sign (−). The use of + and − signs is now preferred. *See also* asymmetry; enantiomorph; optical rotation.

dextrose. A dextrorotatory sugar, also called glucose, which occurs naturally in corn and grapes and is also found in blood. A member of the general class of carbohydrates, it is formed in plants by photosynthesis; it is a monosaccharide having the formula $C_6H_{12}O_6$. Dextrose is an optical isomer of levulose (fructose), which is levorotatory. Dextrose polymers form amylose, a basic constituent of starch. Thus dextrose can be obtained by reaction of starch with water (hydrolysis), which is catalyzed by the enzyme amylase. It is the type of sugar used for intravenous feeding of invalids, and in baby foods, wine manufacture, caramel, and other food flavors. *See also* sugar; optical rotation.

Di Symbol for didymium, a mixture of rare earths.

di-. A prefix meaning "two;" in the names of compounds, it indicates the presence of two atoms of the same element (carbon dioxide, CO_2) or of two identical groups (diethylbenzene, $C_6H_4(C_2H_5)_2$, dicarboxylic acid (oxalic acid, HOOCCOOH). It also appears in more general terms such as dipole (two electric poles), diolefin (two double bonds), dimer (two molecules united), and diatomic (two atoms); the last of these has no connection with diatom or diatomaceous earth, which are named from a biological genus. *See also* dia-; bi-.

dia-. A prefix having several meanings: (a) "passing through," as in dialysis, diaphragm, diameter; (b) "opposite to," as in diamagnetic; (c) "complete" or "throughout," as in diagnosis. It is not a prefix in the word diamond.

diacetone alcohol. A monohydric alcohol, b.p. 169°C (336°F), with formula $CH_3COCH_2C(CH_3)_2OH$. It is derived from acetone by a condensation reaction and has a wide spectrum of solvent applications; these include extraction of numerous organic products, preparation of nitrocellulose lacquers and other cellulosic coating compositions, cleaning mixtures for metals, and stripping dyes from textiles. It has a moderate degree of toxicity and should not be inhaled or used internally. It is also quite flammable and should not be used near open flames or spark sources. *See also* monohydric alcohol.

diallyl phthalate. *See* prepolymer.

dialysis. The diffusion of a substance in solution through a semipermeable membrane; this process permits separation of smaller from larger molecules in a solution and thus might be called molecular filtration. The membranes used are of parchment, collodion, cellophane, or other cellulosic material; many have a pore size of 100 angstroms or less. In nature, plant cell walls act as osmotic diffusion membranes. The rate of diffusion varies with the particle size and concentration of the dissolved substances; sodium chloride passes three times as quickly as sucrose, while macromolecules penetrate extremely slowly, if at all. Colloids were originally distinguished from crystalloids on this basis by Thomas Graham; proteins and other high-polymers tend to be held back or retained by a parchment membrane, whereas salts, sugars, etc., pass through it with varying degrees of ease. *See also* diffusion; colloid chemistry; osmosis; membrane.

diamond. A crystalline allotropic form of carbon and one of the hardest known substances; it is comprised of covalently bonded carbon atoms arranged in stable polyhedral crystals having 8, 12, or 24 sides. There are several types, the purest being used as gemstones, while less valuable forms (known as bort, carbonado, etc.), of much smaller size, are used industrially as cutting surfaces on drilling bits, as dies for wire-drawing, in abrasive wheels, and the like. The gemstone grade occurs in South Africa, while the others are quite widely distributed in South America and elsewhere. Diamond can be made synthetically at high temperature and pressure in an electric furnace, and about one-third of the industrial diamonds now in use are produced in this way. *See also* carbon.

diastase. *See* enzyme; malt;-ase.

diastereoisomer. *See* stereoisomer.

diatomaceous earth. A light, bulky, high-silica material occurring in the southwestern states and said to be the petrified remains of microscopic sea life of an earlier geologic period. The material is also called diatomite and sometimes kieselguhr. It has considerable ability to absorb liquids and thus finds use as a decolorizing and purifying agent. Its most important applications are as a filter medium, as a diluent in many types of industrial mixtures, in paper coatings, and as a carrier for catalysts. Its low conductivity of heat and sound account for its use in both thermal and acoustic insulation. *See also* diluent; earth.

diatomic gas. (1) A molecule composed of two atoms of the same element, which is a gas at room temperature; the elements which normally occur as diatomic gases are oxygen (O_2), hydrogen (H_2), nitrogen (N_2), and the halogens. The molecular form of these elements should be distinguished from their atomic form, which occurs in compounds.

(2) A molecule composed of two atoms, each of a different element, which is a gas at room temperature, e.g., carbon monoxide, hydrogen chloride. These are binary compounds, the elements of which may be replaced by other atoms in a chemical reaction. Unless otherwise stated, meaning (1) is generally assumed. *See also* molecule; gas.

diatomite. *See* diatomaceous earth.

diazonitrophenol. *See* initiating explosive.

diazotization. A chemical reaction used in the production of azo dyes and a wide range of organic intermediates. These contain the diazo group —N≡N—, which carries the color (chromophore). The reaction involves a primary aromatic amine and nitrous acid in a solution of a mineral acid. It is followed by the coupling reaction, in which the diazo group unites with another organic molecule, such as a phenol or amine. A typical product is diazoaminobenzene (C_6H_5N=N—NHC_6H_5). *See also* azo dye; coupling; chromophore.

dibasic. (1) A term used to describe an acid, either inorganic or organic, whose molecule contains two hydrogen atoms that can be replaced by basic elements to form salts. Examples are sulfuric acid (H_2SO_4) and oxalic acid (HOOCCOOH). Dibasic inorganic acids liberate two hydrogen ions when in solution.

(2) An inorganic compound whose molecule contains two atoms of a basic univalent element, for example, dibasic sodium phosphate (Na_2HPO_4).

diborane. A highly flammable and explosive gas composed of boron and hydrogen (B_2H_6); it will ignite at the unusually low temperature of 38°C (100°F). It is stored in cylinders in a cool, dry environment. Carbon tetrachloride and other halogenated extinguishing agents must not be used! It is made (a) by the reaction of hydrogen and boron trichloride, or (b) by the reaction of lithium aluminum hydride and boron trichloride in ether. The gas has a strong, unpleasant smell and is also highly toxic. Its industrial uses are limited to synthesis of organic boron compounds and as a source of boron for doping crystals in semiconductors. It was formerly considered promising for rocket fuels, but this application has not been developed. *See also* hydride; dope; boron.

dibutyl phthalate. A slightly viscous liquid, without odor or color, made by reacting butyl alcohol with phthalic anhydride; its formula is $C_6H_4(CO_2C_4H_9)_2$. It is a plasticizer for nitrocellulose products (lacquers, explosives), as well as for various plastics, resins, and coatings; it also finds use in the perfume industry as a solvent and fixative. Inhalation of the vapor of the pure substance should be avoided. It ignites at 370°C (about 700°F).

dicarboxylic acid. A carboxylic acid in which two carboxyl groups (COOH) are present, for example, oxalic acid (HOOCCOOH), phthalic acid [$C_6H_4(COOH)_2$], and maleic acid (HOOCCH=CHCOOH).

dichlorobenzene. A chlorinated aromatic hydrocarbon ($C_6H_4Cl_2$), having three isomeric forms, depending on the positions at which the chlorine atoms are attached to the benzene nucleus. The meta- (1,3-) and ortho- (1,2-) forms are liquids used as disinfectants and insecticides and to some extent as solvents. The para- (1,4-) isomer is a crystalline solid, familiar as moth balls. All forms emit vapors that are irritant to the eyes and the mucous membranes, and inhalation should be avoided, especially of the ortho- form. These compounds also have some use as chemical intermediates and as analytical reagents. *See also* chlorinated hydrocarbon.

dichlorodifluoromethane. Chlorofluorocarbon gas (CCl_2F_2), b.p. −28°C (−18°F), obtained by catalytic reaction of hydrogen fluoride and carbon tetrachloride. It is not harmful in general, but inhalation of high concentrations may have a narcotic effect; it is also nonflammable. The commercial product is shipped in steel cylinders for a number of industrial uses, especially air-conditioning and refrigeration. There are many similar compounds, commonly called fluorocarbons, but more correctly should be called chlorofluorocarbons. *See also* fluorocarbon; refrigerant.

dichlorodiphenyltrichloroethane. *See* DDT.

dichloroethylene. A liquid chlorinated hydrocarbon (ClHC=CHCl) which is both flammable and irritating to the skin and mucous membranes. It has two isomeric forms: *cis*, b.p. 59°C (138°F), and *trans*, b.p. 48°C (118°F). Its major use is as a solvent in the plastics, dye, perfume, and paint industries and in the manufacture of synthetic organic chemicals. It is a derivative of acetylene. *See also* chlorinated hydrocarbon.

dichloromethane. *See* methylene (1).

dichlorophenoxyacetic acid. *See* 2,4-D; plant growth regulator.

dichroic. A term used in crystallography to denote crystals which refract incident light in two directions, thus displaying two colors when observed from different angles, for example, calcite. *See also* anisotropic; birefringent.

didymium. A mixture of the metals or oxides of several rare earths extracted from monazite sand. Its salts are used to some extent in manufacture of special glasses and in electronic equipment. The nitrate is flammable. Its symbol is Di.

dielectric. Any material which has strong electrical insulating properties. Among the solids of this type are those containing a high percentage of silica, such as glass, diatomaceous earth, mica, etc.; cellulose-containing materials (wood, cotton); and most high-polymers, both natural and synthetic (rubber, plastics). Liquid dielectrics, often called transformer oils, include silicone oils, mineral oils, and chlorinated hydrocarbons (askarels). Values used by engineers and specification writers for defining this property of materials are dielectric constant and dielectric strength, determined by specialized test procedures. *See also* insulator; dielectric constant.

dielectric constant. A value that serves as an index of the ability of a substance to resist the transmission of an electrostatic force from one charged body to another, as in a condenser. The lower the value, the greater the resistance. The standard apparatus utilizes a vacuum, whose dielectric constant is 1; in reference to this, various materials interposed between the charged terminals have the following value: air, 1.00058; glass, 3; benzene, 2.3; acetic acid, 6.2; ammonia, 15.5; ethyl alcohol, 25; glycerol, 56; and water, 81. The exceptionally high value for water accounts for its unique behavior as a solvent and in electrolytic solutions. Most hydrocarbons have high resistance (low conductivity). *See also* polar compound.

Diels-Alder reaction. An important type of organic reaction discovered in 1928 by Otto Diels and Kurt Alder, German chemists, Nobel Prize 1950; it serves as a means of synthesizing such pharmaceutical products as codeine, morphine, cortisone, and reserpine. In simplest terms, it involves the addition at the 1 and 4 positions of an unsaturated molecule of a diolefin in which the double bonds are separated by a single bond (called a conjugated diolefin). The compounds formed contain a hexagonal unsaturated ring, for example, cyclohexene or its derivative. 1,3-Butadiene and cyclopentadiene are often utilized in this type of reaction.

diene. *See* diolefin.

diesel fuel. A petroleum distillate, either straight-run or partially cracked, used as a power source for diesel engines; it is equivalent to the No. 2 grade of domestic fuel oil. It contains a high percentage of aliphatic hydrocarbons. It is suitable for use in heavy automotive equipment and is being increasingly used in automobiles. Its performance is rated by the cetane scale. *See also* cetane number.

diethylamine. An extremely flammable and rather toxic liquid, b.p. 55.5°C (132°F), made by reacting ethyl chloride and ammonia at high temperature and pressure. Its formula is $(C_2H_5)_2NH$. It is used in considerable volume as an intermediate for rubber accelerators and antioxidants and for polymerization retarders, dyes, and miscellaneous organic chemical manufacture. It also has some application as a solvent and laboratory reagent. *See also* amine.

diethyl carbonate. *See* carbonate.

diethyl ether. *See* ethyl ether.

diethyl sulfate. *See* mutagen.

diethyl sulfide. *See* carborane.

diffusion. The mutual permeation of two or more substances due to the kinetic activity of their molecules, so that a uniform mixture or solution results. Diffusion occurs with all forms of matter; it is most rapid for gases, somewhat slower for liquids and for solids in solution. Gases of different densities will mix uniformly counter to gravity; for example, carbon dioxide and air form a uniform blend even though carbon dioxide is considerably denser than air. For gases, the rate of diffusion is inversely proportional to the square roots of their densities. Miscible liquids of different molecular weight diffuse into each other, e.g., alcohol and water. Solids in solution (sugar, salt) move through the solvent until a uniform concentration results, as, for example, salts in ocean water. Two metals in close contact also show a slight tendency to diffuse. Diffusion also takes place when a barrier or film of a microporous solid (metal or cellulosic) is placed between two gases or between two liquid solutions of different concentrations; in the latter case, the rate depends on the relative strength of the concentrations. This phenomenon is involved in dialysis and osmosis, and in the separation of uranium isotopes (gaseous diffusion process). *See also* dialysis; osmosis; Graham's Law.

digestion. (1) The breakdown of cellulosic fibers by the action of heat and chemicals (both acids and alkalies), as occurs in the pulping of wood and in the reclaiming of automobile tires. As a result of digestion with sodium hydroxide solution, for example, the cellulose is separated from the lignin in wood pulp, and the fabric content of tires is separated from the rubber.

(2) Decomposition of solid or semisolid organic wastes such as garbage, cellulosics, sewage, etc., by the action of bacteria, either with or without the presence of air. Air is used in the

activated sludge method of treating sewage wastes and in the composting of garbage and agricultural wastes.

(3) The metabolic sequence of events by which food is decomposed and its nutrient content absorbed by the body. *See also* metabolism.

digitalis. *See* glycoside.

dihydric. An alcohol characterized by the presence of two hydroxyl (OH) groups; this classification is represented by the glycols. The term diol is sometimes used synonymously. *See also* glycol.

m-**dihydroxybenzene.** *See* resorcinol.

p-**dihydroxybenzene.** *See* hydroquinone.

diisocyanate. The essential compound for production of polyurethane resins; it is derived by first nitrating toluene and then reacting the resulting amine with phosgene to form a molecule in which two isocyanate groups (NCO) are attached to the ring at the 2,4 or 2,6 positions. Urethane ($CONH_2OC_2H_5$) and its polymers are formed by a condensation reaction of diisocyanate with, for example, ethylene glycol. A so-called hindered isocyanate may be made by using phenol; this combination gives a urethane which may be kept for some time at room temperature but which will decompose to the isocyanate when heated. *See also* polyurethane.

dilatant. A term used in rheology to describe a phenomenon characteristic of highly concentrated suspensions of solids in liquids, as for instance, rubber-solvent doughs, stiff paint formulations, and the like. If a deforming force is applied quickly, the flow response decreases, that is, the system resists the deforming stress; if the stress is applied gradually, the resistance drops, and flow rate increases. *See also* thixotropy; fluid; liquid.

diluent. A low-gravity material (solid, liquid, or gas) added to a product either (1) to reduce its cost or (2) to lower the concentration of its basic component for some desirable purpose. Examples of (1) are the addition of asphalts, wood flour, etc., to low-grade rubber and plastic products; examples of (2) are the addition of hydrocarbon thinners such as turpentine to paints to lower viscosity and of diatomaceous earth to nitroglycerin to reduce its shock sensitivity (dynamite). The term *extender* is almost synonymous with *diluent,* the chief difference being the specific gravity of the added material; extenders include relatively heavy powders, such

as whiting, barytes, calcium silicate, etc.

dilution ratio. A specific term used by lacquer formulators in reference to nitrocellulose solutions; it is also known as the hydrocarbon tolerance of such solutions. It is defined as the largest number of unit volumes of toluene or naphtha required per unit volume of active solvent to initiate gel formation in a solution containing 8 grams of nitrocellulose per 100 cc of solution. The active solvents referred to are various acetates and ketones. *See also* diluent; kauributanol value.

dimer. A molecule resulting from the combination of two identical molecules called monomers. Straight-chain hydrocarbons and fatty acids are susceptible to dimerization. Many dimers occur naturally, e.g., butene (C_4H_8) is a dimer of ethylene (C_2H_4); others are formed synthetically. *See also* monomer; polymer.

dimethylamine. *See* amine.

dimethylketone. *See* acetone; ketone.

dimethylsulfoxide. An extremely powerful solvent which can be made from dimethylsulfide or recovered from the lignin in pulping waste liquors. It has the formula $(CH_3)_2SO$. It is hygroscopic, almost colorless, and nontoxic. Reactions occur in a solution of DMSO many times more rapidly than in other solvents. It also has the unique property of extremely rapid penetration of both plant and animal tissue, thus greatly aiding the administration of drugs and other medicinals. It was first manufactured on a commercial scale in 1954, its importance being recognized by chemists in the paper industry, who were able to extract it from the lignin in sulfide effluents. Among its other uses are as a spinning solvent for synthetic fibers, an industrial cleaner, and a polymerization solvent.

dinitrobenzene. A cyclic organic compound consisting of two nitro groups attached to a benzene nucleus in either the ortho- (1,2-), meta- (1,3-), or para- (1,4-) positions. The formula thus is $C_6H_4(NO_2)_2$. The compound is made by reacting nitrobenzene with a mixture of sulfuric and nitric acids. It is used in the synthesis of dyes and other organics and as an industrial camphor substitute. It is extremely poisonous and should be kept from contact with eyes and skin. It occurs in the form of a yellowish powder.

diol. *See* dihydric.

diolefin. Any of a series of unsaturated hydrocarbons containing two double bonds; they are also called dienes or alkadienes. Diolefins fall into two groups; those in which the double bonds are conjugated and those in which they are not conjugated. Conjugated diolefins are either straight-chain (1,3-butadiene) or cyclic structures (cyclopentadiene), which readily undergo polymerization. Some occur in nature as plant products (isoprene, turpentine) but most are made by catalytic synthesis from hydrocarbons of petroleum, coal-tar, or natural gas. Unconjugated diolefins do not polymerize easily but react in much the same way as olefins; 1,2-butadiene is an example of this type. *See also* olefin; conjugated compound.

dioxane. A saturated cyclic compound of the ether type, in which the two oxygen atoms are in the 1,4-positions in the ring: $OCH_2CH_2OCH_2CH_2$. Derived by dehydration of ethylene glycol, it is a liquid slightly heavier than water, b.p. 101.3°C (214°F), and has the low ignition temperature of 180°C (355°F). It is also quite poisonous when its vapors are inhaled or when absorbed by the skin. Dioxane is a widely used solvent in the coatings industry and is also a good dispersing and wetting agent. It is volatile and should be kept away from sparks and open flame. The word often appears without the final *e. See also* solvent; ether.

dipentene. A cyclic terpene hydrocarbon b.p. 175°C (347°F), having the formula $C_{10}H_{16}$; it can be regarded as a dimer of isoprene (C_5H_8). It is also known as limonene. It can be obtained by distillation of essential oils and of wood turpentine. Being unsaturated, it oxidizes readily. Dipentene is a combustible, colorless liquid; it is used as a solvent for many materials of similar chemical nature (rosin, rubber, waxes, etc.) and as a dispersing agent in coating formulations, printing inks, and paint driers. *See also* terpene.

diphenylamine. *See* aniline.

diphenyl carbonate. *See* carbonate; polycarbonate.

dipole moment. A force vector exhibited in molecules having closely juxtaposed positive and negative charges, which constitute an electric dipole. The value of a given dipole moment is expressed in electrostatic units (e.s.u.). It is due to variations in the location of nuclear (positive) charges and electronic (negative) charges in the molecule. It occurs in such polar compounds as nitrobenzene, acetonitrile, benzaldehyde, water,

alcohols, and amino acids; hydrocarbon liquids have a very low value, and carbon dioxide and carbon tetrachloride have none. *See also* dielectric constant; polar compound.

direct dye. A water-soluble organic dye, usually a sodium or calcium salt of an azo dye, which is adsorbed directly by the fibers of cotton, rayon, and other cellulosic materials. Alkaline assistants are often added to promote efficient concentration, and detergents are commonly used as well. Such dyes may be introduced in colloidal form, but they will not be effective until a molecular dispersion is achieved (by increasing the temperature of the bath). The alkaline addition retards ionization of the dye molecules, which would prevent satisfactory dyeing action; only the undissociated molecule of the dye can be absorbed by the fibers. *See also* dye.

discharge. (1) In dyeing technology, to cause a color to disappear from a fabric by means of a chemical reaction; substances used for this purpose are called discharging or stripping agents, e.g., titanous sulfate.

(2) In electrochemistry, to withdraw or subtract electrical energy from a system, such as a battery or an electrically stabilized collodial suspension.

disinfectant. A substance having the ability to kill or inactivate bacteria, but usually of too great strength or toxicity for use on living tissue. Among these are certain mercury compounds, phenols, formaldehyde, quaternary ammonium salts, and some synthetic detergents. Other agencies are high temperature and ultraviolet radiation. *See also* antiseptic.

dislocation. *See* lattice.

disorder. *See* order.

disperse dye. *See* anthraquinone.

dispersing agent. *See* surface-active agent.

dispersion. (1) A system comprised of two phases, one of which is in the form of finely divided particles, often in the colloidal size range, distributed throughout a bulk substance; the particles are usually called the internal or disperse phase and the medium in which they are distributed, the external or continuous phase. Under natural conditions the distribution is seldom uniform, but under controlled conditions it can be made so by use of wetting or surface-active agents, such as a fatty acid. There are a number of possible systems: gas/liquid (foam); solid/gas (aerosol); gas/solid (cellular rubber);

liquid/gas (fog); liquid/liquid (emulsion); solid/liquid (paint); and solid/solid (carbon black in rubber). Dispersions of solids in liquids are called suspensions. If the solid or liquid particles are larger than colloidal, they will eventually agglomerate, coalesce, and settle out by gravity. Some dispersions (fat particles in milk) can be stabilized with protective colloids; on the other hand, they can be precipitated by addition of electrolytes, as in water clarification.

(2) In optical terminology, dispersion refers to the retardation of a light ray as it passes through or into a transparent medium; a change of direction is usually involved. The extent of the phenomenon depends on the frequency of the radiation and the refractive index of the medium. Dispersion is a critically important property of optical glass. *See also* refractive index.

displacement series. *See* activity (**1**).

disproportionation. A type of decomposition reaction in which one compound is simultaneously oxidized and reduced; in inorganic compounds this is effected by the interchange of chlorine or other halogen atoms and in organic compounds by the transfer of hydrogen atoms. (The reactions usually require heat and a catalyst). For example, in the reaction $2C_2H_4 \rightarrow C_2H_6 + C_2H_2$, the ethane product is formed by addition of two hydrogens from the ethylene (oxidation), leaving acetylene as the reduced product. Similarly, toluene can be made to yield xylene and benzene: $2CH_3C_6H_5 \rightarrow C_6H_4(CH_3)_2 + C_6H_6$. The latter reaction is also called transalkylation. *See also* oxidation.

dissociation. A specific kind of decomposition occurring (1) when a molecule of a diatomic gas (Cl_2) is separated by heat into its constituent atoms, and (2) when an inorganic compound in aqueous solution separates into charged particles (ions), e.g., hydrochloric acid in water dissociates to form positively charged hydrogen ions (H^+) and negatively charged chloride ions (Cl^-) (electrolytic dissociation). In sense (2), dissociation is synonymous with ionization. *See also* decomposition; ionization; electrolysis.

distillation. Separation of the components of a mixture (usually of liquids or solutions) by heating to the boiling point and condensing the resulting vapor. In multi-component mixtures such as petroleum, the various fractions have different boiling points; thus it is possible to effect separation by condensing the vapor of each component in turn, the lower-boiling fractions distilling off first (fractional distillation). Distillation of petroleum thus yields naphthas, gasoline, kerosine, fuel oils, and ultimately a tarry or waxy residue. depending on the geographical origin of the petroleum. Distillation is also used to purify products such as alcohol.

In nature, simple evaporation from the ocean surface provides pure water vapor, which later condenses as rain (distilled water). Destructive distillation (a form of pyrolysis) is performed on solid materials such as coal and wood by heating them out of contact with air to obtain coal tar, coke, charcoal, etc. Flash distillation is an efficient method of desalting sea water; it involves extremely rapid evaporation and cooling in a multi-stage unit in which the heat removed in the condensation stage preheats the incoming sea water, similar to the action of a countercurrent heat exchanger. *See also* evaporation; fraction; reflux; pyrolysis; azeotrope.

dithiocarbamate. *See* accelerator.

dl-. A prefix indicating a mixture of equal parts of dextro- and levorotatory crystals, which results in optical inactivity. The symbol ± is also used. *See also* meso-(**1**); racemic.

DL. A prefix indicating that a compound contains equal parts of D and L stereoisomers. *See also* racemic; meso-(**1**).

DNA. *See* deoxyribonucleic acid; recombinant DNA.

doctor test. A method used in petroleum technology to determine the content of obnoxious-smelling thiols in gasoline and other distillates. These odors can be diminished (a process called "sweetening" the product) by treating the gasoline with a solution of lead oxide in sodium hydroxide, usually with addition of sulfur (doctor solution). By this means the offensive thiols are changed to disulfides, which have a less noticeable smell.

dodecylbenzene. A synthetic organic compound consisting of a 12-carbon alkyl chain attached to a benzene nucleus, giving the formula $C_{12}H_{25}C_6H_5$; for use as a detergent, it is usually sulfonated and then neutralized with sodium hydroxide to form dodecylbenzenesulfonate $(C_{12}H_{25}C_6H_4SO_3Na)$. The alkyl chain may be either straight or branched, depending on whether it is derived from propylene or ethylene. Since the straight-chain (linear) type (LAS) decomposes more readily, it is preferred for detergent application. Dodecylbenzene is loosely called detergent akylate. *See also* detergent; alkylate (2); alkylbenzene sulfonate.

dolomite. A natural sedimentary rock calcium magnesium carbonate, $CaMg(CO_3)_2$, the most prevalent ore of magnesium. Chemically it is similar to calcium carbonate. It is used as a source of magnesium and magnesium compounds and as a raw material in one form of the process of making wood pulp for paper and other cellulosic products. It is also used as a refractory for furnaces and in ceramics and building materials. When calcined, it forms dolomite lime, $CaO \cdot MgO$, by loss of CO_2.

donor. A nucleophilic atom or ion which supplies an electron pair to another atom, thus creating a covalent bond. The electron-deficient atom, called the acceptor, is said to be electrophilic. Donor and acceptor atoms are involved in acid-base reactions, carbon linkage formation, and coordination chemistry. The term is derived from the Latin word for "give." *See also* nucleophile.

L-dopa. *See* vanillin.

dope. (1) A substance (usually a metal or metallic oxide) added in trace amounts to laser crystals and semiconducting elements to provide essential properties for these applications. For example, neodymium and chromic oxide are used as activators for yttrium-garnet and ruby lasers, and arsenic is added to germanium to increase its semiconducting efficiency.

(2) A carbon-containing additive to dynamite.

(3) *See* drug; hallucinogen.

dosage. (1) Industrially, this term is often used in reference to the percentage of an active ingredient added to a mixture, as, for example, the accelerator dosage in a rubber mixture.

(2) In radiation therapy it means the amount of radiation of any type administered to a patient; the so-called permissible dosage of x-radiation is 0.3 roentgen a week.

(3) In medicine, it refers to the amount of any curative prescribed by a physician.

double bond. A linkage between two multivalent atoms, usually carbon, in which one valence of each is unsatisfied; the two "free" valences are thus available to unite with univalent atoms of other elements such as hydrogen or chlorine under appropriate conditions. Compounds containing double bonds, such as olefins and some fatty acids, are called unsaturated. Double bonds are no stronger than single bonds, but compounds containing them are much more reactive than saturated compounds. *See also* bond; unsaturation.

double salt. *See* salt.

D.P. Abbreviation for degree of polymerization.

drier. A substance having the ability to catalyze the combination of oxygen with the unsaturated linkages of fatty acids in drying oils and to accelerate polymerization of synthetic resins in paints and printing inks. Such substances are metals heavier than sodium or compounds of these with unsaturated acids such as naphthenic, linoleic, etc. The most effective and widely used metals are cobalt and manganese, either as such or in the form of so-called metallic or heavy-metal soaps (cobalt naphthenate, manganese linoleate). *See also* drying (2); naphthenate.

drilling fluid. A concentrated suspension (often called a "mud") used in the rotary method of drilling wells, chiefly for gas and oil. It is comprised of bentonite or other water-binding clay; other materials such as barytes, lignin lignosulfonate, and a variety of similar items are often added for stabilization and fluidizing purposes. The fluid is circulated down the hollow drill pipe and returns to the surface in the space between the drill pipe and the casing or wall of the hole. Its function is to raise the rock cuttings loosened by the drill bit, to cool and lubricate the bit and drill string, and to buoy or decrease the weight of the drill pipe. It also serves to return formation samples to the surface. Three types of fluid are used: water-base, oil-base, and emulsion-base. Each is characterized by being thixotropic, i.e., liquid when agitated or pumped, but a gel when pressure is released; the fluid usually has a density of 2.0 or higher. *See also* bentonite; thixotropy.

drug. A natural or synthetic chemical substance or mixture (also called a pharmaceutical) which has one or more of the following effects: (a) killing or inhibiting the growth of infectious organisms; (b) affecting the activity of a specific organ of the body; (c) stimulating or depressing the central nervous system (narcotics, tranquilizers, and other psychotropic agents). Study of the chemical nature and behavior of drugs is medicinal or pharmaceutical chemistry; the science of drug preparation is pharmacy; pharmacology includes portions of both, with emphasis on the effects on the organism. No drug should be taken without a physician's prescription, as many are extremely poisonous, and some have dangerous side effects. *See also* antibody; chemotherapy; barbiturate; narcotic; hallucinogen.

History. The use of drugs and medicines as curative agents began with the Hindus, who successfully practiced vaccination techniques as early as 550 A.D., and continued with the alchemists in the Middle Ages. They experimented with a wide range of plant principles as well as with arsenic compounds and various metals. The first compilation of drugs, the *Nuremburg Pharmacopeia,* appeared in 1542, just a year after the death of Paracelsus, a German physician and chemist. He modernized the approach of the alchemists to diseases and ridiculed the ancient ideas that were still current in the 16th century. Plant products such as cinchona bark, ipecac, and chaulmoogra oil were used for many years. In 1805, morphine was obtained from the opium poppy, paving the way for the introduction of an extensive series of alkaloid drugs, many of which are still in use.

The first notable breakthrough in disease treatment was Jenner's successful inoculation for smallpox (1775), which initiated the science of immunochemistry. As this disease had ravaged Europe for centuries, Jenner's brilliant work is a landmark in medical history. Pasteur adopted this technique; he was the first to inoculate for anthrax and rabies (1880) and to establish the antigen-antibody relationship. Still more important was his identification of many types of infective microorganisms (germs) as disease-causing agents. Almost simultaneously, Lister introduced surgical antiseptics, such as mercury compounds, carbolic acid, and phenol, and Robert Koch discovered the germs of anthrax, cholera, and tuberculosis.

Chemotherapy may be said to have begun when Ehrlich developed his arsphenamine treatment of syphilis (1910)—so-called "magic bullets" that killed microorganisms with minimum damage to the host. This practice has been broadly applied to cancer therapy in recent years, but without positive results. Another significant event was Funk's discovery (1911) of the ability of certain plant products to cure a disease called beri-beri; these soon were named vitamins. Their classification, metabolic functions, and curative properties were established by many researchers, including McCollum, Szent-Gyorgyi, Sherman, and R. J. Williams (B complex).

The science of endocrinology was developing rapidly at this time. Outstanding was the isolation of insulin for treatment of diabetes by Banting in the early 1920's. Continuing research in this field led to development of many hormones for treatment of a wide variety of disorders, especially the adrenal cortical steroid hormone (ACTH), cortisone, thyroxine, etc.

In 1928, Fleming in England noted the antimicrobial action of certain plant molds that produced penicillin. Some years later, Waksman in the U.S. initiated the development of these mold products, which he called antibiotics. These were so effective against many infectious diseases that they soon displaced the sulfa drugs that had been introduced a few years earlier. Antibiotics have proliferated in recent years and have been the most significant addition to curative drugs in the history of medicine. Coincidental with this development was discovery of the tobacco mosaic virus by Stanley in 1935, which resulted in new concepts of the nature and behavior of infective microorganisms. The early 1940's saw the introduction of antimalarials as substitutes for quinine, and of antihistamines for treatment of allergic diseases.

Significant breakthroughs are by no means confined to the past. Current research on gene-splicing techniques has already resulted in "programming" bacteria to produce insulin and interferon, as well as the creation of wholly new forms of bacteria that have great potential not only in medicine but in chemistry and agriculture. For further details, *see* recombinant DNA.

dry cell.   A primary battery (usually not rechargeable); there are two types, one of which is a self-contained assembly of the component chemicals, including the electrolyte (Leclanche cell), the other reserving the chemicals in dry form and adding electrolyte later. Dry cells have a relatively short life and tend to deteriorate on long standing, especially in a moist atmosphere. They are used chiefly in flashlights and small electrical systems requiring low energy output. *See also* battery.

dry chemical.   A preparation designed chiefly for fire fighting, especially for fires involving flammable liquids, pyrophoric materials, and electrical equipment. Common types contain sodium or potassium bicarbonate, sodium chloride, and free-flowing additives. *See also* extinguishing agent.

dry ice. *See* carbon dioxide.

drying. (1) Evaporative loss of water or other solvent from a substance, usually a solid, with the aid of heat. Many food products and industrial materials are dried as an integral step in processing, e.g., prunes, apricots, etc., some kinds of fish, paper, and textiles. The spray drying of such products as milk and soap to a powdered form can be considered a type of dehydration. Organic coatings (both solvent- and water-based) also dry by evaporation, though polymerization is also involved in the case of paints containing organic solvents. Evaporative drying is not as intensive as dehydration, which removes over 95% of the water. Removal of water by a mechanical device such as a screen or filter is called dewatering.

(2) Polymerization of a vegetable oil or high-polymer base in paints and varnishes, resulting in the formation of a protective film; it is catalyzed by oxygen and is quite distinct from evaporative drying, which occurs simultaneously (but much more rapidly) in paints. *See also* dehydration; freeze-drying; polymerization.

drying oil. A vegetable oil having the property of forming a hard, continuous film when exposed to air; the atmospheric oxygen induces polymerization of the unsaturated fatty acids (glycerides) in the oil, e.g., linoleic and linolenic acids. The most important drying oils are linseed, tung, and dehydrated castor oil. The largest volume use of drying oils is in exterior paints, printing inks, and putty. They have been largely replaced in paints by synthetic latexes, such as acrylic, which polymerize in a similar manner. *See also* polymerization; drying (2); vegetable oil.

dust. An air suspension of solid particles ranging from 10 to 50 microns in diameter. The chemical nature of such particles may vary widely, as represented by coal and oil particulates (smoke), metal and metallic oxide fume, calcium carbonate, and organic materials such as flour, pesticides, etc. Most such dusts, when sufficiently concentrated in air, constitute a serious fire hazard and are capable of causing a severe explosion, if exposed to static sparks or flame under certain conditions. They are also a health risk, particularly if they contain toxic components. Elec-

trostatic precipitators have been used with considerable success in industry, and there are a number of dust-collecting devices available for manufacturing areas where dust may be unavoidable. Silica dust and mine dust are damaging to the lungs and require special precaution. Calcium chloride is effective in laying dust on roads and pavements. *See also* aerosol.

dusting agent. A powdery solid used as an abherent and mold-release agent in the plastics and rubber industries. Typical materials in general use are talc (soapstone), mica, fine sand, slate flour, and the like. Graphite and mica have a flat crystal structure which causes them to act as lubricants, and thus are especially effective in preventing sheets or slabs of hot solid mixtures from sticking together when stacked. *See also* abherent; antiblock agent.

Dy. Symbol for dysprosium, the name being derived from the Greek meaning "hard to get at," referring to its difficulty of separation.

dye. A natural or synthetic organic coloring agent chiefly derived from petroleum or coal tar, with nitrogen and/or sulfur often being present. A vast number of such dyes have been synthesized since William Henry Perkin, an English chemist (1838-1907), developed the original compound mauveine in 1856. One of the largest groups is the azo dyes, but there are many other types including aniline, anthraquinone (alizarin), and sulfonic acid derivatives. Some types are water-soluble, but others are not; there are both acidic and basic dyes. Many dyes require the use of other compounds (called assistants) in order to be effective in combining with the fiber. The chemistry and technology of dyes and dyeing is an extremely complex subject; attention is directed to the cross-referenced entries for further information. Dyes are used primarily to color fibers and fabrics, and the term is conventionally used in this sense. *See also* auxiliary; azo dye; direct dye; fiber-reactive dye; mordant; vat; resist; pigment; Perkin.

dynamite. A mixture of either nitroglycerin or ammonium nitrate with a carbon-containing material such as wood pulp or charcoal. As originally invented and named in 1866 by Alfred B. Nobel, a Swedish chemist and engineer (1833-1896), dynamite was a mixture of nitroglycerin and diatomaceous earth, which greatly decreased the hazard of handling nitroglycerin. It is used chiefly in mining, excavation, and construction work and requires a blasting cap for safe detonation. Its use has decreased in recent years because of the development of more effective types of industrial explosives. Dynamite can be exploded by heat or severe shock, and extreme caution must be used in handling it. *See also* explosive; slurry.

dysprosium. An element.

Symbol	Dy		
State	Solid	Atomic Wt.	162.5
Group	IIIB (Lanthanide Series)	Valence	3
		Isotopes	6 stable
		Atomic No.	66

Closely related to yttrium, dysprosium, m.p. 1407°C (2565°F), is a member of the rare earth family and occurs in gadolinite and several other ores. It is uniquely noncorrosive in air. Relatively stable at room temperature, it reacts rapidly at high temperatures with a number of elements, including halogens, oxygen, boron, hydrogen, and carbon. Its neutron cross-section is quite high (1100 barns), which has led to its use in measuring neutron concentrations and in reactor control. *See also* yttrium.

E

earth. Any of several specific groups of materials, often of metallic nature, some occurring in, or associated with, soils and sands (such as monazite) while others are found in relatively pure deposits. These groups are (1) alkaline earths, which are oxides of such metals as calcium (lime), barium, and strontium; (2) rare earths, which are oxides of a group of elements called the lanthanide series, the name also being applied to the elements themselves; (3) impure oxides of iron such as the pigments known as umber and ochre; (4) aluminum oxides and silicates, which are clays or clay-like materials such as kaolin, fuller's earth, bauxite, etc.; and (5) diatomaceous earth, which is chiefly silica. *See also* *specific entries.*

ebulliometry. Accurate measurement or determination of the boiling point of a liquid under varying conditions of temperature and pressure.

economic poison. *See* pesticide; insecticide.

edetate. *See* ethylenediaminetetraacetic acid.

edible oil. A fatty oil obtained from vegetables, nuts, and seeds, commonly used in food products such as margarine, salad dressings, shortening, etc., either as a liquid or in hydrogenated form. They include corn, soybean, olive, safflower, cottonseed, coconut, peanut, and citrus seed oils; they are comprised largely of oleic, palmitic, linoleic, and other fatty acids. Fish- and shark-liver oils, used as dietary supplements for their vitamin content, and castor oil are also classed as edible oils. Flavoring oils derived from flowers, such as wintergreen, coriander, clove, and citrus peels, are of different chemical composition and are called essential oils, even though they are edible. *See also* vegetable oil; essential oil.

EDTA. *See* ethylenediaminetetraacetic acid.

effect. (1) As a verb, to cause or bring about, as to effect a change of color, state, condition, or property.

(2) As a noun used by chemical engineers, a chamber in a series of evaporation and condensation units which is heated by the latent heat evolved by the condensing vapor from the preceding unit; it is a type of heat exchanger. The number of effects in such a series is usually designated, e.g., double-effect, triple-effect, etc., evaporator.

efflorescence. Loss of combined water molecules by a hydrate when exposed to air, resulting in partial decomposition, indicated by presence of a powdery coating on the material. This action depends on the partial pressure of water vapor in the air. When it is less than the dissociation pressure of the hydrated compound, the latter gives up all or part of its water of crystallization and becomes powdery. This commonly occurs with washing soda ($Na_2CO_3 \cdot 10H_2O$), which loses almost all its water constituent spontaneously.

effluent. Any liquid, slurry, or gas emerging from a pipe or similar outlet. It is usually used in connection with waste disposal of chemical or manufacturing plants, such as stack gases and residual liquid or semiliquid mixtures discharged into rivers or lagoons.

einsteinium. An element.

Symbol	Es	Atomic No.	99
State	Solid	Atomic Wt.	253(?)
Group	IIIB	Valence	3
	(Actinide Series)		

An artificial radioactive element discovered in 1952 as a consequence of the first hydrogen bomb trial. It has been made in extremely small quantities from plutonium and has been used in tracer studies. Named for Albert Einstein. *See also* actinide series.

elaborate. A term used in biochemistry to describe chemical transformations within the organism resulting in the formation of specific types of substances; for example, plants elaborate vitamins and fats, and some poisonous snakes elaborate their own venom. It also refers to formation of metabolic end-products such as purines and uric acid.

elasticity. (1) In materials technology, this term denotes the extent to which a material returns to its original form or shape after being stretched, bent, strained, or otherwise deformed. The work done on the material, or the amount of deformation, is the stress, and the opposing force, or resistance to deformation, is the strain; the ratio of stress to strain is the elastic modulus (sometimes called Young's modulus). Hard materials like metals are extremely elastic, with high modulus and low elongation, whereas softer materials (of which vulcanized rubber is typical) are much less elastic, having a low modulus and high elongation (yield). A material which recovers 100% of its shape after distortion is said to be perfectly elastic (e.g., glass); rubber and many elastomers, on the other hand, tend to be permanently deformed, that is, they may be only 90 to 95% elastic after being stressed to the breaking point, while exhibiting elongation of up to 1000%. *See also* elastomer.

(2) In physical chemistry, this term refers to the fact that atoms and molecules maintain their energy in spite of continual mutual collision; thus, an elastic collision is one in which the particles, for example, of a gas in a closed container, collide with each other or the walls of the container indefinitely with no net energy loss.

elastomer. A term coined about 1935, when synthetic rubber-like materials were introduced on a commercial scale, to describe any high polymer having the essential properties of vulcanized natural rubber, namely, (1) ability to be converted from a thermoplastic to a thermosetting state by cross-linking, and (2) high strength and elongation when stressed, with substantial recovery of its original shape. Among the early synthetic elastomers were polychloroprene, styrene-butadiene copolymer, butyl rubber, and sodium polysulfide polymer. More recent developments produced silicone rubbers, polyurethanes, polyisoprene (an exact duplication of natural rubber), and a number of others. *See also* rubber; high polymer.

electric steel. A specialty high-alloy steel made by heating the components in an electric furnace; it usually contains substantial percentages of nickel, chromium, and molybdenum.

Electrochemical Society. Established in 1902, this society was organized to promote the advance of the science of electrochemistry and related fields. It is comprised of eleven divisions, each devoted to a special branch of electrochemistry, e.g., corrosion, batteries, rare metals, and electrodeposition. It publishes a journal and sponsors books relating to its major interests. Its office is in Pennington, N.J.

electrochemistry. That portion of chemistry concerned primarily with the relationship between electrical forces and chemical reactions. This relationship is fundamental and far-reaching, as the structure of matter is basically electrical. Electrochemistry is directly involved in chemical bonding, ionization, electrolysis, metallurgy, battery science, and corrosion—in short, in any situation in which a chemical change is caused by, or associated with, electrical phenomena. (It does not include electronics, in which electrons are not bound within the atom and in which no chemical changes occur.) Michael Faraday (1791-1867), an English scientist, is generally regarded as the founder of electrochemistry. *See also* electron; electronegative; electrolysis; electroplating.

electrocratic. A term used in colloid chemistry to denote a dispersion of solid particles of colloidal size in a liquid, whose stability is maintained by either positive or negative electric charges on the particles. Since the charges are alike, they repel each other and thus offset the attraction of gravity. Such dispersions, made with colloidal gold, silver, and other metals, have certain medical applications.

electrode. A material used in an electrolytic cell to enable the current to enter or leave the solution. Each cell has a positive electrode (anode) and a negative electrode (cathode). Materials used for electrodes have high conductivity and are exemplified by carbon (graphite), iron, copper, and mercury (calomel). Electrodes are also used in electrodialysis and electrophoretic systems. There are numerous special types of electrodes, such as those used in arc welding, which consist of composite filler materials; the glass electrode, designed for sensitive electrochemical research, which develops a detectable electric potential when placed in a conducting solution; and the hydrogen electrode, which is made of platinum and is placed in a stream of hydrogen. *See also* electrolytic cell; anode; cathode.

electrodeposition. A general term for the coating of an electrode with a metal or other substance as a result of the migration of charged particles, in water solution or suspension, caused by passage of an electric current. The particles may be ions, as in electroplating, large molecules (proteins), colloidal micelles, or still larger units (rubber particles in latex), as in electrophoresis. Thus, the term includes such processes as electroplating, electrocoating, electroforming, electrophoresis, and anodizing. *See also* electrolysis; electroplating; electrophoresis; anodizing.

electrodialysis. The separation of ionizing substances from other materials in a solution, and also from the solvent, by use of one or more semipermeable membranes, and an electric current (electrolysis). This method is being used in some installations for desalination of sea water and for other forms of demineralization. *See also* dialysis.

electroforming. A method of making intricate metal objects, especially those having specifically contoured surfaces, by means of electrolytic deposition. The ions of the coating metal attach themselves to a previously prepared negative mold of the object to be manufactured, to which graphite has been applied to act as a conductor. This method is used in making printing plates, phonograph records, and similar products in which the surface configuration is of primary importance. *See also* electroplating.

electroluminescence. *See* luminescence.

electrolysis. Decomposition of a chemical compound by means of an electric current, the extent being proportional to the quantity of electricity passing through the solution. The ions present move to the oppositely charged electrode, where they are collected; metallic ions discharged on an electrode form a coating, i.e., an electro-

deposit. Water can be separated in this way into its component gases, hydrogen and oxygen. Many industrial operations utilize electrolysis, e.g., metallurgical separations, electroplating, electroforming, anodizing, etc., as well as the production of important inorganic chemicals. Examples of the latter application are (1) production of sodium, chlorine, and sodium hydroxide from sodium chloride, either fused or in solution; (2) production of aluminum from alumina (Al_2O_3) in the presence of cryolite; (3) recovery of sulfuric acid and sodium hydroxide from sulfate paper mill waste. Electrolysis is normally carried out in an electrolytic cell. Its basic principles were stated by Michael Faraday and expanded by Arrhenius. *See also* electrolytic cell; electroplating; Arrhenius; Faraday.

electrolyte. Any ionically bonded substance that will decompose in a solvent and thus increase the electrical conductivity of a solution, or that will act as a conductor between dissimilar metals in a battery or galvanic couple. Materials that commonly serve as electrolytes are salts such as sodium chloride, copper sulfate, aluminum sulfate, ammonium chloride, etc., and acids such as hydrochloric. *See also* electrolysis; electroplating; polyelectrolyte.

electrolytic cell. An electrochemical device in which electrolysis is conducted, typically for the manufacture of chlorine and sodium hydroxide from sodium chloride solution (brine). Its essential features are a graphite anode and a steel or mercury cathode. The types employing steel cathodes (Hooker, Vorce) also utilize a porous asbestos diaphragm as a separation medium and are known as diaphragm cells; the mercury cathode type is represented by the Castner and DeNora cells. The Downs cell is used for manufacture of sodium from a molten mixture of sodium chloride and calcium chloride. When current is passed through an electrolytic cell, the sodium chloride is ionized to Na^+ and Cl^-; it thus acts as an electrolyte, and the ions move to the oppositely charged electrodes. Chlorine then forms at the anode and sodium hydroxide at the cathode, with evolution of hydrogen. Modified electrolytic cells are also used for electroplating. *See also* electroplating.

electromagnetic separation. *See* mass spectrometry; magnetochemistry.

electromotive series *See* activity (**1**).

electron. A particle of negative electricity having a mass that is $1/1837$ of the mass of the hydrogen atom. Its properties combine those of radiation (light waves) and of particles. This double nature of the electron was established in the 1920s by means of quantum or wave mechanics, in which the electron is regarded as a standing wave of energy moving around the atomic nucleus in a so-called orbital. Electrons may occupy from one to seven energy levels (depending on the element involved) which are at varying distances from the nucleus. These are called shells, or groups of orbitals; each shell contains a specific number of electrons ranging from 1 to 32, though in many cases the shells are not filled. Those in the outer shell are known as valence electrons, as they play a predominant part in chemical bonding; for example, carbon has two shells, the inner one being occupied by two electrons and the outer one by four electrons, which represents its valence. Thus, there is a definite relationship between the number of outermost electrons and the chemical activity of an element. No orbital can contain more than two electrons; two electrons in the same orbital are known as paired electrons.

The negative charge on, or represented by, an electron is regarded as being diffuse rather than concentrated, and thus has the appearance of a cloud around the nucleus which becomes gradually less dense as it extends outward, the density being indicated by the square of the wave function.

Electrons were discovered in 1897 by Sir Joseph John Thomson, a British physicist (1856-1940), Nobel Prize 1906, and their behavior in atoms was described in 1913 by Niels Bohr, a Danish physicist (1885-1962), Nobel Prize 1922. The modern wave concept of the electron was developed in the 1920's by several physicists: Louis-Victor, Duc de Broglie, French (b. 1892), Nobel Prize 1929; Werner K. Heisenberg, German (1901-1976), Nobel Prize 1932; and Paul A. M. Dirac, British (b. 1902), and Erwin Schrödinger, German (1887-1961), who shared the Nobel Prize in 1933.

Electrons may be detached from the nucleus of an atom by application of energy; in this free state, they can flow along a conductor or through a vacuum tube or semiconductor in the form of an electric current. Electrons are emitted as beta particles from radioactive nuclei such as radium and uranium.

electronegativity. The extent to which an atom can attract electrons from outside itself and thus become negatively charged; this tendency is due to the attractive force exerted by the nuclei of atoms having vacancies in their outer shells. This attraction makes possible the formation of both covalent and ionic bonds and is thus a fundamental factor in the formation of chemical compounds. Among the less electronegative, i.e., electropositive, elements are the alkali metals; the more highly electronegative elements are the halogens, sulfur, and oxygen. *See also* atom; bond.

electron microscope.   *See* microscopy; resolving power.

electron volt.   A unit of electrical energy used by nuclear scientists in measuring electronic forces. It is defined as the energy an electron receives when it falls through a potential difference of 1 volt, i.e., less than a billionth of an erg. The rupture of a chemical bond yields from 5 to 10 electron volts, whereas the splitting of an atomic nucleus releases about 200 million electron volts. Approved abbreviation is eV for electron volt and MeV for million electron volts.

electrophile.   *See* nucleophile.

electrophoresis.   The movement of charged particles of colloidal size or larger in water suspension, for example, proteins or rubber latex, to the oppositely charged electrode under the influence of an electric current. When electrodes are placed in latex and current is passed, the negatively charged particles of rubber migrate to and coat the anode, that is, they are electrodeposited. Many useful articles are made in this way. Protein molecules, which may have either a positive or a negative charge depending on the pH of their environment, also respond to electrophoresis; research on this behavior has been an important phase of investigation of immunology, especially as applied to allergens. *See also* electrodeposition.

electroplating.   A type of electrodeposition in which a metal or plastic is coated with a film of metal for corrosion protection or for decorative purposes. An electrolytic cell is used in which the metal being coated is the cathode and the metal being deposited is either the anode, or is present as ions in the solution. The solution, or bath, contains ionized salts of the metal being deposited, for example, copper sulfate. As current passes through the solution, the positively charged metal ions are impelled to the cathode, on which they deposit a uniform layer of controllable thickness. The metallic anode may be ultimately consumed in the process. Metals in commercial use for electroplated coatings are copper, chromium, nickel, gold, and silver; less common metals are used for special purposes (rhodium, rhenium, platinum, etc.). The base metal is often steel. *See also* electrolytic cell; electrodeposition.

electropositive. *See* electronegativity.

electrowinning. A term used by metallurgists to refer to separation of a metal from its ores or other impurities by electrolytic means, as in the production and refining of aluminum, sodium, copper, cobalt, and others. It is one of several so-called reduction methods. *See also* extraction **(2).**

element. A unique arrangement of fundamental units of matter having characteristic properties; 106 elements are presently known, of which 92 occur in nature, the others being synthetic. Many more synthetic elements are theoretically possible. The smallest amount of an element that can exist is the atom. The number of protons in the atomic nucleus (atomic number) and the arrangement of electrons around them determine the properties of an element, especially in relation to chemical bonding and reactivity. When arranged in sequence by increasing atomic number, the elements are seen to have recurring similarity of certain properties; it was on the basis of this observation, in 1869 by Dimitri I. Mendeleev (1834-1907), that the Periodic Table was developed. About 75% of the elements are metals, and all those beyond lead are radioactive. All atoms of a given element are identical in every respect except mass; the presence of one or more extra neutrons in the nucleus accounts for the existence of isotopes. Besides the major groups of elements of the Periodic Table, there are several classifications or series based on type or properties; some of these lie within a major group, while others straddle several groups. These are the alkali metals (group IA, except hydrogen); the alkaline earth metals (Group IIA, except beryllium); the transition metals (several groups, including the platinum metals of Group VIII); the halogen elements of Group VIIA; the lanthanide series (rare earth metals); and the actinide series, which includes the transuranic (synthetic) elements. The last two series occupy part of Group IIIB.

Origin. The original synthesis of the chemical elements is thought to have resulted from a huge explosion of undifferentiated energy which astrophysicists believe initiated the universe about 20 billion years ago, at least 7 billion years before the formation of the Milky Way galaxy (the so-called ''big bang'' theory). The sequence of creation of the elements is believed to be hydrogen and helium (within a few minutes of the original cataclysm); then, after a few million years, came carbon (from fusion of helium nuclei), oxygen, magnesium, silicon, sulfur, and iron; after still another interval, the heavy metals were formed by the intense heat resulting from stellar explosions. All this occurred over a period of 5 to 7 billion years. The temperatures involved were inconceivably high, in the range of from 10 to 100 million degrees centigrade. *See also* atom; Periodic Table; isotope; fusion **(1).**

elemental. Consisting of an element only, as distinguished from a compound of an element, e.g., elemental iron is Fe (not Fe_2O_3 or other iron compound). The term *elementary* is often used in this sense but should be avoided because of its more common nontechnical meaning of ''simple'' or ''easy.''

elemi. *See* resin **(1).**

eleostearic acid. *See* vegetable oil.

elutriation. A term used chiefly by metallurgists to denote a method of separating larger and heavier particles from smaller and lighter ones in mixtures such as ores. It is basically a sedimentation operation, in which the mixed particles are subjected to a flow of water in an appropriately designed container; as the water moves upward through the mixture, the heavier particles fall to the bottom of the container, while the water carries the lighter ones over the edge into a collection device. This method is used both in laboratories and in the separation of ores in metallurgical processing. *See also* flotation.

embrittlement. A term used in metallurgy to describe the tendency of a metal or alloy such as steel to become brittle on exposure to a specific environment, with consequent failure known as brittle fracture. The chief agent causing this difficulty is hydrogen, which is liberated in many electroplating and pickling processes or by corrosive acids. It diffuses into the metal surface, where it becomes trapped and eventually ruptures the metal. The extent of embrittlement depends somewhat on the nature and composition of the alloy; titanium and high-strength steels are especially subject to this effect.

emery.   *See* abrasive; aluminum oxide; sapphire.

emission.   *See* spectroscopy.

empirical formula.   The formula obtained by dividing the percentage of each element in a compound by its atomic weight and taking the ratio of the quotients. This procedure gives the *relative* number of atoms present but not necessarily their actual number or their arrangement. For example, benzene (with six atoms each of carbon and hydrogen) has the same relative composition as acetylene (with only two atoms each of the same elements), the empirical formula of both being CH.

emulsion.   A permanent suspension or dispersion, usually of oil or fat particles in water; the reverse also may occur, in which water is dispersed in an oil or fat. These are often designated as o/w (oil-in-water) and w/o (water-in-oil) emulsions, the latter occurring frequently in crude petroleum. The intimate and uniform mixture of naturally immiscible liquids is made possible by the action of an emulsifying agent, commonly fatty acids or long-chain alcohols, at the oil-water interface. A fatty acid is comprised of a hydrocarbon chain which is insoluble in water and a negatively charged (polar) carboxyl radical (COOH)$^-$ which is attracted by water (also polar). As a result, the fatty acid molecule orients itself with the water-soluble carboxyl radical in the water and the insoluble hydrocarbon chain in the oil. The monomolecular film thus formed reduces the repulsion between the two liquids and creates a stable emulsion, as shown schematically below:

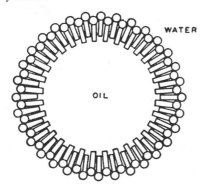

In such natural emulsions as milk and rubber latex, the particles of fat and rubber hydrocarbon are coated with a thin layer of protein, which keeps them from cohering; emulsifying agents of this type are called protective colloids, e.g., alginates, agar, etc., and are used in ice cream, frozen desserts, etc. Polymerization is often carried out with monomers in emulsion form. *See also* interface; surface tension; Langmuir; emulsion polymerization.

emulsion paint.   *See* paint; emulsion.

emulsion polymerization.   An important technique for the manufacture of liquid dispersions of synthetic rubber latexes and paint resins such as acrylic and vinyl coatings. The oil-soluble monomers are emulsified with water at temperatures from about -8 to 60°C (18 to 140°F), the polymerization being activated with catalysts. This method is also applicable to systems involving two or more types of monomers (copolymerization), for example, of styrene and butadiene. *See also* polymerization; copolymer; emulsion.

enamel.   (1) A hard ceramic coating, consisting principally of silicates, which is bonded to a metal surface in the molten state; the ceramic mixture is first applied to the metal in powder form, and the composite is then fired to the point where the coating melts to form a uniform, strongly bonded protective surface when cool. This is called a porcelain enamel.

(2) An organic protective coating similar to a paint, comprised of varnish or of oil-modified alkyd resin and urea or melamine resins, plus necessary pigments; after application, the coated product is exposed to heat or infrared radiation. Such coatings are called baked enamels.

enantiomer.   *See* enantiomorph.

enantiomorph.   One of two or more isomers containing an asymmetric element (usually carbon), which has the effect of causing the plane of polarized light to rotate to either right or left when the substance is placed under crossed nicol prisms. Each enantiomorph (or enantiomer) is a mirror image of its twin, that is, its tetrahedral (pyramidal) structure compares with its reversed image in the same way as right and left hands. These are sometimes called chiral arrangements. *Enantiomorph* is taken from the Greek, meaning "having an opposite form."

The upper illustration is a schematic drawing of a mirror-image molecule, in which C* is an asymmetric carbon and 1, 2, 3, and 4 are different univalent elements or groups. The lower illustration is an actual compound (lactic acid) which displays this arrangement; glyceraldehyde is another such compound. In both cases, structures I and II are enantiomorphs.

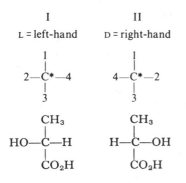

I   II

L = left-hand   D = right-hand

D and L are conventionally used to denote these configurations, the D standing for dextro (right-hand) and the L for levo (left-hand). These designations refer only to the configuration of the molecule and do not indicate the actual rotation; the latter is represented by the symbols + and − (sometimes by *d* and *l*). Some D isomers are levorotatory and vice versa. *See also* asymmetry; glyceraldehyde; optical rotation; chiral.

encapsulation. (1) The coating or covering of an integrated group of small parts such as in an electrical or electronic assembly with a plastic or similar material, often in the form of a foam. The encapsulating material holds the components in position and thus aids the manufacture of complex circuitry for computers, aerospace equipment, etc. The materials used for this purpose are also called potting compounds; among them are polyurethane, epoxy, and polyester resins.

(2) Microencapsulation refers to the formation of very small capsules, 20 to 150 microns in diameter, containing a variety of materials. The capsule is a semipermeable membrane made of gelatin or a polymer, which hinders or prevents release of the contained material until desired. Enzymes are immobilized in such microcapsules until they reach the desired stage or site of an enzymatic process. Medicinals are also microencapsulated for controlled release in the body. The process is also used for adhesives, for carbon or dyes for copy paper, and for perfumes, so that these materials remain stable until the capsules are broken by pressure, heat, or dissolution when the contents are to be released.

endothermic. The term used to characterize a chemical reaction which requires absorption of heat from an external source, as, for example, the formation of carbon monoxide and hydrogen from coke and water:

$$C + H_2O \xrightarrow{\text{heat}} CO + H_2;$$

also the formation of magnetite from iron and water:

$$3Fe + 4H_2O \xrightarrow{\text{heat}} Fe_3O_4 + 4H_2.$$

See also exothermic.

end point. That point in a titration at which no further addition of titrating solution is necessary; it is usually denoted by a change in color of an added substance called an indicator. The end point is often almost identical with the equivalence point, i.e., the neutral point between the titrant and the subject compound. End points are also indicated by instrumental methods. *See also* neutral; indicator.

end-product. (1) As generally used, this term refers to a substance or mixture that is primarily used "as is," without further chemical processing. Examples are carbon black, gasoline, fuel oils, waxes, etc. Many substances, though, are used not only for terminal purposes but as sources or intermediates for other products. For instance, sodium chloride is used both as such and as a source of chlorine; and chlorine also is used as a direct agent (as in bleaching and water treatment) as well as in the manufacture of hundreds of chemicals. The same is true of ethyl alcohol, ethylene dichloride, and many other organics. Thus, many substances are simultaneously end-products and intermediates, depending on their intended use.

(2) The stable element at the conclusion of a radioactive decay series, for example, lead-206.

-ene. A suffix indicating the presence of one or more double bonds in a compound; thus, it appears in the names of alkenes, alkadienes, cycloöl efins, and aromatics. For this reason, the spellings *gasolene* and *kerosene* are less desirable than *gasoline* and *kerosine.*

energy. The fundamental active entity of the universe. Among its more important manifestations are (1) electromagnetic radiation (photons, or radiant energy); (2) the energy of combustion (thermal energy); (3) electrical energy; (4) kinetic energy (the energy of motion); (5)

nuclear energy, the binding force that holds protons and neutrons together in the atomic nucleus; and (6) free energy, a thermodynamic function describing the energy available to a substance for reaction with other substances. The three laws of thermodynamics state the energy conditions which determine the nature and extent of chemical reactions. The first and best-known of these (conservation of energy) is that the total of energy in the universe is constant and cannot be increased or diminished. Einstein was the first to formulate the concept of the equivalence of energy and mass, expressed by the equation $E = mc^2$, from which has developed the science of nuclear physics. *See also* mass; matter; photon.

energy converter. Any element or compound having the ability to convert the radiant energy of sunlight into electrical, thermal or chemical energy. Prominent among them are silicon, selenium and tellurium, as well as the chlorophyll of plants in photosynthesis. *See also* solar cell; chlorophyll.

energy sources. The sources of energy described in this book include both natural materials and man-made devices. Mechanical sources (wind, water, geothermal, oceans, etc.) are not considered, as they involve little or no chemistry.

Non-renewable material sources are those of geologic origin, i.e., petroleum, natural gas, coal, shale oil, oil sands, and uranium, which cannot be replaced once their supply is exhausted. Renewable material sources can be replenished on a predictable basis; they include such cellulosic products as wood, bagasse, agricultural wastes, and residue from forest products industries (shavings, sawdust, bark, etc.), as well as methane from animal manures (biogas) and seaweed products (algae). These are collectively called biomass. The entries listed below give detailed information.

algae	fusion	natural gas
bagasse	fuel cell	nuclear energy
battery	fuel oil	oil sands
biogas	gasoline	peat
biomass	gasohol	petroleum
breeder	gasifiction	shale oil
cellulose	geothermal	solar cell
coal	energy	uranium
ethyl alcohol	hydrogen	wood
fission	methane	

engineering plastic. A type of plastic that is hard and stable enough to be treated as a metal; such materials are capable of sustaining high loads and stresses and are machinable and dimensionally stable. They are used in construction, as machine parts, automobile components, telephone handsets, and numerous other items. Among the more important are nylon, acetal resins, polycarbonate resins, and ABS resins.

enhanced oil recovery. Extraction of oil trapped in rock strata so firmly that it cannot be recovered by conventional pumping. The techniques in present use are as follows: (1) chemical flooding, in which mixtures of strong detergents and salt water are introduced into the formation under high pressure to reduce the surface tension of the oil so that it will flow (a modification of this method, called micellar flooding, involves a second-step addition of a water-soluble high polymer), and (2) hydraulic fracturing, in which a water solution of guar gum (in which bauxite is dispersed) is pumped into the strata; this pressure creates small fissures in the rock formation that are held open by the suspended bauxite particles (called "proppants"), permitting the trapped oil to flow through the apertures.

enhancer. A food additive that brings out the taste of a food without having any taste of its own. Monosodium glutamate is widely used for this purpose. The effective concentration of enhancers is much greater than that of potentiators, being measured in parts per thousand, as compared with parts per billion for potentiators. *See also* potentiator.

enol. A molecular rearrangement characteristic of some aldehydes, ketones, and organic acids, in which a hydrogen atom is removed from the carbon atom nearest to the carbonyl group and united with the oxygen atom of the carbonyl group. Thus, the resulting compound contains both a hydroxyl group (OH) and an unsaturated carbon linkage, as indicated by the name ($-$ene $+$ $-$ol). In this type of rearrangement the hydrogen is especially reactive and is easily replaced with other elements. The reaction is called enolization and occurs in many racemic compounds, e.g., tartaric acid. *See also* racemic.

entrainment. The state of being shut in or occluded, as of small units of gases of liquids trapped within a different phase, for example, air bubbles in a solid or liquid or water droplets in a solid. It also has the sense of "being carried along" when used in reference to liquid droplets present in a stream of vapors, as may occur in evaporation or distillation.

environment. The collective total of external influences to which plants, animals, and man are exposed. This includes many chemical factors such as nutrients, pH, toxic substances, temperature, air, water, waste disposal, pesticides, etc. The study of these and their various maladjustments is called environmental chemistry.

enzyme. Derived from the Greek, meaning "ferment" or "leaven," this term designates a group of organic catalysts formed in the cells of biological systems. Enzymes are characterized by their specificity; each initiates a certain chemical reaction, but is of little significance for other reactions. There are four types: (1) those which effect removal of water, (2) those which aid transfer of electrons, (3) those which transfer radicals, and (4) those which act on a single carbon-carbon bond without removal or transfer of radicals. Enzymes have a vast number of specific functions, among which are synthesis of proteins, conversion of starches to sugars, catalysis of fermentation processes, promotion of digestive reactions, etc. The names of most enzymes terminate in either -ase or -in; examples are rennin, diastase, amylase, pepsin, cholinesterase, and zymase. Enzymes are similar to proteins in their chemical nature and often are closely associated with them (coenzymes). An example is ribonuclease, composed of 124 amino acids arranged in a specific sequence and having four disulfide cross-links. The functioning of nucleotides and the programming of the genetic code are largely controlled by enzymatic activity. *See also* catalyst; coenzyme; yeast.

EPDM rubber. *See* polypropylene.

epichlorohydrin. A poisonous and flammable liquid, b.p. 115°C (239°F), derived indirectly by reacting propylene with chlorine. It is one of the most useful members of the epoxide family of compounds, its major use being the manufacture of epoxy resins. Its formula is

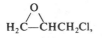

which contains the heterocyclic epoxide group. It also has solvent properties for cellulose de-

rivatives. It should be handled with due caution because of its toxicity. *See also* epoxide.

epimer. One of two isomeric compounds, e.g., sugars such as glucose and mannose, which are the same in all respects except that the order of attachment of the hydrogen atom and the hydroxyl group on one of the carbon atoms is reversed, i.e., —HCOH— and —OHCH—. They occur in six-carbon (hexose) sugars containing four asymmetric carbons. *See also* asymmetry.

epinephrine. *See* adrenaline.

epoxide. An organic compound containing a heterocyclic skeleton made up of two carbon atoms, both of which are attached to an oxygen atom by a single bond, as follows:

Such compounds are quite reactive with oxidizing and reducing agents and with acids, and they also polymerize easily with heat or catalysts, to yield epoxy resins. Epichlorohydrin and ethylene oxide are examples. *See also* epoxy resin; epichlorohydrin.

epoxy resin. A versatile synthetic resin made by reacting an epoxide compound (usually epichlorohydrin) with a hydroxyl-containing substance such as bisphenol A or a polyhydric alcohol (glycerol). Such resins are thermosetting and have a broad spectrum of uses, among the more important of which are as strong-bonding adhesives, effective with metal/ceramic composites, as well as with wood; filament winding; protective coatings; and encapsulation. Epoxy coatings have high resistance to attack by chemicals, corrosion and weathering, and to electricity.

Epsom salt. A hydrated magnesium sulfate ($MgSO_4 \cdot 7H_2O$) used in the textile industry as a fire-retardant agent and dyeing assistant, in leather tanning, and as a bleaching aid. It also has medical applications as an antidote for barbiturates, a laxative, and for reduction of tissue inflammation. Named from Epsom, England.

equilibrium. In chemical engineering terminology, a state of balance in terms of the molecular concentrations of compounds undergoing a reversible reaction. The equilibrium constant, K, is the ratio of the molecular concentration of the reaction products to that of the starting materials. It is strongly affected by temperature, which increases the rate of the reaction. Equilibrium exists when the rates of the two reactions are the same at a given temperature. The concept is useful in chemical engineering calculations for projecting plant-scale reactions.

In physical chemistry, the equilibrium state is said to exist when, in a liquid partially filling a closed container, the number of molecules expelled from the surface by internal molecular impact is equal to the number of molecules attracted back into the liquid at the interface when the air space has become saturated, that is, when it can retain no more molecules of vapor. Thus, there is a reversible physical reaction which reaches an equilibrium condition between two molecular concentrations (of the liquid and the vapor), though no chemical change occurs. *See also* vapor pressure.

equivalent weight.   The weight of an element that will combine with 7.9997 grams (one-half the atomic weight) of oxygen. This weight of oxygen will combine with exactly 1.0079 grams of hydrogen (i.e., the atomic weight of hydrogen). Thus, 1.0079 grams of hydrogen can be taken as equivalent to 7.9997 of oxygen, and hydrogen is said to have an equivalent weight of 1. As hydrogen is the lightest element, the combining power of all other elements is greater than 1. The equivalent weight principle was formerly used in determination of atomic weights, but this method has been replaced by the mass spectrograph. Equivalent weight is also called the combining weight.

Er   Symbol for the element erbium.

erbium.   An element.

Symbol	Er	Atomic No.	68
State	Solid	Atomic Wt.	167.26
Group	IIIB	Valence	3
	(Lan-	Isotopes	6 stable
	tha-		
	nide		
	Series)		

Erbium, m.p. 1522°C (2771°F), is closely related to yttrium and terbium, all of which were discovered in Sweden in gadolinite ore and were later found in other ores. Strongly electropositive, it is a member of the rare earth family; it is readily flammable and has a high electrical resistivity, 87 microhm-cm as compared with copper, 1.67 microhm-cm, at room temperature.

It is used in lasers, magnetic alloys, and similar specialized devices. It forms compounds with the halogens and also appears as the nitrate, oxalate, phosphate, carbonate, and acetate. *See also* yttrium; terbium.

ergosterol.   A member of the biochemically active sterol family of compounds and a precursor of vitamin D_2, or calciferol, to which it is converted by exposure to ultraviolet radiation. It is also known as provitamin D_2. Both the compound and its name are derived from the fungus ergot; it also can be formed from sugars by the action of yeast. Its primary function in the body is to catalyze the deposition of calcium in the bones and teeth (as suggested by the name calciferol). It occurs in yeast and fish-liver oils. *See also* sterol; vitamin.

Erlenmeyer flask.   A useful type of laboratory glassware; it is an open container whose dimensions are, for example, about 8 inches tall, with a relatively narrow neck section about 1½ inches in diameter and 2 inches long, below which the contour becomes cone-shaped. The bottom is flat. It is used for numerous experiments involving liquids, especially titrations and extractive testing. It was named after its inventor.

erythrocyte.   The most vital component of mammalian blood, chiefly composed of the protein-iron complex called hemoglobin and commonly known as the red cell of the blood. Erythrocytes can be removed from blood by centrifugation, to give plasma. *See also* blood.

Es   Symbol for the element einsteinium, the name being assigned in honor of Albert Einstein, an American scientist (German-born) (1879-1955), Nobel Prize 1921.

essential.   (1) An amino acid, vitamin, or fatty acid that is not synthesized by the animal organism but must be obtained externally, that is, by ingestion of plant products of one type or another. *See also* amino acid.

(2) An oil distilled from the flowers, leaves, or stems of certain plants; in a figurative sense it represents the "essence" of the plant and does not imply necessity or need. *See also* essential oil; perfume.

essential oil.   A nonfatty oil with a strong, usually pleasant, odor and taste, obtained from flowers and other parts of plants by solvent extraction or steam distillation. Terpenes are the chief components of many essential oils; others are mixtures of aldehydes, acids, alcohols, and the like, e.g., benzaldehyde and hydrocyanic acid occur in oil of bitter almond. Essential oils are subject to evaporation, in contrast to fixed vegetable oils, which are not. They are obtained from a wide variety of plant life, some having such exotic names as oil bois de rose, neroli oil, ylang-ylang oil, geraniol, rose otto, patchouli oil, and citrus peel oils. They are used in perfumes, odorants, and food flavorings. An exception is turpentine oil, used chiefly as a solvent and paint thinner. The term "essential" refers to the distilled "essence" of a material, not to its importance. Though they are products of vegetation, essential oils are not classified as vegetable oils. *See also* vegetable oil; edible oil.

ester.   A compound that can be regarded as formed by the replacement of the acidic hydrogen of an inorganic or organic acid by an aliphatic, aromatic, or heterocyclic radical. The term usually has the connotation of a substance prepared from a carboxylic acid and an alcohol or phenolic hydroxy compound. The general formula for an ester is RCOOR. While the reaction ROH + RCOOH→RCOOR + H_2O appears analogous to the salt-forming acid-base neutralization of inorganic chemistry, its mechanism is different. By used of a tagged oxygen isotope, it has been shown that the oxygen of the coproduct water comes from the $-OH$ group of the carboxylic acid and not from the alcohol. The reaction is of the condensation type and often requires a catalyst.

Esters are named in terms of the acids and alcohols from which they are formed. Acetic acid yields acetates; fatty acids give glycerides; butyric acid forms butyrates; and carbonic acid gives organic carbonates, such as dimethyl carbonate. Esters are of widespread occurrence and have a broad range of applications. Important types are cellulose esters (acetate, butyrate, propionate) for fibers and plastics; phthalic acid esters for plasticizers; vegetable and animal waxes, which are alkyl esters of monocarboxylic acids; and polyester and alkyl resins, from dicarboxylic acids and dihydric alcohols. Ester formation (esterification) is an important and frequently used reaction in synthetic organic chemistry. *See also* polyester.

ester gum.   An artificial product made by reacting rosin with glycerol or other polyhydric alcohol. It is actually a resin rather than a gum.

It is used as an ingredient of industrial cellulosic lacquers and in special paint formulations. It is soluble in most organic solvents.

esterification.   *See* ester; acetate.

estrogen.   Collective term for naturally occurring steroid compounds formed in the ovary; they are also made synthetically. Estrogens have hormonal activity and are essential for normal female sexual development. Among the more important are estrone and estradiol. They can be obtained from the urine of pregnant animals and can be synthesized from other sterols. Estrogens have applications in the oral contraceptive field and for specialized medical purposes. *See also* antifertility agent.

e.s.u.   Abbreviation for elecrostatic unit.

Et   Symbol often used in chemical formulas for the univalent ethyl group, C_2H_5.

ethanal.   *See* acetaldehyde.

ethane.   A saturated aliphatic hydrocarbon gas, one of the seven basic petroleum-derived gases. It is the second member of the homologous series (paraffins) which starts with methane; its formula is C_2H_6. Like other gases of its type, it is extremely flammable. It is used as a source of ethylene and in general organic synthesis, as a fuel (in liquefied form), and as a refrigerant. It readily combines with chlorine to give, e.g., ethyl chloride. It is not particularly toxic. *See also* ethylene; natural gas.

ethanol.   *See* ethyl alcohol.

ethanolamine.   A syrupy yellowish liquid, b.p. 172°C (342°F), which has a strongly basic reaction, and thus is widely used to remove hydrogen sulfide and other acidic gases from synthesis gas. Its formula is $HO(CH_2)_2NH_2$. It is irritant to the eyes and skin and is considered toxic when inhaled. Ethanolamine and its derivatives di- and triethanolamine result from reacting ammonia with ethylene oxide. Other industrial applications are in the scouring of wool fibers, in dry-cleaning compounds, and for vulcanization of rubber.

ethene.   *See* ethylene.

ether.   A class of organic compounds characterized by the presence of an oxygen atom covalently bonded between two carbon atoms. If the organic groups containing carbon are represented by the letter R, the generalized formula of an ether is ROR. Ethers are derived either by removing water from alcohols (dehydration) or by hydration of olefins by means of a catalyst.

Most common ethers are liquids, and some are extremely flammable. The most prominent one is diethyl ether, $(C_2H_5)_2O$, b.p. 34.6°C (94°F), a valuable anesthetic first used in surgery in 1846; it is also a useful solvent and extraction medium. The ethers of ethylene glycol form a well-known group of useful solvents and plasticizers. There are a few solid ethers of cellulose. The term "petroleum ether" for petroleum-derived naphtha is a misnomer.

ethical drug.   A drug obtainable on prescription, not offered for open sale.

ethyl.   The univalent group, CH_3CH_2—, the second member of the homologous series of paraffinic hydrocarbon (alkyl) radicals; it is derived by dropping one hydrogen atom from ethane, CH_3CH_3, and often appears in formulas as C_2H_5—

When a second hydrogen atom is dropped from ethane, the divalent ethylene group is formed, $—CH_2CH_2—$. The corresponding olefin, $H_2C{=}CH_2$, is also called ethylene, and the two meanings of ethylene are sometimes confused in naming compounds. Like the methyl group, the ethyl group is present in thousands of organic compounds. *See also* ethylene; methyl.

ethyl acetate.   A light, mobile liquid, b.p. 77°C (171°F), resulting from the esterification of ethyl alcohol with acetic acid, catalyzed by sulfuric acid; its formula is $CH_3COOC_2H_5$. It is very flammable and a possible explosion hazard. It is used in the manufacture of smokeless powder and is an excellent solvent in nitrocellulose lacquers; with alcohol, it will also dissolve cellulose acetate. It also has application in the manufacture of pharmaceutical products and as an organic intermediate. *See also* acetate.

ethyl alcohol.   A liquid monohydric primary alcohol, b.p. 78.5°C (173°F), having the formula C_2H_5OH (or CH_3CH_2OH); it is also called ethanol, grain alcohol, or simply "alcohol." It is the most important organic solvent in use today; well over two billion pounds is manufactured annually. Beverage grades are made by fermentation of the sugars in fruits, molasses, and grains. It is classed as a depressant and has a low order of toxicity. Most industrial alcohol is made synthetically by catalytic cracking of hydrocarbons or by the Oxo process. It is used in numerous end-products (detergents, cosmetics, solvents, cleaning preparations) and as an intermediate in the manufacture of organic chemicals. Recent production from agricultural wastes

has made possible its expanding use as a motor fuel additive (gasohol); it may eventually replace gasoline, as is already the tendency in Brazil. Denatured grades contain certain noxious or toxic additives (often methyl alcohol) to prevent internal use. Ethyl alcohol is flammable and should be protected from ignition sources. *See also* denaturant; Oxo process; fermentation; gasohol.

ethylamine.   A flammable and toxic liquid, b.p. 16.6°C (62°F), having the formula $CH_3CH_2NH_2$ and made by reacting ammonia with ethyl chloride; also called aminoethane. Though it has solvent properties, its chief uses in the chemical industry are as an intermediate for the synthesis of dyes and and related organic compounds. It should be handled with caution. *See also* amine.

ethylbenzene.   *See* aluminum chloride; Friedel-Crafts reaction.

ethycellulose.   A thermoplastic product, insoluble in water; it is made by replacing about half of the hydroxyl groups of cellulose with ethoxy groups (OC_2H_5), derived from ethyl alcohol or similar compounds. It is thus a cellulose ether. Its major uses are in coatings for a broad range of industrial products (paper, textiles, wire and cable), and as an adhesive and binding additive in printing inks, pigments, and similar materials.

ethyl chloride.   A saturated chlorinated hydrocarbon, C_2H_5Cl, b.p. 12°C (54°F), gaseous at room temperature but manufactured and transported in compressed form as a liquid. Like other compounds of this class, it is quite poisonous and extremely flammable and should be handled with caution and protected from exposure to static sparks or other flame source, which may cause explosion. It has solvent properties for organic materials and for certain elements such as sulfur and phosphorus; it is used in making tetraethyllead and as an insecticide base. *See also* chlorinated hydrocarbon.

ethylene.   (**1**) An unsaturated aliphatic hydrocarbon (olefin), ethylene is obtained by thermal cracking of petroleum gases (butane, ethane, etc.), a process known as pyrolysis. Ethylene is one of the most prolific sources of synthetic organic chemicals and plastics. Its formula, $H_2C{=}CH_2$, is a reactive structure of far-reaching importance which occurs in many products besides ethylene, e.g., butadiene and isoprene. Ethylene (also called ethene) is a flammable and explosive gas, b.p. $-103.9°C$ (-155°F), from which a number of basic petrochemicals are derived by catalytic processes, for example, ethyl alcohol, ethylene oxide, ethylene dichloride, and ethylene chlorohydrin. These in turn are intermediates for a wide range of synthetic organics, typical of which is the ethylene glycol family. Ethylene is also the parent substance of many elastomeric and plastic products, e.g., polystyrene, polyethylene, ethylene-propylene rubbers, and polyester resins. Many of these can be cross-linked to form thermosetting plastics.

(2) The divalent group —CH_2CH_2—, formed when a hydrogen attached to each carbon atom of ethane, CH_3CH_3, is replaced by another element or group, as in ethylene dichloride, $ClCH_2CH_2Cl$. *See also* polyethylene; ethylene glycol; ethyl.

ethylene bromide. *See* bromine.

ethylene chlorohydrin. Made by reacting ethylene with hypochlorous acid, this compound is an extremely toxic liquid which is readily absorbed by the skin, sometimes with lethal effect. Its formula is $Cl(CH_2)_2OH$, b.p. 128.8°C (262°F); it may be regarded as an alcohol in chemical constitution. It is used as a solvent for various cellulosic plastics and in the synthesis of other organic compounds including ethylene glycol and ethylene oxide. Great caution should be exercised in handling this material.

ethylenediamine. *See* ligand.

ethylenediaminetetraacetic acid. Often referred to by its abbreviation EDTA, this compound is one of the best-known and most effective complexing agents, coordinating strongly with metal ions to form chelates. It is commercially obtainable in the form of various salts (edetates) as, for example, tetrasodium EDTA. It forms stable compounds with metal ions and thus has the effect of deactivating them. It coordinates through no less than six linkages—two nitrogen atoms and four carboxyl groups:

EDTA forms soluble complexes with ions of calcium, magnesium, iron, etc., and is used as a water-softening agent and detergent; it also has applications in electroplating, preparations for rust and scale removal, and in decontamination of radioactive surfaces. It is made synthetically in the form of water-soluble crystals. *See also* chelate; coordination compound; sequestering agent.

ethylene dichloride. A saturated chlorinated hydrocarbon liquid made by catalytic reaction of ethylene with chlorine, giving the formula $ClCH_2CH_2Cl$, b.p. 83.5°C (182°F). It is both flammable and poisonous and must be handled with caution. It is an important starting material for making vinyl chloride and other chlorinated organics and is also used for degreasing metals and in paint and varnish removers.

ethylene glycol. A dihydric alcohol with formula CH_2OHCH_2OH, b.p. 197.2°C (387°F); it is a hygroscopic liquid of medium viscosity obtained from ethylene, chiefly by catalytic oxidation. Its major use is as a permanent antifreeze; of secondary importance is its use as a polyester fiber intermediate. It is also used in hydraulic fluids, heat-transfer agents, and humectants for cellophane, fibers, leather, and adhesives. The proprietary "Cellosolve" groups of mono- and dialkyl ethers of ethylene glycol are important solvents derived from ethylene glycol. Ethylene glycol is combustible and sufficiently toxic to preclude its use in foods. Its many ether derivatives are excellent plasticizers, emulsifiers, and dispersing agents. There are a number of condensation polymers of similar properties and applications, among which are di-, tri-, and polyethylene glycols. *See also* glycol.

ethyleneimine. A strongly reactive liquid, b.p. 57°C (135°F), having the formula $H_2C=NHCH_2$; the unsaturated carbon-nitrogen linkage accounts for the reactivity of compounds of this type. It is corrosive to tissue and damaging to lungs when fumes are inhaled. It is a useful organic intermediate and has a few specialized uses in refining of hydrocarbons and in some types of protective coatings. *See also* imine.

ethylene oxide. A reactive intermediate made by catalytic oxidation of ethylene; it is a flammable and explosive gas, b.p. 10.73°C (51.3°F), and is also quite toxic. An ether of the epoxy type, also called oxirane, its formula is H_2COCH_2(or CH_2CH_2O). Its chief industrial importance is as a source of the ethylene glycol group of compounds, as well as of polyester resins of various types. It is also an effective detergent, fumigant, and disinfectant. When reacted with cellulose, it forms the water-soluble polymer hydroxy-ethylcellulose. *See also* ether; epoxide.

ethyl ether.   The most important of the ether group of compounds (generalized formula ROR), ethyl ether has the formula $C_2H_5OC_2H_5$, or $(C_2H_5)_2O$; it is also called diethyl ether and often simply ether. It is a highly volatile and extremely flammable liquid boiling at 34.6°C (94°F). When exposed to air, it readily forms toxic peroxides which are an explosion hazard; thus, it must be stored in air-tight containers and must never be used where there is a slightest chance of ignition from a static spark or other heat source, no matter how transitory. Its use as an anesthetic has declined considerably in favor of less dangerous materials. It finds application in the chemical industry as a solvent for oils, rubber, perfumes, and other natural organics, as well as for nitrocellulose in the manufacture of smokeless powder. *See also* ether.

ethyl mercaptan.   *See* odor.

ethyl methyl ketone.   *See* methyl ethyl ketone.

Eu   Symbol for the element europium, the name being adapted from Europe.

europium.   An element

		Atomic No.	63
Symbol	Eu	Atomic Wt.	151.96
State	Solid	Valence	2,3
Group	IIIB	Isotopes	5 stable
	(Lan-		
	tha-		
	nide		
	Series)		

The most reactive of the rare earth family of metals, europium, m.p. 826°C (1519°F), is obtained from monazite sands and bastnasite ore. Its high neutron cross-section (4600 barns) makes it usable as a control element in nuclear reactors; this use virtually disappeared in 1965 when its unique ability to activate yttrium compounds in red television phosphors led to its almost exclusive use for this purpose. It forms such compounds as the oxide, sulfate, nitrate, chloride, and oxalate. *See also* yttrium.

eutectic. (1)   A term used in metallurgy and glass technology to indicate the lowest melting point attainable with a solution of two or more components by varying the percentage of the components. For example, an alloy or solid solution which melts at a lower temperature than any

other alloy of the same metals is called a eutectic alloy. The few eutectic alloys that exist form a group of so-called fusible alloys, some of which melt as low as 60°C (140°F). *See also* alloy; fusible alloy.

(2)   A term used in the physical chemistry of solutions, e.g., of salts in water, to designate the point at which two or more components precipitate together in a constant ratio as the temperature of a saturated solution is changed (usually lowered) or as the concentration is changed by removal of solvent (e.g., evaporation of water).

eutrophication.   A term which has come into prominence in recent years as a result of concern over water pollution. It refers specifically to the "fertilization" of lakes, rivers, and coastal bays resulting from disposal of phosphate-bearing detergents in waste waters. The phosphates stimulate the growth of algae to the point where most of them die from lack of light and thus pollute the water with their decomposition products, removing oxygen in the process with the resulting death of fish. *See also* algae.

evaporation.   Expulsion of molecules from the surface of a liquid (or solid) due partly to the kinetic forces exerted by molecules within the substance and partly to the chemical nature of the substance, which largely controls its surface tension. The rate of expulsion increases as the temperature is raised, and is greatest at the boiling point. Low-boiling organic liquids have a greater tendency to evaporate (i.e., they are more volatile) than water, since their surface tension is lower. The term evaporation is essentially synonymous with vaporization. Evaporation takes place naturally from exposed water surfaces and is the first step in the distillation process which produces rain.

Industrially, evaporation has many applications: (1) distillation of petroleum, alcohol, and other organic materials; (2) purification of sea water; (3) concentration of solutions and recovery of salt, soda, ash, etc.; (4) refining of sugar; (5) drying of various materials (food products, paper, textiles). The energy required to convert a liquid to vapor at the boiling point is retained by the vapor (latent heat of vaporization) and is released when condensation occurs; this principle is utilized in designing multiple-unit evaporation systems. Solids such as ice, naphthalene, etc., also evaporate below their melting points (sublime) but at a comparatively slow rate. *See also* distillation; effect; surface tension; sublime.

evolution.   The act of giving off or emitting a gas or vapor as the result of a chemical reaction; for example, acids evolve hydrogen when reacted with a metal or metal-containing compound; carbon dioxide is evolved by combustion. In biology and mathematics the term has quite different meanings.

excited state.   A higher than normal energy level (vibrational frequency) of the electrons of an atom, group, or molecule, resulting from absorption of photons (quanta) from a radiation source (arc, flame, spark, etc.) in any wavelength of the electromagnetic spectrum. X-ray, ultraviolet, visible, infrared, microwave, and radio frequencies are used for excitation in various types of spectroscopy. When the energizing source is removed or discontinued, the atom or molecule returns to its normal or stable state either by emitting the absorbed photons or by transferring the energy to other atoms or molecules. The increased vibrational activity of the atom or molecule yields line or band spectra characteristic of its structure, thus permitting identification. Photochemical reactions are induced by excited chemical entities, which are also responsible for the phenomenon of luminescence (phosphorescence and fluorescence). *See also* spectroscopy; photochemistry; absorption **(2)**.

exothermic.   The term used to characterize a chemical reaction which evolves heat as it proceeds, as, for example, the formation of carbon dioxide from carbon and oxygen: $C + O_2 \rightarrow CO_2$ + heat; also the reaction between metallic sodium and water to form sodium hydroxide: $2Na + 2H_2O \rightarrow 2NaOH + H_2$ + heat. *See also* endothermic.

exotic.   A term occasionally used to describe or indicate materials of particularly high efficiency, derived from an unusual source for a given application; for example, when boron-derived fuels for rocket engines were developed, they were known collectively as exotic fuels. A number of new and unusually powerful solvents have also been so described. There are also so-called exotic metals, e.g., titanium, hafnium, zirconium, etc.

expander.   **(1)** In storage batteries, a composition made up of carbon black, barium sulfate, and wood lignin; when added in the amount of about 1 percent to the lead oxide used to coat the negative plates of the battery, this expander composition maintains the capacity of the plates and helps to extend the active life of the battery.

(2) An agent added to blood plasma, e.g., polyvinylpyrrolidone, or substituted for it in the treatment of wounds, burns, etc. Such materials are called plasma volume expanders.

experiment.   An objective trial undertaken in order to discover or establish facts. Its essential parts are (1) statement of the problem or the fact to be ascertained; (2) setting up the equipment so as to exclude all irrelevant or chance factors; (3) observing and recording the events; (4) use of one of more controls; and (5) interpretation of results. Experiments are the lifeblood of chemistry, and some laboratory training is essential to an effective chemical education.

Experimentation is one of the most basic aspects not only of chemistry but of all natural sciences. Leaders in experimentation through the centuries were Archimedes, Roger and Francis Bacon, Copernicus, Galileo, Franklin, Faraday, Curie, and Edison. Famous experiments in the history of science include Archimedes' discovery of the principle of specific gravity in his bathtub; Galileo establishing the uniform acceleration of falling bodies by dropping objects of different weights from the tower of Pisa; Francis Bacon stuffing a fowl with snow (food preservation); Franklin's key-and-kite identification of lightning as a form of electricity; Newton's discovery of the spectra of sunlight by passing it through a glass prism; and the Michelson-Morley experiment (1887) by which the speed of light was determined.

explosive.   An unstable chemical compound of which nitrogen usually is the active constituent, which decomposes more or less violently when shocked or ignited, particularly when confined. Low explosives, typified by black powder and smokeless powder, burn or deflagrate rapidly, with evolution of gases. High explosives detonate, with production of a rapidly traveling shock wave of destructive violence. Initiating explosives are those used in blasting caps to detonate dynamite, TNT, etc., which are less shock-sensitive than many types. Silver acetylide and copper acetylide are instances of explosives that do not contain nitrogen. While explosions can occur by ignition of dusts and vapors of organic liquids, these are not classified as explosive materials. *See also* deflagration; detonation; high explosive; initiating explosive.

explosive limits. The range of concentration of a flammable gas or vapor in air (in percent by volume) in which combustion or explosion can occur. A concentration below the lower limit is not sufficient to support combustion, and above the upper limit the mixture is too "rich" to burn (as when a carburetor is flooded). Flammable or explosive limits for some common substances in air are as follows:

	Lower (%)	Upper (%)
carbon disulfide	1	50
benzene	1.5	8
methane	5	15
butadiene	2	11.5
butane	1.9	8.5
propane	2.4	9.5
natural gas	3.8	17
hydrogen	4.1	75.5
ether	1.73	23.3
ethanol	4.16	19
acetylene	2.5	80
hydrogen sulfide	4.3	46

extender. *See* diluent.

extinguishing agent. Any compound used to put out fires. Water is, of course, the best-known; it not only has a powerful cooling effect but also serves to keep oxygen away from the flame. It is especially effective in the form of spray or mist, and as such it can be used to extinguish fires in heavy oils. A direct stream of water should never be used to fight fires involving gasoline, hydrocarbon liquids, or any other material of specific gravity less than 1.00. Carbon dioxide is often used on these and other fires. Dry chemical (a mixture of sodium chloride and sodium bicarbonate) is useful for extinguishing fires in metals and electrical equipment. Fires in hydrocarbon liquids are best handled by use of chemical foam (carbon dioxide in water). Such toxic halogenated hydrocarbons as carbon tetrachloride and methyl bromide should be avoided. *See also* flame retardant.

extraction. (1) The removal of soluble components from a solid or liquid mixture by means of an appropriate solvent, the substance removed being the extract. Numerous solvents are used in analytical chemistry, pharmacy, etc.; for example, a sample of vulcanized rubber can be analyzed for specific components by extracting it with acetone or chloroform. Solvent extraction is an industrial method of removing impurities from liquids and also of obtaining essential oils from plant sources. Steam distillation is also an extraction method.

(2) In metallurgy, the term is applied to the mining of metals and separation of pure metals from their ores (sometimes also called winning).

extrusion. A fundamental mechanical operation of the process industries, applied in one variation or another in the manufacture of a wide range of plastics and rubber articles, synthetic fibers and filaments, plastic film and sheet, wire and cable coating, some metals, and such food items as spaghetti, etc. The base material may be a low-viscosity liquid such as molten glass or a polymer dispersion passed through a spinneret; a viscous liquid polymer as used in injection molding; or a semisolid mass such as a rubber or plastic composition. In the case of high-viscosity materials, the process involves passing the mass over a revolving screw of varying pitch in an enclosed chamber; this forces it through a metal form or die to impose the required shape. Film is made by extrusion through a slit of predetermined width and thickness. Polymer fibers are usually hardened by passing them through a bath of formaldehyde or similar agent. Extrusion involves application of rheological principles of some complexity; viscosity, temperature, and rate are critical factors. *See also* liquid; rheology.

F

F **(1)** Symbol for the element fluorine, the name being derived from the Latin word for *flow*.

(2) Abbreviation for Fahrenheit temperature scale.

factice. A synthetic softener made by reacting a vegetable oil (usually corn oil) with sulfur chloride; it is a brown to light-tan semisolid material used in special types of rubber and plastic products to impart a soft, smooth feeling, as well as to make art gum erasers. It is also called rubber substitute and vulcanized oil.

factor. A term used chiefly by biochemists to indicate any member of a biologically active complex, especially if its exact chemical nature is unknown or if its function in cellular metabolism has not been elucidated. Several of the B complex vitamins were originally referred to as "factors" until their identity had been established by research. *See also* Rh factor.

Fahrenheit. The temperature scale commonly used in the U.S., though centigrade is preferred by scientists. On the Fahrenheit scale, the freezing point of water is 32° and the boiling point 212° at sea level (1 atmosphere, or 760 mm of mercury). To convert °F to °C, first subtract 32 from the Fahrenheit temperature and then multiply the balance by 5/9 or 0.55. The temperature value for both Fahrenheit and centigrade scales is the same at -40°. This scale was devised by Gabriel Daniel Fahrenheit, a German physicist (1686-1736). *See also* centigrade.

fallout. Radioactive fission products resulting from nuclear and thermonuclear explosions; these may drift for hundreds of miles from the explosion site, contaminating the earth's surface as they go. Much potentially dangerous radioactivity has already been added to the environment by nuclear testing, especially in the form of strontium-90. No adequate protection from fallout is known. No fallout occurs from controlled fission reactions in atomic power plants, whose deleterious products are neutralized within the plant area or shipped elsewhere for treatment or disposal.

Faraday, Michael (1791-1867). A native of England, Faraday did more to advance the science of electrochemistry than any other scientist. A profound thinker and accurate experimentalist and observer he was the first to propound correct ideas as to the nature of electrical phenomena, not only in chemistry but in other fields. His contributions to chemistry include the basic laws of electrolysis, electrochemical decomposition (the basis of corrosion of metals) as well as of battery science and electrometallurgy. His work in physics led to the invention of the dynamo. Faraday was in many respects the exemplar of a true scientist, combining meticulous effort and interpretive genius.

fast. **(1)** A term used to describe a dye or pigment whose color is not impaired by prolonged exposure to light, steam, high temperature, or other environmental conditions. Inorganic pigments are normally superior in this respect to organic dyes.

(2) In nuclear technology, the term describes neutrons moving at the speed at which they emerge from a ruptured nucleus, as opposed to "slow" or thermal neutrons whose speed has been reduced by impinging on a moderating substance such as graphite.

fat. A solid to semisolid glyceride of a fatty acid, of high nutritional value. Fats are chemically classified as lipids. Natural fats are produced by both plants and animals; the former are represented by coconut and cocoa "butters," and the latter by lard, tallow, and milk-derived butter. Chemically, such fats have the same composition as oils of the same origin; all are glycerides of stearic, palmitic, oleic, and similar fatty acids, varying only in the extent of their saturation, the animal fats being the more highly saturated. Thus, the only difference between a vegetable or animal fat and a similarly derived oil is that fats are solid and oils are liquid at room temperature. Some fats show traces of crystal formation on cooling. Synthetic fats are made by catalytic hydrogenation of vegetable oils to a solid consistency, chiefly for culinary purposes. Some fish oils are also hydrogenated for use as dispersing agents. Fats can be hydrolyzed or "split" by reacting the glyceride with water at about 260°C (500°F) to form glycerol and free fatty acids. *See also* glyceride; hydrolysis; soap; fatty acid; lipid.

fatliquor.   A mixture of neatsfoot oil and water, emulsified with a fatty acid or sulfonated oil; the neatsfoot oil is derived from water-extraction of animal tendons. Fatliquors are used primarily in the leather industry to lubricate skins and hides, which tend to become hard and dry as a result of tanning. This treatment enables the finished leather to bend easily without cracking and to withstand the stresses it receives during fabrication into shoes and other products.

fatty acid.   An aliphatic acid, either saturated or unsaturated, whose molecule consists of an alkyl chain containing from 1 to over 30 carbon atoms, terminating in a carboxyl group (COOH). The simplest fatty acids are acetic and formic; those of intermediate molecular weight (from 4 to 20 carbon atoms) are derived from glycerides by hydrolysis; still higher forms are found in waxes. Fatty acids are in the general classification of lipids. Natural fats contain many unsaturated fatty acids having up to six double bonds along the chain (polyunsaturates), and many isomeric structures are possible. Saturated types occur in animal fats and some nut fats (coconut); unsaturated acids occur chiefly in such vegetable oils as castor, linseed, and safflower. Most fatty acids are mixtures of several types of glycerides. They are produced commercially by catalytic hydrolysis ("splitting") of the fat or oil to separate the acid from it (Twitchell process), with glyc-

erol as a by-product. The best known saturated fatty acids are stearic, palmitic, butyric, and lauric; among the unsaturates are oleic, linolenic, linoleic, and ricinoleic. Some fatty acids are synthesized within the animal and are thus called non-essential; others that are nutritionally necessary but are not formed within the organism must be obtained from external sources and are hence called essential. These are arachidonic, linoleic, and linolenic acids. Fatty acids are used primarily as raw materials for soap manufacture, as dispersing and activating agents in the rubber industry, in synthetic detergents, and in candle making. *See also* glyceride; lipid.

fatty alcohol.   An aliphatic alcohol containing from 8 to 20 carbon atoms in the chain; they are derived from natural sources by reduction, hydrolysis, or hydrogenation and may be saturated (octyl and stearyl) or unsaturated (oleyl and linoleyl). Those having more than 12 carbons in the molecule are solids. Fatty alcohols are used as solvents, detergents, and in pharmaceutical products. *See also* alcohol; cholesterol.

Fe   Symbol for the element iron, taken from the Latin, *ferrum,* meaning "iron." This form is conventionally used in the names of most iron compounds (ferric, ferrous).

feedstock.   A term used in the petroleum and synthetic organic chemicals industries to refer to petroleum or petroleum-derived hydrocarbons (liquids or gaseous) from which gasoline and petrochemicals are produced by cracking, reforming, and similar operations. It is also sometimes called the charging stock. It should not be confused with animal feedstuffs.

feldspar.   A clay-like material containing potassium as well as aluminum and silica. It is somewhat coarser and harder than clay and is often used as a component of abrasives. It also has application as a flux in ceramic products and as a filler in compositions for paving, roofing, and insulation. It is obtained in various states in the South and Southwest. *See also* spar (**1**).

fenchyl alcohol.   A secondary monohydric alcohol having a saturated cyclic structure, with formula $C_{10}H_{17}OH$. It is optically active and exists in several forms, one of which is solid at room temperature. Though known to occur in pine oil, the commercial grade is made synthetically. It has limited use as an odorant in detergents and antiseptic sprays and as a solvent.

fermentation.   An anaerobic, energy-yielding

reaction brought about or catalyzed by enzymes or bacteria and involving the decomposition of carbohydrates (sugars and starches) to alcohol and carbon dioxide. The ethyl alcohol in spiritous liquors is derived in this way from cereals and fruits. The fundamental reaction is:

$$C_6H_{12}O_6 \rightarrow 2C_2H_5OH + 2CO_2$$

Yeast contains enzymes which activate this reaction, and the bacteria of some molds induce the fermentation responsible for growth of antibiotics. Bacterial action also accounts for the souring of milk, spoilage of vegetables, and similar types of organic decomposition. Proteins can be obtained from petroleum by means of bacteria-induced fermentation. Several vitamins, including riboflavin and vitamin B_{12}, are also fermentation products. *See also* enzyme; bacteria; Pasteur.

Fermi, Enrico   (1901–1954). Italian physicist who later became a U.S. citizen. He developed a statistical approach to fundamental problems of physical chemistry, based on Pauli's exclusion principle. He discovered induced or artificial radioactivity resulting from neutron impingement, as well as slow or thermal neutrons. He was Professor of Physics at Columbia (1939) and was awarded the Nobel Prize in Physics in 1938. He was the first to achieve a controlled nuclear chain reaction, directed the construction of the first nuclear reactor at University of Chicago (1942), and worked on the atomic bomb at Los Alamos. He also carried on fundamental research on subatomic particles using sophisticated statistical techniques. Element 100 (fermium) is named after him.

fermium.   An element.

Symbol	Fm	Atomic No.	100
State	Solid	Atomic Wt.	254(?)
Group	IIIB	Valence	3
	(Actinide		
	Series)		

Discovered in 1953 and named after Enrico Fermi, an Italian scientist (1901-1954), Nobel Prize 1938, fermium is one of the artificially made radioactive members of the transuranic series, which is part of the actinide series of elements. The 254 isotope has a half-life of 3.24 hours; though the 257 isotope half-life is 97 days, its production is not feasible. The use of fermium is confined to tracer work.

ferric, -ous.   Adjectival forms derived from the Latin word, *ferrum,* meaning "iron," and indicating, respectively, the oxidation states of $+3$ and $+2$ for the metal in a given compound, e.g., ferric chloride ($FeCl_3$) and ferrous chloride ($FeCL_2$). An alternative nomenclature recommended for such compounds is iron (III) chloride and iron (II) chloride. Undoubtedly because of possible confusion with nonscientific meanings, the word "ironic" is not used in chemistry; however, iron oxide is accepted as a synonym for ferric oxide (Fe_2O_3). *See also* -ic; -ous.

ferricyanide.   *See* blue.

ferriferrocyanide.   *See* iron blue.

ferrite.   A solid compound consisting of the iron oxides Fe_2O_3 and FeO, the latter being replaceable by oxides of several other transition metals such as copper, nickel, or cobalt. Their characteristic property is their magnetic moment, for which they are used in computers, tape recorders, and many related communication devices where magnetic properties are essential. A simple example of a ferrite is the mineral magnetite, whose composition is $FeO \cdot Fe_2O_3$. Ferrites are manufactured by ceramic methods involving grinding, pressing, and sintering. Some types substitute a nonmagnetic oxide component, e.g., sodium, zinc, or aluminum. *See also* magnetochemistry.

ferroalloy.   An alloy containing varying proportions of iron and some other metal, such as molybdenum, manganese, titanium, etc., commonly added to steel for purposes of strength, heat resistance, and the like. Such alloys are used as a means of introducing the specific element into the "melt" and have no independent applications. They are also called ferrous alloys.

ferrocene.   An organometallic complex in which an iron atom is coordinated with a cyclic unsaturated structure (cyclopentadiene); it has the formula $Fe(C_5H_5)_2$. Discovered in 1951, it was the first of a broad group of organometallic "sandwich" compounds, so-called because the metal atom is located halfway between the two organic rings. Ferrocene is quite stable, melts at 173°C (343°F), resists pyrolysis at 400°C (752°F), and reacts in the same way as aromatic compounds. It reduces smoke emanation from fuel oils and has some applications as a catalyst and knock suppressor in gasoline. *See also* metallocene; coordination compound.

ferrocerium. *See* misch metal.

ferromagnetism. *See* garnet; ferrite.

ferrous sulfate. *See* copperas.

fertilizer. An agricultural chemical (which may be an element, compound, or mixture) containing one or more soil nutrients. Fertilizers are usually made artificially and applied either as solids or solutions. The elements essential for soil fertility are chiefly nitrogen, phosphorus, and potassium; several others are of less importance (sulfur, calcium, and trace elements). Fertilizers used in largest tonnage are ammonium nitrate, superphosphate (20% phosphoric acid), and potassium chloride. Controlled-release fertilizers are made by impregnating the fertilizer pellets with sulfur to permit gradual release of the fertilizing elements under varying moisture conditions. Some fertilizers are obtained from organic sources, such as processed municipal wastes, various types of manures, and tankage. The chemical characteristics of a soil are important in the selection of an optimum fertilizer. *See also* tankage.

fiber. An elongated crystalline structure ranging in length from a few millimeters to several feet (for some woody fibers) and in diameter from 1 to 1,270 microns (0.05 in.). Their strength varies greatly, from upwards of one million psi for single-crystal metal fibers (often called whiskers) to about 40,000 psi (theoretically) for cotton. Many common natural fibers are of vegetable origin (cotton, flax, ramie, jute, etc.); there are also animal fibers (wool, mohair, silk) and natural inorganic fibers such as occur in asbestos. Semisynthetic fibers are rayon (cellulose-derived), glass, boron nitride, graphite, and a number of other materials. Synthetic fibers are essentially plastics passed through fine nozzles in liquid form into a hardening bath; these can be cut to any length or manufactured in continuous lengths called filament. Polyamides (nylon), polyesters, and polyethylene are among the more familiar types. A filament is an extremely long or continuous fiber. *See also* whiskers; cellulose.

fiber, optical. *See* optical fiber.

fiber-reactive dye. An organic dye, often of the azo type, which can form chemical bonds with cellulose and thus become an integral part of the fiber. Such dyes were developed by synthetic

means about 40 years ago. They can also be used with a number of man-made fibers.

fibrinogen. A major protein component of blood plasma; when activated by the enzyme thrombin at the onset of bleeding, the fibrinogen undergoes chemical modification to fibrin; this in turn triggers the protective agglutination or clotting of blood. Fibrinogen is water-soluble and has a fibrous structure; it is synthesized in the liver. Blood clotting is an involved series of reactions in which fibrinogen is a critical factor.

filament. *See* fiber.

filament winding. An application of reinforced plastics in which a filament is passed through a bath of liquid thermosetting resin and wound under tension onto a form of appropriate shape for the component being manufactured. The resulting composite is then heat-treated. Filaments (or fibers) used are usually glass, though other types (graphite, boron, epoxy resin) are used for special purposes. Filament-wound structures have many industrial and structural applications.

filler. A fine-ground solid, usually inorganic in nature, used in considerable tonnages in the plastics and rubber industries, partly to save costs and partly for the heavy, deadening effect they produce in such items as floor coverings, gaskets, heels, etc. The materials used include whiting (calcium carbonate), barytes (barium sulfate), certain types of clays, wood flock, etc. They have neither reinforcing nor coloring effect. *See also* clay; barium sulfate.

film. A completely uniform and continuous layer or sheet of a material, ranging in thickness from molecular or colloidal dimensions to about one-hundredth inch. Films may exist independently, as typified by soap bubbles, toy balloons, etc., or they may form a surface coating on another substance, as a monomolecular film of oil or fatty acid on water or an electrodeposited metal serving as a protective coating. Plastic films are widely used for sealing and packaging food products to keep out moisture. It is essential that a film be free from the smallest holes or cracks. The term has specific application in the photographic industry. Thin films of atomic dimensions have important uses in electronics. *See also* monomolecular; thin.

filter aid. Any material used in a finely divided state to effect filtration of a liquid. Sand can serve as a filter in water purification; in the laboratory, such materials as clays, diatomite, coarse cellulosic structures, and glass fibers are used. These are sufficiently porous to allow passage of water and other liquids of similar viscosity, the foreign particles adhering to the surfaces of the filter aid. *See also* filtration.

filtrable virus. *See* virus.

filtration. An operation involving liquid-solid separation, used in chemistry and chemical engineering; it consists essentially of either gravity or forced flow of a liquid containing dispersed solid particles through a structure having a pore size small enough to retain the particulates. Special papers, cloth, fine-mesh wire screens, and membranes of collodion or other plastic are all used. In water purification, a bed of fine sand is effective; in the paper industry, the formation of the sheet on the fourdrinier wire is a practical example. Filter papers, glass fiber mats, and similar materials are used for laboratory filtration. *See also* filter aid.

fine chemical. A chemical produced in comparatively small quantities and in a relatively pure state. Pharmaceutical and biological products, perfumes, photographic chemicals, and reagent chemicals are examples.

fireclay. *See* clay; refractory.

Fischer, Emil (1852–1919). German organic chemist, recipient of Nobel Prize in chemistry (1902) for his original research in the chemistry of purines and sugars. He was Professor of Chemistry at University of Berlin (1882), succeeding Hoffmann. He synthesized fructose and glucose and elucidated their stereochemical configurations; also established the nature of uric acid and its derivatives. Additional work included enzyme chemistry, proteins, synthetic nitric acid, and ammonia production.

Fischer, Hans (1881–1945). German biochemist who studied under Emil Fischer. He was awarded the Nobel Prize in chemistry in 1930 for his synthesis of the blood pigment hemin. He also did important fundamental research on chlorophyll, the porphyrins, and carotene.

Fischer-Tropsch process. A process for the manufacture of liquid hydrocarbon fuels developed in Germany in the 1930's. It is basically a catalyzed reaction of carbon monoxide and hydrogen, yielding aliphatic hydrocarbons and oxygen-containing organics. The gaseous mixture of hydrogen and carbon monoxide can be obtained from petroleum or coal (water gas or synthesis gas). A number of metallic catalysts have been successfully used. Variations of this process are being applied to coal gasification.

fish by-products. A number of useful materials other than human food obtained from fish: (1) Fish meal is steamed, pressed, and dried processing scrap, used largely in animal and pet foods; it is flammable when stored in bulk. (2) Fish glue is an adhesive derived from the skins of the more common fish; one of its uses is in photographic work, as it can be light-sensitized with ammonium dichromate. (3) Fish-liver oils, chiefly from cod, halibut, and shark, are a valuable source of vitamin A. (4) Oils obtained from whole fish or extracted from body and head cavities have good drying properties; they are used in paints, alkyd resin mixtures; and (when hydrogenated) for cooking and as an industrial substitute for stearic acid. Oils derived from whales (both sperm and blubber oil) and from porpoises are excellent lubricants for precision machinery; in recent years, their use has been discouraged for ecological reasons. *See also* jojoba.

fission. A nuclear reaction brought about by direct impact of a neutron upon the uranium isotope 235 or upon the artificial element plutonium (239). This causes the nucleus to split or divide into two fragments (fission products); for example, U-235 yields radioactive isotopes of barium and krypton. More important is the tremendous energy release, amounting to about 200 million electron volts for each nucleus split, which represents a portion of the binding energy of the nucleus. In addition, fission produces two or three fast neutrons per nucleus, each of which is then able to impinge upon and divide another nucleus. In this way, a progressive and self-sustaining chain reaction occurs. For effective fission, the fast neutrons must be retarded by contact with a moderating substance such as graphite; in this condition, they are called slow or thermal neutrons. Fission reactions can be closely controlled; their most important use is for electric power generation. *See also* chain reaction; binding energy; reactor; mass defect.

fixative. (1) In dyeing technology, either an adhesive substance which holds a dye in contact with a fiber, or a metallic complex (mordant) which chemically combines with the dye and thus binds it to the fiber.

(2) In the perfume industry, a substance that retards the evaporation rate of the volatile com-

ponents and thus stabilizes the odor; a number of natural materials have been used as perfume fixatives (ambergris, musk), but these have now largely been replaced by synthetics.

(3) A quick-drying cellulosic solution applied as a spray to pencil and ink drawings to maintain distinctness of the lines.

(4) A reagent, or mixture of reagents, used to preserve organs and organisms intended for microscopic examination.

fixed. As applied to nitrogen, this term refers to atmospheric nitrogen that has been made available to plants by soil bacteria or has been combined by synthetic processes into useful chemical compounds, especially ammonia. As applied to oils, it refers to vegetable oils of high fatty acid content and low volatility. *See also* nitrogen fixation; fatty acid.

flame retardant. A substance applied to or incorporated in a combustible material to reduce or eliminate its tendency to ignite when exposed to a low-energy flame such as a match or cigarette. There are three methods of application: (1) as a coating or surface finish (non-durable, readily removed); (2) in solution form to penetrate the fibers (semi-durable, reasonably stable); and (3) as an integral part of the polymer structure of a synthetic fiber (durable, not removable). The latter method provides permanent protection, as it not only makes the material self-extinguishing, but cannot be leached out by laundering or drycleaning. Substances commonly used in methods (1) and (2) include such inorganic salts as ammonium sulfamate, zinc borate, and antimony oxychloride; chorinated organic compounds such as chlorendic anhydride; alumina trihyrate; and certain organic phosphates and phosphonates. Major uses of flame-retardant chemicals are in carpets, rugs, upholstery, plastics used in construction, and miscellaneous wearing apparel. Certain types of fibers (polyamides and aramids) are inherently flame-retardant, e.g., nylon.

flammable material. Any substance (solid, liquid, or gas) which will catch fire instantly on exposure to a spark or flame and which will burn rapidly and continuously. Many solids which approach colloidal size in one or more dimensions are dangerous fire hazards (suspended dusts, fibers, films, etc.); others will take fire spontaneously at room temperature (pyrophoric materials). Organic liquids having a flash point less than 27°C (80°F) comprise a broad group of flammable materials. An important factor in the case of solids is the state of subdivision of the material; fine fibers and tiny particles of solids may burn with explosive rapidity under certain conditions due to the large surface area exposed to air. *See also* combustible material; flash point.

flash. A term used to describe several quite different physicochemical operations in the general sense of "extremely rapid" or "almost instantaneous." In flash photolysis, for example, an actual flash of light is used to initiate a reaction; flash distillation and flash evaporation involve very fast conversions of liquid to vapor, taking place in thousandths of a second. Flash pasteurization involves heating the product, e.g., milk, for only 15 seconds at 70°C (158°F). Flash pyrolysis converts waste solids to usable materials, including fuel oil, by applying a temperature of 480°C (900°F) for 30 minutes. *See also* flash point; photolysis.

flash point. The temperature at which an organic liquid evolves a high enough concentration of vapor at or near its surface to form an ignitable mixture with air. Flash points range from far below zero for volatile liquids like ethyl ether to several hundred degrees for heavy, nonvolatile liquids such as vegetable oils, petroleum, etc. A flash point from 27 to 38°C (80 to 100°F) is considered by most authorities to be the dividing line between flammable and combustible liquids. *See also* flammable material.

flat. A term used in the coatings industry to indicate a paint or similar coating material which contains enough coarse solid particles to give a slightly roughened surface when applied; the effect of this is to decrease direct reflectivity by scattering light rays at angles, rather than reflecting them in a direction normal to the surface. The most effective flatting agents are silica, diatomite, and low percentages of heavy-metal soaps.

flatting agent. *See* flat.

flavonoid. *See* pigment.

flavor. (1) Approximately synonymous with taste. Though much research has been expended on analyzing this sense, no correlation between flavor and chemical type or structure has been determined. The basic tastes are bitter, sweet, salty, and sour; beyond this, the problem is physiological and psychological, rather than chemical. Taste, like odor, is a property widely used by chemists for identification purposes.

(2) A substance having a strong characteristic taste, often obtained by extraction from the leaves, twigs, or flowers of plants; many are now made artificially. Among the more common natural flavoring materials are vanillin and chocolate (from beans), oils of wintergreen, peppermint, clove, and cinnamon (from plant leaves or blossoms), oil of bitter almond (from nuts), and citrus flavors (from seeds or rinds of oranges, lemons, etc.). Over 1,500 flavoring materials are known. *See also* potentiator; enhancer.

flaxseed oil. *See* linseed oil.

Fleming, Sir Alexander (1881-1955). A Scottish biochemist and bacteriologist who, in 1928, discovered the bactericidal properties of penicillin produced from the mold *Penicillium notatum*. From this observation, which was somewhat fortuitous, has developed the broad spectrum of antibiotics. He received the Nobel Prize in medicine in 1945. Another outstanding researcher in this field was Selman A. Waksman (1888-1973) of Rutgers University, who was awarded the Nobel Prize in medicine in 1952 for his discovery of streptomycin; he introduced the term antibiotics.

flint. **(1)** A term used in glass technology to denote a type of optical glass characterized by a high refractive index and high light dispersion properties; it normally contains lead oxide, which increases the refractivity. Lead-containing glass has the brilliance characteristic of genuine cut glass.

(2) A naturally occurring hard crystalline silica similar to quartz, used as an abrasive material for spark ignition, in pulverizing devices, and in powder form as an extender in various materials.

floatation. *See* clay; air; classification.

flocculation. A type of aggregation of finely divided solid particles dispersed in a liquid or semiliquid matrix; under certain conditions they tend to clump together to form small clusters or "flocks" similar in appearance to tufts of fiber. This often occurs with colloidal dispersions of carbon black, clays, etc., in rubber and plastic mixes, drilling muds, and paints. The effect can often be reversed by agitation whereas coagulation cannot. The word is derived from the Latin word for tuft of wool. *See also* aggregation; coagulation.

flock. **(1)** A finely divided, low-gravity pulverized material made from vegetable fibers; it is flammable when suspended in air as a dust. Its uses are as a diluent in rubber, plastics, and paint mixtures. Cotton linters can be considered as a type of flock.

(2) *See* flocculation.

flotation. A method used in extractive metallurgy to separate mineral values from gangue in ores. Since the two components of the ore mixture (metals and small rocky particles) are often closely similar in specific gravity, effective separation requires a counter-gravity effect. This is obtained by adding water and wetting agents to the ore and beating or stirring the mixture to incorporate air; the unwanted siliceous material, being hydrophilic, is wet by the water and precipitates, while the mineral particles, which are not wetted, are buoyed up by the foam created by stirring. Thus, they rise to the top, where they can be skimmed or poured off. Foaming agents are often used to facilitate the operation. Ore flotation should not be confused with air floatation, by which clays and other particulates are classified by suspending them in a stream of air. *See also* elutriation.

flow diagram. A graphic or pictorial representation used by chemical engineers to indicate the various successive steps in the production of a chemical. It is sometimes also called a flow sheet, the term *flow* in this instance referring to the progress of the various reactants through the system. Such diagrams also indicate quantities, energy input and output, by-products, waste, etc., as well as efficiencies and other relevant production data. Flow diagrams are essential in setting up a pilot plant, as well as for full-scale operations.

fluid. Any material or substance that changes shape or direction uniformly, in response to a force imposed upon it. The term applies not only to liquids but also to gases and to finely divided solids. Fluids are broadly classified as Newtonian and non-Newtonian depending upon their obedience to the laws of classical mechanics. *Fluid* is derived from the Latin word for *flow,* as are also the related words *flux* and *flue.* The science dealing with the mechanical behavior of fluids is called rheology. *See also* liquid; gas; hydraulic; fluidized bed.

fluidization. *See* fluidized bed.

fluidized bed. A device originally developed for catalytic cracking of petroleum gases; it consists of a container partially filled with finely divided solids through which a uniform stream of hydrocarbon vapors is passed upward through the bottom of the container. The catalyst is incorporated in the bed of solids, and as the gases move uniformly through the solids, they attain a random motion, and the bed tends to expand. Thus, the surface of the bed assumes the appearance of a boiling liquid, which gave rise to the name "fluidization" for this method. The cracking reaction occurs at the surface of the solid particles. This is an example of heterogeneous catalysis. Other applications of fluidization include the oxidation of naphthalene to form phthalic anhydride, the decomposition of limestone to produce lime and carbon dioxide, coal gasification, and the roasting of sulfide ores to produce metallic oxides and sulfur dioxide. *See also* cracking; catalyst.

fluorescence. A kind of luminescence in which the emanation of light continues only as long as the excitation lasts and in which the emission of radiation follows the excitation almost instantaneously. Among inorganic substances having the property of fluorescence are sodium vapor and numerous compounds of zinc, cadmium, magnesium, and calcium with sulfur, tungsten, silicon, boron, and others. Such materials (called phosphors) are utilized in fluorescent lights, television tubes, and similar devices. Organic substances which fluoresce are anthracene, fluorescein, and other fused ring structures, used in fluorescent dyes and as tracer materials. *See also* luminescence; phosphorescence; phosphor.

fluoridation. Incorporation of an extremely low percentage (usually about 1 part per million) of a fluoride such as sodium fluoride in municipal drinking water supplies to prevent or retard the development of tooth decay, especially in children. This is well below the level at which a toxic effect is possible. Fluoridation is gradually increasing in spite of opposition in some areas.

fluoride. Any compound containing fluorine may be designated as a fluoride, though many have alternate names as well; for example, sodium silicofluoride is also known as sodium fluosilicate, and fluoborates are sometimes called borofluorides. Fluorine readily reacts with metals, as well as with non-metals (silicon, hydrogen, carbon, and boron). Inorganic fluorides and many organic fluorides are quite poisonous, and are often extremely corrosive materials, some of the most hazardous being hydrogen fluoride and

uranium hexafluoride (used in the production of U-235). The group of organic fluorides called fluorocarbons have low toxicity and are not flammable. The most unusual fluorides are those recently made by inducing two supposedly inert gases (krypton and xenon) to form compounds with fluorine—a feat which has important possibilities. Several metallic fluorides are used in toothpastes in low concentrations, and sodium fluoride (1 ppm) is used to fluoridate drinking water. *See also* fluorine; fluorocarbon; fluoridation.

fluorine. An element.

Symbol	F	Atomic No.	9
State	Gas	Atomic Wt.	18.9984
Group	VIIA	Valence	1

Isolated in 1886 by the French chemist Henri Moissan (1852-1907), fluorine is the most reactive member of the halogen family. It exhibits strong attraction for electrons. Its outer shell contains seven electrons, which accounts for its tendency to acquire an eighth electron to complete the shell; for this reason, it is extremely electronegative. Its oxidation state is -1. Thus, it combines easily with other elements (except helium, neon, and argon) to form fluorides, fluorocarbons, etc. It is a vigorous oxidizing agent and therefore is a fire hazard in the presence of combustible materials. Fluorine occurs naturally in the ores cryolite (Na_3AlF_6), fluorapatite, and fluorspar (CaF_2) and is also made by electrolytic processes from hydrogen fluoride. It is available commercially as a compressed gas, b.p. $-188°C$ ($-308°F$), and as a liquid. It is used for manufacturing of fluorine-containing chemicals, including refrigerants, propellants, and polymers. Fluorine is a dangerously toxic and corrosive substance and should be handled and stored with great caution. It has the highest bactericidal activity of all the halogens; it is classed as a bone-seeking element, which accounts for its use in drinking water and toothpastes (as fluoride). *See also* fluorspar; fluoride; fluorocarbon; Moissan.

fluorocarbon. Any of a broad group of organic compounds analogous to hydrocarbons, in which all or most of the hydrogen atoms of a hydrocarbon have been replaced by fluorine; some types also contain chlorine, and these are called chlorofluorocarbons. Both aliphatic and aromatic fluorocarbons are in extensive commercial use. The saturated aliphatic type comprises the series of refrigerants developed in the 1930's by Thomas Midgley, Jr., an American chemist (1889–1944); these are characterized by low toxicity and nonflammability and are designated by various trademarks combined with a numbering system, e.g., "Freon" 11 and 12.

The unsaturated aliphatic fluorocarbons such as vinyl fluoride ($H_2C\!\!=\!\!CHF$) and tetrafluoroethylene ($F_2C\!\!=\!\!CF_2$) are the basis of an important series of heat-resistant thermoplastic polymers. The aromatic fluorocarbons, e.g., hexafluorobenzene (C_6F_6), are valuable intermediates from which many useful compounds can be made. *See also* fluoride; fluorocarbon resin; refrigerant; Midgley.

NOTE: Use of chlorofluorocarbons as aerosol propellants has been discontinued because of the possibility that they may deplete the atmospheric ozone layer.

fluorocarbon resin.   Any of a group of organic polymers containing fluorine and having a number of exceptional properties, among which are relatively high heat resistance, permitting them to be used at temperatures of 260°C (500°F) or more; resistance to electricity and to attack by chemicals, oxidation, and severe environmental conditions; they are also nonflammable and nontoxic. Examples are polyvinylidene fluoride, polytetrafluoroethylene, and polychlorotrifluoroethylene. All are thermoplastic and are available in several forms. Their uses include numerous types of high-temperature service (insulation and nonstick coatings for cooking utensils, steam packing, etc.). *See also* fluorocarbon.

fluorspar.   The most important of the three fluorine minerals, the others being cryolite, Na_3AlF_6, and fluorapatite, $CaF_2 \cdot 3Ca_3(PO_4)_2$. It has the formula CaF_2 and thus is a chemical compound; m.p. 1360°C (2480°F). It occurs in the U.S. and Canada and is used as the principal source of fluorine compounds. It also has application as a flux in metallurgy and ceramic technology and as a component of abrasives and special cements. *See also* spar.

flushed color.   A pigment dispersed in oil, varnish, etc., the transfer from the water phase to the oil phase having been effected without drying and subsequent grinding of the dry pigment. Flushed colors are ready for use as prepared.

flux.   The literal meaning of this term is "flow;" thus, it is used by physicists in the sense of a neutron flux or a light flux, referring to the rate of flow or emanation of radiation from a given source. Chemists use it to mean any material or substance that will reduce the melting or softening temperature of another material when added to it; it is not the name of any particular group of substances. Many substances may act as a flux on one material when added to it in small proportions and yet may have their own melting point decreased when the situation is reversed. Fluxing agents are chiefly used in ceramic glazes and in ore smelting.

fly ash.   Fine particulates introduced into the air by burning pulverized coal; they can be recovered within the stack by electrostatic precipitation, as an aid to air purification. The collected ash has value as a cement additive, especially for use in oil-well casings. Fly ash has been tried as a means of combating oil spills on sea water. In Europe, it has been used as a source of the semiconducting element germanium.

Fm   Symbol for the element fermium, named for the Italian physicist Enrico Fermi.

foam.   A liquid or solid in which a gas is more or less uniformly distributed, the entrapped vacuoles ranging in size from colloidal to optically visible. The gases present are usually air or carbon dioxide. Examples of liquid foams are the froth on beer and ale (proteins and fats being excellent foam-formers), soapsuds, fire-extinguishing foams, shaving creams, etc. Solid foams are exemplified by flexible and rigid plastic foams, cellular rubber, and in the food field, by bread and other baked products. Special blowing agents (nitrogen) are used in plastics; baking powders and yeast provide the carbon dioxide for edible solid foams. Metals and ceramics can also be foamed. *See also* blowing agent.

folic acid.   A member of the vitamin B complex, folic acid plays an essential part in biological oxidations, in conjunction with enzymes. Several of its derivatives, e.g., folinic acid, have a similar function. Liver, yeast, and fruit are rich sources of these substances, which are vital to the metabolism of living organisms. The formula of folic acid is $C_{19}H_{19}N_7O_6$, and it is made synthetically. It has medical use in treatment of vitamin deficiency and certain blood diseases. *See also* B complex.

food.   Any substance that acts as energy source for the growth of plants and animals by chemical reaction and assimilation. Water and carbon dioxide are plant foods which are converted into carbohydrate energy sources by photosynthesis. (Water and oxygen are classed as nutrients rather than as foods by some authorities.) Other essential plant foods are nitrogen and phosphorus compounds, from which amino acids and proteins are synthesized. The basic food requirements of animals and man are fats, carbohydrates, proteins, vitamins, and minerals, all of which are obtained by the ingestion of plants or their assimilated forms (meats, milk). The energy stored in these is made available by oxidation, hydrolysis, and similar decomposition reactions. *See also* nutrition; nutrient; digestion (3).

food additive.   Any substance purposely added to a foodstuff or mixture of foodstuffs with the intention of providing a specific property or function. All substances that affect flavor (salt, sweeteners, spices, synthetic or natural flavoring agents), color, and texture modify the properties of taste, appearance, and feeling, respectively; substances that aid in processing and preservation are functional in nature, i.e., they have a specific physicochemical action in the food product. Among this group are leavening agents (baking powders), emulsifiers, antioxidants, bacteriostats, gel-formers, and the like. Additives are not used for nutritional purposes, even though some have nutrient value, e.g., sugar, starches, and fats used in low percentage as sweeteners, thickeners, and shortening, respectively. Vitamins and amino acids added as fortifiers should more properly be considered as dietary supplements. The toxicity of all food additives is closely supervised by the FDA, which periodically conducts tests and issues lists of approved items, called GRAS (Generally Recognized as Safe). No additives that can be proved to induce cancer in test animals are permitted.

Some authorities recognize a group of so-called unintentional additives, which are actually contaminants; these include materials that find their way into foods by accident, such as insecticides, fumigants, packaging materials, etc. *See also* fortification; GRAS.

foots.   A term used chiefly in the soap industry to describe the residual combination of fatty acids and impurities which results from the purification of oils or fats with an alkali. Less specifically, the solids which precipitate from impure oils as sediment or dregs, analogous to the term lees in wine-making.

forensic chemistry.   The legal aspect of chemistry, applying to both civil and criminal cases, as well as to violations involving Federal or state regulations, e.g., transportation of hazardous materials, air and water pollution, food and drug control, etc. Important subdivisions of forensic chemistry are crime detection and the entire body of patent law as it applies to chemical products and processes.

formaldehyde.   An aliphatic aldehyde which is one of the most important industrial chemicals; its officially approved name is methanal, with formula CH_2O. The pure substance is a poisonous gas, b.p. $-21°C$ ($-6°F$), with an obnoxious odor. Formaldehyde is so soluble in water that it is regularly marketed for industrial use in solutions ranging in strength from 37 to 50%. Formaldehyde is made by catalytic oxidation of either methanol or hydrocarbon gases in the vapor phase. It easily polymerizes to form paraformaldehyde (by hydration); polymers of high molecular weight, especially those resulting from reaction with phenol or urea, were among the first commercially made plastics. Another polymer group derived from formaldehyde includes the polyoxymethylene acetal resins.

Formaldehyde is a peculiarly versatile chemical used in high tonnage volume for many types of industrial service. It readily combines with alcohols, ketones, and other aldehydes to form a host of useful products; in organic synthesis, it permits addition of methyl groups to molecules. Pentaerythritol and hexamethylenetetramine are other important derivatives. An important application of recent years is its use in permanent-press and wash-and-wear fabrics.

Formaldehyde and many of its derivatives are quite toxic, and care should be observed in handling them. *See also* aldehyde.

formamide.   *See* amide.

formic acid.   The simplest carboxylic acid with formula HCOOH, m.p. 8.4°C (47°F). The term is derived from the Latin word for ant *(formica,)* since this acid has been found to be produced by these insects, as well as by bees. It is produced industrially from sodium formate and sulfuric acid, the sodium formate being the result of reaction of carbon monoxide and caustic soda. Formic acid is a poisonous liquid which is extremely irritating to the skin and eyes and can be a strong allergen. It is used in textile dyeing, leather tanning, electroplating, and in pesticide manufacture. It is also a reducing agent.

formula, chemical. The chemical constitution of a substance indicated by the use of accepted written symbols. Formulas may be of several kinds, some indicating only the chemical makeup of a substance and others its physical structure. Other types are general in nature showing the constitution of a class of chemical compounds (generic formula). Various graphic devices are used to indicate three-dimensional configurations. *See also* empirical formula; configurational formula.

formula, product. A list of the ingredients and their amounts or percentages required in an industrial product. Such formulas (or recipes) are mixtures, not compounds; they are generally used in such industries as adhesives, food, paint, rubber, and plastics. Selection of ingredients to provide optimum specific properties for a given end use or for specification requirements is made by experienced technologists.

fortification. A term used in the food industry to describe the addition of essential nutritional factors (vitamins, minerals, proteins, etc.) to certain food products to replace losses incurred in the processing of raw materials, and to meet standard nutritional requirements. It is practiced especially with milled wheat flours, prepared mixes, meat and fish by-products, margarine, and infant foods. In cases where the food is baked after fortification, a large percentage of the vitamins so added may be inactivated by heat. *See also* food additive.

fossil. Derived from the Latin word for *dig,* this term is used in chemistry to indicate that a material is composed of geologically ancient animal or vegetable matter which has lain underground from a depth of a few feet to a mile or more. Thus, fossil fuels include coal (vegetable) and petroleum (animal); there are also fossil waxes (ozocerite), fossil resins (amber), and silicated (petrified) woods. Peat is a semifossilized carbonaceous material in the process of becoming lignite, or brown coal.

foundry sand. (1) A mixture of silica and a binding material, used in metal-forming molds in the dry-sand casting operation; the binder may

be sodium silicate, casein, or a thermosetting resin. The mixture is heated to form a hard shell for the molding of units having thin or intricate shapes.

(2) Sand used in a metal casting operation without a binder, often called greensand. It is used for molding objects of simple design, without thin projections.

fourdrinier. The best-known and most widely used papermaking machine; it was the first straight-line continuous production assembly for any major industrial material. It was patented in England in 1807 by two brothers of that name and was first operated in the U.S. in 1827. A typical fourdrinier is over 100 feet long and 6 feet wide, delivering finished sheet (web) up to 72 inches wide at high speed. Its essential components are: (1) the head box containing the "furnish," a dilute suspension of pulp in water; (2) the moving "wire" or metal screen onto which the furnish is flowed uniformly from the head box and on which the sheet is formed within seconds by filtration; (3) a series of pressure and suction rollers which remove additional water, the semi-dried web being carried on felt pads as it moves through the system; (4) the drying section, comprised of a double bank of rotating steam-heated drums over which the sheet passes; and (5) a four- or five-roll calender which irons and polishes the sheet. The entire process reduces the water content of the product from about 95% to 5%. Continuous operation at high speed for weeks at a time is possible with web thicknesses ranging from tissue to moderately heavy paperboard. *See also* paper.

Fr Symbol for the element francium, named for the native country of its discoverer, Marguerite Perey (1910–1975).

fraction. Any homogeneous component of a mixture which can be separated in relatively pure form by a number of operations, e.g., chromatography and distillation (for liquids) and electrophoresis, filtration, centrifugation, and crystallization for solids. One of the most widely used types of separation into fractions is distillation, especially of petroleum; the many components of petroleum differ widely in boiling point, which makes it possible to boil off and condense each in turn, thus obtaining naphthas, straight-run gasoline, kerosine, and fuel oils of various grades. The operation is called fractionation. *See also* distillation; filtration; rectification.

fragrance.   Any substance having a pleasant smell; it is virtually synonymous with "perfume," but may be considered to include odoriferous materials that are not used specifically for perfumery but for masking unpleasant odors or for imparting a "natural" smell to such items as leather substitutes and other imitative products. Examples are pine oil and balsam. *See also* perfume; deodorant; odor.

francium.   An element.

Symbol	Fr	Atomic Wt.	223(?)
State	Solid	Atomic No.	87
Group	IA	Valence	1

Francium is a radioactive metal having no stable form. Its existence as a decay product of actinium was discovered in 1939 by the French scientist Marguerite Perey (1910–1975). Its longest-lived isotope (223) is the only one that occurs naturally, with a half-life of 21 minutes. It forms no compounds. Francium is the heaviest of the alkali metal group.

Frasch process.   *See* sulfur.

free energy.   *See* energy.

free radical.   A reactive portion of a stable molecule occurring as a result of rupture of one or more single bonds as by thermal decomposition or ionizing radiation. For example, several free radicals can be formed by breaking the carbon-to-hydrogen linkages in methane (CH_4), namely CH_3, CH_2, and CH. All such radicals have one or more free electrons, which may be either paired or unpaired. Because of their extreme reactivity, the life of a free radical is very short, on the order of 0.1 second. The original evidence for the existence of free radicals and the elucidation of the chain mechanisms by which they initiate such reactions as oxidation, polymerization, etc., was obtained by photochemists about 50 years ago. *See also* chain reaction (**1**).

freeze-drying.   A unique method of dehydrating solids or solid/liquid suspensions of organic materials or mixtures. It is especially valuable in the pharmaceutical and medical fields for preservation of vaccines, blood plasma, and similar materials which are readily degraded even by normal room temperature. The method has also been applied to some food products, notably coffee. The material is frozen under vacuum and kept at between −20 and −40°C (−4 and −40°F); at this temperature the ice sublimes, removing about 99% of the water. *See also* drying; dehydration.

freezing point.   *See* melting point.

Friedel-Crafts reaction.   A type of reaction involving anhydrous aluminum chloride and similar metallic halides as catalysts, originally studied by Charles Friedel, a French chemist (1832–1899), and James Mason Crafts, an American chemist (1830–1917), in 1887-8 during joint research in France; it has been developed since then for many important industrial uses, exemplified by the condensation of ethyl chloride and benzene to form ethylbenzene and the manufacture of acetophenone from acetyl chloride and benzene.

frit.   Fragmented or finely ground glass particles used as components of ceramic glazes, porcelain enamels, etc. It is sometimes mixed with fluxes and colorants. Frits used as components of glazes are usually made by special methods to make possible the use of water-soluble substances (alkalies and borates) in the glaze.

FRP.   Abbreviation for fiberglass-reinforced plastic. *See* reinforced plastic.

fructose.   A simple levorotatory sugar having the formula $C_6H_{12}O_6$. It occurs naturally in honey and can be derived from the Jerusalem artichoke; also made artificially by hydrolysis of the reserve polysaccharides in the bulbs of various plants, or from the so-called invert sugars obtained by hydrolysis of sucrose. Though it is sweeter than other natural sugars, it does not require insulin for metabolism within the body and is thus a useful diabetic food. *See also* dextrose; sugar; sucrose.

fuel.   Any substance that evolves heat as a result of combustion can be used as a fuel. The most common materials used for heating and electric power production are the so-called fossil fuels (coal, petroleum, and the gases or liquids derived from them). Economic feasibility is usually the critical factor in selection of a fuel for a given purpose. Many organic waste products are used as fuels in locations where they are a disposal problem, e.g., sugarcane waste, corn shucks, and other farm wastes. Rocket fuels are a special group, requiring nearly instantaneous combustion and high temperatures; among them are hydrogen peroxide, hydrazine, and boron hydride. Quite distinct from combustible fuels are the nuclear fuels whose energy is derived from controlled disintegration of atomic nuclei; they are used exclusively in electric power reactors. *See also* fuel oil; boron hydride; combustible material; fission.

fuel cell.   A primary power-generating device based on the production of an electric current by the oxidation of a gaseous fuel, usually hydrogen. An electrochemical reaction takes place at the electrodes and will continue as long as fuel is supplied. An electrolyte is necessary to act as a conducting medium. The electrodes are of porous carbon to permit diffusion of the gases, e.g., hydrogen and oxygen, into the cell. The oxidation of hydrogen to water at the cathode yields electrons, which pass through an external circuit and reenter the cell at the anode; the water evaporates. Fuel cells of many types have been developed, chiefly for use in spacecraft; their efficiency is over twice that of a steam engine, and they can operate indefinitely. The first oxygen-hydrogen fuel cell was constructed in 1839 by the English physicist, William Robert Grove (1811–1896), who in 1854 proposed the idea that a fuel cell be used for the direct conversion of the heat of combustion of coal into electrical energy. *See also* battery.

fuel element.   *See* reactor.

fuel oil.   Liquid petroleum fractions or residual refinery products used for heating and power generation. There are several grades designated by numbers. No. 1 is a relatively light distillate used for vaporizing burners; No. 2, called diesel oil, is used for domestic heating and as a fuel for trucks, locomotives, and some automobiles; No. 4 is a light residual grade used for heating of large buildings such as churches, hospitals, etc.; Nos. 5 and 6 are heavier residual types called bunker oils, which must be preheated before use. All grades have flash points above 40°C (104°F), which places them above the generally accepted range of flammable liquids. *See also* fuel; petroleum.

fugitive.   A term used in chemistry and chemical technology to denote an ingredient which tends to be partially deactivated during or immediately after processing. This occurs with certain organic dyes, rubber accelerators, softeners, etc., and may be due to adsorption, neutralization, heat, or other physicochemical action.

fuller's earth.   A finely divided clay-like material capable of adsorbing colors and odors. It is "mined" in the southeastern U.S. and in eastern Canada. Its uses are similar to those of other types of clays: to remove impurities from various liquids, as a filler in rubber and plastic mixes, and in oil-well drilling fluids. It is also an excellent filter aid. The name originated from its early use in removing natural oils from wool and other textiles (fulling). *See also* clay; bentonite; diatomaceous earth; earth.

fumaric acid.   *See cis-.*

fume.   Derived from the Latin word for *smoke,* this term has several distinct, though related meanings: (a) the vapors given off by strong inorganic liquids or by evaporation of organic solvents; (b) the combination of smoke and gases resulting from incomplete combustion, as from a gasoline engine (exhaust fumes); (c) the submicroscopic particles emitted by heated metals or metallic oxides (usually in the singular, as metal fume). *See also* smoke; fuming.

fumigant.   A pesticide or bactericide applied either indoors or in a closely restricted locality in the form of a vapor or gas. Common fumigants are para-dichlorobenzene (fabrics), sulfur dioxide (grains), and hydrogen cyanide (tree crops). Phenol and naphthalene, as well as sulfur compounds, are used for disinfection of sickrooms. *See also* pesticide; insecticide; disinfectant.

fuming.   A term used in inorganic chemistry to describe the property characteristic of highly active liquids which evolve visible smoke-like emanations on contact with air. Most familiar are the forms of nitric and sulfuric acids designated as "fuming." These are not pure concentrated acids; low percentages of nitrogen dioxide and water are present in fuming nitric acid, and fuming sulfuric acid contains sulfur trioxide. Hydrofluoric acid (a mixture of hydrogen fluoride and water) also fumes. Pure compounds in which fuming occurs are fluosilicic acid and hydrazine.

fundamental particle.   Ultimate units occurring in the atom or formed by impact of high-energy particles from an outside source. For the chemist, the most important fundamental particles are the proton, the neutron, the electron, and the photon; the last of these has no mass at all when at rest and is the basic unit of radiation. Protons have a mass of 1 and a positive charge; neutrons have a mass of approximately 1, with no charge; and electrons have a vanishingly small mass and a negative charge. Besides these, there are mesons of various designations, neutrinos, positrons, etc., which are of more direct concern to the physicist than to the chemist. *See also* proton; electron; neutron; photon; particle.

fungicide.   Any substance that kills or inhibits the growth of spores and fungi. The latter are plantlike organisms which do not form chlorophyll. Older fungicidal agents for agricultural purposes included sulfur, lime, and their combinations; inorganic copper compounds such as Bordeaux mixture and copper oxychloride, the latter widely used to control fungi in grapevines; and formaldehyde for grain smuts. Hypochlorite solutions have long been used as antifungal agents in swimming pools and in water-cooled heat-exchangers, as well as in locker-room showers to prevent athlete's foot. A number of dithiocarbamate fungicides were introduced in 1940, among them ferbam, ziram, nabam, and quinones such as dichlorobenzoquinone. Mercury compounds have not been used since 1971 because of their toxicity to humans. Various metallic naphthenates have been used to impregnate tent fabrics and clothing for the military. Fungus infections in humans are a complicated and serious medical problem as many types are extremely persistent and require sophisticated medical treatment. *See also* fungus; herbicide.

fungus.   A broad group of plantlike organisms which reproduce by means of spores and which do not form chlorophyll. Authorities are not agreed as to whether fungi should be considered plants or unicellular bacteria; they include molds, yeasts, smuts, rusts, mushrooms, etc., and often act as parasites on living organisms. Thus, they are responsible for many infective diseases of plants, animals, and man, e.g., athlete's foot and similar skin infestations. They also cause spoilage of stored grains and other food products and the deterioration of such cellulosic products as paper, clothing, tent fabrics, cordage, etc., in the hot, damp environment of the tropics. Soil sterilants and fungicides of various types are used to combat them in agriculture. *See also* fungicide; yeast; mold.

Funk, Casimir   (1884–1948). Born in Poland and later an American citizen, Funk, in 1911, isolated a food factor extracted from rice hulls, which he found to be a cure for a disease due to malnutrition (beriberi). Believing this to be an amine compound essential to life, he coined the name "vitamine," from which the final *e* was later dropped. The various types and functions of vitamins were not differentiated till some years later as a result of the work of Elmer V. McCollum (1879–1967), Albert Szent-Gyorgyi (1893-), Nobel Prize 1937, Roger J. Williams, (b. 1893), and others. Funk's discovery should rank high among the great biochemical advances of the early 20th century.

furan resin.   A thermosetting polymer made from furfuryl alcohol in two stages: it is first partially polymerized to a viscous liquid, to which such fillers as asbestos, wood fibers, etc., are added; it is then further polymerized to a hard solid form at the site of its use by adding an inorganic acid which acts as a catalyst. The chief characteristic of furan resins (besides their relatively low cost) is their stability toward chemicals such as acids, solvents, and the like. Thus, they are widely used for equipment housings, vat linings, and molding forms.

furfural.   A heterocyclic aldehyde, b.p. 162°C (324°F), which is the parent substance of a group of compounds including furan, furfuryl alcohol, hydrofuramide, furoic acid, etc., all of which are characterized by an unsaturated five-membered ring in the form of a pentagon, in which oxygen occupies the apex position, as follows:

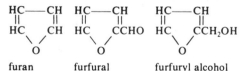

furan   furfural   furfuryl alcohol

Furfural is made by distillation of bran, corn cobs, wood, and similar agricultural waste products containing pentosan sugars, in the presence of sulfuric acid. Besides its use as a source of the derivatives mentioned above, it is a solvent in nitrocellulose lacquers, in the refining of petroleum products, and is also a replacement for formaldehyde in the leather industry and as a fungicide. A group of polymers known as furan resins is derived from it. It is quite toxic and a strong skin and eye irritant. *See also* furan resin; furfuryl alcohol; heterocyclic.

furfuryl alcohol.   A monohydric heterocyclic alcohol, b.p. 170°C (338°F), in which an oxygen atom appears in the unsaturated five-membered ring; the formula is $C_4H_3OCH_2OH$. It can be made by reacting furfural with sodium hydroxide without a catalyst, or directly from furfural by catalytic hydrogenation. It readily polymerizes in the presence of acids to form furan resins, which is its largest commercial use; it also has application as a dispersing agent and solvent for dyes, as an ingredient of adhesives and sealants, and as a source of tetrahydrofurfuryl alcohol by hydrogenation. *See also* furfural.

furnace black. *See* carbon black.

fuse. The primary meaning of this term as used in chemistry is to melt or soften from a solid to a liquid or semiliquid state; closely associated with this meaning is the idea of joining or uniting. The word has a number of derived forms in chemical literature, each with a rather specific meaning, as follows: (1) two metals may be *fused* together, i.e., joined by melting, as in welding; (2) *fusible* alloys are low-melting alloys which fuse (melt) on slight heating; (3) *fused* salts are melted or molten salts; (4) nuclear *fusion* is the union of light atomic nuclei to form heavier nuclei at high temperature; (5) *fused* rings occur in many cyclic structures, where two or more rings have a side in common. (In the last two the joining concept has displaced the melting concept.)

Nonchemical meanings are: (1) a strip of conductive metal that melts when subjected to too high a voltage, and (2) a cord or strip of flammable material used to detonate explosives.

fused salt. A melted (liquefied) halide or nitrate salt of a metal such as sodium, potassium, barium, or calcium; these are useful for electrolytic processes because of their high conductivity. They are also used for the heat treatment of metals at high temperatures and as coolants for nuclear reactors. Their heat-transfer capacity is similar to that of water. They are also called molten salts.

fused silica. *See* quartz.

fusel oil. A mixture of alcohols existing as impurities in ethyl alcohol derived by fermentation (yeast enzymes) of molasses and other materials rich in sugar; after separation from ethyl alcohol, its major components are various amyl alcohols (except normal primary), propyl alcohol, and low percentages of other alcohols and esters. So-called refined amyl alcohol is produced from fusel oil by distillation. It is used as a solvent and for the manufacture of various ethers and esters. Fusel oil is rather toxic and combustible.

fusible alloy. Any of a number of mixtures of metals such as bismuth, tin, lead, cadmium, etc., which melt at a relatively low temperature [260°C (500°F) or less]. The best known is Wood's metal, used as heads on automatic fire-extinguishing water lines, for vent control on tanks for compressed gases, and for similar uses. *See also* eutectic.

fusion. (1) A thermonuclear reaction in which positively charged hydrogen nuclei (protons) unite or fuse to form helium, with evolution of energy. A temperature of several million degrees is necessary to overcome the repulsion of the positive charges. This process occurs in the sun, where it is catalyzed by carbon. Uncontrolled fusion has been achieved in the hydrogen bomb, the initiating temperature being supplied by a fission reaction. Research is now being devoted to developing a controlled and sustained fusion reaction, using the hydrogen isotopes deuterium and tritium. The most efficient experimental reaction is, for each fusion event, $D + T + e \rightarrow {}^4He + n + 17.5$ MeV, where e represents at least 44 million °C.

Fusion has two great advantages over fission as an energy source: (1) it utilizes water and readily available lithium as its raw materials instead of scarce and costly uranium; (2) it produces no long-lived radioactive fission products, though it does yield high-energy neutrons. Large-scale development should not be expected for the indefinite future.

(2) Conversion of a solid substance to its liquid state, commonly called melting. *See also* fuse; heat of fusion.

G

g **(1)** Abbreviation of gram.

(2) (italic) Acceleration due to gravity.

Ga Symbol for the element gallium, the name being derived from the Latin, *Gallia,* the original name for the area that later became France.

gadolinium. An element

Symbol	Gd	Atomic No.	64	
State	Solid	Atomic Wt.	157.25	
Group	IIIB	Valence	3	
	(Lan-thanide Series)	Isotopes	7 stable	

Gadolinium is a member of the rare earth family of metals, occurring in low percentages in monazite and bastnasite ores. It is a heavy crystalline metal, m.p. 1312°C (2394°F), not unlike steel in appearance. It has the highest absorption capacity for neutrons of any metal, with a capture cross-section of 45,000 barns; thus, one of its major uses is in nuclear shielding and reactor control. It is also used in laser garnets as a substitute for yttrium. The selenide is used for thermoelectric power generation. The metal is flammable in dust or powder form.

galactomannan. *See* guar; mannan.

galactose. A member of the monosaccharide family of sugars having the formula $C_6H_{12}O_6$. It is a constituent of lactose, as well as of a number of polysaccharides, such as agar, gum arabic, carrageenan, and pectins. It is readily absorbed in metabolism and is converted into glucose by the liver. *See also* sugar; carbohydrate.

galacturonic acid. A dextrorotatory carboxylic acid found principally in the pectin group of compounds occurring in plants. The molecules are of large (colloidal) dimensions, most of the carboxyl groups being esterified with methyl groups to form so-called methoxyl units ($COOCH_3$). The pectin structure is composed of chains of such molecules, which provide the gelling property characteristic of these substances. *See also* pectin.

galena. *See* glance; lead.

gallium. An element.

Symbol	Ga	Atomic Wt.	69.72
State	Liquid	Valence	2,3
	(Solid)	Isotopes	2 stable
Group	IIIA		
Atomic No.	31		

Gallium has a melting point of 29.8°C (85°F), below which it is a solid resembling zinc in appearance; the liquid form looks much like mercury. Thus, gallium is one of only three metals that are liquid at this temperature. Gallium metal has no industrial uses, but several of its compounds (arsenide, antinomide, and phosphide) have semiconducting properties and are used in electric and solid state equipment (lasers, microwave generators, and electroluminescent devices).

gallium arsenide. *See* arsenic; gallium.

galvanizing. Application of a protective layer of zinc to a metal, chiefly steel, to prevent or inhibit corrosion. The length and extent of the protection are dependent upon the weight or thickness of the zinc coating. It is applied by (a) the hot-dipping process by passing the metal through molten zinc, or by (b) electrodeposition of zinc by a plating process, usually onto strip or wire. The former method gives a unique mottling or "spangle" pattern on the surface, formed as the zinc crystallizes, while the latter results in either a white matte surface or one which is bright and reflective. Galvanized metal is used chiefly on inexpensive products for utilitarian service (pipe, nails, wire, containers, etc.) and represents one of the largest uses of zinc. The process derives its name from Luigi Galvani, an Italian scientist (1737–1798). *See also* protective coating; sacrificial; zinc.

gamma   (γ). A term or symbol with meanings analogous to those of alpha, i.e., to designate locations of substituents in a compound, or a particular form or modification of a protein (gamma globulin), or metal crystal. It also identifies the most intense form of short-wave radiation released by radioactive atomic nuclei and by fission. *See also* alpha; beta; gamma radiation.

gamma radiation.   A form of highly energetic radiation emanating from unstable atomic nuclei and characterized by ultrashort wavelengths (about 1 angstrom unit). It is emitted slowly by decaying radioactive nuclei (radium) and very rapidly in the fission process. Gamma radiation is highly dangerous to body tissues and may cause genetic damage or even death on prolonged exposure. Cobalt-60 has been used in industry as a source of gamma rays for initiating the synthesis of ethyl bromide and the cross-linking of polyethylene and other high polymers. Gamma radiation is used for spectroscopic analysis and in medicine (cancer treatment). *See also* ionizing radiation; spectroscopy.

gangue.   A term used in extractive metallurgy to refer to the waste rocky or siliceous content of metal ores which is separated by flotation or elutriation. It is also a component of slag, which separates from the ore on smelting. Ore residues are sometimes called tailings; the latter term is also used in other industries for any residual processing waste.

garnet.   The natural form (gemstone) is a silicate compound containing iron and aluminum, $Fe_3Al_2(SiO_4)_3$. Synthetic garnets contain iron oxide and one or more rare-earth oxides which have specific magnetic properties (ferromagnetism), making them suitable for use in microwave equipment, lasers, and similar devices. Such garnets, or garnet-ferrites, are structurally analogous to natural silicate garnets; their unique magnetic function was discovered in 1956. They often contain yttrium and aluminum oxides. *See also* ferrite; ruby; sapphire.

gas.   A state or phase of matter in which the molecules are so widely dispersed under standard conditions of temperature and pressure that they cannot exert sufficient mutual attraction to form a denser (liquid) phase. A vapor is the gaseous phase of a substance that is normally a liquid. Gases behave like fluids; any gas can be liquefied when subjected to sufficient pressure and low temperature. The molecules of a gas are in continuous motion, exerting pressure equally in all directions as a result of their kinetic energy. The volume of a confined gas is inversely proportional to the pressure applied to it if the temperature remains constant (Boyle's Law). A perfect or ideal gas is one that conforms exactly to this and other gas laws (although no gas behaves in this ideal manner).

Gases may be composed of (1) monatomic molecules of the noble gases, e.g., helium; (2) diatomic molecules of hydrogen (H_2), oxygen (O_2), nitrogen (N_2), fluorine (F_2), and chlorine (Cl_2); (3) molecules of compounds, such as hydrogen sulfide (H_2S), carbon dioxide (CO_2), carbon monoxide (CO), nitrous oxide (N_2O), ammonia (NH_3), methane (CH_4), acetylene (C_2H_2), and some other hydrocarbons; (4) mixtures of these, such as air, natural gas, water gas, etc. Not all gases are light; carbon dioxide and chlorine are much heavier than air; nonetheless, it is a characteristic of gases to diffuse into each other to form uniform mixtures, regardless of their weight. Equal volumes of the same or different gases contain the same number of molecules (Avogadro's Law).

NOTE: Use of the word *gas* in the sense of gasoline or domestic fuel is improper in technical communication, though acceptable in everyday speech.

See also liquid; fluid; Avogadro number; vapor pressure.

gaseous diffusion process.   *See* diffusion; uranium.

gas hydrate.   A type of hydrate in which a molecule of a gas is enclosed (without chemical bonding) in a multisided "cage" of water molecules; this results in an insoluble crystalline clathrate or inclusion compound in which the combining ratio of gas to water molecules varies from about 1 to 18 for propane to as low as 1 to 6 for some other gases. Gas hydrates form in pipelines, causing considerable operating difficulty. The structure has been found useful in separating fresh water from brine; a salt-free cage compound is obtained by use of propane-seawater mixtures. Many gases can form hydrates of this kind, including the noble gases, fluorocarbons, acetylene, carbon dioxide, cyclopropane, ethyl chloride, and ethane. *See also* hydration; clathrate.

gasification. Production of gaseous or liquid hydrocarbon fuels from coal (1) by direct addition of hydrogen to form methane (hydrogasification); (2) by reacting steam with hot coal (800°C) in the presence of air or oxygen to form carbon monoxide and hydrogen (synthesis gas), followed by a methanation reaction; (3) by underground combustion. In methods (1) and (2), the coal enters the reaction sequence in finely divided form; hydrocarbon gases, naphtha, and fuel oils of various grades can be produced by several modifications, e.g., the Fischer-Tropsch process. In method (3), air is pumped down to previously ignited coal seams and the combustion products evolved are collected.

Many variations of methods (1) and (2) have been researched in the U.S. since 1945, but none was economically competitive with the then low-priced crude oil. Several have advanced to the pilot stage, and projections of cost, investment, and installation for large-scale operation have been made. Environmental and geographic considerations have also been investigated. The controlling factor for future development is the price of crude oil.

All three methods mentioned above are technically feasible. During World War II, hydrocarbon fuels were produced from coal on large scale in Germany by both hydrogasification and Fischer-Tropsch methods. More recently, coal conversion has been successfully achieved in Scotland and elsewhere by the Lurgi process and in South Africa by the SASOL method, both of which are based on the Fischer-Tropsch reaction. *See also* Fischer-Tropsch process.

gas laws. *See* Boyle's Law; Gay-Lussac's Law; Charles' Law.

gasohol. A fuel, chiefly for internal combustion engines, composed of a 9/1 ratio by volume of unleaded gasoline and water-free ethyl alcohol.

Its octane number is essentially equivalent to that of gasoline. The alcohol is derived from fermentation of the sugar and starch in some agricultural crops, such as corn, thus substituting a renewable raw material for increasingly scarce petroleum. Studies have shown that, at present, the energy consumed in raising and harvesting of the crops, plus that needed for fermentation, distillation and drying of the ethyl alcohol, is somewhat greater than the energy value of the alcohol. This difference can be supplied by energy from coal and biomass. Ethyl alcohol from other raw material sources, such as glucose derived from the enzymatic decomposition of cellulose, also can be used to make gasohol. Increasing amounts of gasohol are becoming available, and this trend is expected to increase as petroleum products become more scarce and expensive.

gasoline. A refined mixture of petroleum hydrocarbons having an octane number ranging from about 60 to 100 or more, used primarily as a fuel for automobiles and airplanes. Its vapors are flammable, and its partial combustion products toxic in high concentrations. Most high-octane gasoline is made by catalytic cracking, platinum being one of the most effective catalysts. Branched-chain paraffins (isooctane), cycloparaffins, and aromatic compounds predominate. Straight-run gasoline is obtained by direct distillation of petroleum, and casinghead (natural) gasoline from various hydrocarbon gases present in natural gas; these two types are not cracked and have low octane number. They are used for blended gasolines and low-power fuels for motor boats, etc. They can be upgraded by catalytic reforming. Antiknock gasoline of adequate otcane number is now in general use without addition of tetraethyllead, which is being gradually replaced by methyl tert-butyl ether. Control of toxic emission products is under active study. *See also* cracking; reforming; octane number; knock.

gastric juice. Also referred to as digestive secretion, gastric juice is a mixture of hydrochloric acid and the enzyme pepsin produced in the stomach by a nerve reaction of the conditioned reflex type. Both the acid and the enzyme are secreted by the glands of the stomach; it is reported that as much as 700 cc of gastric juice is made available for digestion during one meal. The pH of gastric secretion is close to 2.0 (the value for 0.01N hydrochloric acid) or nearly 1000 times as acidic as urine. Secretion of gastric juice is the essential step in the metabolic breakdown of foods in the digestive process, though carbohydrate decomposition begins with the saliva. *See also* digestion (3); pepsin.

Gay-Lussac, Joseph Louis (1778–1850). A French chemist and physicist noted for the brilliance and accuracy of his reasoning and experimental work. He contributed greatly to the knowledge of gases in his discovery (1808) of the law of combining volumes and his independent discovery (1802) of the law of Charles, the relationship of temperature to the volume of gases. It can be stated as follows: the volume of a confined gas at constant pressure is proportional to its absolute temperature. Gay-Lussac also determined the fact that the volumes of gases involved in a chemical change are always in the ratio of small whole numbers. Avogadro accepted this law, and from it the famous Avogadro hypothesis evolved.

The work of Gay-Lussac in chemistry was extensive, resulting in the discovery, with Louis-Jacques Thenard, of boron, which he named, and a variety of compounds, such as boron trifluoride, chloric acid, and dithionic acid ($H_2S_2O_6$). He identified iodine as an element, named it, and studied its properties. He investigated the relationship of acids and bases and introduced many analytical techniques, such as the use of litmus as an indicator. Among his many contributions to industrial chemistry were improvements in the production of sulfuric acid. Much of the progress of chemistry in the early 19th century is associated with his career.

Gd Symbol for the element gadolinium, named for J. Gadolin, an 18th-century Finnish chemist.

Ge Symbol for the element germanium, named from the ore germanite in which it was discovered, which in turn was undoubtedly named for Germany.

gel. A term used in colloid chemistry to denote a structure formed by substances of high molecular weight, in which the macromolecules consist of long chains or fibers. Substances which produce gels readily are common soap, certain proteins (collagen), and carbohydrates such as starches, complex sugars, and related compounds (pectin). They absorb many times their own weight of water, with consequent swelling and set to a viscous or semisolid jelly (gel) when

cool. The macromolecules become entangled with one another as their motion slows with decreasing temperature, water being occluded in the interlocked network so formed; when warmed, such gels resume their liquid form and are thus called reversible. Some heavy-metal soaps have a similar tendency. Gels such as dextran can be used as sieves to separate molecules of other substances, an operation called gel filtration. *See also* gelatin; agar.

gelatin. A purified form of collagen used as a gel-forming agent and protective colloid in ice cream, desserts, confectionery, and jellied meat products; it also has major use as a coating for photographic film and pharmaceutical capsules. It is strongly hydrophilic, absorbing many times its weight of water; it is composed of amino acids and various degradation products of collagen. A white, amorphous solid when dry, it becomes liquid when mixed with water and warmed; as it cools, it forms a gel which has excellent film-forming properties. The term is also used to refer to any material of similar nature, for example, gelatin dynamite (nitroglycerin with a low percentage of nitrocellulose). *See also* collagen; gel; agar.

gem-. A prefix used in naming organic compounds; derived from the Latin word for *twins (gemini)*, it indicates a carbon atom to which two identical groups are bonded. It is not of common occurrence.

gen-. A Latin root having several related meanings, as follows: (1) "to be born" or "to create," as in a number of chemical terms where it appears as a suffix with the meaning "maker," e.g., hydrogen (water maker), oxygen (acid maker), glycogen (sweet maker), and halogen (salt maker); antigen and carcinogen are other examples of this meaning; (2) "race" or "kind" (in the sense of family relationship), as in gene, genus, generic, generation, homogeneous; (3) "origin" or "beginning," as in genesis, genius, regenerate.

gene. A protein-nucleic acid complex present in the chromosomes of all organisms and having the ability to direct the transmission of hereditary characteristics from one generation to another. This is accomplished through the agency of deoxyribonucleic acid (DNA), which programs the sequence of amino acids in protein synthesis. *See also* nucleic acid; genetic code; deoxyribonucleic acid.

gene splicing. *See* recombinant DNA.

genetic code. A natural computer code provided in the genes by deoxyribonucleic acid and activated by ribonucleic acid, two extremely complex substances that are present in the nuclei of all living cells. The code programs not only the specific amino acids required in a protein, but also their sequence in the molecular chain of which the protein is composed, some of which contain several hundred amino acid units. In this way, the code determines the type of cells necessary for the many functional structures of a developing organism (eyes, blood, hair, brain, etc.) and also the transmission of hereditary traits. *See also* nucleic acid; deoxyribonucleic acid; ribonucleic acid (RNA); gene.

Geneva. A system of naming organic compounds proposed in 1892, which has received considerable acceptance among chemists. It was amended in 1930 by the International Union of Pure and Applied Chemistry. Among the changes introduced by the Geneva system were the naming of alcohols, e.g., methanol for methyl alcohol; use of the suffix *-ene* to characterize aromatic and unsaturated aliphatic hydrocarbons; and substitution of alkane for paraffin, alkene for olefin, alkyne for acetylenic hydrocarbons, and thiol for mercaptan. Some of the reforms proposed have never been accepted, e.g., phene for benzene and hydrol for water, regardless of their logical correctness. *See also* IUPAC; nomenclature.

geochemistry. A branch of chemistry dealing with the chemical composition and distribution of elements in the earth's crust, including oceans and atmosphere. It is based largely on the detailed analysis of minerals and on the application of the chemistry of crystals to mineral formation. Industrially, one of its most valuable contributions has been to petroleum exploration. It has also shed important light on the abundance of elements and the age of the earth. Geochemistry is directly involved with study of the moon's surface and elucidation of the information gained from rocks obtained on the moon.

geometric isomer. A type of stereoisomer in which a compound containing one or more double bonds can appear in two different spatial configurations, though they are otherwise chemically identical, i.e., they are *cis-trans* isomers, differing only in the location of one or more substituent groups in relation to the carbon chain.

Such isomerism occurs in many natural products, for example, natural rubber and gutta percha, where the methyl group occurs on opposite sides of the carbon chain. This can be illustrated for the compound butene-2 ($CH_3HC=CHCH_3$): in the *cis* form, the two hydrogens are on one side of the carbon chain and the two methyl groups on the other; in the *trans* form, the hydrogens and methyl groups are diagonally opposite each other, i.e., a methyl group has exchanged places with a hydrogen atom.

$$H-C-CH_3 \qquad CH_3-C-H$$
$$\parallel \qquad\qquad\quad \parallel$$
$$H-C-CH_3 \qquad H-C-CH_3$$
$$cis \qquad\qquad\quad trans$$

See also isomer; stereoisomer; *cis-*.

geothermal energy. Superheated water and steam trapped in rock strata in areas characterized by volcanic activity or by intrusions of molten magma. Its temperatures range from 150 to 300°C. It escapes either from natural surface vents (geysers, fumaroles, hot springs) or from boreholes drilled through the strata. A contributing source of heat is the natural radioactivity of rocks in the earth's upper mantle. The only geothermal power plant in the country, located at The Geysers in central California, produces about 400,000 kilowatts from steam at 177°C and 100 psi. Geological formations appropriate for geothermal heat are so rare that this form of energy will never be more than a minor factor in U.S. resources.

geraniol. An unsaturated primary alcohol, b.p. 230°C (446°F), derived from terpenes (citronella and palmerosa essential oils); it is optically active, with a smell suggestive of geraniums. It is used to impart odor to soaps and as a food additive (flavoring). The formula is $C_9H_{15}CH_2OH$. It is the major component of geranium oil. *See also* terpene; essential oil.

germanium. An element.

Symbol	Ge	Atomic Wt.	72.59
State	Solid	Valence	2,4
Group	IVA	Isotopes	5 stable
Atomic No.	32		

Germanium, m.p. 937°C (1720°F), is a rare element present in low percentages in a few widely scattered ores and in coal in trace amounts. It is usually classed as a nonmetal; its structure is similar to that of carbon. Its distinguishing feature is its ability to act as a semiconductor; it forms perfect crystals and can be made in extremely pure form. Germanium can be readily ''doped'' with impurities on a controlled basis, an essential feature of semiconduction. For these reasons it has been ideal for the study of solid state physics, which made possible the discovery of transistors in the late 1940's. It is also used in special alloys and in brazing. *See also* semiconductor; nonmetal.

getter. A metal used in vacuum tubes and similar evacuated devices to absorb (scavenge) residual molecules of oxygen, nitrogen, and other gases. Rare earth metals and misch metal are commonly used.

ghatti. *See* arabic gum.

gibberellic acid. One of a class of plant hormones or growth regulators called gibberellins; it is a cyclic, unsaturated compound ($C_{19}H_{22}O_6$), found in the seeds of many varieties of plants. It is effective in accelerating the germination of seeds under certain conditions when applied externally but has little effect on root growth. *See also* auxin.

Gibbs, Josiah Willard (1839–1903). The father of modern thermodynamics. During his life-long post as professor of mathematical physics at Yale, he stated the fundamental concepts embraced by the three great laws of thermodynamics, especially the nature of entropy and the phase rule. A theorist rather than an experimenter, Gibbs was the first to expound with mathematical rigor the ''relation between chemical, electrical and thermal energy and capacity for work.'' It has been said that throughout his adult life, Gibbs did nothing but think. The results have established him as one of the foremost creative scientists in history.

glacial. A term applied to a number of acids (acetic, phosphoric) which, in a highly pure state, have a freezing point slightly below room temperature. For example, glacial acetic acid is 99.8% pure and crystallizes at 20°C (62°F).

glance. A mineralogical term meaning brilliant or lustrous, used to describe hard, earth-derived materials that have a bright, reflecting surface or fracture. It is applied to hard asphalts, which are often called glance pitch, as well as to galena (PbS) or lead glance, an ore of lead.

glass. A semisynthetic, amorphous material

classed as a ceramic. It is a vitreous liquid which is not fluid except at high temperatures; it cools from the molten state without forming the crystal structure characteristic of solids, even though it is mechanically rigid. Glass does not have a specific melting point, but softens gradually on heating. Glass occurs in nature in limited amounts as the volcanic glass, obsidian. The basic constituents of glass are silica, soda ash, and lime, to which oxides of lead, cobalt, lithium, etc., are often added. These are heated to melting in a furnace, from which the melt is withdrawn by ladles and poured onto tables (for flat products) or removed in small amounts and formed into specific shapes by manipulation. Superior grades (plate glass) are annealed in a long oven called a lehr, where it is subjected to slow and carefully controlled cooling to eliminate internal stresses and distortions. There are many special-purpose glasses, which are described in the cross-referenced entries. The simplest glass is sodium silicate, or water glass. *See also* crown glass; flint; optical glass; photochromic; water glass; borosilicate; glass fiber; metal glass.

glass-ceramic. A mixture based on the silica-soda-lime composition of regular glass (an amorphous or noncrystalline material) but also containing an ingredient which causes crystal formation by a nucleating process; the glass is thus devitrified, that is, it becomes a crystalline solid rather than a liquid. Such products are resistant to extreme heat and can be used for laboratory hoods, restaurant ranges, etc. The nucleating agent is titanium dioxide. *See also* nucleation; vitreous.

glass electrode. *See* electrode.

glass fiber. An extruded filament of glass, often only a few microns in diameter, which can be twisted into yarns and woven into fabrics. It is used as such in insulation bats, filter aids, tire cord, drapes, etc., and also in composites with phenol-formaldehyde and other resins (glass-reinforced plastics). Such composites have high strength, light weight, and excellent corrosion and shock resistance. *See also* reinforced plastic.

glassine. A type of paper in which the cellulose fibers have been subjected to extensive breakdown by mechanical action during manufacture and to which a small amount of paraffin is usually added to achieve a high degree of proofness to water vapor. It is translucent, flexible, and of very low friction coefficient and is used widely for food packaging and other applications requiring moisture exclusion. Its non-stick property makes it a useful household item. *See also* paper.

Glauber's salt.   *See* sodium sulfate.

glaze.   A glassy coating fired on the surface of ceramic ware and clay products, either to prevent absorption of liquids or (more often) to provide a smooth, attractive appearance and feel. Many glazes are colored and thus offer the possibility of unusual decorative effects. It is applied by dipping, brushing, or spraying before firing, the glaze being tightly bound (fused) to the base material by heat. The composition of glazes is similar to that of glass, i.e., a mixture of silicates and metallic oxides; some types also contain frit. *See also* glass; frit.

globulin.   *See* antibody.

glucose.   *See* dextrose; glycoside.

glue.   A type of adhesive based on proteins derived from animal and some vegetable sources; it lacks the strength and durability of synthetic-resin adhesives. There is still considerable industrial use of casein-based glues and fish glue; other types made by extracting the collagen from slaughter-house waste are used chiefly for household purposes. *See also* adhesive; fish by-products.

glyceraldehyde.   A crystalline compound formed by the metabolism of sugars; other than its value in nutritional and biochemical research, it has few applications of major importance. It is also called glycerose; it is an aldose compound containing both an aldehyde and an alcohol grouping. The structure of the molecule is of significance, as its optically active enantiomorphic forms are used as the basis for designating left- and right-handed configurations in other compounds having asymmetric carbon atoms. The conventional symbols are small capital D (for dextro) and L (for levo); these are not the same as the italic *d* and *l* often used to indicate optical rotation; they refer only to the configuration of the glyceraldehyde isomers shown below, in which the central carbon atom is asymmetric (C*):

$$
\begin{array}{cc}
\text{L} & \text{D} \\[1em]
\text{CHO} & \text{CHO} \\
| & | \\
\text{HO}-\text{C*}-\text{H} & \text{H}-\text{C*}-\text{OH} \\
| & | \\
\text{CH}_2\text{OH} & \text{CH}_2\text{OH}
\end{array}
$$

See also enantiomorph; asymmetry; optical rotation.

glyceride.   An ester formed by reaction of glycerol, a trihydric alcohol, with a fatty acid; when all three hydroxyl groups of the glycerol are replaced, the product is called a triglyceride, but if only one or two are involved, the products are known as mono- and diglycerides, respectively. If more than one type of fatty acid is present, as is usually the case in natural fats, the product is a mixed triglyceride. Glycerides are the major components of animal and vegetable fats and oils and may be saturated or unsaturated. They are readily hydrolyzed to fatty acids and glycerol. Monoglycerides are excellent emulsifying agents. Unsaturated glycerides are often hardened to make margarines and shortenings by catalytic addition of hydrogen. *See also* fatty acid; hydrogenation; ester.

glycerin.   *See* glycerol.

glycerol.   A trihydric (or polyhydric) alcohol, also called glycerin; it has the formula $CH_2OHCHOHCH_2OH$ (three hydroxyl groups). It is a rather thick, syrupy liquid which is hygroscopic and has a sweetish taste. It occurs widely in esterified form in the fatty acids present in animal and vegetable fats and oils; these esters are called glycerides. When a fatty acid is treated with a metal hydroxide to form a soap, glycerol is the by-product; this reaction is a major source of the commercial product. It is also made by hydrolysis of fats with steam. A completely synthetic process involves chlorination of propylene followed by hydrolysis and treatment with hypochlorous acid.

Glycerol has no toxic or flammable hazards and is widely used in industry in the manufacture of alkyd resins and nitroglycerin, as a plasticizer for cellophane and many composite materials, and as a humectant for tobacco. More limited uses are as an antifreeze, a solvent in cosmetics, and in the manufacture of ink rollers for printing presses.

glycerose.   *See* glyceraldehyde.

glyceryl phthalate.   A type of alkyd resin made from glycerol, phthalic anhydride, and a drying oil, used in high-grade paints. *See also* alkyd resin.

glycine.   An amino acid found in many natural proteins; it is not regarded as an essential nutritive factor. The formula is NH_2CH_2COOH (sometimes called aminoacetic acid); it is the only amino acid which contains no asymmetric carbon atom. It has some applications as a feed supplement and as an additive to inhibit rancidity of animal and vegetable fats and oils (not over 0.01%). It is conventionally abbreviated Gly. *See also* amino acid.

glyco-. The Greek word for sweet, appearing in many chemical terms for carbohydrate compounds derived from sugar or starch and for alcohols having a characteristically sweet taste, e.g., glycogen, glycerol, glyceride, glycol, etc.; glucose has the same derivation.

glycogen. A starch polymer which is classed as a polysaccharide and is found chiefly in the liver of animal organisms where it acts as a reserve fund of energy. It can be broken down to simpler forms by the action of enzymes, a reaction called glycolysis.

glycol. One of a group of alcohols containing two hydroxyl groups (dihydric or diol), each attached to a different carbon atom in the molecule. They are liquids of medium to high viscosity and high boiling point [well above 100°C (212°F)]; the high molecular weight polyglycols are waxy solids at room temperature. Many of the simple glycols become glasslike solids below 0°C (32°F). All are combustible, but none is flammable; their toxicity varies from moderate (ethylene glycol) to negligible (propylene glycol).

Glycols are stable materials which absorb and retain moisture and are thus much used as humectants. Other major applications are as solvents for dyes and in protective coatings, as plasticizers in plastics, as permanent antifreeze compounds, and as hydraulic fluids. The more important members of this group are ethylene glycol and propylene glycol, together with their respective condensation polymers, esters, acetates, etc. The propylene group finds use as stabilizers and emulsifiers in food products. *See also* ethylene glycol.

glycolysis. Enzymatic decomposition of sugars, starches, and other carbohydrates, with release of energy, a type of reaction occurring in yeast fermentation and in certain metabolic processes. Lactic acid is one of the products formed. The root word is *glyco-* (''sugar''), not *glycol* (''alcohol'').

glycoside. Any of a number of organic plant products whose molecules are composed of an alcohol or alcohol-like group bound to a sugar

(saccharide). When the sugar present is glucose, the compound is called a glucoside. Glycosides can be hydrolyzed into their sugar and nonsugar components. Many sugars occur only in the form of glycosides; among these are the so-called steroid glycosides, such as those obtained from digitalis, which are used in pharmaceutical preparations for specific medical purposes.

gold. An element.

Symbol	Au	Atomic No.	79
State	Solid	Atomic Wt.	196.9665
Group	IB	Valence	1,3

Gold, m.p. 1063°C (1945°F), is one of the transition metals and is heavier than both lead and mercury. Its specific gravity is 19.3, compared to 11.3 for lead and 13.5 for mercury. It has only one stable form. Though not particularly reactive, it readily forms compounds with the halogen elements. It ranks third (after silver and copper) in electrical conductivity. Its softness (ductility) enables it to be extended into sheets (leaf) as thin as 50 angstrom units, which have become extremely useful in surgical techniques. Even in thin layers, gold is highly reflective to radiation and is used in space applications, such as the face masks of astronauts' suits. It can also be prepared in colloidal suspension and mixed with molten glass to give the red color of ruby glass. A well-known suspension containing gold is called purple of Cassius, in which an oxidized tin compound acts as a protective colloid. Other industrial uses of gold are in electrical contacts, electronic devices, and spinneret nozzles. White gold is an alloy of gold and palladium containing about 60% gold. Gold has many artificially radioactive isotopes, some of which are used in medicine.

graft polymer. A type of ''tailor-made'' composite high-polymer in which one or more molecular chains of one type of polymer are attached to the main chain of another type. These structures result from free radical initiation at active points on the main chain; their composition and structure are subject to close control so that copolymers of many types and properties can be constructed by chemical ''grafting'' of subchains to the backbone of the major component. Styrene, vinyl acetate, and acrylonitrile have been attached in this way to polyethylene and other polymers. A simple diagram is as follows:

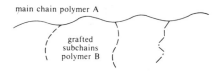

main chain polymer A

grafted
subchains
polymer B

See also block polymer.

Goldschmidt process.   The process invented in 1905 by the German chemist, Hans Goldschmidt (1861–1923), in which finely divided aluminum reacts with a metal oxide when ignited, to form pure metal and aluminum oxide. Also called the thermite process or aluminothermy, the exothermic reaction is rapid, self-sustaining, and very vigorous, the molten metal and alumina reaching a temperature of about 3000°C (5432°F). It is used to produce high-purity metals such as chromium, nickel, vanadium, manganese, and uranium from their oxides and is also used to make molten iron from iron oxide, to be used for joining (welding) massive metal parts such as rails and shafts. *See also* thermite.

Goodyear, Charles   (1800–1860). Born in Woburn, Mass., Goodyear was the first to realize the potentialities of natural rubber. Frustrated by its lack of stability to temperature and other weaknesses in the uncured state, he experimented with additives such as magnesium and sulfur. The discovery of vulcanization in 1839 was not accidental, as is often stated, but the result of intelligent trials and correct evaluation of their results. Though Goodyear's patents were contested by Hancock in England, he well merits the credit for making rubber usable in countless ways, and for helping to make the automobile possible.

Gordon Research Conferences.   *See* American Association for Advancement of Science.

Graham, Thomas   (1805–1869). Born in Scotland, Graham is famous for his basic studies in diffusion which led to the development of colloid chemistry. He was the first to observe a marked difference in the rate of passage of certain types of substances through a parchment membrane. Those which readily crystallize, like sugar, pass rapidly through the membrane, but gelatinous types are "slow in the extreme." The latter, which comprise albumin, starch, gums, etc., he designated as colloids, and their solutions as colloidal solutions. The former, which he called crystalloids, form "true" or molecularly dispersed solutions. *See also* colloid chemistry.

Graham's Law.   At constant conditions of temperature and pressure, the rate of diffusion of a gas is inversely proportional to the square root of the density of the gas. Based on 1833 experiments of Thomas Graham, a Scottish chemist (1805–1869). *See also* diffusion.

grain.   (1) A unit of weight equivalent to about 1/15.5 gram, or 1/7000 pound. It is used in weighing textile fibers and filaments and for specifying the amounts of drugs used in prescriptions and medical preparations.

(2) The seeds of wheat, corn, oats, rye, etc., or an alcohol made from them by fermentation.

(3) The crystalline particles of which metals are composed.

(4) The side of a hide or skin which has been dehaired.

(5) The orientation of pulp fibers in paper.

grain alcohol.   *See* ethyl alcohol.

gram.   A standard unit of mass (weight) equivalent to 1/453.59 pound, or 15.4 grains. It is the weight of 1 milliliter (ml) of water at 4°C and 1 atmosphere pressure. A kilogram is 1000 grams; a milligram (mg) is 1/1000 gram; and a microgram (μg) one-millionth gram. One milliliter so closely approximates 1 cubic centimeter that, for all but the most accurate purposes, it may be said that 1 cc of water weighs 1 gram.

gram atomic weight.   *See* atomic weight.

gram molecular weight.   *See* molecular weight; mole.

graphite.   A crystalline allotropic form of carbon in which the atoms are arranged in parallel tiers, 3.4 angstrom units apart, within which the bonds are extremely strong in a lateral direction, but between which the vertical attraction is comparatively weak. The crystals are hexagonal plates having a soft, fatty feeling and made up of thin sheets that can be separated easily. Graphite occurs naturally, but most is made synthetically by high-temperature treatment of amorphous carbon. It is used as a refractory in foundries; for electrodes, brushes, and other electrical equipment; and in steel making. It is a good solid lubricant and also is employed as the "lead" in pencils, for crucibles, and for other heat-resistant laboratory equipment. Highly purified graphite is an effective moderator in nuclear reactors.

GRAS.   These initial letters stand for "Generally

Recognized as Safe,'' referring to those food additives that meet the requirements of the Food and Drug Administration. Materials on this list are considered to be ''safe for their intended use.'' This list is under constant testing and review; changes in the status of these materials are reported in the ''Federal Register.'' *See also* food additive.

gravimetric.   A term used by analytical chemists to denote methods of quantitative analysis which depend on weighing the components of the sample. Various methods of separation are required to obtain the individual components, for example, precipitation. *See also* titration.

grease.   (1) A mixture comprised of a petroleum-derived oil thickened until it becomes a semi-solid, by means of soaps of heavy metals such as calcium, aluminum, lead, etc.; copper phthalocyanine, rosin, and colloidal silica are also used. The high viscosity of these greases enables them to withstand the severe mechanical action of machine parts, gears, and bearings even at temperatures up to 260°C (500°F). Corrosion inhibitors and other special-purpose additives may also be included.

(2) A cholesterol fatty-acid ester with an unpleasant odor produced by the hair follicles of some animals, especially sheep and goats. This material must be separated from the wool before the latter can be used. The unrefined product, called degras, is used as a leather softener and in printing inks; refined wool grease is the basis of lanolin, used in ointments and cosmetics.

Greek letters.   *See* alpha; beta; gamma.

green.   (1) A light-stable, nonfading inorganic pigment (chrome green) made by combining iron blue and chrome yellow from calcined chromium oxide, or by chlorination of phthalocyanine blue. Guignet's green is made by heating a mixture of potassium chromate and boric acid.

(2) Paris green (which has several other names ending in green) is an extremely toxic compound of copper and arsenic still used to a limited extent in mosquito control and as a wood preservative. It is copper acetoarsenite, $(CuO)_3As_2O_3 \cdot Cu(C_2H_3O_2)_2$.

(3) Green liquor. *See* liquor.

(4) A term used in industrial parlance to mean uncured, for example, leather, rubber, tobacco, cheese, etc. It has no reference to color. It is undoubtedly used in this sense in the word *greensand,* referring to an untreated sand, packed around a metal mold.

(5) Green soap (*q.v.*).

greenhouse effect.   Absorption of infrared radiation by the carbon dioxide and water vapor in the atmosphere, which retards dissipation of heat from the earth's surface. This phenomenon significantly affects global temperature. The CO_2 content of the air is not constant; at present, it averages 0.033%. Authoritative estimates indicate that the combustion of fossil fuels increased it by 10 to 12% between 1900 and 1950, causing a rise of about 1°C in *average* global temperature. Presumably, an equivalent rise will have occurred by 2000, for a total of 2°C for the century. Coal and coal-derived fuels produce substantially more CO_2 on combustion than either oil or natural gas. Environmentalists warn that, since the temperature increase is considerably greater at the poles than near the equator, sufficient melting of polar ice may eventually occur to raise the level of the oceans enough to threaten coastal cities.

greensand.   *See* green (4).

green soap.   A liquid, water-soluble soap made by reacting potassium hydroxide with various vegetable oils; it has a slightly greenish color and flows readily through dispenser orifices at room temperature. It is used chiefly for shampoos and as a hand cleaner in public washrooms.

grex.   A term used in the textile industry to designate the weight per unit length of a filament, i.e., its diameter or ''fineness;'' a filament is called 1 grex if 10,000 meters of it weigh 1 gram. *See also* denier.

Grignard.   A series of powerful organometallic reagents named after their discoverer (1900), Victor Grignard, a French chemist (1871–1935), Nobel Prize 1912. Each is a compound made up of three members: an organic group (alkyl, aryl), a halogen atom other than fluorine, and magnesium; it is prepared by adding an ether solution of the organic halide to finely divided magnesium. The formula of a simple Grignard reagent (methyl magnesium iodide) is CH_3MgI. Typical synthetic organic applications are: (with water) hydrocarbons, (with aldehydes) secondary alcohols, and (with ketones) tertiary alcohols. The last two reactions are in two steps, the second of which requires the presence of water. Many other syntheses are possible with these versatile reagents. They require caution in preparation and handling because of their high flammability, especially when in contact with water.

groundwood. *See* lignin; paper; wood.

group. (1) A number of elements having similar electronic structure and chemical properties as classified according to the Periodic Law and appearing in vertical columns in the Periodic Table. There are nine major groups, seven of which are divided into two subgroups, designated A for the major group elements and B for transition elements. The two groups that are not so divided are VIII (transition elements only) and the noble gas group (sometimes called Group 0). Thus, the numbered groups in a widely used type of Periodic Table are as follows: IA, IIA, IIIB, IVB, VB, VIB, VIIB, VIII, IB, IIB, IIIA, IVA, VA, VIA, VIIA, and noble gas. *See also* Periodic Table.

(2) Two or more associated elements which tend to remain together during a reaction. The term *group* properly refers to these only when no dissociation is involved. Thus, there are inorganic groups such as OH (hydroxyl), CO_3 (carbonate), SO_4 (sulfate), as well as organic groups such as COOH (carboxyl), CO (carbonyl), CH_3 (methyl) and homologous univalent alkyls, and C_6H_5 (aryl) groups. Alkyl groups may acquire an unpaired electron when they break away from their molecules, in which case they are called free radicals. Besides the above, there are a number of combinations of elements that are characteristic of large classes of compounds; these are also called groups, e.g., COH (alcohol or methanol), CHO (aldehyde), $CONH_2$ (amide), CH_3CO (acetyl), NH_4 (ammonium), and SH (sulfhydryl). *See also* radical.

growth. (1) In biochemistry, the continuous process of cell division and reproduction, characteristic of all living organisms. The basic phenomenon is considered to be osmosis, by which nutrients are transferred through cell walls and tissue structures; it is thus essential to the metabolic functioning of the organism.

(2) In crystallography, the process of crystal formation and development by nucleation and accretion. Crystals of many kinds are artifically produced for a variety of uses (e.g., lasers) by vapor condensation, electrodeposition, or rapid cooling of a saturated solution. *See also* osmosis; nucleation.

guaiac. A natural resin obtained from trees in the Caribbean islands and Central America; it is

insoluble in water but soluble in organic solvents and sodium hydroxide solution. It was formerly used as a varnish base, but its application is now chiefly as an additive in edible fats and oils; it contains 15% vanillin. *See also* resin (1).

guanidine. *See* accelerator.

guanine. A basic component of nucleic acids, usually obtained by hydrolysis of yeast and also to some extent from beet sugar; its formula is $C_5H_5N_5O$, which is characteristic of the purines. It is a metabolic product formed in the pancreas of some birds and animals and is a component of the fertilizer nitrates of Chile and Peru, obtained from the feces of sea birds (guano). It also occurs in the scales of fish and when recovered can be used for manufacture of imitation pearls.

guar. A polysaccharide (galactomannan) occurring in the seeds of a plant grown in southern Asia and technically classified as a mucilage. It is strongly hydrophilic, absorbing many times its own weight of water; this property is responsible for its use as a paper coating material, textile size, thickening agent in cosmetic products, ointments, food products, and in hydraulic fracturing of oil strata. It is now being cultivated in the southwestern U.S. *See also* mucilage; mannan.

guayule. A rubber-like hydrocarbon obtained from a shrub grown extensively in Mexico and the southwestern U.S.; it is well adapted to dry, semi-arid environments. The latex is contained in microscopic individual cells in every part of the plant except the leaves; it is separated by crushing and parboiling the entire plant. This coagulates the latex, which is then removed by grinding in caustic solution. Almost all the hydrocarbon content is recoverable as a possible substitute for both natural and synthetic rubbers. Still more important is the possibility, being currently investigated, that the hydrocarbons of this plant family may provide a source of liquid fuels and chemical feedstocks. *See also* rubber.

Guignet's green. *See* green (1).

gum. (1) As a noun, this term refers to either a natural water-soluble carbohydrate polymer (polysaccharide) occurring in globules on the trunks of various types of tropical trees, or to chemically modified starch or cellulose products made synthetically. The water-absorption properties of gums make them useful as thickeners and stabilizers in food products, as coating agents for paper and textile products, as adhesives for stamps and labels, in pharmaceutical preparations, in confectionery, and as protective colloids. The best-known natural gums are arabic and tragacanth; the synthetic type is represented by cellulose gum (carboxymethylcellulose) and polyvinyl alcohol.

(2) As an adjective (gum rosin, gum turpentine, gum benzoin, gum camphor, etc.), the word appears to indicate only that the material is derived from trees or shrubs, since both the chemical and physical properties of these mixtures are quite different from those of true gums, e.g., they are not polymeric molecules but are mixtures of simple substances; thus, they are more accurately classified as resins.

(3) In petroleum technology, the term refers to a thick hydrocarbon oxidation product formed as residue in gasolines not properly inhibited by antioxidants. *See also* resin; mucilage; carboxymethylcellulose.

guncotton.   *See* nitrocellulose.

gutta percha.   The geometric isomer of natural rubber, which is *cis*-polyisoprene; the molecular formula of the two substances is the same (C_5H_8), the only difference being that in gutta percha, the methyl group is on the opposite side of the hydrocarbon chain, in the *trans* position. Unlike rubber, gutta percha is definitely crystalline, but it can be vulcanized with sulfur; it lacks the resilience and elastic recovery of rubber, and for this reason it has little commercial value. Some applications are in cable covers, electrical insulation, and dentistry. *See also cis-.*

gypsum.   *See* calcium sulfate; plaster.

H

H Symbol for the element hydrogen, the name being derived from the Greek, meaning "water maker."

Haber, Fritz (1868–1934). Born in Breslau, Germany, Haber's notable contribution to chemistry was his development (with Bosch) of a workable method of synthesizing ammonia (NH_3) by the water gas reaction from hot coke, air, and steam; the gas mixture obtained includes nitrogen from the air as well as hydrogen from the steam. It was the first successful attempt to "fix" atmospheric nitrogen in an industrial process. This discovery was developed to production scale about 1912; it enabled Germany to manufacture an independent supply of explosives for World War I. Haber was awarded the Nobel Prize in 1918. He left Germany to escape persecution in 1933. *See also* ammonia; Haber-Bosch process.

Haber-Bosch process. The first commercially feasible method of synthesizing ammonia, originally developed in Germany about 1912 by the chemists, Fritz Haber (1868–1934), Nobel Prize 1918, and Carl Bosch (1874–1940), Nobel Prize 1931. Both this process and various modifications of it now used are the major source of ammonia worldwide. It involves the use of water gas or synthesis gas made by reacting steam with hot coke or natural gas in the presence of air, which yields a mixture containing hydrogen, carbon monoxide, and considerable amounts of nitrogen and carbon dioxide. This mixture with additional steam is subsequently passed over a catalyst at about 500°C (932°F) to convert the CO and water to hydrogen and CO_2; all the CO_2 is removed, leaving the hydrogen and nitrogen

in a 3/1 ratio. These then are catalytically reacted at high temperature and pressure to form ammonia. Other sources of nitrogen, e.g., from air liquefaction, and hydrogen, e.g., from electrolytic cell gases, can also be used for production of ammonia. *See also* ammonia; synthesis gas.

habit. A term used in crystallography to denote the type of geometric structure a given crystalline material invariably forms, e.g., thin plates or sheets, fibers, cubes, rhomboids, etc. The detailed morphology of crystals is a science in itself, with a multitude of classes and subclasses, but the crystal habit of a given substance under comparable conditions of formation is quite constant and predictable. *See also* crystal.

hafnium. An element.

Symbol	Hf	Atomic Wt.	178.49
State	Solid	Valence	2,3,4
Group	IVB	Isotopes	6 stable
Atomic No.	72		

Hafnium, m.p. 2230°C (4046°F), is a ductile, corrosion-resistant metal, similar chemically and physically to zirconium, with which it occurs in the major ore, zirconium silicate. It is separated from zirconium by chlorination and solvent extraction, followed by reduction of hafnium chloride with magnesium (Kroll process). Its relatively high capacity for absorbing neutrons makes it a useful metal for nuclear reactor control; it also has limited application in electronic devices. While harmless from the toxicity standpoint, it is a fire hazard in powdered form. It is classed as a rare metal. *See also* Kroll process.

half-life. (1) In physical chemistry, the time required for an unstable element or nuclide to lose one-half of its radioactive intensity in the form of alpha, beta, and gamma radiation. Half-lives of radionuclides range from microseconds to millions of years; that of radium is 1620 years. *See also* decay; radiocarbon dating.

(2) In chemical kinetics, the time required for a reaction to be half-completed or for one-half of a reactant to be consumed; it is also called half-period.

Hall, Charles Martin (1863–1914). A native of Ohio, Hall invented a method of reducing aluminum oxide in molten cryolite by electrochemical means. This discovery made possible the large-scale production of metallic aluminum and resulted in formation of the Aluminum Co. of America. The process requires a relatively high power input. Hall is generally considered as the founder of the aluminum industry. *See also* aluminum; Hall process.

Hall process. The basic method for the separation of aluminum from alumina (Al₂O₃), whose most important source is bauxite, an earthy material rich in aluminum oxide (alumina). The process involves electrolysis in which the alumina is mixed with fused cryolite (Na₃AlF₆), a sodium-aluminum fluoride found in Greenland. It requires a high power input which approaches 4% of the electricity produced in the United States. It was developed in 1888 by Charles Martin Hall, an American chemist (1863–1914), and was the beginning of commercial aluminum production. In a remarkable coincidence, it was also invented independently in 1888 by Paul Louis Toussaint Héroult, the French chemist whose life span equalled that of Hall, 1863–1914. The process is therefore also called the Hall-Héroult process. *See also* aluminum; bauxite.

hallucinogen. Any of a group of cyclic organic compounds which cause various aberrations of perception and feeling, as well as undesirable physical reactions. Many of these compounds are from natural sources and encompass six distinct types. The best-known of these are derivatives of lysergic acid, chiefly the diethylamide (LSD-25), which is one of the most powerful; mescaline, a phenylethylamine; and the tetrahydrocannabinol group occurring in marijuana and its concentrated form, hashish. *See also* psychotropic agent.

halocarbon. *See* halogen.

halogen. This term, whose literal meaning is "saltmaker," refers to the five elements of Group VIIA of the Periodic Table. In the order of their activity these are fluorine, chlorine, bromine, iodine, and astatine; they are all electronegative and thus are strong oxidizing agents. They are classified as nonmetals. Their inorganic compounds readily form negatively charged ions in solution. They also combine with organic compounds to form a multitude of useful products, among which are fluorocarbons, chlorinated hydrocarbons, and plastics of various kinds; all these are collectively known as halocarbons. For further information, see the respective elements.

hand. A term used chiefly in the textile industry to describe the feeling or texture of a fabric; it is largely subjective, being based on sensory perception rather than on measurable properties. For example, a piece of woven fabric, especially when coated with a finishing agent, may have a soft hand, a stiff hand, etc. In the leather industry, the term *plumpness* is often used to describe a soft, pliable tanned product.

handedness. *See* chiral; enantiomorph.

hard acid. *See* acid.

hardness. (1) The most familiar property of solid materials, which is difficult to define accurately. It is measured by determining the extent to which a material resists the force applied by a needle-like instrument under a specific load. Various hardness testers of this type are used for metals, glass, plastics, rubber, etc. Among the more important are Brinell and Rockwell testers (for steel and iron) and the Shore Durometer (for rubber and plastics). The hardness of minerals is indicated by the Mohs scale.

(2) The amount of inorganic matter contained in a sample of water; the contaminant is usually a calcium or magnesium compound picked up by the water from the rocks and soil over which it has flowed. The mineral content is very small and is usually given in parts per million. Water that is free of such compounds or that contains a vanishingly small amount is called soft.

(3) In reference to radiation, hardness refers to wavelengths beginning about the middle of the x-ray spectrum and moving downward through the gamma and cosmic ranges. These have extremely high energy and are called "hard" radiation. *See* x-ray.

(4) The term is also applied to Lewis acids

and bases having certain properties. *See* acid; base.

hard water. *See* calcium carbonate; boiler scale; hardness **(2)**.

hardwood. *See* wood; pulping.

hazardous wastes. *See* waste control.

Hb Abbreviation for the iron-protein complex hemoglobin.

He Symbol for the element helium, the name being taken from the Greek word for *sun,* where it was first identified spectroscopically.

heat. A mode of energy associated directly with and proportional to the random molecular activity (motion) of a material system. It maintains or raises the temperature of the system and can be transferred to another body (gas, liquid, or solid) by radiation, convection, or conduction. It can also be converted to other forms of energy, e.g., electricity and motion. Heat is also defined as energy which is transferred under a temperature gradient or difference from one body to another. Heat is variously generated, e.g., by chemical reaction, flow of electricity, friction, nuclear fission and fusion, etc. *See also* temperature.

heat of combustion. The heat evolved when a definite quantity of a substance is completely oxidized (burned); usually stated for a standard state of 25°C and one atmosphere pressure for reactants and products in kilocalories per gram-formula weight of the substance. It is applied most commonly to the combustion of fuels and organic compounds but may also be applied to many inorganic substances, e.g., the burning of sulfur to form sulfur dioxide. It is a special type of heat of formation and is closely related to the heating value of fuels of all sorts, in which case the value is commonly expressed as BTU per pound of fuel.

heat of crystallization. The heat evolved or absorbed when a crystal forms from a saturated solution of a substance; usually expressed in kilocalories per gram-formula weight of the crystalline substance. In water solutions, the crystal formed often contains water of crystallization, e.g., $Na_2SO_4 \cdot 10H_2O$. For engineering calculations, the value is often taken as the negative of the heat of solution of the crystalline form, thereby neglecting the much smaller heat of dilution, which is the actual difference between the heat of solution and the heat of crystallization.

heat of dilution. The heat evolved per mole of solute when a solution is diluted from one specified concentration to another.

heat of formation. The heat evolved or absorbed when a compound is formed in its standard state from its elements in their standard states, all at a specified temperature and pressure; usually expressed as kilocalories per gram-formula weight of the compound.

heat of fusion. The heat required to convert a substance from the solid state to the liquid state with no temperature change, it is also called the latent heat of fusion; usually measured at the melting point of the substance, i.e., the temperature at which the solid and liquid forms are in equilibrium and usually expressed as the heat required to convert a definite weight of the substance into a liquid, e.g., calories per gram, or kilocalories per gram-formula weight.

heat of hydration. (1) The heat associated with the hydration or solvation of ions in solution, usually expressed in kilocalories per gram-formula weight or per mole of the ion involved.

(2) The heat evolved or absorbed when a hydrate of a compound is formed, i.e., the difference between the heats of solution of the anhydrous and the hydrated forms of the compound, e.g., Na_2SO_4 and $Na_2SO_4 \cdot 10H_2O$; usually expressed as kilocalories per mole of the compound. *See also* hydration.

heat of reaction. The heat evolved or absorbed when a chemical reaction occurs, in which the final state of the system is brought to the same temperature and pressure as that of the initial state of the reacting system. The value usually is accompanied by a chemical equation expressing the reaction and the number of gram-moles involved in it. The heat change is usually expressed in kilocalories per gram-formula weight of the product or products.

heat of solution. The heat evolved or absorbed when a substance is dissolved in a solvent. The total or integral heat of solution is the heat change resulting when one mole of solute dissolves in a definite amount of solvent. Most values are of this type. The partial or differential heat of solution is the heat change involved when one mole of solute dissolves in such a large amount of solution of definite concentration that the concentration is not appreciably altered. Both types are expressed in calories per gram or in kilocalories per gram-formula weight of solute.

heat of sublimation.   The heat required to convert a unit mass of a substance from the solid to the vapor state (sublimation) at a specified temperature and pressure, without the appearance of the liquid state. It is equal to the sum of the heats of fusion and vaporization for the substance; usually expressed in kilocalories per gram-formula weight of the substance.

heat of transition.   The heat evolved or absorbed when a unit mass of a given substance is converted from one crystalline form to another crystalline form. It can be calculated from the respective heats of formation of each crystalline form and is usually expressed in kilocalories per gram-formula weight of the substance.

heat of vaporization.   The heat required to convert a substance from the liquid to the gaseous state with no temperature change; also called the latent heat of vaporization. It is usually measured at the normal boiling point of the substance, i.e., the temperature at which the liquid boils at one atmosphere pressure. It is usually expressed as the heat required to convert a definite weight of the substance into a gas, e.g., kilocalories per gram-formula weight.

heat exchange.   The use of the heat developed in one operation to activate a second operation by means of a heat-transfer agent. A common type of heat exchange is carried out in evaporator units (called effects) set up in series arrangement, in which the latent heat of vaporization released on condensation in the first unit is passed on to the next, where it is utilized as a heat source. A similar procedure uses the heat developed in a nuclear reactor to make steam for electric power generation. *See also* heat transfer; effect.

heat transfer.   The absorption of heat by a fluid medium which carries it away from the source and releases it either to the air, as in an automobile, or to another liquid (heat exchange). Heat-transfer agents are often called coolants, since that is their usual function. The efficiency of a coolant depends largely on its heat capacity, that is, its ability to quickly absorb and release thermal energy. Water is most commonly used where freezing is not a problem; for low-temperature conditions, ethylene glycol, alcohol, and various proprietary mixtures are used. Liquid sodium and lithium are used in cooling some types of nuclear reactors. Air and other gases also act as heat-transfer agents for special types

of service. *See also* heat exchange.

heavy.   As used in chemistry, this word has several meanings: (1) in the phrase "heavy-metal" soaps, it refers to metals of atomic weight greater than sodium (22.99) which form soaps on reaction with fatty acids, e.g., aluminum, lead, cobalt, and others; (2) the term *heavy isotope* refers to deuterium, the isotope of hydrogen having an atomic weight of 2.016, which occurs in heavy water; (3) the term *heavy chemical* is applied to chemicals produced on a very large scale, without reference to their actual weight, e.g., soda ash, borax, sulfuric acid, etc.

heavy chemical.   *See* heavy.

heavy hydrogen.   *See* deuterium; heavy water.

heavy-metal soap.   *See* soap; drier; grease (**1**); heavy.

heavy water.   A form of water also known as deuterium oxide (D_2O), in which the symbol D_2 stands for two atoms of heavy hydrogen (H^2), an isotope of hydrogen containing one proton and one neutron. The specific gravity of heavy water is 1.1, and its boiling point at normal atmospheric pressure is 101.4°C (214.7°F). Its occurrence in H_2O is in the ratio of approximately 1 to 6000 parts. It was used as a moderator in early types of nuclear reactors but has no other important application. *See also* deuterium.

Heisenberg, Werner K.   (1901–1976). A native of Germany, Heisenberg received his doctorate from the University of Munich in 1923, after which he was closely associated for several years with Niels Bohr in Copenhagen. He was awarded the Nobel prize in physics in 1932 for his brilliant work in quantum mechanics. In 1946, he became Director of the Max Planck Institute. His notable contributions to theoretical physics, best known of which was the Uncertainty Principle, imparted new impetus to nuclear physics and made possible a better understanding of atomic structure and chemical bonding. *See also* uncertainty principle; orbital.

helium.   An element.

Symbol	He	Atomic Wt.	4.00260
State	Gas	Valence	0
Group	Noble Gas	Isotopes	2 stable
Atomic No.	2		

Helium (from Greek, meaning "the sun") was so named because of its original discovery in the sun by spectrographic analysis (1868); it was not found on earth until 1895. It is a component of natural gas, from which it is separated by compressing the gas at low temperatures until all the other components have been liquefied. Helium does not become liquid until a temperature of 4.2 Kelvin (about $-265°C$) is reached and is thus a cryogenic liquid, which remains fluid at temperatures near absolute zero. The nuclei of helium atoms are called alpha particles and are emitted during radioactive decay of unstable nuclei. Both gaseous and liquid forms can penetrate even dense solids three times as rapidly as air. Though helium cannot enter into chemical combination and thus forms no compounds, it has unusual properties as a liquid which have been the subject of much physical research. The chief uses of helium are inflation of weather balloons, leak detection in high-vacuum equipment, arc-welding, as inert medium for semiconductor crystal growth, in rocket and space vehicle launching, and in breathing equipment for divers and in space vehicles. It is noncombustible and nontoxic.

hemicellulose. A fraction of cellulose containing relatively few monomer units, compared with normal cellulose; as the name suggests, it is a sort of "half-cellulose." It is soluble in mild solutions of sodium hydroxide and is a minor component of wood; it is sometimes also called beta- or gamma-cellulose. High percentages occur in corn cobs and in other agricultural wastes. *See also* cellulose.

hemin. *See* porphyrin.

hemoglobin. The iron-protein complex which is the characteristic component of red blood cells (erythrocytes). Its chief function is the transport of oxygen in the body. It exists in the form of a molecule of high molecular weight, each containing one atom of iron; the symbol Hb is often used. Hemoglobin is able to pick up oxygen from the air inhaled into the lungs; it also picks up carbon monoxide very much more readily, resulting in dangerous and sometimes fatal poisoning from exposure to atmospheres containing more than 1% carbon monoxide. *See also* blood.

heparin. A nitrogen-containing organic compound of polymeric nature, occurring in the tissues of animals and man; its major function is to prevent or retard coagulation of the blood, and it is used in medicine for this purpose, i.e., where blood clots in the circulatory system are

considered likely to occur. It is also used as a rodent poison, as large dosages cause fatal hemorrhage.

heptamethylnonane. *See* cetane number.

heptane. A hydrocarbon liquid of paraffinic structure having seven carbon atoms in the chain; its formula is C_7H_{16}, or $CH_3(CH_2)_5CH_3$. It has long been used as a basis for establishing gasoline octane ratings, since its own octane number is zero, i.e., it produces extremely high knock when used as a motor fuel. Heptane, like most petroleum distillates, is used for synthesis of organic chemicals and to some extent as a solvent. It is dangerously flammable. *See also* octane number.

heptene. *See* Oxo process.

herbicide. A type of pesticide, including so-called weed-killers, silvicides and defoliants, which kills shrubs, small trees, grasses, etc. There are both organic and inorganic herbicides; the latter is typified by common salt, sodium borate, and various arsenical compounds and the former by 2,4-D and similar chlorinated compounds and by the defoliant picloram. Some herbicides will kill only the weeds while leaving the desirable plants unharmed (selective); others (called silvicides) are especially effective on heavy growths such as trees and shubbery (ammonium sulfamate). *See also* pesticide; defoliant; silvicide.

hetero-. A prefix having the meaning "different" or "unlike," as in *heterogeneous* ("different kinds") and *heterocyclic* ("different atoms in the ring").

heterocyclic. An unsaturated cyclic compound containing one or more atoms other than carbon as part of the ring structure. The rings may be hexagonal or pentagonal, the latter appearing in the furan, purine, and pyrrole families of heterocyclics. Schematic structures would be as follows, where X stands for nitrogen, oxygen, or other element:

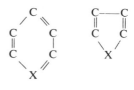

heterogeneous. Descriptive of a mixture or solution of two or more substances or materials, regardless of whether or not their dispersion in one another is uniform. Examples are (for gases) air at sea level; (for liquids) gasoline; (for solids) marble, alloys; (for colloidal dispersions) milk, rubber latex. Heterogeneous mixtures can be separated by physical means, e.g., evaporation, distillation, filtration, sedimentation, etc.; they are not chemically combined but are mixtures of different substances, as indicated by the derivation of the term, from *hetero* (different) and *gen* (kind). In heterogeneous catalysis, the catalyst and the feedstock constitute two separate phases, e.g., as when a hydrocarbon gas is passed over a fixed bed of catalyst. *See also* homogeneous; mixture; compound.

HETP.   An abbreviation used in chemical engineering; it stands for "height equivalent to a theoretical plate" in a fractionating (distillation) column. The value in a given case is obtained by dividing the height of the column or tower by the number of theoretical plates present. *See also* theoretical plate.

hexachlorophene.   A bacteriostat derived by reacting formaldehyde with trichlorophenol with sulfuric acid as a catalyst. Its formula is $(C_6HCl_3OH)_2CH_2$. Formerly a widely used additive to toothpastes and various cosmetic preparations, such as soap, where its bacteriostatic action gave it deodorant properties, its use has been prohibited by FDA except by prescription. *See also* antiseptic.

hexafluorobenzene.   *See* fluorocarbon.

hexafluoroethane.   *See* fluorocarbon.

hexamethylenediamine.   *See* nylon.

hexamethylenetetramine.   A nitrogen-containing heterocyclic compound derived from formaldehyde by condensation with ammonia; its formula is $(CH_2)_6N_4$. It is used as a reagent in organic synthesis, as a catalyst in the curing of two-stage phenol-formaldehyde (novolak) resins, and as a rubber accelerator. On reaction with nitric acid, it produces the powerful explosive RDX, introduced in World War II. It is classed as a flammable solid and is a skin irritant on prolonged contact.

hexose.   *See* sugar; epimer.

Hf   Symbol for the element hafnium, the name being derived from the Latin word for *Copenhagen,* where it was first isolated.

Hg   Symbol for the element mercury, derived from the Latin *hydrargyrum,* meaning "liquid silver" (from its appearance and behavior).

hiding power.   The extent to which a pigment is able to conceal or "cover" another color when imposed over it; it is also sometimes called covering power. It depends on various factors, chiefly refractive index and fineness of subdivision. Carbon black and titanium dioxide are black and white pigments of maximum hiding power. The term is used chiefly in the paint industry, mostly in reference to white pigments. *See also* pigment; opacity.

high explosive.   An explosive whose shattering power, or brisance, is due to detonation brought about by heat or mechanical shock. Some are much more readily detonated than others; among the more sensitive are mercury fulminate and nitroglycerin; TNT and ammonium nitrate are much more stable. *See also* explosive.

high flash (hi-flash) naphtha.   *See* naphtha.

high polymer.   A natural or synthetic polymer of high molecular weight (i.e., 5000 or more). Such polymers may contain as many as several thousand monomer units, the molecular weights sometimes running into the millions. Units of this size are called macro-molecules and are in the colloidal size range; they include such biochemically active agencies as enzymes and viruses. High-polymer molecules are chain-like structures that may be linear, coiled, helical, or cross-linked. Among the natural high polymers are carbohydrates (starches, cellulose, polysaccharides), proteins, and certain unsaturated hydrocarbons; the exact number of units present is not definitely known. These are the basis for many types of plastics and fibers: (1) cellulose and its derivatives (rayon, cellophane, nitrocellulose, carboxymethylcellulose); (2) proteins (casein, soybean); (3) and hydrocarbons (rubber and related materials.)

Synthetic high polymers (synthetic resins) are now in general more widely used for plastics and fibers than are the natural types; they are made by an almost infinite variety of catalytic reactions between amides, amines, alcohols, acids, ketones, halogens, etc., and petroleum-derived compounds such as benzene and its derivatives (phenol, cumene, xylene, etc.), ethylene, propylene, acetylene, and others. Those having rubber-like properties are called elastomers. An exceptional type is represented by silicone resins, in which carbon is replaced by silicon in the chain. In many of these synthetic materials, the molecular weight (degree of polymerization) and the structure (tacticity) can be selectively controlled by catalysts, short-stopping agents, and the like, so that products of predetermined properties can be "tailor-made" for specific uses. *See also* monomer; polymer; plastic (1); resin (2); stereospecific; elastomer.

hindrance.   Partial or complete blocking of a reaction resulting from the three-dimensional structure of a molecule, often referred to as steric hindrance. For example, differences in the geometric arrangement of alkyl groups in saturated acids may cause substantial retardation of their rates of esterification; trimethylacetic acid esterifies at only a small fraction of the rate of acetic acid because the reaction is hindered by three methyl groups attached to the first carbon atom. *See also* stereochemistry.

histamine.   *See* antihistamine.

histochemistry.   A branch of biochemistry dealing with the detailed structure and chemical composition of the tissues of plants and animals. Since this involves the study of the cells of which the tissues are composed, there is some overlap between histochemistry and cytochemistry. Histochemistry is concerned with the part played in tissue structure by its many component chemical substances, e.g., metals, cellulose, proteins, pigments, enzymes, etc. *See also* cytochemistry.

history of chemistry.   The following resume lists the significant developments in chemical theory and practice since its beginnings. (For further details, see individual entries.)

Democritus (465 B.C.). First to conceive matter in the form of particles, which he called atoms.

alchemists (ca. 1000–1650). Attempted to (1) change lead and other base metals to gold; (2) discover a universal solvent; and (3) discover a life-prolonging elixir. Used plant products and arsenic compounds to treat diseases.

Boyle, Sir Robert (1637–1691). Formulated fundamental gas laws. First to conceive the possibility of small particles combining to form molecules; distinguished between compounds and mixtures; studied air and water pressures, desalination, crystals, and electrical phenomena.

Priestley, Joseph (1733–1804). Discovered oxygen, carbon monoxide, and nitrous oxide.

Scheele, C. W. (1742–1786). Discovered chlorine, tartaric acid, sensitivity of silver compounds to light (photochemistry); and oxi-

dation of metals.

LeBlanc, Nicholas (1742–1806). Invented a process for making soda ash from sodium sulfate, limestone, and coal.

Lavoisier, A. L. (1743–1794). Discovered nitrogen; studied acids and described composition of many organic compounds. Generally regarded as the father of chemistry.

Volta, A. (1745–1827). Invented the electric battery, a series of "piles" or stacks of alternating layers of silver and zinc, or copper and zinc, separated by paper soaked in brine (electrolyte). *See* activity (1).

Berthollet, C. L. (1748–1822). Corrected Lavoisier's theory of acids; discovered bleaching power of chlorine; studied combining weights of atoms (stoichiometry).

Jenner, Edward (1749-1823). Discoverer of vaccination for prevention of smallpox (1776).

Dalton, John (1766–1844). The first great chemical theorist; proposed atomic theory (1807); stated law of partial pressure of gases. His ideas led to laws of multiple proportions, constant composition, and conservation of mass.

Avogadro, A. (1776–1856). Proposed principle that equal volumes of gases contain the same number of molecules. The number (6.02×10^{23} for 22.41 liters of any gas) is a fundamental constant that applies to all chemical units.

Davy, Sir Humphry (1778–1829). Laid foundation of electrochemistry; studied electrolysis of salts in water and other electrochemical phenomena; isolated sodium and potassium.

Gay-Lussac, J. L. (1778–1850). Discovered boron and iodine; studied acids and bases and discovered indicators (litmus); improved production method for sulfuric acid; did basic research on behavior of gases versus temperature and on the ratios of gas volumes in chemical reactions.

Berzelius, J. J. (1779–1848). Classified minerals chemically; discovered and isolated many elements (Se, Th, Si, Ti, Zr); coined the terms *isomer* and *catalyst;* noted existence of radicals; anticipated discovery of colloids.

Faraday, Michael (1791–1867). Extended Davy's work in electrochemistry; developed theories of electrical and mechanical energy; electrolysis; corrosion; batteries; and electrometallurgy.

Wohler, F. (1800–1882). First to synthesize an

organic compound (urea, 1828) (a rearrangement reaction). This discovery was the beginning of synthetic organic chemistry.

Goodyear, Charles (1800–1860). Discovered
vulcanization of rubber (1844) by sulfur, inorganic accelerator, and heat. Hancock in
England made a parallel discovery.

Liebig, J. von (1803–1873). Fundamental investigation of plant life (photosynthesis) and
soil chemistry; first to propose use of fertilizers. Discovered chloroform and cyanogen
compounds.

Graham, Thomas (1805–1869). Studied diffusion of solutions through membranes; established principles of colloid chemistry.

Pasteur, Louis (1822–1895). (1) First to recognize infective bacteria as disease-causing
agents; (2) developed concept of immunochemistry; (3) initiated heat-sterilization of wine
and milk (pasteurization); (4) observed optical
isomers (enantiomers) in tartaric acid.

Lister, Joseph (1827–1912). Initiated use of antiseptics in surgery, e.g., phenols, carbolic
acid, cresols.

Kekule, A. (1829–1896). Laid foundations of
aromatic chemistry; conceived of 4-valent
carbon and structure of benzene ring; predicted isomeric substitutions (o-, m-, p-).

Nobel, Alfred (1833–1896). Invented dynamite,
smokeless powder, blasting gelatin. Established international awards for achievements
in chemistry, physics, and medicine.

Mendeleev, D. I. (1834–1907). Discovered periodicity of the elements and compiled the first
Periodic Table.

Hyatt, J. W. (1837–1920). Initiated plastics industry (1869) by invention of Celluloid (nitrocellulose modified with camphor).

Perkin, Sir W. H. (1838–1907). Synthesized first
organic dye (mauveine, 1856) and first synthetic perfume (coumarin). His work on dyes
was continued and expanded by Hofmann in
Germany.

Beilstein, F. K. (1838–1906). Compiled "Handbuch der Organischen Chemie," a multi-volume compendium of properties and reactions
of organic chemicals.

Gibbs, Josiah W. (1839–1903). Stated three
principal laws of thermodynamics; expounded
nature of entropy and phase rule and the relation between chemical, electric, and thermal
energy.

Chardonnet, H. (1839–1924). First to produce
a synthetic fiber (nitrocellulose) with properties similar to rayon.

Boltzmann, L. (1844–1906). Developed kinetic
theory of gases; their viscosity and diffusion
properties are summarized in Boltzmann's Law.

Roentgen, W. K. (1845–1923). Discovered x-
radiation (1895). Nobel Prize, 1901.

LeChatelier, H. L. (1850–1936). Fundamental
research on equilibrium reactions (Le-
Chatelier's Law), combustion of gases, and
metallurgy of iron and steel.

Becquerel, H. (1851–1908). Discovered radioactivity, deflection of electrons by magnetic
fields, and gamma radiation. Nobel Prize, 1903
(with Curies).

Moisson, H. (1852–1907). Developed electric
furnace for making carbides and preparing
pure metals; isolated fluorine (1886). Nobel
Prize, 1906.

Fischer, Emil (1852–1919). Basic research on
sugars, purines, uric acid, enzymes, nitric acid,
ammonia. Pioneer work in stereochemistry.
Nobel Prize, 1902.

Thomson, Sir J. J. (1856–1940). Research on
cathode rays resulted in proof of existence of
electrons (1896). Nobel Prize, 1906.

Arrhenius, Svanté (1859–1927). Fundamental
research on rates of reaction versus temperature, expressed by the Arrhenius equation,
and on electrolytic dissociation. Nobel Prize,
1903.

Hall, Charles Martin (1863–1914). Invented
method of aluminum manufacture by electrochemical reduction of alumina. Parallel discovery by Héroult in France.

Baekeland, Leo H. (1863–1944). Invented phenol-
formaldehyde plastic (1907), the first completely synthetic resin (Bakelite).

Nernst, Walther Hermann (1864-1941). Nobel
Prize, 1920, for his work in thermochemistry;
did basic research in electrochemistry and
thermodynamics.

Werner, A. (1866–1919). Introduced concept of
coordination theory of valence (complex
chemistry). Nobel Prize, 1913.

Curie, Marie (1867–1934). Discovered and isolated radium; research on radioactivity of
uranium. Nobel Prize, 1903 (with Becquerel)
in physics; in chemistry, 1911.

Haber, F. (1868–1934). Synthesized ammonia
from nitrogen and hydrogen, the first indus-

trial fixation of atmospheric nitrogen (the process was further developed by Bosch). Nobel Prize, 1918.

Rutherford, Sir Ernest (1871–1937). First to prove radioactive decay of heavy elements and to carry out a transmutation reaction (1919). Discovered half-life of radioactive elements. Nobel Prize, 1908.

Lewis, Gilbert N. (1875–1946). Proposed electron-pair theory of acids and bases; authority on thermodynamics.

Aston, F. W. (1877–1945). Pioneer work on isotopes and their separation by mass spectrograph. Nobel Prize, 1922.

Fischer, Hans (1881–1945). Basic research on porphyrins, chlorophyll, carotene; synthesized hemin. Nobel Prize, 1930.

Langmuir, Irving (1881–1957). Fundamental research on surface chemistry, monomolecular films, emulsion chemistry. Also electric discharges in gases, cloud seeding, etc. Nobel Prize, 1932.

Staudinger, Hermann (1881–1965). Fundamental research on high-polymer structure, catalytic synthesis, polymerization mechanisms, resulting eventually in development of stereospecific catalysts by Ziegler and Natta (stereoregular polymers). Nobel Prize, 1963.

Fleming, Sir Alexander (1881–1955). Discovered penicillin (1928); initiated antibiotics. Nobel Prize, 1945. The science was developed in the U.S. by Selman A. Waksman.

Moseley, Henry G. J. (1887–1915). Discovered the relation between frequency of x-rays emitted by an element and its atomic number, thus indicating its true position in the Periodic Table.

Adams, Roger (1889–1971). Noted educator and contributor to industrial research in catalysis and structural analysis. Priestley Medal.

Midgley, Thomas (1889–1944). Discovered tetraethyllead and antiknock treatment for gasoline (1921) and fluorocarbon refrigerants; early research on synthetic rubber.

Banting, Sir Frederick (1891–1941). Isolated the insulin molecule. Nobel Prize, 1923.

Chadwick, Sir James (1891–1974). Discovered the neutron (1932). Nobel Prize, 1935.

Urey, Harold C. (1894–1981). Discovered heavy isotope of hydrogen (deuterium). Nobel Prize, 1934. A leader of the Manhattan Project. Made original contributions to theories of the origin of the universe and of life processes.

Carothers, Wallace (1896–1937). Polymerization research resulting in synthesis of neoprene (polychloroprene) and of nylon (polyamide).

Heisenberg, W. K. (1901–1976). Research in quantum mechanics resulting in development of the orbital theory of chemical bonding. Stated Uncertainty Principle. Nobel Prize, 1932.

Fermi, Enrico (1901–1954). First to achieve a controlled nuclear fission reaction (1939); basic research on subatomic particles. Nobel Prize, 1938.

Lawrence, Ernest O. (1901–1958). Invented the cyclotron, in which first synthetic elements were created. Nobel Prize, 1939.

Libby, Willard F. (1908–1980). Developed radiocarbon dating technique based on carbon-14. Nobel Prize, 1960.

Ho   Symbol for the element holmium, named after Stockholm, Sweden.

Hofmann, August Wilhelm   (1818–1892). German organic chemist who studied under Liebig. While professor of chemistry at the Royal College of Chemistry in London, he did original research on coal-tar derivatives which later led him into a study of organic dyes. Perkin, who first synthesized the dye mauveine in England, was a student of Hofmann. When the latter returned to Germany, he continued his work in the field of dyes, which became the basis of German leadership in synthetic dye manufacture which continued until World War I.

hole.   A term used in semiconductor technology to refer to an energy deficit in a crystal lattice due to (1) electrons ejected from unsatisfied covalent bonds at sites where an atom is missing (vacancy) or (2) to electrons supplied by atoms of impurities present in the crystal, e.g., arsenic or boron. The free electrons from these two sources move through the crystal leaving energy deficits which are positively charged; these deficit sites, or holes, are also considered to move as they become alternately filled and vacated by electrons; thus, a flow of positive electricity results. *See also* semiconductor; impurity; vacancy.

holmium.   An element.

Symbol	Ho	Atomic No.	67
State	Solid	Atomic Wt.	164.9304
Group	IIIB	Valence	3

Holmium, m.p. 1470°C (2678°F), is a rare earth metal prepared by reducing holmium fluoride with calcium. It is strongly electropositive; it has a high magnetic moment and electrical resistivity. It also has scavenging properties. There are no important industrial uses of holmium, though it is of considerable theoretical interest. It has only one stable form. *See also* lanthanide series.

holocellulose. The carbohydrate component of wood; depending on the species and botanical nature of the wood, its holocellulose content varies between 67 and 80%, the remainder being lignin. Holocellulose is not soluble in water; it is composed of alpha-cellulose (insoluble in strong caustic) and hemicellulose (soluble in weak caustic). Alpha-cellulose is the basis of paper manufacture. *See also* cellulose; paper.

homo-. A prefix having the meaning of "the same," as in the terms *homogeneous* ("the same kind"), *homologous* ("the same proportion"), *homopolymer*, etc.

homocyclic. Any organic molecule which has a ring or cyclic structure in which the ring contains only one element, which is usually carbon. This is true of cycloparaffins, cycloolefins, benzene and its derivatives, and cyclic terpenes. The term carbocyclic is also used for rings composed only of carbon. *See also* heterocyclic.

homogeneous. (1) Derived from *homo* ("the same") and *gen* ("kind"), this term is properly applied to chemical elements and compounds but not to mixtures or solutions. For example, pure water is homogeneous, whereas gasoline is not; nitrogen is homogeneous, but air is not. (Those which are not homogeneous are heterogeneous.) A compound can be subdivided or decomposed only by a chemical or electrochemical reaction, the products of which are different from the starting substance, whereas mixtures can be separated into their components by physical means such as evaporation, distillation, filtration, etc.

(2) The term is loosely (but improperly) used to describe mixtures of two or more liquids (solutions) which are uniformly dispersed in each other, so that samples taken at random have the same percentage composition. Under these conditions, a solution of water and alcohol, for example, is said to be homogeneous, regardless of the fact that it is comprised of two compounds that can easily be separated by heating. The term is similarly applied to colloidal dispersions.

(3) A homogeneous reaction is one involving only a single phase of matter, as in certain types of catalysis in which the catalyst and the reacting substances are both liquids. *See also* heterogeneous; homogenization; mixture; uniform dispersion.

homogenization. Reduction of the sizes of solid or semisolid particles in aqueous suspension to colloidal dimensions by mechanical action, the purpose being to stabilize the suspension so that the particles will neither rise to the surface nor precipitate. This is performed, for example, on pigments used in latex dispersions. A protective film of a hydrophilic nature (gelatin, casein) is first formed on the particles by wetting them in a water solution of these materials. The coated particles are then passed through a homogenizer, or "colloid mill," which exerts a strong shearing force that reduces the particles to uniform diameter. So-called homogenized milk is made in this way, except that its fat particles have their own protein coating. The operation may also be applied to paints and similar solid-liquid dispersions. The products are not homogeneous in the strict sense of the word. *See also* homogeneous; heterogeneous.

homologous series. A related succession of organic compounds, each containing one more carbon atom and two more hydrogen atoms than the one before it in the series. For example, the paraffin (alkane) hydrocarbons form a homologous series:

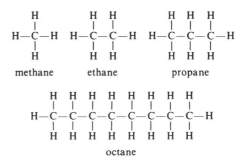

methane ethane propane

octane

Thus, for the paraffin series, the generic formula is C_nH_{2n+2}. Similarly, the olefin (alkene) series has the generic formula C_nH_{2n}, and the aliphatic alcohols, $C_nH_{2n+2}OH$.

homopolymer. A high polymer made by polymerizing only one monomer substance, as compared with a copolymer, which is made from two or more different monomers. Examples of homopolymers are natural rubber, whose monomer is isoprene, and polyethylene, whose monomer is ethylene. *See also* monomer.

hormone. One of a number of complex organic substances which control or regulate various metabolic processes in plants and animals. In man they are formed in the so-called ductless glands (adrenal cortex, pituitary, pancreas, thyroid) and are secreted directly into the blood stream. Examples are adrenaline, ACTH, thyroxine, insulin, cortisone, estrogen, and testosterone. Some are steroids, others amino acids, proteins, and polypeptides; most have been synthesized, but some are made from the specific organs of animals. They regulate such metabolic functions as blood-sugar level (insulin), reproductive functions (estrogen), metabolic rate (adrenaline), and other body processes. They have often been described as chemical messengers. Plant hormones are preferably called plant growth regulators. *See also* insulin; plant growth regulator; auxin.

hot-melt. In the adhesives and sealants fields, a hot-melt refers to a material which is easily liquefied by a relatively low temperature so that it can be poured or spread and which sets quickly on cooling to form a strong and permanent bond. Many hot-melt adhesives contain such synthetics as polyethylene; hot-melt sealants for highway construction are usually composed of combinations of asphalt, rubber, coal-tar pitch, etc. Low-melting waxes are used in packaging.

humectant. A hygroscopic liquid used to maintain constant moisture content in materials with which it is mixed or associated. Common types are glycerol, some glycols, sorbitol, etc. Humectants are used chiefly in tobacco, as a top dressing on leather and plastic products, in textile products, and in cosmetic preparations (skin creams and lotions). When not actually mixed with the materials, they are able to keep the humidity reasonably constant in small containers for tobacco, etc. *See also* hygroscopic.

humidity, relative. The ratio of the amount of water vapor present in air at a given temperature to the maximum that can be held by air at that temperature, i.e., saturation. Since the latter is always represented by 100, the ratio is expressed as a percentage. The saturation capacity of air for water vapor varies greatly with temperature, showing a sharp increase at about 24°C (75°F). Thus, 50% relative humidity at 21°C (70°F) is quite pleasant, whereas at 32°C (90°F) it is highly uncomfortable. Humidity control is of the utmost importance in many industries, such as textile, leather, food, etc.

humus. That portion of a soil comprised of semi-decomposed organic matter (dead vegetation), the modification resulting primarily from the action of bacteria. The lignin, cellulose, and proteins interact to form a diverse range of substances, including aromatic compounds, amino acids, polysaccharides, and organic phosphorus compounds. The part played by microorganisms in humus formation is critical, as they not only effect the decomposition of the vegetation but structurally modify its chief components (lignin and carbohydrates); they also are able to synthesize the numerous aromatic substances found in humus.

Hyatt, John Wesley (1837–1920). Born in Baltimore, Hyatt is generally credited as being the father of the plastics industry. In 1869 he and his brother patented a mixture of cellulose nitrate and camphor which could be molded and hardened. Its first commercial use was for billiard balls. The trade-mark "Celluloid" was the first ever applied to a synthetic plastic product; because of its great flammability hazard, it is little used today.

hybrid. *See* resonance (1).

hydrate. *See* hydration (1).

hydration. (1) As used by inorganic chemists, this term refers to combination of one or more molecules of water with a crystalline inorganic compound, usually a salt, without rupture of the H—OH bond. Such compounds are called hydrates, the structure being indicated by a centered dot, e.g., $Na_2SO_4 \cdot 10H_2O$. When the water (often called water of crystallization) is removed by heating or other means, the compound is called anhydrous. The same substance may be capable of forming several hydrates, which are designated mono-, tetra-, decahydrates, etc., according to the number of water molecules present. A common group of hydrates are the alums (named from aluminum); the best-known of these is $KAl(SO_4)_2 \cdot 12H_2O$. Hydration occurs readily in solution, where the negative portion of the water molecule is attracted to positively charged metal ions. The term is also used to refer to a chemical reaction in which water combines with another substance to form a "hydrated" derivative, e.g., lime; in this case, the reaction is $CaO + H_2O \rightarrow Ca(OH)_2$.

(2) In physical chemistry, hydration (or solvation) refers to the strong affinity of water molecules for ions of dissolved substances, an interaction which is the fundamental cause of electrolytic dissociation. Ions and other charged particles thus acquire a tightly held film of water. Such particles are said to be solvated. This phenomenon may lead to hydrate formation, as defined in (1); it also has important effects in stabilization of colloidal solutions. *See also* solvolysis; solvation.

hydraulic. (1) In chemical usage, this term refers to the hardening or setting action induced by the addition of water to a limestone or cement which contains varying proportions of alumina and calcium silicate. In fine-ground form, the material will set even out of contact with air, the chemical action being a type of hydration. Some heat is evolved in the process. Portland cement is the most familiar example.

(2) In engineering terminology, this term describes fluids used to exert pressure in mechanical systems (braking devices and automobile transmissions). Water is often sufficient for this purpose in industrial equipment such as hydraulic presses; in smaller and more specialized mechanisms, materials of higher viscosity, such as petroleum oils, ethylene glycol, and phosphate esters are used. *See also* (1) cement, hydraulic; (2) brake fluid.

hydraulic fracturing. *See* enhanced oil recovery.

hydrazine. A liquid reaction product of ammonia, chlorine, and caustic soda; its formula is N_2H_4, and it boils at 113.5°C (236°F). It is extremely irritant and corrosive and may explode when heated or by oxidation. A high-energy substance, it is used as a propellant for rockets, as well as in a number of chemical applications such as polymerization initiator, inhibitor of oxidation and metal corrosion, "getter" for carbon dioxide, and in the manufacture of various pharmaceutical products. It is a powerful reducing agent, especially of metallic salts. A highly reactive compound, it readily forms hydrazides with many acids and esters, as well as forming

hydrazones when reacted with carbonyl compounds.

hydride. A compound formed by combination of hydrogen either with a single metal (lithium hydride, LiH), with two metals (sodium aluminum hydride, $NaAlH_4$), with a metal and a nonmetal (lithium borohydide, $LiBH_4$), or with metals plus an organic goup. The latter are called organometallic hydrides. Hydrides are hazardous materials which may catch fire when rubbed or agitated. They are powerful reducing agents used in chemical conversion reactions; they are also employed as sources of hydrogen and as catalysts. The boron hydrides, or boranes, have been researched extensively because of their high energy capability; they evolve hydrogen in the presence of water and are spontaneously flammable. The gas arsine (AsH_3) is extremely poisonous. The organometallic hydrides are assuming prominence and are competitive with the inorganic type.

hydriodic acid. *See* iodine; aqueous.

hydro-. A prefix derived from the Greek word for *water;* in many chemical terms such as *hydrocarbon* and *hydrogenation,* it merely indicates the presence of hydrogen. Closely related is the prefix *hygro,* meaning "dampness" or "moisture," as in *hygroscopic* ("absorbing moisture").

hydroboration. The reaction of boranes (boron hydrides) with unsaturated aliphatic compounds (olefins or acetylenes) to produce organoboranes; when followed by oxidation, this process is useful in the synthesis of alcohols and other organic compounds.

hydrocarbon. Any compound composed of carbon and hydrogen; hydrocarbons may be gases (methane), liquids (benzene), or solids (waxes). Their chemical structure may be straight- or branched-chain, alicyclic, or aromatic. Their chief sources are coal, natural gas, petroleum, and plant life. They have multitudinous uses as heating fuels, solvents, gasoline, and as starting materials for synthetic chemicals. The single most versatile hydrocarbon from the chemical point of view is probably benzene; petroleum is a mixture of many hydrocarbon compounds. *See also* aliphatic; aromatic; paraffin; olefin; acetylene.

hydrochloric acid. *See* hydrogen chloride.

hydrocracking. *See* hydrogenation.

hydrocyanic acid. *See* hydrogen cyanide.

hydrocolloid. A hydrophilic colloidal material used largely in food products as emulsifying, thickening and gelling agents. They readily absorb water, thus increasing viscosity and imparting smoothness and body texture to the product, even in concentrations of less than 1%. Natural types are plant exudates (gum arabic), seaweed extracts (agar), plant seed gums or mucilages (guar gum), cereal gums (starches), fermentation gums (dextran), and animal products (gelatin). Semisynthetic types are modified celluloses and modified starches. Completely synthetic types are also available, e.g., polyvinylpyrrolidone. Most are carbohydrate polymers, but a few, such as gelatin and casein, are proteins.

hydrodealkylation. A term used in petroleum technology to describe a method of removing methyl groups (or other alkyl groups) from a hydrocarbon compound and replacing them with hydrogen. This is effected by treating the compound with hydrogen at high temperatures and pressures. For example, toluene ($CH_3C_6H_5$) can be converted to benzene (C_6H_6) in this way. The process is also useful in improving the quality of more complex petroleum distillation products. *See also* hydrogenation.

hydrofluoric acid. *See* hydrogen fluoride.

hydroforming. *See* reforming.

hydrogasification. *See* gasification.

hydrogen. An element.

Symbol	H	Atomic Wt.	1.0079
State	Gas	Valence	1
Group	1A	Isotopes	
		(stable)	$H^1;H^2$ (deuterium)
		(radioactive)	H^3 (tritium)
Atomic No.	1		

Hydrogen is by far the most abundant element in the universe. The word *hydrogen* literally means ''water-maker,'' since water is an oxide of hydrogen. Atomic hydrogen, made up of one proton and one electron, is represented by H; molecular hydrogen is a diatomic molecule represented by H_2; the isotope deuterium (one proton and one neutron in the nucleus) is represented by H^2; or D; the artificial radioactive isotope tritium (one proton and two neutrons) is indicated by H^3 or T. Hydrogen is the lightest of all elements and is extremely flammable over a concentration range of 4 to 75% by volume.

It is formed readily by the action of steam on iron, on hydrocarbons, or on carbon monoxide, as well as by many other chemical reactions including electrolysis of water. The latter method requires too much electricity to be commercially useful. Several experimental techniques for obtaining hydrogen from water are being explored; e.g., thermochemical decomposition, which involves addition to water of calcium, mercury, and bromine at 500 to 800°C (Euratom plan), and decomposition by solar radiation. Photolytic decomposition has been achieved experimentally. As hydrogen is a major potential energy source, there is continuing active research on its use as a fuel.

Hydrogen is the characteristic element of acids and is a constituent of water, hydrocarbons, and carbohydrates. Among its many chemical uses are production of ammonia and methyl alcohol, numerous hydrogenation reactions (especially of fatty acids) and oxidation-reduction reactions. It is essential in many catalytic reactions in petroleum refining (hydrocracking, hydrodealkylation, etc.). It is also used in welding and in meteorological balloons. It can be liquefied under pressure at −252°C (−420°F); in this form it has rocket fuel capability, and is used as a low-temperature research tool (cryogenics). Its heavy isotopes are used in thermonuclear experimentation. *See also* tritium; fusion.

hydrogenation. A reaction of great value and importance in organic chemistry in which hydrogen is added to the molecules of hydrocarbons, fatty acids, and the like, usually by means of a catalyst. Gasoline may be made by hydrogenation of coal, with tin or iron as catalyst, and many vegetable and fish oils are converted to solid form by hydrogenation with a nickel catalyst. Destructive hydrogenation, in which organic molecules are split into smaller ones capable of reacting with hydrogen, occurs in catalytic hydrocracking of gasoline (hydrogenolysis). Reactions of hydrogen with inorganic materials are not considered as hydrogenation.

hydrogen bond. A type of chemical bond (also called a bridge bond) in which the hydrogen atom of a hydroxyl group is attracted by the electronegative oxygen of a similar molecule. This attractive force effects an association between the molecules of liquids such as water and alcohols, holding them together almost as if they were one huge molecule. Hydrogen bonding at least partially accounts for the fact that alcohols and water do not evaporate rapidly. Bonding of this type is relatively weak; it may be represented by

$$\overset{H}{:\underset{\cdot\cdot}{O}:}\;H:\overset{H}{\underset{\cdot\cdot}{O}:}\;H:\overset{H}{\underset{\cdot\cdot}{O}:}\;H$$

hydrogen chloride. A gaseous compound of hydrogen and chlorine (formula HCl), b.p. $-85°C$ ($-121°F$), which is a powerful irritant to tissue and thus requires extreme care in handling. It is produced by the direct union of chlorine and hydrogen or by distillation of HCl from hydrochloric acid, a water solution containing 18 to 38% HCl. It is used for numerous reaction products in the chemical industry, such as the manufacture of vinyl chloride and other olefin-derived chlorides. Hydrochloric acid is made by the action of sulfuric acid on sodium chloride; as a by-product of the substitution chlorination of hydrocarbons, e.g., $C_6H_6 + Cl_2 \rightarrow C_6H_5Cl + HCl$; and by the thermal dehydrochlorination of organic chlorides such as the conversion of ethylene dichloride to vinyl chloride by the reaction: $C_2H_4Cl_2 \rightarrow CH_2CHCl + HCl$. Hydrochloric acid can be converted to the more valuable product, chlorine, by electrolysis and by reaction with oxygen (Deacon process). It has manifold applications in chemistry and industry, among which are treatment of oil wells to decompose subsurface calcium carbonate structures, metal cleaning and pickling, laboratory reagent, pharmaceuticals, etc. It is a strong and reactive acid which is corrosive to skin and body tissues. In the pharmaceutical trade, it is often called muriatic acid, but this term is not used by chemists. *See also* chlorine; ethylene dichloride; vinyl chloride; Deacon process.

hydrogen cyanide. A highly toxic, flammable, and explosive gas [about $27°C$ ($80°F$)] and liquid (below $80°F$). It is obtained from the reaction of ammonia, air, and hydrocarbon gases. When dissolved in water, it is known as hydrocyanic acid or prussic acid; the formula in either case is HCN. Solutions up to 10% strength are used commercially and are stabilized with phosphoric acid to prevent the violent reaction which occurs on autopolymerization. The acid is found in nature in a few plant products, chiefly bitter almond, from which it is removed before use. It is classed as a protoplasmic poison and should be handled with extreme caution. In spite of this fact, it is used to fumigate fruit trees. Its main uses are in the manufacture of synthetic chemicals such as acrylonitrile, acrylic resins, etc. *See also* cyanogen.

hydrogen fluoride. A compound of hydrogen and fluorine (formula HF) which is extremely corrosive to organic tissue; any direct contact or inhalation can cause burns and internal injury. It is produced in tonnage quantities by the reaction of sulfuric acid and fluorspar (CaF_2). It is gaseous above $19.5°C$ ($70°F$) and liquid below that point. Its major uses are as a catalyst and fluorinating agent. Solution of the gas or liquid in water from 20 to 70% produces hydrofluoric acid, one of the strongest acids known. Its ability to attack glass accounts for its use in roughening and etching panels, bottles, graduates, and other glass products; it is also used in metal treatment and as a fluorine source. It should be handled with proper protective equipment.

hydrogen iodide. *See* iodine; aqueous.

hydrogenolysis. *See* hydrogenation.

hydrogen peroxide. The best known and most widely used of the peroxide compounds, which are characterized by the presence of two linked oxygen atoms; the formula is H_2O_2, the structure being represented by H—O—O—H. It is liquid at room temperature, with a freezing point about the same as water. It is usually handled commercially in aqueous solutions from as low as 3% (for antiseptic use) to 90%. Unless inhibited with acetanilide, it decomposes in the presence of impurities, evolving atomic oxygen. For this reason it is an active oxidizing agent; organic materials will ignite on contact with concentrated H_2O_2, and solutions of such materials in the concentrated peroxide may explode. It has been used for military propulsion purposes (rockets, etc.), but its chief commercial applications are as a bleach in the textile and paper industries; in a number of important reactions in the manufacture of synthetic chemicals; as a blowing agent; and as an initiator of polymerization, the latter effect being due to formation of free radicals during decomposition. *See also* peroxide.

hydrogen sulfide. A flammable and poisonous gas (formula H_2S) with an obnoxious odor; it is a natural decomposition product of protein-containing substances and decaying organic tissue. It is made for industrial use by reaction of ferrous sulfide with sulfuric acid and is also obtained from coke-oven gas. It is a toxic hazard in viscose rayon manufacture, as a result of the treatment of cellulose with carbon disulfide and sodium hydroxide. It is also present in refinery gases and natural gas, from which it is recovered. Its industrial uses are limited; it has some application as a reducing agent, in the manufacture of thiophene (with acetylene), and as a source of elemental sulfur.

hydrolysis. A chemical reaction in which water reacts with another compound so that both it and the compound are cleaved or split; the water decomposes to its ions, hydrogen and hydroxyl, each of which reacts with a portion of the cleaved compound. In inorganic chemistry, it refers to hydrolysis of salts of a weak acid and a strong base to give a basic solution, e.g., $CH_3COONa + H_2O \rightarrow NaOH + CH_3COOH$, or of salts of a strong acid and a weak base to give an acidic solution, e.g., $SnCl_4 + 4H_2O \rightarrow Sn(OH)_4 + 4HCl$. Hydrolysis also refers to the interaction of water with certain molecules, such as phosphorus trichloride: $PCl_3 + 3H_2O \rightarrow P(OH)_3 + 3HCl$. The rate of reaction with water is generally increased with the addition of an acid or base to increase the concentration of hydrogen or hydroxyl ions. In organic chemistry, many types of hydrolysis are important; among them are: (1) formation of a fatty acid and an alcohol by hydrolysis of an ester ($CH_3COOC_2H_5 + H_2O \rightarrow CH_3COOH + C_2H_5OH$); (2) conversion of starch to sugar by enzyme-catalyzed hydrolysis; (3) hydrolytic separation (cleavage) of proteins to amino acids; (4) fat-splitting and saponification reactions. Catalysts are normally required to accelerate hydrolysis; enzymes and bacteria are frequently involved in such catalysis. *See also* solvolysis; saponification.

hydronium ion. An ion (H_3O^+) formed by the transfer of a proton (hydrogen nucleus) from one molecule of H_2O to another; a companion ion (OH^-) is also formed; the reaction is $2H_2O \rightarrow H_3O^+ + OH^-$. Formation of such ions is very infrequent, resulting from the interaction of water molecules in a ratio of about 1 to 550 million.

hydroperoxide. An organic peroxide having the generalized formula ROOH, in which R represents an organic group (either alkyl or aryl). An example is ethyl hydroperoxide (C_2H_5OOH).

Methyl and ethyl hydroperoxides are unstable and thus are strong oxidizing agents and explosion hazards; those of higher molecular weight are more stable. Hydroperoxides can be derived by oxidation of saturated hydrocarbons or by alkylating hydrogen peroxide in a strongly acidic environment; they are used as polymerization initiators. *See also* peroxide.

hydrophilic. (1) With the literal meaning of "water-loving," this term refers to substances that tend to absorb and retain water; this results in swelling, often accompanied by formation of gels. This property is characteristic of some proteins (collagen, gelatin), water-soluble gums (arabic, tragacanth), pectins, and starches, as well as of some types of clay such as bentonite. Hydrophilic agents are widely used as stabilizers and thickeners in food products, pharmaceuticals, and coatings for textiles, paper, etc. The term should not be confused with hygroscopic. *See also* gel; bentonite; hygroscopic; hydrocolloid.

(2) The term is also applied to the surfaces of solids which are readily wetted by water or on which water is adsorbed easily. Thus, wetting is important, e.g., in detergency and in the secondary recovery of oil which involves the displacement of oil from sand by water.

hydrophobic. Literally meaning "water-hating," this term describes substances which repel water and thus are difficult to "wet" or emulsify. Most notable of these are fats, waxes, and oils, but this property also is found in many solids in finely divided form, e.g., carbon black, magnesium carbonate, and metal powders. Rubber is notably hydrophobic, as are many synthetic polymers. In liquids, this property is often due to differences in surface tension. *See also* surface tension; wet.

hydroponics. The growing of plants (flowers and vegetables) in aqueous solutions of nutrients, without the aid of soil; also called water culture and soilless growth. This method has been successfully used on small islands where neither space nor soil for agriculture is available. A high yield of vegetables can be obtained in a small area if the operation is conducted by experienced personnel.

hydroquinone. An unsaturated cyclic compound, m.p. 170°C (338°F) containing two hydroxyl groups; the synonym para-dihydroxybenzene explicitly describes its structure, represented by the formula $C_6H_4(OH)_2$:

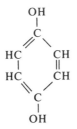

It is made in several ways, (1) by reducing para-benzoquinone with sulfur dioxide, (2) by formation of quinone from aniline by oxidation, followed by reduction. Hydroquinone is a strong reducing agent and thus is an effective photographic developing agent; it is also used as an oxidation inhibitor and stabilizer in organic materials such as rubber, paints, oils, etc.

hydrosol. *See* organosol.

hydroxyaniline. *See* aminophenol.

hydroxyapatite. *See* phosphorus.

hydroxyethylcellulose. *See* ethylene oxide.

hydroxylamine. *See* oxime.

hydroxyl group. A combination of one atom each of oxygen and hydrogen (OH), which tends to behave as a single atom in chemical reactions. It has a valence of 1 and occurs in many organic and inorganic compounds. In organic compounds, this group is characteristic of all alcohols, including phenol, cresol, resorcinol, cholesterol, etc., its presence being signified by the suffix *-ol*. Common examples of hydroxyl-containing inorganics are water (HOH), sodium hydroxide (NaOH), and hydrated lime [calcium hydroxide, $Ca(OH)_2$]. When a hydroxyl-containing inorganic molecule ionizes in solution, the group assumes a negative charge (OH^-) and is then called a hydroxyl radical. *See also* group; radical.

hygroscopic. A term used to describe solid or liquid materials which pick up and retain water vapor from the air. Some solids having a high internal surface area are made especially for this purpose; by adsorbing vapor molecules they can maintain a small volume of air virtually free from moisture, for example, in food packages and laboratory desiccators. Examples of these are silica gel and activated carbon. Hygroscopic liquids, such as glycerol and propylene glycol, are used as humectants. Many materials that are more or less hygroscopic, e.g., paper and textiles, should be stored under conditions of controlled humidity. Several inorganic compounds are very hygroscopic. Among them are calcium and magnesium chlorides and sodium and potassium hydroxides, which will literally "melt" or dissolve in the water they absorb from the air when exposed. Such materials are called deliquescent, and must be carefully stored and handled. *See also* desiccation; humectant; hydrophilic; deliquescent.

hypo-. (1) A prefix used in inorganic chemistry to denote acids and salts whose central multivalent positive element is in its lowest oxidation state; it is also the acid or salt containing the least oxygen. Examples are hypochlorous acid, HOCl, whose Cl has a valence of $+1$, and hypophosphorous acid, H_3PO_2, whose P has a valence of $+1$. Salts of these acids also carry the prefix *hypo*.

(2) Other meanings of *hypo* vary somewhat but have the general sense of "below," "less than," and "underneath," which appear in such terms as *hypodermic* ("underneath the skin") and *hypothesis* ("less than certain"). It is the opposite of *hyper*.

(3) *Hypo* is often used as a slang term in the photographic industry to refer to sodium thiosulfate, a solution of which is a developing assistant. This compound is erroneously called sodium hyposulfite.

hypochlorous acid. A weak acid, HOCl, but a strong oxidizing agent, it is the first of the series of chlorine-oxygen acids. While the formula HClO is also used for it, HOCl corresponds to its structural formula, H—O—Cl, similar to that of the other chlorine acids containing chlorine, i.e.,

H—O—Cl=O for chlorous acid; $H—O—\overset{\overset{O}{\|}}{\underset{\underset{O}{\|}}{Cl}}$

for chloric acid; and $H—O—\overset{\overset{O}{\|}}{\underset{\underset{O}{\|}}{Cl}}=O$

for perchloric acid. In these the chlorine has the positive valences of $+1$, $+3$, $+5$ and $+7$, respectively. Hypochlorous acid exists in solution when chlorine added to water hydrolyzes: $Cl_2 + H_2O \rightarrow HCl + HOCl$. When an alkaline compound, such as NaOH or $Ca(OH)_2$, is present in the solution, the corresponding sodium or calcium hypochlorite is formed, and it is in this form that the material is most useful. The acid and its salts are widely used in the bleaching of paper, textiles, and other materials; they are also used for water purification and for sterilization. The oxiding value of the Cl^{+1} of HOCl changing to Cl^{-1} is the same as that of Cl_2 changing to $2Cl^{-1}$ in such applications. Chlorohydrins can be made by the addition of HOCl to the double bond of alkenes. *See also* chlorohydrin; bleaching; chlorine.

hysteresis.   The dissipation of energy in one form or another in a material, when the forces acting on it are reversed. This applies, for example, to the magnetic induction of iron and iron alloys which does not respond reversibly to change in the magnetizing force, as well as to the resilient energy of rubber and other elastomers, some of which is "lost" when the deforming force is reversed, the energy decrement appearing as heat. The term actually means a "lagging behind."

I

I Symbol for the element iodine, the name being derived from the Greek word for *violet,* referring to the color of its fumes.

-ic. A suffix used in naming inorganic compounds to indicate that the multivalent positive element in the compound is in its highest valence state. It is used (a) to designate that the metal in a compound has its highest oxidation number, e.g., ferric chloride ($FeCl_3$), ferric oxide (Fe_2O_3) or ferric sulfate [($Fe_2(SO_4)_3$)], in which the iron valence is 3; and (b) to indicate that the central positive valence element in an acid is in its maximum valence state, e.g., sulfuric acid (H_2SO_4), in which the sulfur has a valence of 6. In the first case, a more accurate terminology is to use the symbol or name of the metal followed by a roman numeral in parentheses to show the valence state, e.g., Fe(III) or iron (III) chloride. *See also* -ous.

Iceland spar. *See* calcium carbonate; Nicol; spar.

-ide. A suffix used in naming compounds comprised of two elements; in such names, the first element, being electropositive, retains its name without change, but the second, being electronegative, utilizes the suffix *-ide* as a modification of the elemental name. Examples are: sodium hydroxide, magnesium chloride, hydrogen sulfide, etc. Similarly, oxygen is modified to oxide, fluorine to fluoride, phosphorus to phosphide, and carbon to carbide.

ideal. Descriptive of a substance or system which exists only in theory, that is, which is in exact accord with the principles governing its behavior. The term *perfect* is also used in the same sense. For example, an ideal or perfect gas would be one that precisely obeys the laws of Boyle and Charles, and an ideal solution would be one that behaves in accordance with Raoult's law. Actually, there are no instances where this occurs, as all materials and systems deviate slightly from ideal behavior. The concept is essentially mathematical, enabling scientists to reach important theoretical conclusions of broad general application.

identity period. In any polymeric substance, the characteristic repeating unit, or monomer, which may occur an indefinite number of times in the natural or synthetic molecule. Its length is usually measured in angstrom units. The brackets in the following structure indicate the identity period of anhydroglucose units in cellulose:

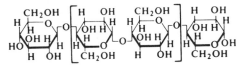

ignition point. *See* autoignition point.

imide. An unsaturated organic acid whose formula includes nitrogen and in which two double bonds are present, e.g., succinimide.

imine. A nitrogen-containing organic substance having a carbon-nitrogen double bond

Such compounds are highly reactive, even more so than the carbon-nitrogen triple bond characteristic of nitriles.

immiscible. A term used to describe substances of the same phase that cannot be uniformly mixed or blended. It usually applies to liquids, such as oil and water, but may also refer to solids in powder form which differ widely in some physical property, such as magnesium carbonate and barium sulfate. In the case of liquids, immiscibility is due to differences in their chemical nature, which can often be overcome by addition of a third substance called an emulsifier or detergent. *See also* miscible; compatibility.

immunochemistry. The chemistry of antibodies and antigens and their disease-protective mechanism in animals and man. It primarily involves the study of specialized proteins, structural antagonism, allergies, and related responses. Its original applications were made by Jenner (1775) and Pasteur (1880). *See also* serum **(2)**; antibody.

impalpable. A term used to refer to a powder composed of particles that are too small to be sensed as such by manipulation with the fingers. Such powders usually have a slippery feeling, such as talc and mica, and thus have definite lubricating properties.

impermeable. Relatively resistant to the passage of a gas, vapor, or liquid. For example, many synthetic rubbers are said to be impermeable to air, by which is meant that air will pass through them much less slowly than through natural rubber. The same applies to moisture vapor-proof plastics for food packaging. There are few materials that are absolutely impermeable to gases; glass most nearly approaches this, but helium will migrate through it at cryogenic temperatures.

impingement. A term used in the chemical industry to refer to a carbon black made by combustion of natural gas in an open burner, the carbon given off being collected on a metal surface (called a channel) immediately above the burner. The carbon particles strike, or impinge on, the metal and are later removed for use as "channel black" or "impingement black."

impregnation. Application of a liquid or semiliquid material to a more or less porous or absorptive substrate, for example, an asphaltic compound to paper or a soft rubber mixture to fabric. The effect is one of complete penetration of the substrate fibers, rather than of a surface coating.

impurity. An extremely low percentage of an extraneous substance either naturally present in a material or added to it accidentally or intentionally. Often it is impossible to completely eliminate naturally occurring impurities by any separation method, but there are many processes that are effective in reducing them to an acceptable minimum. So small are the amounts involved that impurities are usually indicated in parts per million (ppm). Common examples are calcium carbonate in hard water, strontium-90 in milk, mercury compounds in fish, and insecticide residues in foods. Of special importance is the fact that not all impurities are undesirable; their presence may have a beneficial effect, and they are often added intentionally. Examples are (1) arsenic or boron in semiconductors, (2) carbon in iron or steel, (3) doping agents in laser crystals, and (4) fluoride in drinking water. *See also* contaminant; additive.

In Symbol for the element indium, the name being assigned because the lines of its spectrum are the color of indigo.

incident. The literal meaning of this term is "falling on;" it is conventionally used to refer to light rays and their angle of contact as they impinge on a solid or liquid surface.

inclusion complex. *See* clathrate.

indanthrene blue. *See* vat.

Indian red. *See* iron oxide.

indicator. An organic compound which, when in contact with another substance, mixture, or solution, indicates its pH (acidity or basicity) by a change in color; it also may indicate whether or not a redox reaction has gone to completion and the extent of a precipitation. Many indicators are organic dyes (Congo red, methyl orange, methylene blue, litmus) which change color within a specific range of pH, as from red to yellow or blue to red. Others are derivatives of phthalic anhydride (phenolphthalein), which are colorless under acid conditions and red under basic conditions. Starch is used for iodine determination. The most common use of indicators is in titrations. *See also* pH; titration; scavenger.

indium. An element.

Symbol	In	Atomic Wt.	114.82
State	Solid	Valence	1,3
Group	IIIA	Isotopes	2 stable
Atomic No.	49		

Indium is a very soft, silvery metal, m.p. 156.6°C (314°F), always occurring as a component of ores of other metals; it is corrosion resistant at room temperature and thus is used as a coating for bearings. It alloys readily with many other metals, the products being used for brazing and soldering. It is also useful in semiconductor and electronic technology. In reactions, indium is able to form covalent rather than ionic bonds. It has no toxic properties. Indium is produced in considerable volume in Japan and to a lesser extent in British Columbia and Peru.

indoleacetic acid.   *See* auxin; plant growth regulator.

industrial alcohol.   This term includes four items (1) pure ethyl alcohol purchased by chemical manufacturers for nonbeverage use, with or without payment of tax, depending on the circumstances of its use (scientific research, government agencies, etc.); (2) completely denatured alcohol; (3) specially denatured alcohol; and (4) proprietary solvents made from these. *See also* denaturant.

industrial carbon.   A general term considered to include all forms of carbon used for industrial purposes with the exception of fuels; it does not include graphite and diamond but does encompass carbon black, charcoal, activated carbon, and similar products. Most of the uses of these amorphous carbons are based on their strong adsorptive capacity, electrical conductivity, reinforcing ability, and pigmenting property. *See also* carbon black; activated carbon.

industrial chemistry.   For many years this term was used to describe a practical mixture of general chemistry and chemical engineering, applying to large-scale manufacturing in which chemical reactions are involved; it embraced the so-called process industries, including paint, plastics, textiles, inorganic and synthetic organic chemicals, leather, paper, rubber, etc. In recent years it has fallen into disuse in educational circles; though it is no longer generally regarded as an independent discipline, several universities offer courses in this subject.

industrial diamonds.   *See* diamond.

-ine.   When used as a suffix, *-ine* is generally characteristic of nitrogenous substances, e.g., amine, benzidine, purine, thiamine, serine, lysine, etc. The spelling *benzine* for *benzene* is thus inadmissable. In names of halogen ele-

ments, the terminal *-ine* is not a suffix but is an integral part of the word.

inert.   Having little or no chemical affinity or activity. The six elements comprising the noble-gas group of the Periodic Table (sometimes called Group 0) are either wholly or relatively inert. The first three (helium, neon, argon) have a valence of 0 and do not enter into any chemical combination. The last three (krypton, xenon, radon) have several valences but are capable of only limited compound formation. Thus, these gases are collectively called "noble." Less specifically, nitrogen and carbon dioxide are often referred to as "inert atmospheres," as they are unreactive under normal conditions; they are used to protect certain highly reactive substances from contact with atmospheric oxygen. In the same sense, inert solids are generally considered to be clays, asbestos, talc, sand, etc. The phrase *inert ingredients* on labels usually refers to water, the most common example of an inactive solvent. *See also* noble.

inflammable.   This term is not acceptable to safety authorities; the proper word is flammable.

infrared spectroscopy.   An instrumental technique of analytical chemistry in which substances of any type (gases, liquids, or solids) are identified by their tendency to absorb light (energy) of infrared wavelength (2 to 40 microns). The resulting absorption spectrum makes it possible to analyze complex mixtures quantitatively, to identify impurities and other unknown components, and to determine the various kinds of atomic groups in a molecule; in short, the infrared spectrum offers an infallible molecular "fingerprint." The ability to detect trace quantities of impurities on a continuous basis makes this method useful as a quality control tool in many industries, e.g., steel making. Standard infrared spectra of a wide range of pure substances have been prepared on punched cards.

Substances also give off radiation in the infrared at extremely high temperatures. Study of these emission spectra yields important information to aerospace engineers and astronomers. *See also* spectroscopy; radiation.

inhibitor.   Any substance that retards or reduces the rate of a chemical reaction; the opposite of a catalyst. Antioxidants are well-known inhibitors of oxidation reactions in foods, rubber, and other organic materials. Protective coatings of various kinds inhibit the corrosion of metals. Spontaneous polymerization reactions are prevented in reactive monomers, e.g., acrylic acid, by incorporation of suitable inhibitors during storage and shipment. Salicylic and benzoic acids are often added to food products to prevent rancidity. The formation of unwanted foams can be suppressed by certain inhibiting agents (higher alcohols), and there are several benzenoid compounds which prevent deterioration of a substance due to ultraviolet light by absorbing the radiation (U.V. inhibitors). Organic phosphorus esters are inhibitors of cholinesterase formation in the body. *See also* retarder; cholinesterase.

initiating explosive.   An explosive of high shock sensitivity used to cause detonation of a less sensitive explosive such as dynamite or TNT, usually in the form of blasting caps or similar devices. Initiating explosives regularly used are mercury fulminate, lead azide, and diazonitrophenol. *See also* initiator; explosive.

initiator.   A substance exhibiting unusual chemical activity, which enables it to set in motion, or induce, a chemical reaction such as polymerization, oxidation, or detonation. Well-known initiators are peroxides, which evolve atomic (nascent) oxygen, organic ions (carbanions and carbonium ions), and free radicals. Benzoyl peroxide is an instance of an organic initiator. *See also* free radical; chain reaction; peroxide.

ink.   (1) Writing: water solutions of appropriately colored dyes or pigments having consistencies varying from free-flowing liquids to semisolid pastes. Tannic acid is present in small amounts to act as mordant in fixing the colorant to the paper.

(2) Printing: mixtures of linseed or similar drying oil or a resin with a suitable pigment; metallic soaps may be included to accelerate oxidative drying. The composition is similar to that of exterior paints. Carbon black, iron blue, lithol red, indigo, and phthalocyanine pigments are used as colorants.

inorganic.   This term refers to a major and the oldest branch of chemistry and is concerned with substances which do not contain carbon, with several exceptions: such binary compounds as the carbon oxides, the carbides, and carbon disulfide; such ternary compounds as the metallic cyanides, metallic carbonyls, phosgene ($COCl_2$), carbonyl sulfide (COS), etc.; and the metallic carbonates, such as calcium carbonate and so-dium carbonate. Primarily, the field of inorganic chemistry centers around the elements other than carbon and their compounds, including much of physical chemistry, atomic structure, chemical bonding, electrochemistry, and coordination chemistry. Another view is that inorganic chemistry includes all substances except the hydrocarbons and their derivatives; however, organic chemistry is also concerned with compounds containing such elements as nitrogen, oxygen, sulfur, etc., in addition to hydrogen and carbon, e.g., the carbohydrates. There is no sharp dividing line between inorganic and organic chemistry in view of the many borderline cases such as organometallic compounds, electroorganic reactions, etc.

Regarding the importance of inorganic chemistry, R. T. Sanderson has written: "All chemistry is the science of atoms, involving an understanding of why they possess certain characteristic qualities and why these qualities dictate the behavior of atoms when they come together. All properties of material substances are the inevitable result of the kind of atoms and the manner in which they are attached and assembled. All chemical change involves a rearrangement of atoms. Inorganic chemistry [is] the only discipline within chemistry that . . . examines specifically the differences among all the different kinds of atoms." *See also* organic.

insecticide.   A chemical compound that is lethally toxic to insects either by ingestion or by body contact; it is applied to vegetation, crops, and insect-breeding areas either as liquid spray or as dry powder. There are five major types: (1) inorganic metallic compounds, for example, Bordeaux mixture (copper sulfate-lime); (2) natural organic compounds, such as rotenone; (3) chlorinated hydrocarbons, typified by DDT; (4) organic phosphorus esters (parathion and its derivatives); (5) methylcarbamates, such as carbaryl (l-naphthyl N-methylcarbamate). Of these, the fourth is by far the most toxic to man, and such insecticides must be used by trained personnel who are familiar with their risks. The chlorinated hydrocarbons are highly persistent and though their toxicity to man is much lower, their presence in foods is officially regulated. They are also ecologically damaging to certain species of fish and birds; agricultural use of several types (including DDT) has been prohibited in the U.S. Fuel oil used on ponds for mosquito control and hydrogen cyanide in orchard sprays may also be considered insecticides. *See also* pesticide.

instantizing. *See* agglomeration.

instrumental analysis.   As ordinarily used, this term refers to methods of analytical chemistry in which rather sophisticated instruments are required. Most important are those involving light or radiant energy (optical methods), which are used in the many types of spectroscopy (infrared, x-ray, NMR, etc.), and those utilizing electrical measurements, including potentiometry, coulometry, magnetic susceptibility, and mass spectrometry. *See also* analysis; spectroscopy.

instrumentation.   Application of specialized devices, often electronic, to measure continuously, and in some cases control, such phenomena as temperature, chemical composition, light, electrical activity, and radioactivity. Many such devices, e.g., the thermostat, operate on the feedback principle; others, such as photoelectric cells and automatic thickness gauges, are activated by interruption of the current supply. These instruments thus permit continuous automatic control of specific conditions or properties. The term includes the broad array of instruments for measuring absorption and emission of radiant energy in analytical processes (spectrometers, scintillation counters, and the like). *See also* instrumental analysis.

insulator.   Derived from the Latin word for *island,* this term refers to any material which can inhibit the radiation of heat (thermal insulator) or the passage of an electric current (electrical insulator or dielectric). Among the thermal insulators are such materials as glass fiber, cellulosic fibers, magnesium silicate, etc. Air is an excellent thermal and electrical insulator. Commonly used electrical insulators are paper and other solid forms of cellulose, glass, rubber, mineral oils, silicone oils, and askarel. The liquid types are known as transformer oils, the last two being synthetic products. *See also* dielectric; refractory.

insulin.   An important hormone whose function in the body is to control the metabolism of starches and sugars. It is a protein of comparatively low molecular weight and was first isolated by the Canadian physician, Frederick G. Banting (1891–1941), Nobel Prize 1923, in the early 1920s. The name is derived from *insula* (Latin word

for *island*) because the hormone is secreted from the so-called isles of Langerhans in the pancreas. It is used in medical control of diabetes. Chemical synthesis of this complex molecule was a notable achievement. It has also been made by gene-splicing techniques. *See also* hormone; recombinant DNA.

interface.   The area of contact or boundary layer between two forms of matter or two phases of a dispersion, such as: (1) solid particles in contact with a gas or vapor; (2) solid particles in contact with a liquid; (3) liquid droplets dispersed in another liquid; (4) solid surfaces in contact with each other, as in protective coatings. Interfaces are physicochemically active areas at which a considerable force exists because of the adsorptive power of finely divided solids and the surface tension of liquids. Together these comprise what is known as interfacial tension. This affects the foregoing situations as follows: (1) as a result of strong physical or chemical adsorption of gases at the interface, solid particles (especially metals) are able to act as catalysts; (2) solid particles of such materials as clays are not "wetted" by liquids of high surface tension (water) and do not form uniform dispersions without the aid of a wetting agent; (3) in liquid-liquid dispersions (oil-in-water), the interfacial tension holds the droplets apart; thus, a detergent is necessary to decrease the tension enough to permit miscibility, as in emulsions. The study of interfacial phenomena comprises a large part of the science of colloid chemistry. *See also* surface tension; intermetallic compound; detergent; emulsion; catalysis; surface-active agent.

interferon.   An antiviral protein produced in body cells in response to virus infection. Discovered in 1957, it is a product of the infected cell, rather than of the disease-inducing virus. It can be formed by any type of infected cell, and is non-specific in its protection; i.e., it is effective against many viruses, but only in the cells of the organism that produced it. A synthetic interferon molecule has been produced, by gene-splicing methods (recombinant DNA), that will greatly increase the supply of this disease-fighting material and thus make it available for treatment of cancer and virus infections. *See also* recombinant DNA.

intermediate.   As originally used in organic dye chemistry, this term referred to a compound midway in the reaction sequence between the starting material and the terminal product. For example: (1) benzene may be reduced to aniline, the parent substance of a large family of dyes; (2) oxidation of naphthalene gives phthalic anhydride, from which are produced plasticizers, dyes, resins, and pharmaceuticals. In these examples, aniline and phthalic anhydride are intermediates. The situation is similar with phenol and other aromatic compounds. In current terminology, the meaning has been extended to include many straight-chain (acyclic) materials which themselves have many important end uses, as well as being sources for useful derivatives, e.g., ethyl alcohol, chloral, etc. Thus, the term intermediate is much less specific now than formerly and is often loosely used for any organic compound from which many others can be made.

intermetallic compound. A term used in metallurgy to describe an alloy formed by the heat-induced diffusion of two metals at the interface between them; for example, when one metal is coated with another for protective purposes, an intermetallic layer often is formed. This occurs in the process known as cementation. The alloys are not true compounds in the chemical sense but are actually mixtures or solid solutions of two or more metals. *See also* cementation; alloy.

International Union of Pure and Applied Chemistry (IUPAC). A voluntary, nonprofit association of national organizations representing chemists in 45 member countries. It was formed in 1919 with the object of facilitating international agreement and uniform practice in both academic and industrial aspects of chemistry. Examples are nomenclature, atomic weights, symbols and terminology, physicochemical constants, and certain methods of analysis and assay. Reports on such subjects are presented at the biennial conferences of the Union by the numerous Commissions dealing with them, discussed, and, after approval, are published. Its offices are in Basle, Switzerland.

interstitial. (1) In a crystal lattice, an atom of another element (impurity), which causes a defect or dislocation, for example, the presence of an atom of carbon or nitrogen in an iron crystal, or of arsenic in a semiconductor.

(2) In biochemical terminology, a substance (usually water) occurring between the cells of other structural units in a tissue.

interstitial compound. A metallic compound of a transition metal containing carbon, boron, nitrogen, or hydrogen, wherein the interstices between the atoms of the metal lattice are occupied by one of these elements to form carbides, borides, nitrides, or hydrides, respectively. Such compounds are frequently nonstoichiometric in that they do not have definite formulas.

inversion. In chemistry, this term refers to a reversal in the direction of optical rotation of a carbohydrate molecule as a result of acid hydrolysis; it may be either from right (dextro) to left (levo), or the opposite. Sucrose, the common form of sugar, is dextrorotatory; it can be readily hydrolyzed to glucose and fructose. The resulting mixture is levorotatory and thus is called *invert* sugar. It is commercially used in confectionery and other food products because it does not crystallize as easily as do other forms of sugar and tends to minimize loss of moisture. Inversion of sucrose can also occur by fermentation in the presence of the enzyme invertase. *See also* optical rotation; sugar.

invertase. *See* sucrose.

invert sugar. *See* inversion; sucrose.

in vitro. An expression adopted from the Latin, literally meaning "in glass," that is, an experiment (usually of a biochemical nature) that is carried out in laboratory equipment rather than on living organisms. The opposite meaning is given by the expression *in vivo*.

iodine. An element.

Symbol	I	Atomic No.	53
State	Solid	Atomic Wt.	126.9045
Group	VIIA	Valence	1,3,5,7

Iodine, m.p. 113.6°C (236.5°F), is one of the less active members of the halogen family of elements and is classed as a nonmetal. It has only one stable form. It is quite poisonous in the concentrated state, and even the 3% solution used as an antiseptic is dangerously toxic if swallowed, though it can be applied topically without risk. It is extracted from Chilean nitrate, which contains about 0.2%, from natural brines in Michigan, and to a limited extent from seaweed. It is used in iodized salt for protection against goiter, as a reagent chemical, as a catalyst, as an indicator of chemical unsaturation, and in organic synthesis. One of its important compounds is hydrogen iodide which in water solution is called hydriodic acid.

iodine number. The amount of iodine, in percent by weight, that is chemically added to a given quantity of an unsaturated material, especially one in which the double bonds are not conjugated. This technique for determining the extent of chemical unsaturation is used primarily for fatty acids (vegetable oils), rubber, and similar materials. The iodine number (or value) increases rapidly with the number of double bonds present. Historically, the names of Wijs, Hubl, and Hanus are associated with this procedure, which dates back to the late nineteenth century.

ion. An atom (H, Ag, Cl), group (OH, SO$_4$), or molecule (O$_2$) that has either lost one or more electrons and thus become positively charged (cation) or gained one or more electrons and thus become negatively charged (anion). The properties of ions are quite different from those of the neutral units from which they are derived, because of the presence of the electric charge. In a solution, they are closely associated with molecules of the solvent; they also may exist in solids and gases. Positive ions are formed in some gases by electric discharge or electromagnetic radiation. Strong ionizing radiation (x- and gamma rays) can remove electrons from atoms and molecules of organic substances, with lethal effects on living organisms. Short-lived, highly reactive organic ions can be formed with carbon; these act as initiators in organic synthesis. *See also* ionization; carbonium ion; ionizing radiation.

ion exchange. Substitution of one ion, either positive (cation) or negative (anion), for another of the same charge when an ion-containing solution is passed into a molecular network having either acidic or basic substituent groups which can be readily ionized. The ions in the solution attach themselves to the network, replacing the free or mobile ions contained therein. Thus, there are two major kinds of ion-exchange reactions, depending on whether the substituent groups of the network are acidic or basic: (1) cation exchange, including replacement of a cation by hydrogen, and (2) anion exchange, in which a hydrogen ion and anion are removed. There are a number of modifications and variants of these types.

Many insoluble substances, both inorganic and organic, which have an interlocking molecular structure (cage) can act as ion exchangers, e.g., aluminosilicates (clays), cross-linked organic polymers (ion-exchange resins), sulfonated coal and cellulose, and, to some extent, proteins and nucleic acids. Ion exchange is a reversible phenomenon with many large- and small-scale uses: water softening (zeolite), chromatographic analysis, elimination of ionic impurities (demineralization), rare-earth separations, acid removal, etc. Because of their predominantly siliceous nature, soils exhibit ion-exchange activity, which has an important bearing on fertilizer application. *See also* molecular sieve; zeolite.

ionic atmosphere. A term used by physical chemists to characterize a diffuse population of ions surrounding a charged particle in a solution; the particle may be an ion, a molecule, or a colloidal micelle. Most of the ions in the "atmosphere" have a charge opposite from that of the particle they surround, and the net strength of the charge diminishes with their distance from the central particle.

ionic bond. A type of chemical bond (often called electrostatic) in which atoms of different elements unite by transferring one or more electrons from one atom to the other to form an ionizing or polar compound. For example, an element such as sodium which has but one electron in its outer shell readily links with chlorine, which has seven; this provides a total of eight electrons (characteristic of argon), which is energetically ideal. The sodium loses its electron and thus becomes positively charged, while the chlorine gains the electron and becomes negatively charged. Bonds of this kind are less stable than covalent bonds; the compounds formed will dissociate into ions when dissolved in water, as a result of the high dielectric constant of the water, which reduces the strength of the bond to the point of rupture. Ionic bonding and dissociation can be represented as follows:

$$\text{Na} + \text{Cl} \longrightarrow \text{NaCl} \xrightarrow{\text{H}_2\text{O}} \text{Na}^+ + \text{Cl}^-$$

| sodium | chlorine | sodium chloride | sodium ion | chloride ion |

See also ionization.

ionization. A chemical change by which ions are formed from a neutral molecule of a solid, liquid, or gas. The most common type of ionization occurs when an ionically bonded inorganic compound, such as NaCl or H$_2$SO$_4$, is dissolved in water (or other solvent); the molecule separates, or dissociates, into two ions, the metallic ion being positively charged by loss of an electron and the nonmetallic ion being negatively charged by gaining an electron (electrolytic dissociation). The degree of dissociation varies with the type of compound, the solvent, and the temperature. Gaseous molecules can be ionized by electric energy.

Compounds that ionize in solution are called electrolytes as they carry an electric current and thus greatly increase the conductivity of the solvent. Ionization is most effective in water which, because of its high dielectric constant, lowers the ionic bonding forces in the solute molecules enough to cause separation of their constituent atoms or groups. Ion formation produces a notable rise in the boiling point and a depression of the freezing point of water. An electric current passed through a solution containing ions causes them to move to the oppositely charged electrode; this effect is the basis of many industrial electrochemical operations, such as electroplating and the manufacture of sodium hydroxide and chlorine. *See also* ion; electroplating; electrolysis; electrolyte.

ionizing radiation. Electromagnetic radiation, e.g., cosmic, gamma, and hard x-rays, which has sufficient energy to cause formation of free radicals in substances with which it comes in contact, by removal of electrons (ionization). This causes degradation of organic matter, especially of body tissue, which may result in death if the absorption is prolonged. Less severe exposures are likely to bring about damaging changes in DNA, affecting the genes, and to have deleterious effects on bone marrow and protein structures. The gamma rays emanating from nuclear fission are thus a threat to entire populations, wholly apart from the blast effect. *See also* radiation.

ionomer. A type of cross-linked noncrystalline copolymer of unsaturated hydrocarbons, such as ethylene, and unsaturated acid salts; these form aggregates on moderate heating and bring about cross-linking, but at higher temperatures, they are weakened to the point that the material becomes thermoplastic. Both ionic and covalent bonds are present. Such resins contain positively and negatively charged groups which impart unique physical properties. They are strong, transparent, and resistant to oils and abrasion. One of their major uses is as packaging films for food and confectionery products.

Ir   Symbol for the element iridium, the name being derived from the Greek word for *rainbow,* referring to the numerous colors it displays when attacked by acids.

iridium.   An element.

Symbol	Ir	Atomic Wt.	192.22
State	Solid	Valence	1,2,3,4,6
Group	VIII	Isotopes	2 stable
Atomic No.	77		

A member of the so-called platinum group of metals, iridium is the heaviest of all elements, with a density, 22.65 g/cc, twice that of lead. Its m.p. is 2443°C (4430°F), and it is extremely resistant to corrosion and other forms of chemical attack. It occurs chiefly in Canada, Alaska, U.S.S.R., and South Africa. In finely divided form it is used as a catalyst, especially for petroleum products; its major industrial application is as an alloying element with platinum to produce electrical contacts, resistance thermal devices, and the like. Its electrical conductivity is about twice that of platinum. With its numerous valences, it is quite reactive, forming many oxides, halogen salts, and sulfides, as well as more complex compounds.

Irish moss.   *See* carrageenan.

iron.   An element.

Symbol	Fe	Atomic Wt.	55.847
State	Solid	Valence	2,3
Group	VIII	Isotopes	4 stable
Atomic No.	26		

A strongly reactive transition metal, m.p. 1536°C (2797°F), iron occurs widely (especially in Minnesota and Alabama), usually in the form of oxides (hematite, taconite); it is present in most soils in a concentration of from 1 to 6%. Iron ores are reduced by heating (smelting) with coke and calcium carbonate to yield pig iron, primarily used for steel manufacture. Iron has important electrochemical and magnetic properties; it will remove hydrogen from water and thus is easily oxidized to form rust (corrosion), a reaction in which water acts as catalyst. Finely divided and purified iron is used as a chemical reagent and catalyst, as well as in magnets. Its physical properties can be modified by heat treatment (tempering) and by the presence of low percentages of carbon. Biochemically, it is essential in life processes and is the primary constituent of red blood cells (hemoglobin).

iron blue.   A blue pigment whose molecule is made up of iron in combination with carbon, nitrogen, and hydrogen (ferriferrocyanide). It is also called Prussian blue. In spite of the chemical name, it is not poisonous; it is insoluble in water and organic solvents but is attacked by alkaline media. Iron blue is used chiefly as a colorant in paints, baked enamels, printing inks, and laundering additives. *See also* phthalocyanine.

iron oxide.   Any of a number of iron-oxygen compounds (also called ferric or ferrous compounds) in which the iron appears in several oxidation states, e.g., FeO, Fe_2O_3, Fe_3O_4, or mixtures of these. In finely divided form they have colors ranging from dark reddish black to dull red. The last is the best-known (Fe_2O_3) and is variously called burnt sienna, Turkey red, Indian red, red oxide, and rouge; it is widely used as an industrial pigment, as a polishing abrasive, and as a catalyst. Special preparations of iron oxide are iron sponge and iron mass, in which the fine-ground oxide is mixed with wood particles. It is a generally used colorant in plastics, rubber, paint, cosmetic preparations, etc., and in the removal of sulfides from gases. A synthetic type is used as an abrasive for polishing glass (especially mirrors). The darkest iron oxide, which has magnetic properties, is used in recording tapes and electronic devices.

iron sponge.   *See* sponge; iron oxide.

irradiation.   Subjection of a substance to radiation of a specific type or wavelength in order to initiate a chemical change. There are several kinds of irradiation: (1) ultraviolet irradiation of milk to increase its vitamin D content (no longer widely practiced); (2) gamma-ray irradiation of ethylene and hydrogen bromide to synthesize ethyl bromide; (3) cross-linking of polyethylene and vulcanization of rubber by high-energy irradiation. *See also* radiation.

irreversible.   A chemical or physicochemical change of state that is permanent and cannot be reversed. Examples are: (1) A chemical reaction that can proceed in only one direction, i.e., to the right, giving a product that is stable and that cannot revert to the original constituents; most reactions are of this type. (2) A colloidal system that cannot be restored to its original form after coagulation or precipitation, for example, the hardening of egg white or milk protein by heat,

the formation of butter from milk (mechanical action), and the coagulation of rubber latex by acid (chemical action). *See also* reversible.

iso-.   A prefix used in chemical and biological terms meaning "the same," for example, isotope, isomorphic, isomer. When used in combination with the name of an aliphatic compound, it means that the molecular structure of the compound is a branched chain, for example, isopentane, isooctane, isopropyl alcohol. *See also* isotope; branched chain.

isobutane.   *See* alkylation; isomerization.

isobutene.   *See* butyl rubber.

isobutyl alcohol.   A primary monohydric alcohol, b.p. 107°C (225°F), derived chiefly as a by-product of the synthesis of methyl alcohol from carbon monoxide and hydrogen; it is also obtained from fusel oil. Its formula is $(CH_3)_2CHCH_2OH$; it has a branched-chain structure and is isomeric with *n*-butyl alcohol. It is quite flammable but has only moderate toxicity. Being similar to butyl alcohol in structure and properties, its applications are of the same type; it has been widely used in hydraulic fluid systems and as a solvent in paints and coating resins. Its major use is in the manufacture of isobutyl acetate for nitrocellulose lacquers. *See also* butyl alcohol.

isocyanate.   *See* diisocyanate.

isoelectric point.   The pH at which the positive and negative charges on the particles of a colloidal solution cancel each other so that the particles become electrically neutral. This condition is brought about by altering the original pH by addition of acids or alkalies to the solution. The pH at which the isoelectric point occurs varies with the chemical nature of the particle; for example, it is about 6.0 (faintly acid) for glycine and near 10 (moderately alkaline) for lysine. At the isoelectric point, particles will not migrate in an electric field; the system becomes unstable, and coagulation or precipitation occurs, unless the change of pH is great enough to cause reversal of the charge by "jumping" the isoelectric point.

isolation.   In chemistry, the identification and separation of a particular element or compound, often the first step in large-scale production or laboratory synthesis. Usually complex analytical techniques, extensive fractionation, and purification procedures are required, as in the isolation of radium by the Curies.

isomer. One of two or more compounds having the same molecular weight and formula but often having quite different properties and somewhat different structure. A vast number of isomers are possible in organic chemistry, e.g., in alcohols, benzene derivatives, and petroleum hydrocarbons. The following example indicates the nature of such closely related substances, the difference between them in this case being the position of the chlorine atoms attached to the benzene ring. Dichlorobenzene has the formula $C_6H_4Cl_2$, but it has three distinct configurations, all having identical molecular weights:

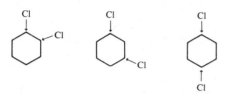

ortho (1,2-) liquid   meta (1,3-) liquid   para (1,4-) solid

Xylene has a similar isomeric arrangement, but with CH_3 groups instead of chlorine. Special types of isomers called stereo (3-dimensional) isomers are optical (occurring primarily in sugars) and geometric (found in polymer structures). *See also* stereoisomer.

isomerization. A method used in petroleum refining to convert straight-chain to branched-chain hydrocarbons or alicyclic to aromatic hydrocarbons, to increase their suitability for high-octane motor fuels. For example, butane (a gaseous paraffin hydrocarbon, $CH_3CH_2CH_2CH_3$) can be slightly modified in structure by catalytic reactions to give the isomeric isobutane ($CH_3CH_3CHCH_3$) used as a component of aviation fuel. Similarly, methylcyclopentane can be isomerized to cyclohexane, which is then dehydrogenated to benzene. Isomerization techniques were introduced on large scale during World War II. *See also* isomer.

isoprene. The monomer unit of rubber hydrocarbon (polyisoprene); it belongs to the classification of terpenes. It is a liquid, b.p. 34°C (93°F), having the molecular formula C_5H_8 and the structural formula

$$CH_2{=}\overset{\overset{\displaystyle CH_3}{|}}{C}{-}CH{=}CH_2.$$

It also is the chemical basis for many other terpene hydrocarbons containing from two to eight C_5 units ($C_{10}H_{16}$, $C_{15}H_{24}$, etc.). All these are formed in plants, occurring in essential oils, turpentine, pinene, and carotenoids. The synthesis of isoprene, which led to the manufacture of synthetic "natural" rubber, was achieved about 1940. *See also* terpene.

isopropyl alcohol. A secondary monohydric alcohol, b.p. 82.4°C (180°F), derived from the petroleum gas propylene by reaction with sulfuric acid. Its formula is $(CH_3)_2CHOH$; it is highly flammable and hazardous to handle near sparks or other ignition source. Its chief commerical use up to 1970 was in the manufacture of acetone, but this application has been declining in recent years. It also has a broad spectrum of other uses: as a solvent for lacquers and paints, for solvent extraction, dehydration, and deicing, and for the manufacture of glycerol. It is not regarded as a toxic hazard.

isotactic. A type of three-dimensional, geometrically symmetrical polymer (called stereospecific) whose structure is characterized by a primary chain or "backbone" of carbon atoms with their accompanying hydrogens, the other constituents (R = Cl, CH_3, C_6H_5, etc.), being located either above or below the primary chain. The following schematic structure suggests the isotactic arrangement, though it is oversimplified since in the actual molecule the carbon atoms are not in the same plane:

```
   H   R   H   R   H   R   H   R   H   R
   |   |   |   |   |   |   |   |   |   |
 —C— C— C— C— C— C— C— C— C— C—
   |   |   |   |   |   |   |   |   |   |
   H   H   H   H   H   H   H   H   H   H

   H   H   H   H   H   H   H   H   H   H
   |   |   |   |   |   |   |   |   |   |
 —C— C— C— C— C— C— C— C— C— C—
   |   |   |   |   |   |   |   |   |   |
   R   H   R   H   R   H   R   H   R   H
```

Such polymers are made with stereospecific catalysts developed by Ziegler and Natta and are distinguished from the older type by having a crystalline rather than an amorphous structure. They are made from propylene, styrene, and other olefinic monomers; isotactic polypropylene has been notably successful in the fiber field. *See also* stereospecific; tacticity.

isotonic.   Having equal osmotic pressure (of solutions); an isotonic salt solution is one containing the same concentration of sodium chloride as occurs naturally in body fluids (about 1%). For this reason, such a solution can be used effectively to restore osmotic balance in tissues when too much water has been lost from the body as a result of injury. It is also called physiological salt solution.

isotope.   Any of two or more forms of an element whose weights differ by one or more mass units as a result of variation in the number of neutrons in the nuclei. The literal meaning of this term is "the same place" (in the Periodic Table). The basic work on isotopes was done by the British chemists, Frederick Soddy (1877–1956), Nobel Prize 1921, who suggested the word in 1913, and Francis W. Aston (1877–1945), Nobel Prize 1922, who invented the mass spectrograph (1920). Hydrogen has two natural isotopes; as one hydrogen atom in 6500 contains a neutron, its nucleus has a mass of 2; it is called heavy hydrogen (deuterium) and is designated H^2. Its atomic number remains the same; all its other properties are identical with those of normal hydrogen, H^1, and its position in the Periodic Table is unchanged. In uranium, the 235 isotope occurs once for every 140 heavy isotopes (238). Most elements have from two to eight or more natural isotopes; carbon has six (10, 11, 12, 13, 14, 15), 10, 11, 14, and 15 being radioactive. Artificial radioactive isotopes of many elements can be created by bombardment of atomic nuclei with neutrons.

A few elements have only one stable atomic form, e.g., manganese, sodium, beryllium, iodine; this form is commonly referred to as an isotope, though strictly it should not be, since there are no other forms that are like it. Such an element can correctly be said to have *no* isotopes, because at least two forms must exist to comply with the definition. *See also* radioactive isotope.

isotropic.   *See* anisotropic.

-ite.   A suffix indicating an intermediate oxidation state of the central element of a metallic salt, analogous to *-ous* for acids, e.g., sodium sulfite (Na_2SO_3), which contains one less oxygen atom than the sulfate, and in which the sulfur has a valence of four, compared with a valence of six in the sulfate.

IUPAC.   Abbreviation for International Union of Pure and Applied Chemistry (*q.v.*).

J

japan. A type of varnish originally developed by Japanese craftsmen for use on ceramics and other objects of art. It contains a natural resin, a drying oil, various pigments, and lead oxide; after baking, it forms a hard, durable, and shiny coating.

Japan wax. A mixture of glycerides obtained from a plant grown in the Far East; it has a pronounced odor similar to tallow and a light yellowish color. Its uses are similar to those of beeswax.

jelly. (1) A modified (transliterated) form of *gel,* this term refers specifically to the colloidal dispersions formed by warming and then cooling mixtures of pectins, sugar, and water. The so-called high-ester pectins have a recognized jelly grade when over 50% sugar is used; low-ester pectins form gels of the gelatin type.

(2) This term is also used in nontechnical designations such as mineral jelly or petroleum jelly as alternative names for petrolatum. *See also* gel; gelatin; pectin.

Jenner, Edward. *See* antibody; drug; immunochemistry.

JET. Abbreviation of Joint European Torus, a device for experimental fusion research now in operation at Oxford, England. Similar installations are located in the U.S. at Princeton, La Jolla, and Cambridge (MIT).

jet fuel. Any of a number of combustible mixtures used in turbine-propelled aircraft; some types are blends of gasoline and light petroleum fractions, others are specially prepared kerosines. Their composition and properties are closely controlled by specifications. *See also* kerosine.

jojoba. An evergreen desert shrub of the southwestern U.S. and northern Mexico. From it are obtained a thick, waxy liquid used as a substitute for sperm oil in transmission and high-pressure lubricants. Wax from this plant is used as a substitute for carnauba and beeswax. Yield of oil from seeds is about 50%.

jute. A woody fiber, grown chiefly in India, whose strands are exceptionally long, in some cases as much as 10 feet. They are prepared by retting (water-soaking) the stems of the plant in which they occur. Jute has a higher lignin content than other vegetable fibers (about 20%), the balance being cellulose. It is used in twine and rope making, burlap manufacture, etc. *See also* fiber.

K

K (**1**) Symbol for the element potassium; it stands for the Latin *kalium* (''alkali'').

(**2**) Abbreviation for the Kelvin temperature scale.

kaolin. The most widely used industrial type of clay (aluminum silicate); it is relatively free from impurities and has small and uniform particle size; its neutral color and high heat resistance have led to its use as the primary ingredient of fine ceramic ware. For this reason, it is often called china clay; it is also known under the proprietary name of ''Dixie'' clay. It occurs in the southeastern states, as well as in England and France. Besides its use in ceramics, it has wide application as a filler and reinforcing agent in plastics and rubber, in paper coatings, and as a catalyst carrier; it is also used in cosmetic preparations, as an anticaking agent, and as an abherent. *See also* clay.

kapok. A light, fluffy vegetable fiber grown in various tropical areas. Though high in cellulose, it is not suitable for textile processing but is useful for sound insulation and stuffings for life preservers, cushions, etc.

karaya gum. A hydrophilic vegetable gum composed of complex sugars of high molecular weight. Like others of its type, it occurs on tropical trees in the form of hard clumps on the bark, probably resulting from mechanical injury sites from which the sap has emerged. Karaya absorbs water and swells to form a colloidal gel structure; for this reason, it is used as a stabilizer and thickener in food products, as well as in pharmaceuticals and textile and paper coatings. *See also* hydrophilic; protective colloid.

kauri-butanol value. A determination used as an index of the solvent power of a hydrocarbon liquid. Kauri is a natural resin that is completely soluble in butyl alcohol (butanol) but is not soluble in hydrocarbons. Thus, when a test sample of a hydrocarbon mixture is added to a standard kauri-butanol solution, a portion of the resin comes out of solution and appears as a cloudy suspension (turbidity). The amount of test solvent required to induce this is called the kauri-butanol value of that solvent. *See also* dilution ratio.

kcal. An abbreviation often used (without period) in chemistry and nutritional science for kilogram calorie, a unit of heat equivalent to 1000 small calories. The abbreviation Cal is also used with the same meaning. A small (or gram) calorie is the amount of heat required to raise 1 gram of water 1 degree centigrade (1.8 degrees Fahrenheit).

Kekulé, August (1829–1896). Born in Darmstadt, Germany, Kekulé laid the basis for the ensuing development of aromatic chemistry. His idea of a hexagonal structure for benzene in 1865 was a monumental contribution to theoretical organic chemistry. ''This had been preceded in 1858 by the remarkable notion that carbon was tetravalent and that carbon atoms could be joined to each other in molecules. The theory of the benzene ring has been called the 'most brilliant piece of scientific prediction to be found in the whole field of organic chemistry,' for besides promulgating the idea, he had predicted the number and types of isomers which might be expected in various substitutions on the ring.''

kelp. Any of many brown algae (seaweeds) which grow freely in seawater and are especially plentiful off the west coast of the United States and in the northeast Atlantic. Some varieties grow to very large size, 50 feet or more in length. It is harvested in some countries and used for fertilizer, cattle feed, and as a source of alginic acid, iodine, and potassium carbonate. *See also* algae.

Kelvin. A temperature scale named after Lord Kelvin, an English physicist (1824–1907) of the late nineteenth century. Absolute temperatures are denoted in degrees Kelvin by adding 273 to the centigrade temperature; thus, 25 degrees centigrade would be expressed as 298 degrees Kelvin.

keratin. A class of natural fibrous proteins occurring widely in animals and man; they are characterized by their high content of several amino acids, especially cystine, arginine, and serine. They are generally harder than the fibrous collagen group of proteins; the softer keratins are components of the external layers of skin, wool, hair, and feathers, while the harder types predominate in such structures as nails, claws, and hoofs. The hardness is largely due to the extent of cross-linking by the cystine molecules, which makes them difficult to dissolve. *See also* wool.

kerogen. The bituminous component of shale oil, which has the approximate composition: carbon 75%, hydrogen 10%, nitrogen 2.5%, sulfur 1.5%, and oxygen 9%. On refining, it yields about 18% gasoline, 30% kerosine, 25% gas oil, 25% lube oils of various weights. *See also* shale oil.

kerosine. A mixture of hydrocarbons obtained from petroleum by fractional distillation; it is the fraction boiling between 177 and 300°C (350 and 572°F). It is a free-flowing, oily liquid of higher viscosity than gasoline. Its various uses are as a booster fuel for trucks and other heavy vehicles, as a fuel for aircraft engines of the jet type, as an insecticide (sprays and mosquito control), and as a paint and varnish solvent. It has the low flammable concentration in air of 0.7 to 5% and should be used with care. The term is derived from the Greek word for *wax* (the distillation end-product of paraffinic hydrocarbons).

ketene. A class of organic compounds characterized by the presence of two double bonds. The term also applies to the compound which is the most important member of this class; its formula is $H_2C{=}C{=}O$. Others are derived by replacement of the CH_2 group with one or more methyl groups (CH_3) to form methylketene, dimethylketene, etc. These are all quite reactive compounds. Ketene itself is a severely toxic gas made by heat treatment (pyrolysis) of acetone or acetic acid. It has numerous applications in organic synthesis, especially as an acetylating medium.

ketone. A class of unsaturated and reactive compounds whose formula is characterized by a carbonyl group to which two organic groups are attached; these groups may be either paraffinic (alkyl) or aromatic (aryl). The best-known example of this class is acetone, or dimethylketone, $(CH_3)_2C{=}O$. Other important ketones are methyl ethyl ketone, acetophenone, and cyclohexanone. The reactions of ketones are generally similar to those of aldehydes. Their major uses are in the solvent field, especially in the plastics, textile, paint, and lacquer industries. They are particularly good solvents for cellulose derivatives. *See also* acetone.

kier-boiling. Boiling cotton fibers, frequently under pressure, in an aqueous solution of a medium-to-strong alkali, e.g., sodium hydroxide, to remove resinous impurities; it is one form of the process known as scouring. (A kier is a large tub or vat.)

kieselguhr. *See* diatomaceous earth.

kinematic viscosity. A relationship defined as the ratio of the viscosity of a fluid in centipoises to its density; the value is expressed in centistokes. Determination is made by noting the time required for the test fluid to pass through an aperture of small diameter. *See also* viscosity; poise.

kinetics. As applied to chemistry, kinetics is the study of matter as conceived of discrete moving particles and the forces resulting from their behavior at various temperatures. The term is derived from the Greek word for *moving*. The basic principles of kinetics are emobdied in the gas laws of Boyle, Charles, Dalton, and Avogadro. This is one of the two fundamental approaches to the problems of rate changes in chemical phenomena, the other being thermodynamics.

Thermodynamics determines the driving potential for a system to change from one state to another, i.e., whether a chemical reaction will proceed and how far it will proceed. Kinetics determines the rate at which a system changes and what resistance must be overcome to allow the change to occur. In chemical reactions, the latter is indicated by the activation energy needed to initiate the change.

Kjeldahl. An analytical method for determination of nitrogen in certain organic compounds. It involves addition of a small amount of anhydrous potassium sulfate to the test substance and heating the mixture with concentrated sulfuric acid and a catalyst; ammonium sulfate is formed. Ammonia is distilled off after alkalizing with sodium hydroxide, the nitrogen present being determined by titration.

knock. Undesired ignition of a portion of a motor fuel in the cylinder head due to spontaneous oxidation reactions rather than to the spark. This is not only noisy but causes loss of power, particularly in high-compression systems. It is reduced by the presence of branched-chain hydrocarbons in the fuel and virtually eliminated by anti-knock agents such as tetraethyllead and manganese tricarbonyl. The use of these is declining because of their toxicity, their reputed contribution to air pollution, and their adverse effect on catalytic exhaust gas converters. In their place, increasing use is being made of nonmetallic antiknock agents, such as methyl *tert*-butyl ether (MTBE) or a mixture of methanol and *tert*-butyl alcohol. *See also* octane number; gasoline; methyl *tert*-butyl ether.

Kr Symbol for the element krypton, the name being derived from the Greek word for *hidden*.

kraft. This term, adapted from the German word for *strength*, refers to a type of paper made from pine and noted for its overall utility where strength is a factor. It is comparatively inexpensive to manufacture by the sulfate process. Kraft has become the leading tonnage paper in the United States, the mills being chiefly in the South, where slash pine is plentiful. *See also* paper.

Krebs cycle. Also called the citric acid or tricarboxylic acid cycle (TCA cycle), this complex series of metabolic transformations of the major nutrients (carbohydrates, proteins, and fats) converts the intermediate pyruvic acid ($CH_3COCOOH$) into carbon dioxide and water through formation of the phosphorus-containing acetyl coenzyme A. A number of enzyme-induced reactions occur, involving acetylation, dehydrogenation, decarboxylation, and reoxidation, in which citric, succinic, malic, and fumaric acids play a part. The Krebs cycle is a multi-faceted stepwise biological oxidation process, the energy of oxidation being retained as adenosine triphosphate. *See also* citric acid.

Kroll process. The most important process for the production of pure titanium metal by the reduction of titanium tetrachloride with molten magnesium; it takes place under a protective inert atmosphere of argon or helium, which prevents harmful contamination of the titanium by oxygen, nitrogen, water, and carbon. Named for its inventor, William J. Kroll, a Luxembourg metallurgist, who developed the process in the United States (1946). The exothermic reaction gives off enough heat to maintain the temperature at about 800°C (1472°F) for the reaction: $TiCl_4 + 2Mg \rightarrow Ti + 2MgCl_2$. The process is also used to produce pure hafnium and zirconium from their tetrachlorides. *See also* titanium; hafnium; zirconium.

krypton. An element.

Symbol	Kr	Atomic Wt.	83.8
State	Gas	Valence	2
Group	Noble Gas	Isotopes	6 stable
Atomic No.	36		

Natural krypton is a gas occurring in trace percentages in air, from which it is extracted by fractionation processes. It was formerly classed as an inert gas but later was found to form a few transient compounds with fluorine. It is more than twice as dense as air and can be prepared in liquid and solid forms at very low temperatures. It is used in fluorescent light tubes, in flash equipment for photography, and to a limited extent in lasers. The radioactive krypton-85 is used in analytical determinations. *See also* inert; noble.

L

L-. A prefix conventionally appearing as a small capital, indicating the molecular structure or configuration of an optically active substance; it stands for "levo," the left-handed enantiomorph of an optical isomer. It does not mean that the substance is levorotatory. *See also* glyceraldehyde; enantiomorph.

l-. A prefix indicating that a compound is levorotatory, i.e., that it turns the plane of polarized light to the left as it passed through the crystal. A minus sign ($-$) is preferably used instead of the letter. Do not confuse with L, which refers only to molecular configuration.

La Symbol for the element lanthanum, the name being derived from a Greek word, meaning "to lie concealed."

labeled atom. *See* tracer element.

lacquer. A type of organic coating in which rapid drying is effected by evaporation of solvents, as distinguished from a paint, where drying is largely due to polymerization of a resin or oil. Nitrocellulose lacquers are the most important industrially; their largest use is for automobiles, furniture, and other products requiring a decorative and permanent finish. The rate of drying is rapid enough to permit economical large-scale operation. Other components of nitrocellulose lacquers (in addition to colorant) are plasticizers, alkyd or other resins, and solvents (both active and latent). *See also* paint; latent solvent; plasticizer.

lactalbumin. A protein occurring in milk; it is generally considered to act as a protective colloid which coats the particles of dispersed fat, preventing them from cohering. When the coating is ruptured by churning or other agitation, the lactalbumin becomes part of the serum (whey), of which it comprises from 10 to 12% in cow's milk. *See also* protective colloid.

lactic acid. A carboxylic acid, m.p. 18°C (64.4°F), derived either by hydrolyzing lactonitrile or by bacterial fermentation of lactose (milk sugar). It is this reaction that causes the souring of milk, during which the casein coagulates to form a curd. The formula is $CH_3CHOHCOOH$. Its chief uses are to perform various functions in food chemistry (preservative, flavor potentiator, acidulant); it also is used in the textile and leather industries and is available in tonnage quantities. *See also* lactose.

lactone. An ester formed by an intramolecular chemical change or rearrangement in certain types of carboxylic acids containing hydroxyl groups or halogen atoms, often on evaporation of aqueous solutions. The water (or halogen compound) formed by interaction of the OH and COOH groups within the acid molecule is split out, and a heterocyclic lactone structure results, as follows:

$$CH_2{-}CH_2{-}CH_2 \atop {\mid \qquad\qquad\quad \mid} \atop O\ \ \vdots H\ \ HO\ \vdots CO \longrightarrow {CH_2{-}CH_2{-}CH_2 \atop {\mid \qquad\qquad\quad \mid} \atop O{-}\!\!-\!\!-\!\!-\!\!-\!\!-CO} + H_2O$$

hydroxybutyric acid   butyrolactone

Examples are formation of butyrolactone from hydroxybutyric acid (above) and valerolactone from hydroxyvaleric acid (these products can also be synthesized). Lactones are useful organic intermediates and also have application in perfumery and flavorings. They occur naturally in some plants, where they are instrumental in odor formation.

lactose. A hydrated form of sucrose occurring in the milk of mammals ($C_{12}H_{22}O_{11} \cdot H_2O$), the name being derived from the Latin word for *milk*. Cows' milk contains about 5%. It is a disaccharide obtained commercially by separating the whey from skim milk, followed by concentrating until crystallization takes place. It is used primarily in the manufacture of penicillin; yeast and edible protein can also be derived from it, as well as lactic acid, alcohol, and riboflavin. All these products involve fermentation reactions. *See also* milk; lactic acid.

ladder polymer. A cross-linked high polymer comprised of double chains or strands of monomers or similar molecular groupings joined at regular intervals by hydrogen (or chemical) bonds, thus simulating a ladder-like structure. Such polymers occur in nature in the proteins, where polypeptide chains are crosslinked with hydrogen bonds; in deoxyribonucleic acid, the interlocked chains have a curved or helical conformation. The tanning of leather and the vulcanization of rubber approximate this mechanism. Some inorganic silicon compounds also exhibit a ladder structure, as do many synthetic organic polymers:

$$
\begin{bmatrix}
& C_6H_5 & \\
& | & \\
-\!\!\!- & Si\!\!-\!\!O & -\!\!\!- \\
& | & \\
& O & \\
& | & \\
-\!\!\!- & Si\!\!-\!\!O & -\!\!\!- \\
& | & \\
& C_6H_5 &
\end{bmatrix}
$$

lake. A class of water-insoluble organic pigments made by precipitating a soluble dye on an absorptive inorganic base (aluminum hydrate). Soda ash, calcium salts, and other metallic salts can be used as precipitants. Lakes are limited in their service aspects because of their sensitivity to light; their chief uses are in printing inks, interior paints, etc., where there is minimum exposure to sunlight, and in food colorants. *See also* dye; pigment.

laminate. A composite structure made up of alternating strips or layers of the same or different materials, which are joined tightly with adhesives. Plywood is a familiar type of laminate in which the wood is placed with the grain at right angles on successive layers to achieve maximum strength, the bonding agent usually being a phenolic resin. Similar structures are made with glass fibers (reinforced plastics). Adhesive-bonded paper and cloth are also used. Laminates of this type usually exhibit high strength and electrical resistance and are widely used in the construction industries. Rubber/fabric laminates, used in tire carcases, and textile fibers bonded with adhesive, as in nonwoven fabrics, are other examples of laminates. *See also* composite; reinforced plastic.

Langmuir, Irving (1881–1957). A brilliant American physical chemist who, after graduating from Columbia University, obtained his doctorate under Walter Nernst in Germany. Employed by the General Electric Research Laboratories from 1909 until his retirement in 1950, he received the Nobel Prize in chemistry in 1932 for his fundamental research in the field of surface chemistry. He was the first eminent scientist employed in an industrial laboratory to be so honored. His investigation of monomolecular films, especially the behavior of organic liquids such as alcohols and fatty acids on water, whereby the water-soluble end of the molecule is attracted downward into the water, while the hydrocarbon portion extends upward, led to the development of modern knowledge of emulsification and detergency. He also did pioneer research in electric discharges in gases at low pressure and high temperature, which led to the modern incandescent light bulb filled with inert gas; on atomic hydrogen, e.g., the atomic hydrogen welding torch; and on cloud-seeding techniques.

lanolin. A mixture of cholesterol esters occurring in the sebaceous glands of sheep and similar animals, which exudes into the wool. The crude product is known as degras; the type resulting from tanning chamois leather with various fish oils is called moellon degras. When removed by scouring and purified, the product becomes lanolin (from Latin word for *wool*). One grade contains 25 to 30% water; the other is virtually free of water. Lanolin is a well-known ingredient of ointments and cosmetic creams; it is also used in soaps and as a softener for textiles, leather, rubber mixtures, and the like. The original degras is a dark-brown semisolid with an objectionable odor; the lanolin of commerce is an almost colorless viscous liquid without odor.

lanthanide series. The sequence of elements having atomic numbers from 57 to 71, occurring in Group IIIB of the Periodic Table. They are more commonly known as rare earths. The spelling recommended by IUPAC is lanthanoid, but this has not been generally followed by authorities in the United States. The lanthanide series elements are: lanthanum, cerium, praseodymium, neodymium, promethium, samarium, europium, gadolinium, terbium, dysprosium, holmium, erbium, thulium, ytterbium, and lutetium. *See also* rare earth.

lanthanoid series *See* lanthanide series.

lanthanum.   An element.

Symbol	La	Atomic Wt.	138.9055
State	Solid	Valence	3
Group	IIIB	Isotopes	2 stable
Atomic No.	57		

The first of the rare-earth metals comprising the lanthanide series of the Periodic Table. It is recovered from monazite sands and the ore bastnasite by action of sulfuric acid. It undergoes considerable surface attack in moist air, and in finely divided form it will catch fire at room temperature, i.e., it is pyrophoric. Lanthanum melts at 920°C (1688°F). It has some application in high-energy fuels and as a reducing agent, but its principal use is in alloys such as misch metal. Lanthanum compounds are used in glass and ceramics, as phosphors in fluorescent lamps, and as analytical reagents. *See also* lanthanide series; misch metal.

LAS.   Abbreviation for linear alkyl sulfonate. *See* detergent; dodecylbenzene.

laser.   This term is made up from the initial letters of "light amplification by stimulated emission." The emission referred to is that of a controlled beam of high-energy radiation of constant wavelength and frequency (called coherent light). It results from photon-initiated atomic vibrations in certain crystals, such as garnet and ruby when activated by trace additions of other metals (yttrium, chromium), in gases (argon, carbon dioxide), and in other materials (plastics, glasses). The exciting energy is due to photons from an external source (flash lamp) which initiates continuously increasing emission of photons within the laser crystal; in chemical lasers it is due to the bond rupture of a chemical reaction. Laser beams are exceedingly powerful and highly accurate cutting agents; adequate protection for the eyes is essential. They have been adapted to surgical techniques, diamond and metal cutting, missile guidance, and other high-precision work. One of the most successful applications of lasers to chemistry has been the enrichment of uranium isotopes for preparation of more effective reactor fuels. There are also developing applications in spectroscopic analysis and thermonuclear research. Fundamental research on the effect of laser light on chemical bonding and in the kinetics of gas-phase reactions is continuing.

latent heat.   The quantity of energy, in calories per gram, absorbed or given off as a substance undergoes a change of state, that is, as it changes from liquid to solid (freezes), from solid to liquid (melts), from liquid to vapor (boils), or from vapor to liquid (condenses). No change in temperature occurs. Water has unusually high latent heat values: the latent heat of fusion (melting) of ice is 80 cal per gram, and the latent heat of condensation of steam (latent heat of vaporization of water) is 540 cal per gram. The considerable energy delivered by steam condensation is utilized not only for power generation but for heating a variety of chemical plant equipment, such as dryers, evaporators, reactors, and distillation columns. *See also* evaporation; heat; heat of fusion; heat of vaporization.

latent solvent.   An inactive solvent (invariably an alcohol) used in conjunction with a true solvent for nitrocellulose, usually in the ratio of 1 volume alcohol to 2 volumes of active solvent. Alcohols alone are not good solvents for nitrocellulose but are less costly than active solvents such as ketones, acetates, and glycols and are effective when blended with them in nitrocellulose lacquers. For this reason, their solvent power is said to be latent, i.e., hidden. An example is isopropyl alcohol. The term *cosolvent* is often used synonymously. *See also* solvent; dilution ratio.

latex.   An aqueous suspension of proteins and resins occurring in some plants, trees, and shrubs, such as sapodilla, guayule, and especially the tree *Hevea braziliensis*, from which natural rubber latex is obtained. The latter contains about 33% rubber hydrocarbon particles dispersed in a watery serum and readily visible in an optical microscope; these are coated with a protein layer which keeps them from cohering (protective colloid). Synthetic rubber latex is made by emulsion polymerization. Precipitation of the particles is brought about by adding acetic or formic acid. Many types of rubber products are made directly from prevulcanized latex by electrodeposition or by dipping. The pH of natural latex is 6.8, but as it develops acidity on standing, ammonia is usually added as a preservative.

lattice. The stable geometric arrangement of atoms in a crystal (metals, nonmetals, and their compounds). The lattice structure constitutes a submicroscopic three-dimensional grating or network of atoms, which refracts incident light of short wavelength. Measurement of the angles of this refraction has made it possible to determine with x-rays the arrangement of atoms in a crystal with a high degree of accuracy. The science of crystallography is largely based on such measurements. Vacancies or dislocations may occur in lattices due to lack of enough atoms to satisfy all the covalent bonds; such imperfections have a marked effect on the electrical and optical properties of the crystal. *See also* crystal; x-ray; semiconductor.

laughing gas. *See* nitrous oxide.

lauric acid. A 12-carbon fatty acid, m.p. 44°C (111°F), occurring chiefly in coconut and laurel oils; it is obtained by fractional crystallization of the plant oil after solvent extraction. The formula is $CH_3(CH_2)_{10}COOH$. It is used as a dispersing agent in rubber and plastic formulations, cosmetic preparations, synthetic food flavors, and detergents.

Lavoisier, Antoine Laurent (1743–1794). French chemist generally regarded as the "father" of chemistry. His "Traité Elementaire de Chimie" (1789) listed 30 elements, clarified the nomenclature of acids, bases, and salts, and described the composition of numerous organic substances. He erroneously believed that oxygen is the characteristic element of acids. However, his fundamental work on combustion, as a result of which he identified and named nitrogen (azote), and on the separation of hydrogen from water by a unique reduction experiment carried out in a heated gun barrel, earned him a leading position among early chemists. *See also* Mendeleev.

Lawrence, Ernest O. *See* cyclotron; accelerator.

lawrencium. An element.

Symbol	Lr	Atomic Wt. 257
Group	IIIB	
	(Actinide	
	Series)	
Atomic No.	103	Valence 3

Lawrencium is an experimentally developed radioactive element named for E.O. Lawrence (1901–1958), inventor of the cyclotron; it was identified in 1961 by irradiating californium with boron ions. Its half-life of 8 seconds is too short for further determinations; and its existence is of only theoretical importance.

LD_{50} test. *See* toxicity.

leach. *See* lixiviate.

lead. An element.

Symbol	Pb	Atomic Wt. 207.2
State	Solid	Valence 2,4
Group	IVA	Isotopes 4 stable
Atomic No.	82	

A soft, heavy metal, m.p. 327°C (621°F), obtained chiefly from the sulfide ore galena. Lead has such high density that it is almost impermeable to high-energy radiation; thus, a major use is to protect personnel from sources of x-rays, gamma rays, etc., in the form of barriers of various thicknesses. Other important applications of metallic lead are as grids in storage batteries, solders, vibration absorbers in heavy construction, type-metal alloys, piping, and cable sheathing. The metal is not toxic in bulk form and is safe to handle and process. However, fine powders and fumes are damaging if inhaled, and the presence of lead in any form in foods is strictly forbidden. Lead compounds are poisonous; their presence in exhaust fumes of autos (from tetraethyllead) has been considered to be an air pollution factor, and their use in gasoline has been greatly reduced. Use of lead compounds in house paints has been virtually discontinued because of their toxicity when ingested by children. *See also* litharge; tetraethyllead.

lead azide. *See* azide; detonation.

lead chromate. An inorganic yellow pigment

(also called chrome yellow) which is a finely divided solid having the formula $PbCrO_4$, the color being due to the chromium. Like most inorganic pigments, it does not fade appreciably on exposure to light and heat, but it is quite toxic and cannot be used in interior paints or in products that come into contact with foods. It is used as a colorant in rubber and plastic products, printing inks, baked enamels, and engineering paints for steelwork, cement, etc. Combined with iron blue, it gives chrome green pigment;

lead naphthenate. See antifouling paint; naphthenate.

leather. The skin or hide of an animal treated with a chemical (tanning agent) which protects the protein (collagen) fibers from degradation due to bacteria, increases the strength, and maintains the flexible, microporous nature of the material indefinitely. The tanning process involves cross-linking and coordination reactions, depending on the type of agent used (vegetable or metallic). Well-made leather can remain in serviceable condition for well over a century. Its uses are manifold, ranging from formal gloves to transmission belting. Many of its applications have been preempted by plastics in recent years, particularly for soling of shoes. See also tanning; poromeric.

leavening agent. See baking powder; yeast.

Le Blanc, Nicholas (1742–1806). French inventor of the first successful process for making soda ash. His patent was confiscated by the Revolutionist government and the process was used widely for years without either acknowledgment or remuneration. His original formula was 100 parts salt cake, 100 parts limestone, and 50 parts coal.

Le Chatelier, Henri Louis (1850–1936), French chemist and metallurgist who served as professor of chemistry at the Ecole des Mines (1877–1919), at the College de France (1877–1908), and at the Sorbonne (1907–1925). His pioneer research on equilibrium reactions related chiefly to industrial problems; he made vital contributions to such subjects as the chemistry of the setting of calcium sulfate (plaster of Paris) and cement; reactions in the blast furnace; and combustion of gases, which resulted in the development of the oxy-acetylene torch. In 1884, prior to receiving his doctorate in physical and chemical science (1887), he announced his famous principle (see next entry). He developed several re-search instruments, e.g., high-temperature thermocouples and the dilatometer, founded and edited the Revue de Metallurgie, and published and lectured extensively.

Le Chatelier's Law. If a stress, such as change of temperature, pressure, or concentration, is applied to a system in equilibrium, the equilibrium is shifted in such a manner as to undo the effect of the stress.

lecithin. A complex organic substance formed in the cells of plants and animals from fatty acids, glycerol, phosphoric acid, and choline; it is closely related to the lipids, differing from them chiefly because of its phosphorus content. It is considered to be a member of the phosphatide family (also called phospholipid). The lecithin molecule contains fatty acid groups, nitrogen, and phosphorus. It has important functions in the cellular metabolism of plant and animal organisms. Lecithin is obtained commercially from vegetable oils and egg yolk and is also made synthetically. It is widely used in the food industry as an emulsifier and nutritional additive; there are many minor uses, including cosmetic preparations, dietetic foods, animal feeds, and special soaps. See also lipid; phosphatide.

Leclanche cell. See dry cell.

lees. See foots.

legal chemistry. See forensic chemistry.

leucocyte. A white blood cell. See blood.

level. (1) To promote maximum uniformity of a dissolved dye by means of an additive in the bath which acts as a wetting or dispersing agent. For sulfonic acid dyes, lignin is effective, as it forms lignosulfonates in the dye solution.

(2) Of a paint, to brush or spray evenly over a prepared surface, while maintaining uniform thickness and covering power. Good leveling ability is related to proper viscosity: the paint is neither too thin to remain in place nor too thick to flow or spread evenly.

levigate. To grind solid materials into powders of approximately uniform particle size, either in the dry form or mixed with liquid; the term is also used in the broad sense of effecting a uniformly dispersed mixture of solids, pastes, or gels by mechanical agitation.

levulose. See fructose.

Lewis acid. Any molecule or ion (called an electrophile) that can combine with another molecule or ion by forming a covalent bond with two electrons from the second molecule or ion. An acid is thus an electron acceptor. Hydrogen ion (proton) is the simplest substance that will do this, but many compounds, such as boron trifluoride, BF_3, and aluminum chloride, $AlCl_3$, exhibit the same behavior and are therefore properly called acids. Such substances show acid effects on indicator colors and when disolved in the proper solvents.

Lewis, Gilbert N. (1875–1946). American chemist, native of Massachusetts; professor of chemistry at M.I.T. from 1905 to 1912, after which he transferred to University of California at Berkeley. His most creative contribution was the electron-pair theory of acids and bases which laid the groundwork for coordination chemistry. He was also a leading authority on thermodynamics, and his textbook on this subject, written with Merle Randall and published in 1923, became world-famous.

Li Symbol for the element lithium, the name being derived from the Greek word for *stone*, doubtless because of the hardness of the mineral in which it was discovered. Actually, lithium is a very soft metal.

Liebig, Justus Von (1803–1873). German chemist who founded the *Annalen*, a world-famous chemical journal. He was a great teacher of chemistry, training such men as Hofmann, who did basic work on organic dyes. Liebig contributed original research in the fields of human physiology, plant life, and soil chemistry and was the discoverer of chloroform, chloral, and cyanogen compounds. He was the first to recommend addition of nutrients to soils and thus may be considered the originator of the fertilizer industry.

ligand. Derived from the Latin word for *bind*, a ligand is a molecule, ion, or atom that is capable of furnishing or donating one or more pairs of electrons to a transition-metal ion, thus forming a coordination compound. Common substances that act as ligands are ammonia, water, and chlorine. When organic nitrogen compounds behave as ligands, a heterocyclic ring is formed, and the product is called a chelate. The number of points of attachment of the ligand to the metal is called its coordination number and is normally 2,4, or 6. Ligands which coordinate by donation of electrons from more than one atom are known as bidentate (e.g., ethylenediamine), tridentate, and so on, the multidentate type being the most stable. Ligands may be neutral, positively charged, or negatively charged. *See also* coordination compound; chelate.

light. *See* radiation.

light hydrocarbon. A gaseous or liquid aliphatic or aromatic hydrocarbon obtained from natural gas by fractionation or cracking and used in the manufacture of synthetic organic chemicals; they include ethane, methane, propane, ethylene, acetylene, benzene, toluene, etc. Some are used as liquefied petroleum gases.

light metal. This term is generally used to refer to metals having a specific gravity of less than 3 that are capable of being used for construction purposes; these are aluminum, magnesium, and beryllium. Many others have low specific gravity, especially the alkali metals, but they are too reactive and much too soft for any but chemical use.

light oil. A coal-tar or petroleum fraction obtained by distillation, which boils between 110 and 210°C (230 and 410°F). It contains a number of important organic solvents, including phenol and cresol, which can be separated from it by further fractionation processes.

light water. Ordinary water, as distinct from heavy water (deuterium oxide). It is widely used as a moderator in nuclear reactors.

lignin. The noncellulosic portion of wood, of which it constitutes from 25 to 35%. It is an amorphous organic polymer whose function is to hold the cellulose fibers together. In papermaking, it must be removed from high-grade products, as it causes yellowing and hardening of the sheet; it is permitted in newsprint and in some other coarse papers made from groundwood. After removal from the wood in the pulping process, the recovered lignin has some commercial uses as an extender in plastics, as a binder in heavy emulsions and drilling muds, and as a leveling agent in dye baths. It is a source of vanillin; phenol and benzene can be made by fluidized bed cracking of lignin. *See also* lignosulfonate.

lignite. *See* coal; fossil.

lignosulfonate. A waste product from pulp digesters in which the lignin of the wood is removed in the sulfite and sulfate pulping processes; it is a component of the so-called black liquor. It can be recovered and used in the manufacture of vanillin, as well as in ore flotation and as a dispersing aid in heavy emulsions and slurries. It is sometimes called lignin sulfonate.

ligroin (pronounced lig-ro-in). A low-boiling petroleum-derived solvent, sometimes also called petroleum ether; the latter name is not recommended since the material is not an ether in the chemical sense. The origin of the word is not known. Ligroin is highly flammable and should be used with care.

lime. *See* calcium oxide; hydration; hypochlorous acid.

limestone. *See* calcium carbonate, calcium oxide.

limonene. *See* dipentene.

linoleic acid. *See* drying oil; drier; fatty acid; polyunsaturate.

linolenic acid. *See* drying oil; fatty acid.

linseed oil. The most widely used drying or oxidizing oil in such products as exterior paints, printing inks, putty, and (formerly) linoleum. It is obtained by crushing seeds of the flax plant and thus is also called flaxseed oil. It is a mixture of saturated and unsaturated fatty acids, the latter being oleic, linoleic, and linolenic. It dries to a hard, resistant film as a result of polymerization caused by air oxidation. It may be thickened to increase its drying properties by blowing air through it at about 93°C (200°F). Heating it to about 260°C (500°F), often with admixed metallic driers, causes still further increases in viscosity. This product is called heat-bodied or (erroneously) "boiled" linseed oil (actually it has no boiling point). Other names used are stand oil and resinoid. The meal left from the presscake is used as an animal feed. *See also* vegetable oil; resinoid.

linters. A collective term for short fragments of cotton fiber which remain attached to cottonseed after the first trip through a gin but are removed by further ginning. They are exceedingly flammable because of their fine state of subdivision and are a source of cellulosic derivatives used in the manufacture of explosives and rayon.

lipid. A broad class of organic compounds formed in the cells of plants and animals as metabolic products; they include fats, waxes, lecithins (phosphatides), sterols, carotenoids, and fat-soluble vitamins. Fatty acids and their glycerides occur in many lipids found in animal and vegetable oils. One of the most biologically important lipids is cholesterol. Lipids are transported in the blood in the form of lipoproteins, which are synthesized in the liver. The term is derived from the Greek word for *fat* and is also

spelled with a final *e*. *See also* fatty acid; cholesterol.

lipoprotein. *See* lipid.

liquation. Fractionation of an ore or alloy whose components have appreciably different melting points, by application of heat to the mixture; the lower-melting components are thus removed, leaving the pure metal (in the case of an ore) or the base metal (in the case of an alloy or bimetallic structure).

liquefied petroleum gas. Any of a group of hydrocarbon fuels which constitute a portion of the broader group of compressed gases. Liquefied petroleum gases (usually abbreviated LPG) include butane, butenes, isobutane, propane, and propylene. All have ignition points above 430°C (806°F), but extremely low flash points, below −70°C (−100°F). They are handled in appropriate metal cylinders or similar containers designed for portable household or mobile fuel supply. If a container leaks or is accidentally broken, the liquid vaporizes immediately and constitutes a fire and explosion hazard in the presence of the slightest spark or any type of open flame. Neither the liquids nor the gases they evolve are toxic unless the concentration exceeds 1000 parts per million in air. LNG stands for liquefied natural gas. *See also* compressed gas.

liquid. (divided liq-uid). A type of fluid which is a state of matter intermediate between gaseous and solid; liquids differ from gases because of their greater density and more complex molecules, and from solids because of their random (amorphous) molecular arrangement, which is the distinctive nature of the liquid state. Almost all liquids contract on freezing. The most notable exception is water, which expands about 10%; others that expand on solidification include gallium and bismuth. Otherwise, the physical properties of liquids vary widely. Water is unique in its ability to exist in all three states of matter at 0.01°C. This is known as the triple point of water and was the basis for establishing the centigrade temperature scale. Three metals are liquid at or near room temperature (mercury, gallium, and cesium).

Two classes of liquids are distinguished on the basis of their response to an applied force: (1) Newtonian liquids, which flow immediately when stressed, the rate of flow being directly proportional to the stress, and assume any shape imposed upon them; and (2) non-Newtonian liquids, which do not flow until a force of critical magnitude is applied. Examples of this are thick colloidal aggregates such as putty and, under certain conditions, asphalt and rubber. These exhibit what is called plastic flow. Many common materials resembling solids are either combinations of amorphous and crystalline phases or change from one to the other when under stress. Cellulose has been found to consist of crystalline structures in an amorphous matrix; unvulcanized rubber is amorphous when unstretched but develops crystallinity when stretched. Glass is technically a Newtonian liquid of such high viscosity that it is rigid. Thus, liquids are often difficult to differentiate from solids, as both conditions can exist in the same material. *See also* fluid; gas; amorphous; liquid crystal; dilatant.

liquid crystal. A state of matter intermediate between amorphous and crystalline, sometimes designated by the terms *mesophase* and *mesomorphic*. The most common examples are cholesterol derivatives. Materials of this nature combine the properties of liquids and crystals, that is, they have fluidity but at the same time have optical properties associated with crystal structures. They are known to occur in such biological forms as muscles and nerves. Liquid crystals have some applications in medicine, as well as in electronics, especially in color television tubes. *See also* liquid.

liquor. In chemical technology, any aqueous solution of one or more chemical compounds. In sugar manufacturing, it refers to the sirups obtained from various refining steps (mother liquors). The paper industry uses this term extensively as follows: (a) black liquor is liquid digester waste (also called spent sulfate liquor) containing sulfonated lignin, rosin acids, and other waste wood components, from which tall oil is made; (b) green liquor is a solution made by dissolving chemicals recovered in the alkaline pulping process in water; (c) white liquor is made by adding caustic soda to sodium sulfide solution. In dyeing technology, red liquor is a synonym for mordant rouge (aluminum acetate).

liter. A standard unit of volume for gases and liquids; it is the volume occupied by 1000 grams of water at a pressure of 1 atmosphere at 4°C (39°F). It is equivalent to about 1.05 quarts. A milliliter is 1/1000 liter. Avogadro's number of molecules of any gas (i.e., one gram-mole) occupies 22.4 liters. *See also* gram; milliliter.

lith-. A base word borrowed from the Greek term for *stone*; its literal meaning is apparent in such words as *lithography* ("drawing on stone") *lithosphere* ("earth's crust"), but is less obvious in a number of other terms in which it appears (*lithopone, litharge, lithium*).

litharge. The most common type of lead oxide (PbO), a heavy (sp. gr. 9.53) reddish-yellow powder made by heating lead in air. Though it shares the toxic properties of lead, it is quite safe to work with, as its weight precludes any substantial amount of inhalation. Its largest uses are in coating the grids of storage batteries and in the manufacture of red lead (Pb_3O_4), a major component of undercoat paints for protecting metals from corrosion. Mixed with glycerin, it forms an acid-resistant cement used for ceramic products; it also has minor use as a rubber accelerator. *See also* lead.

lithium. An element.

Symbol Li	Atomic Wt. 6.941
State Solid	Valence 1
Group IA	Isotopes 2 stable
Atomic No. 3	

The first member of the alkali-metal series, lithium, m.p. 179°C (335°F), has some unusual properties. It has the lowest specific gravity (0.54) of any of the solid elements and is also one of the softest, rating 0.6 on the Mohs scale. It has the highest melting and boiling points of the alkali metals; though it is the least reactive of these, it is nonetheless a reactive element, forming many inorganic compounds and a few organic salts. When bombarded with neutrons, it yields the hydrogen isotope tritium. It evolves heat in contact with air ar room temperature and ignites spontaneously at about 177°C (350°F) . Elemental uses of lithium include alloying agent with many metals with which it forms intermetallic compounds; scavenging and deoxidizing molten metals; and heat-transfer agent. It is available in the form of ribbon and wire for laboratory use and must be shipped in airtight containers under kerosine. Important compounds are lithium aluminum hydride, a versatile reducing agent; lithium chloride, used in air-conditioning and in carbonated beverages and mineral waters to inhibit loss of carbon dioxide; and lithium soaps for lubricating greases.

lithium chloride.   *See* lithium.

lithium hydride.   *See* hydride.

lithopone.   A heavy white pigment prepared by the reaction $ZnSO_4$ + BaS 1 $BaSO_4$ + ZnS, composed of about 1 part zinc sulfide and 3 parts barium sulfate; its coloring or hiding powder is about the same as that of zinc oxide but much less than that of titanium pigments. It was once the most widely used pigment in the paint and rubber industries but has been largely replaced by titanium oxide.

litmus.   *See* indicator.

liver.   A term used in the paint and printing industries to describe the irreversible coagulation caused by accidental polymerization of the binder. This makes the ink or paint useless; it can be prevented by use of inhibitors or antioxidants.

lixiviate.   A term usually applied to ore processing, where the water-soluble components are dissolved out by washing the ore with water after roasting. It is also used to describe the removal of soluble salts from any solid, for example, the solvent action of seawater on rock or soil. Originally, it referred to the extraction of water-soluble components from wood ashes. The term *leach* has essentially the same meaning.

LNG.   Abbreviation for liquefied natural gas.

locust bean.   *See* mucilage.

long-oil.   A term applied in organic coatings technology to a paint or varnish containing an unusually high proportion of drying oil, which makes it suitable for prolonged use in outdoor service, for example, spar varnish. Opposed to this are "short-oil" varnishes intended for interior exposure, as on furniture, floors, etc.

low explosive.   *See* explosive; deflagration.

LPG.   Abbreviation for liquefied petroleum gas.

Lr   Symbol for the element lawrencium, named in honor of Ernest O. Lawrence (1901–1958), the American physicist who invented the cyclotron, Nobel Prize 1939.

Lu   Symbol for the element lutetium, the name being derived from the Latin term for *Paris,* where its discoverer, G. Urbain, was born.

lubricant.   An inclusive term embracing a wide variety of substances (solid, semisolid, liquid, and even gaseous) whose molecules (or crystals) are able to pass over one another freely when slight shearing force is applied. Materials having as low a viscosity as air and water have limited lubricating ability; the crystals of graphite, talc, and mica, being arranged in parallel planes, enable these solids to be used in antifriction bearings. The majority of lubricants are oils of various types derived from petroleum and greases made by compounding mineral oils with soaps of heavy metals. The oils are used chiefly to lubricate fast-moving parts, as in automobile cylinders, whereas greases are applied to slower-moving members, such as axles, gears, and rotors on heavy machinery. There is also a wide variety of synthetic lubricants based on silicone compounds, polyglycols, phosphate esters, and chlorofluorocarbons.

lubricity.   A term used to denote the effectiveness of a lubricant in reducing the friction between two given surfaces, as determined by a control or by a standard value. The term *oiliness* is sometimes applied to lubricating oils in the same sense.

luminescence.   Radiation of any wavelength emitted by the atoms or molecules of a substance (solid, liquid, or gas) as a result of absorption of exciting energy (photons, electrons, chemical reactions) and subsequent release of some of that energy on return of the excited atoms to their normal state, without accompanying evolution of heat. The wavelength ranges include the entire visible spectrum (reds, greens, blues, etc.) and extend into the ultraviolet and infrared. There are several types of luminescence, designated by the nature of the energy source: electroluminescence is induced by an electric impulse or current; chemiluminescence results from chemical bond formation, electron capture, or oxygen elimination; bioluminesence occurs in organisms (insects, marine fauna) by oxidation catalyzed by the enzyme luciferase; photoluminescence is due to shortwave radiation (photons). Two distinct aspects of luminescence are fluorescence and phosphorescence. *See also* fluorescence; phosphorescence; phosphor; spectroscopy; photochemistry.

lutetium.   An element.

Symbol	Lu	Atomic Wt.	174.97
State	Solid	Valence	3
Group	III B (Lanthanide Series)	Isotopes	1 stable
Atomic No.	71		

Lutetium is a soft metal, m.p. 1675°C (3047°F), of the rare-earth group, recovered from monazite sands in which it occurs in concentrations of as low as 0.003%. It is quite similar to erbium and holmium. Because it is expensive to produce, it has no industrial applications but is of considerable theoretical interest. *See also* lanthanide series.

lyophilic.   Having strong affinity for a liquid; the term is usually used to describe a finely divided or colloidal substance in a solvent dispersion. The opposite term is **lyophobic,** meaning a finely divided material that repels liquids. In both cases, if water is the liquid, the more exact term is *hydrophilic* (or *hydrophobic*); if an organic solvent is involved, the respective terms are *organophilic* and *organophobic*. *See also* hydrophilic; hydrophobic.

lysergic acid.   *See* hallucinogen.

lysine.   An amino acid which is not synthesized within the animal body and is therefore regarded as essential for proper nutrition. It is a component of many proteins, from which it can be extracted; it can also be made by fermentation of various carbohydrates, as well as from caprolactam. It is used in numerous foods and pharmaceutical products. The abbreviation is Lys. *See also* amino acid.

-lysis.   A suffix commonly used in chemical terminology derived from the Greek, meaning "to free" or "to loosen," i.e., *analysis* is the "loosening" of a problem or a material. When referring to substances, the meaning is more clearly expressed by the term *decomposition of,* as in *hydrolysis, glycolysis, solvolysis*. *See also* catalysis.

M

M Abbreviation for molar solution.

m- Abbreviation for the chemical prefix *meta*; abbreviation for molal solution.

m Abbreviation for meter.

macromolecule. A large molecule whose size may be in the lower colloidal range (0.5 to 1 micron). Such molecules are usually composed of either a sequence of chemically similar units linked together in chains of known length, or of an indefinite number of identical repeating units which may or may not be cross-linked. An example of the first type is a protein in which a known number of amino acids are arranged in specific sequence; in some cases (DNA), the number of units is 3000 or more; ribonuclease contains 124. Examples of the second type are cellulose (an unknown number of carbohydrate units joined by oxygen atoms) and rubber (an unknown number of unsaturated hydrocarbon units whose chain can be crosslinked by added sulfur). Synthetic macromolecules have structures analogous to these and can also be cross-linked. The molecular weight of macromolecules may run to several million. *See also* polymer; protein; molecule.

madder. *See* lake.

magnesia. *See* magnesium oxide.

magnesite. *See* magnesium carbonate.

magnesium. An element.

Symbol	Mg	Atomic Wt.	24.305
State	Solid	Valence	2
Group	IIA	Isotopes	3 stable
Atomic No.	12		

Magnesium m.p. 650°C (1202°F), is a light, strongly electropositive metal obtained from seawater by electrolysis of molten magnesium chloride (Dow process); in another widely used method, magnesium oxide is reduced with ferrosilicon at high temperature. Coincidentally, both these methods went into large-scale use in 1941. Magnesium occurs in the ores dolomite and magnesite, as well as in seawater. The bulk metal will burn when heated to its melting point; but in thin flakes or powder form, it ignites readily and burns with a bright flame, being used in flares and incendiary equipment for this reason. Biochemically, it is of great importance, as it is the characteristic element in chlorophyll; it also has metabolic functions in the body. It is a useful construction metal where weight is a factor; thus magnesium alloys are used in airplanes, auto pistons, and portable equipment. Magnesium also has applications in the battery field, in organometallic chemistry, and as a reducing agent. *See also* Grignard; chlorophyll; Kroll process.

magnesium carbonate. Though it occurs in nature as the ore magnesite, commercial magnesium carbonate ($MgCO_3$) is either mined as a natural material or made by carbonation of magnesium oxide or magnesium hydroxide with carbon dioxide. The product is very light, fluffy powder which flies about in the air and is extremely bulky, a cubic foot weighing only 4 pounds. It is used in foods and cosmetics to prevent caking, in rubber and plastics as a stiffening agent, and as a component of insulating materials. As its refractive index is indentical with that of natural rubber, it can be used in quite high percentage in translucent rubber products. It is strongly hydrophobic (water-repellent).

magnesium chloride. A deliquescent white powder produced from seawater and used as a source of magnesium (by electrolysis). $MgCl_2$ melts at 708°C (1297°F) and its hydrate, $MgCl_2 \cdot 6H_2O$, loses water above 100°C (212°F), but also hydrolyzes by the reaction: $MgCl_2 \cdot 6H_2O \rightarrow MgO + 2HCl + 5H_2O$, so the anhydrous form cannot be made by drying the hydrate. It is a useful fireproofing and extinguishing material and also has numerous applications in the textile industry. It has some use as a catalyst and precipitant for solids, as well as in special cements, and has mild disinfectant action.

magnesium oxide. This compound is also called magnesia and has the formula MgO; it is made by heat treatment (calcining) of other magnesium compounds. Its chief use is in furnace linings and insulating materials; it has a few specialized applications in cements, as an accelerator in low-grade rubber products, and as a food additive. Fumes from the heated material are irritant.

magnesium silicate. *See* asbestos; silicate.

magnesium sulfate. *See* Epsom salt.

magnetic compound. *See* ferrite.

magnetic susceptibility. *See* magnetochemistry.

magnetite. A black naturally occurring oxide iron ore, formula Fe_3O_4, or more exactly $FeOFe_2O_3$ since it is a mixed oxide or a spinel mineral. Characterized by its magnetism, it is also known as magnetic iron ore or lodestone. It was probably the basis of the first compass; when a piece was floated in a dish of mercury, one end would always point north.

magnetochemistry. A specialized branch of chemistry devoted to the study of the effects of magnetic forces or fields upon chemical substances. The two most important considerations are magnetic moment (the turning force exerted on an atom, ion, or molecule) and magnetic susceptibility (the extent to which a substance can be magnetized). If a substance turns away from a magnetic force, it is said to be diamagnetic; if it turns toward the force, it is called paramagnetic. Magnetic behavior studies have been useful in physical chemical research, especially in connection with coordination phenomena, spectrographic analysis, and catalytic properties.

The fact that ions are attracted into a curved path when they enter a magnetic field at high velocity makes possible the separation of ions which have nearly the same mass; electromagnetic separation of isotopes is utilized in isolating U-235 and is also used in mass spectrometry for analytical purposes. Magnetic fields of high intensity are used to contain plasmas in fusion research. *See also* nuclear magnetic resonance; mass spectrometry; tokamak.

magnetohydrodynamics. *See* plasma (1).

Maillard reaction. *See* browning reaction.

maleic acid. An unsaturated dicarboxylic acid, m.p. 130°C (266°F), made by catalytic oxidation of benzene or butylene; its formula is $HOOCCH{=}CHCOOH$. It is an important chemical intermediate for the manufacture of such acids as malic, fumaric, tartaric, and lactic, all of which are used in the food industries, and also has application in the dyeing of textile fibers. The related compound, maleic anhydride, is still more useful, though it is quite poisonous. It is a component of various tonnage synthetic resins (polyesters, alkyds, and permanent-press types) which have a broad spectrum of uses. Malic acid, made by heating maleic acid with steam under pressure, also occurs naturally in such fruits as apples, from which the name is derived; it is also a dicarboxylic acid, formula $COOHCH_2CH(OH)COOH$, but is not the same as maleic, in spite of the similarity in spelling.

maleic anhydride. *See* maleic acid.

malic acid. *See* maleic acid.

malt. The product derived from barley or similar grain as a result of controlled germination followed by heating or drying; it is the basic material of the brewing industry. Malt contains several enzymes: diastase, which hydrolyzes the starch to sugars which are later fermented to alcohol; maltase, which converts maltose to glucose; and amylase. Malt concentrates rich in these enzymes are used in the food industry. Malt also contains considerable protein and has high nutrient value. Malt extract is called wort. *See also* brewing.

maltose. A sugar whose molecule is composed of two saccharide units and is dextrorotatory; it occurs in barley and other grains and can be produced by enzymic decomposition of starch. It is a reducing sugar which is hydrolyzed to glucose during the digestive process. Its chief use is in infant food preparations. The formula is $C_{12}H_{22}O_{11}$. *See also* malt; sugar.

manganese.   An element

Symbol	Mn	Atomic No.	25
State	Solid	Atomic Wt.	54.938
Group	VIIB	Valence	2,3,4,6,7

Manganese-bearing ores are scarce in the United States, most of the country's requirements being imported from Brazil and Zaire. In recent years, an untapped supply of manganese "nodules" has been found on the ocean floors. The metal is derived commercially by electrolysis of manganous sulfate and has only one stable form among its four allotropes. Its multiple valences make manganese a versatile metal for both analytical and industrial procedures; it can act either as a reducing agent or an oxidizing agent depending on its valence. Manganese has a density of 7.44 g/cc and melts at 1245°C (2273°F). It is subject to corrosion and is combustible at high temperatures. Its principal uses are in manufacture of manganese salts, as a deoxidizer and purifier in steel manufacture, and as an alloying agent in steel (ferromanganese) and in aluminum and magnesium alloys. Manganese is an essential trace element in plant nutrition. Its fumes have toxic effects on prolonged inhalation, and it is flammable in powder or dust form.

manganese dioxide.   The most industrially important manganese compound; MnO_2 is obtained from natural sources in Africa and is also made electrolytically. It is used in large quantities in dry cell batteries; it is also a strong oxidizing agent, as in the manufacture of hydroquinone from aniline; is an effective catalyst, and a versatile reagent.

manganese linoleate.   *See* drier.

Manhattan project.   *See* uranium.

mannan.   A polymer of the carbohydrate mannose and classified as a polysaccharide; the mannose units are linked at the 1,4 positions in the molecule. The simple mannans occur in softwoods (pine, spruce, etc.), as well as in some nuts. A subgroup, called galactomannans, is a combination of galactose and mannose, the two being isomeric sugars with slight differences in molecular linkage; they bear about the same relationship as cellulose and starch. Galactomannans are found in the seeds of such plants as guar, locust, and other rare species. Mannans are used in textile and paper sizings, leather tanning, and as food additives. *See aslo* polysaccharide.

mannose.   *See* mannan.

marihuana.   *See* hallucinogen.

marsh gas.   *See* methane.

mass.   As generally used, the term *mass* refers to the amount of material substance present in a body, irrespective of gravity; for example, a given unit of mass is the same on the moon as on the earth. The term *weight* is a measure of gravitational attraction and is substantially less on the moon than on the earth. A more fundamental meaning is based on the ultimate equivalence of mass and energy; thus, mass can be defined as a certain amount of energy which is more or less highly concentrated, as represented by a proton and an electron, respectively. The basic unit of mass is the proton, which has a mass number of 1. *See also* energy; mass defect; mass number.

mass action (law).   In its simplest terms, this law states that the rate (speed, velocity) of a chemical reaction is directly related to the concentrations of the reactants, if the temperature remains constant. Another name for this principle is the law of molecular concentration.

mass conservation (law).   In every chemical reaction or chemical change, the mass (weight) of the substances formed is precisely equal to that of the reacting substances, regardless of any alteration in their form or other properties. This implies the principle of balance in a reaction, which means that the products formed contain the same number of atoms of each element as were present in the reacting substances.

mass defect.   The difference in mass between an actual atomic nucleus and the sum of the masses of its constituent protons and neutrons when added as individual units. The mass of a naturally occurring nucleus is found to be very slightly less than that of the sum of its constituents when taken separately; this discrepancy is accounted for by Einstein's equivalence equation $E = mc^2$, which states that mass can be transformed into energy. This refers to the binding energy which holds the nucleus together, a portion of which is released when fission takes place. *See also* binding energy; fission.

massecuite   (mass-kweet). A term derived from the French, literally meaning a "cooked mass," it is used in sugar technology to denote the thick sirupy mixture of sucrose crystals and molasses resulting from concentration and boiling of the mother liquor. The crystals are separated from the liquid molasses by centrifuging.

mass number. A whole number representing the total number of protons and neutrons in a given atomic nucleus; it should not be confused with atomic number, which is the number of protons only. For example, hydrogen has a mass number of 1 (one proton, no neutrons), an atomic number of 1, and an atomic weight of 1.008; helium, with two protons and two neutrons, has a mass number of 4, an atomic number of 2, and an atomic weight of 4.0026. Carbon has a mass number of 12, an atomic number of 6, and an atomic weight of 12.01115 (an average including all isotopic forms); the mass of the 12 isotope is the international standard for calculation of atomic weights.

mass spectrometry. Analysis of substances or mixtures of substances by electromagnetic separation of their ions. A given sample is vaporized and then ionized by electrons, either as the entire molecule or as constituent atoms. The charged particles are accelerated by an electric impulse and then are passed through a magnetic field, which has the effect of changing their path from a straight line to a curve. Since some particles have greater mass than others, the effect is a separation of the particles to an extent dependent on their ratio of mass to charge. Their impact on a photographic plate thus yields a spectrum from which the constituents can be determined. Isotopes and substances present in a mixture in extremely small amounts can be identified in this way. The mass spectrograph is designed especially for mass determinations, as in study of isotopes, whereas the mass spectrometer measures the ion population. Mass spectrometry has been especially useful in analyzing mixtures of petroleum hydrocarbons and in detection of tracer elements; recently developed techniques are used in chromatography. *See also* analysis; magnetochemistry; isotope.

material. An ambiguous term which should be avoided in chemical usage except in such conventionally established phrases as *raw material* and *material balance*. Its meaning is imprecise since it can refer to either an element, a compound, or a mixture; it also has a number of nonscientific meanings which can become a source

of confusion, e.g., the material of a report, etc. *See also* substance, mixture.

material balance. A term used by chemical engineers to denote an essential requirement for the design of processing equipment, namely, a list of all the substances introduced into a reaction and all the substances which leave it in any given period of time. The two sums must be precisely equal, by the law of conservation of mass. *See also* mass conservation (law); balance **(1)**.

matter. A broad term customarily defined as anything that has mass and occupies space, thus distinguishing it from electromagnetic radiation (light) and thermal activity (heat), which are forms of energy. It is not much used by chemists except in reference to the "three states of matter," i.e., solid, liquid, and gas. Additional states of matter have been proposed from time to time but have never been widely accepted, e.g., the colloidal state, the vitreous or glassy state, and the plasma state.

mauveine. *See* dye.

Md Symbol for the element mendelevium, named in honor of the Russion chemist Dmitri I. Mendeleev (1834–1907), originator of the Periodic System.

Me Symbol often used in chemical formulas for the methyl group, CH_3.

mean free path. A term used by physical chemists to denote the average distance traversed by an atom, molecule, or other particle without collision with another atom or molecule. This distance is much greater in gases than in liquids because of the much lower population of atoms or molecules. It is determined mathematically on the basis of probability assumptions.

medicinal chemistry. A subdivision of chemistry which deals with the effects of drugs and pharmaceuticals on the human body and on various infective organisms and with the synthesis of compounds specifically for certain diseases, such as antimalarials and antihypertensive agents. It is also concerned with such topics as immunology, hormone activity, etc. *See also* clinical chemistry.

megaton. One million tons; used to indicate the blast effect of a nuclear weapon, it refers to 1,000,000 tons of TNT.

melamine resin. A thermosetting resin made by a condensation reaction of a melamine derivative, trimethylolmelamine,

$$C_3N_3(NHCH_2OH)_3,$$

with formaldehyde. Such resins are prepared in the form of molding powders which can accept colorants and are thus used for production of decorative plastics. They are often extended with wood flour and inorganic fillers and may be used in combination with paper, textiles, and similar materials, either as coatings or impregnants. They are resistant to both heat and abrasion. *See also* amino resin.

melting point. The temperature at which the forces which unite the crystals of a solid are ruptured, resulting in a change from the crystalline to the amorphous state. This change involves absorption of a characteristic amount of heat for each solid substance, called the heat of melting or fusion; it is 80 calories per gram for ice. Synonymous with the term melting point is freezing point, where the change is from amorphous to crystalline, the only difference being that heat is removed from instead of absorbed by the substance. Thus, the melting and freezing points of a substance are identical, e.g., for water (ice) they are 0°C (32°F).

membrane. An extremely thin, porous layer or film of material, either natural or synthetic; the thickness is approximately 100 angstrom units, the diameter of the pores ranging from as little as 8 angstroms for natural cellular membranes up to 100 angstroms for manufactured membranes. Membranes which have pores so small that only the molecules of solvents, ions, and a few substances of very low molecular weight can pass through are called semipermeable; these are characteristic of living cell tissue and are an essential feature of osmosis. Membranes made of cellophane, collodion, parchment, and certain types of plastics are used in many industrial procedures (waste liquor recovery, desalination, and chemical separation techniques). All types are virtually impermeable to macromolecules and colloidal particles. *See also* osmosis; dialysis; colloid.

Mendeleev, D.I. (1834–1907). Born in Siberia, Mendeleev in 1869 made one of the greatest contributions to chemistry of any scientist by establishing the principle of periodicity of the elements. His first Periodic Table was compiled on the basis of arranging the elements in ascending order of atomic weight and grouping them by similarity of properties. So accurate was

Mendeleev's thinking that he predicted the existence and atomic weights of several elements that were not actually discovered till years later. The original table has been modified and corrected several times, notably by Moseley, but it has accommodated the discovery of isotopes, rare gases, and synthetic radioactive elements. Its importance in the development of chemical theory can hardly be overestimated. *See also* Becquerel; Moseley; Periodic Table.

mendelevium. An element.

Symbol	Md	Atomic No.	101
State	Solid	Atomic Wt.	256(?)
Group	IIIB		
	(Actinide		
	Series)		
		Valence	2,3

A synthetic radioactive element discovered in 1955 and made by bombardment of einsteinium in a cyclotron. The half-life of the 256 isotope is 1.5 hours and that of the heaviest isotope (258) is 2 months. The element is named for Dmitri Mendeleev, discoverer of the Periodic System. It can be produced only in quantities measured in numbers of atoms. *See also* actinide series.

menthol. *See* antiseptic; terpene; alicyclic.

mercaptan. *See* thiol.

mercerizing. In textile technology, the process of passing cotton fibers through a 20% solution of sodium hydroxide, followed by washing with water. This treatment not only increases the strength of the fibers but develops controlled shrinkage and elasticity. It has been applied to the manufacture of modern stretch fabrics. Mercerizing is the first step in the manufacture of viscose rayon, in whcih alkali cellulose is formed by reaction of the hydroxyl groups of cellulose fibers with the sodium hydroxide. *See also* viscose process.

mercuric oleate. *See* antifouling paint.

mercury. An element.

Symbol	Hg	Atomic Wt.	200.59
State	Liquid	Valence	1,2
Group	IIB	Isotopes	7 stable
Atomic No.	80		

Mercury is the only metal that is liquid not only at room temperature but down to −39°C (−38°F); it forms a monatomic vapor and can be compressed to a soft solid state. It is unique among liquids in its very high surface tension which causes it to fragment into globules rather than to flow; this property is responsible for the trivial name quicksilver. Elemental mercury is light gray in color, with a specific gravity 13.6 and high electrical conductivity. It is used in thermometers and barometers, in special lamps, as a component of dental amalgams, as electrodes in electrolytic cells, and in electric switches. It may have toxic effects as a result of long contact with the skin or inhalation of vapor; while oral intake should be avoided, it is not serious in small amounts. Disposal of mercury in streams and coastal waters is an ecological hazard, due to bacterial conversion of the element into toxic compounds, such as methyl mercury. *See also* mercury compounds.

mercury compounds Both organic and inorganic compounds of mercury have considerable use as disinfectants, fungicides, seed sterilants, textile preservatives, and as ingredients of antifouling paints. Their effectiveness in these applications is due to their extreme toxicity. Their presence in lakes, rivers, and even the ocean in trace concentrations has resulted partly from drainage of agricultural mercurials from the soil and partly from discharge of metallic mercury wastes from industrial processing (manufacture of chlorine and some types of plastics); in the latter case, the element is converted to compounds by bacterial action. It is probable that the use of these materials will diminish in the future as a result of increased ecological concern. However, methods of separating and collecting the mercury from waste effluents are being introduced.

mercury fulminate. *See* high explosive; initiating explosive.

mescaline. *See* hallucinogen.

mesh. The number of apertures per linear inch in a woven or electroformed metal screen or sieve made especially for the laboratory testing of high-gravity dry powders for fineness and impurity content. Such screens are available in mesh sizes ranging from 100 to 400. Fineness specificaitons are often given in terms of these sizes. *See also* particle size.

meso **(1)** A prefix denoting that the compound lacks optical activity because it contains two asymmetric carbon atoms which rotate plane-polarized light by the same amount in opposite directions, resulting in zero rotation. This is called internal compensation and occurs, e.g., in tartaric acid. *See also* racemic.

(2) A prefix indicating that certain groups or substituents occupy a middle or central position in the molecule.

mesomerism. *See* resonance **(1)**.

meson. *See* fundamental particle; nucleonics.

messenger. A term often used to describe the function of an active biochemical agent or complex. For example, it is used to characterize the behavior of hormones, which regulate many essential metabolic functions; it also refers to one type of ribonucleic acid (RNA), which carries essential genetic coding information from DNA molecules to the ribosomes, on the basis of which proteins are built up from various amino acid sequences. *See also* hormone; amino acid; ribonucleic acid; genetic code.

meta-. A prefix derived from the Greek word for *beyond*; it is used in naming isomers of aromatic compounds and denotes location 3 in a benzene ring. Thus, if substituent atoms or groups replace hydrogen at locations 1 and 3, the compound is a meta- isomer; often the numerals 1,3- are used to indicate such substitution. The prefix is also used in inorganic nomenclature to indicate a lower state of hydration, e.g., metaphosphoric acid. *See also* ortho-; para-; isomer.

metabolism. The chemical and physiochemical changes occurring in a plant or animal in response to intake of nutrients. Plant metabolism involves the assimilation of the major nutrients, e.g., nitrogen, potassium, phosphorus, trace elements, water, carbon dioxide, etc., and their conversion into cellulose, fats, starches, sugars, and proteins within the plant. In animals the entire assimilative process serves to extract the energy values from nutrients provided directly or indirectly by plants in a complex series of steps, from carbohydrate breakdown in the saliva to absorption of glucose and amino acids in the intestines. Digestion of foods is aided by enzymes (pepsin), by hydrochloric acid in the stomach, and by bile formed in the liver.

metabolite. An active growth factor in organic systems typified by vitamins, enzymes, proteins, and their various complexes. Substances that are able to inhibit the activity of metabolites because of their chemically similar structure are called antimetabolites or structural antagonists. *See also* antagonism; prosthetic.

metal. All metals are elements; they rarely occur in elemental form in nature, but rather as compounds mixed with rocky material (ores) from which they are extracted by heat and chemical processing. Metals comprise about 75% of the elements; all but three (mercury, gallium, and cesium) are solid at room temperature. They display a broad range of physical properties with respect to conductivity, weight, strength, corrosion resistance, hardness, etc.; most have a crystalline structure and a characteristic sheen or luster. The atoms of metals are joined by bonds having free electrons, which account for their electrical and thermal conductivity. The most electrically conductive are silver, copper, gold, and aluminum.

Most metals are quite reactive chemically, though a few are not; they form positive ions and exert electrochemical potential, lithium and potassium having the highest potential. Many metals oxidize (corrode) readily; a few, such as chromium, gold, tantalum, and vanadium, are quite resistant to corrosion. Low percentage content of other elements may have a profound influence on the properties of metals; for example, the effect of carbon in steel is so marked that the carbon content must be controlled to within hundredths of a percent. A few metals are toxic (beryllium, barium, cadmium) or combustible (magnesium, sodium) in elemental form; many finely divided metals are flammable when dispersed in air.

Some metals exert a strong influence on the metabolism of plants and animals, where they have essential nutrient and other biochemical functions; among these are iron, cobalt, copper, potassium, magnesium, sodium and zinc, usually in trace concentrations. Magnesium is the central element in chlorophyll, as iron is of the pigment hemin.

Metals fall into several groups: (1) alkali metals (Group IA); alkaline-earth metals (Group IIA, except beryllium); rare-earth metals (lanthanide series); radioactive metals (polonium, radium, and actinide series); and the transition metals (atomic numbers 21–29, 39–47, 72–79). Some authorities include the lanthanide and actinide series in the transition metal group; the platinum group (atomic numbers 44, 45, 46, 76, 77, 78) are also transition metals. *See also* transition element; rare metal; lanthanide series; corrosion; trace element.

metal carbonyl. *See* carbonyl.

metal glass. A product made by subjecting a molten metal or alloy to extremely rapid cooling (quenching), thus preventing formation of a crystalline structure. This results in a vitreous material that is predominantly amorphous and has unique behavior in respect to corrosion resistance, conductivity, and magnetic properties.

metallic bond. A strong electrostatic force, existing between atoms in metallic crystals, which binds the crystal together in a close-packed structure; in this type of bonding, the electrons are distributed among all the available vacancies rather than being paired between specific atoms. Thus, a metallic bond is the attraction between a positively charged metal ion and the unlocalized body of electrons which envelope it. This imparts a number of properties characteristic of metals, such as electrical and thermal conductivity, ductility, hardness, etc. When sufficient energy is supplied, the bonds rupture and the crystal melts.

metallic soap. *See* soap; drier; grease (1).

matallize. *See* ABS resin.

metallocene. Any of a number of organometallic complexes in which a transition metal atom is coordinated with cyclopentadienyl rings (C_5H_5); the metals most often involved are iron (ferrocene) and cobalt (cobaltocene), though the structure can be formed by a number of others. These are sometimes called sandwich structures, since the metal atom lies midway between the two organic rings. Metallocenes were introduced in the early 1950s and have been used as antiknock agents, catalysts, and reducing agents. *See also* ferrocene.

metallography. Study of the properties of metals and alloys as related to their physical structure, involving knowledge of crystal formation and structure, solid solution theory (phase rule), and microscopy at all levels.

metalloid. An obsolete term formerly used to refer to a small group of solid elements which have strong semiconducting properties. These are now preferably called nonmetals or semiconductors. *See also* nonmetal; semiconductor.

metallurgical coke. *See* coke.

metallurgy. The body of scientific knowledge relating to the recovery, properties, and uses of metals. It is generally considered to comprise the following subdivisions: mining and ore separation (extractive metallurgy); alloying, fabrication, and testing (physical metallurgy); and manufacture and fabrication of metal powders (powder metallurgy). Application of the principles of chemical analysis and separation techniques and of electrochemical phenomena are also involved in these subdivisions. *See also* ore dressing; metal; powdered metals.

meter.   A standard unit of length equivalent to 39.375 inches, i.e., 3 feet, 3⅜ inches. A centimeter is one-hundredth meter, or about 0.4 inch. A millimeter is one-thousandth meter, and a micron one-millionth meter. The abbreviation for meter is m.

methadone.   *See* narcotic.

methanal.   *See* formaldehyde.

methanation.   A reaction by which methane is formed from the hydrogen and carbon monoxide derived from coal gasification. It requires a catalyst, e.g., nickel, and temperatures in the range of 500°C. The reaction is

$$CO + 3H_2 \xrightarrow{Ni} CH_4 + H_2O.$$

In one process the reaction is performed in an adiabatic fixed-bed reactor. *See also* gasification.

methane.   The simplest hydrocarbon and the first member of the paraffin (alkane) series; formula CH_4. It is a degradation product of carbonaceous materials and thus occurs in association with petroleum and coal, as well as in bogs and marshes; it is often called marsh gas. Methane is found in natural gas, of which it constitutes from 75 to 85%, and can be recovered from it by various sorption methods. It can also be made by gasification of coal. Its major uses, other than as a fuel, are in synthetic organic chemistry as a source of many tonnage chemicals, including methyl alcohol, methyl chloride, and numerous petrochemicals. When cracked by treatment with steam, it yiedls a mixture of carbon monoxide and hydrogen called synthesis gas. When chlorinated, methane yields carbon tetrachloride and numerous other chlorinated hydrocarbons. Natural gas is frequently used for such processes without prior separation of the methane. Methane is extremely flammable and may cause explosions in confined areas; it is not toxic but is classed as an asphyxiant gas. *See also* synthesis gas; natural gas.

methanol.   *See* methyl alcohol.

methanol group.   The characteristic atomic grouping found in monohydric alcohols; it con-

sists of a carbon atom, with three open valences, bonded to a hydroxyl group:

$$-\overset{|}{\underset{|}{C}}-OH.$$

The three valences may be attached to hydrogen atoms, giving methyl alcohol, to alkyl groups (R), or to a combination of H's and R's. The carbon atom of this group is often referred to as the methanol carbon. *See also* monohydric alcohol.

methyl.   The univalent group CH_3-; it is the simplest alkyl group, having one less hydrogen atom than methane. A methyl group occurs at both ends of paraffinic hydrocarbon molecules having two or more carbon atoms in the chain, e.g., ethane, H_3CCH_3, and in at least one position in every other alkyl group, for example, propyl, $CH_3CH_2CH_2-$; tert-butyl, $(CH_3)_3C-$; etc. It is also present in many olefinic compounds, such as propylene, $CH_3CH=CH_2$; in alicyclic compounds, e.g., methyl cyclohexane; in aromatic compounds, such as toluene and xylene; and in heterocyclic compounds, e.g., methyl furane. Probably the most common alkyl group, it is present in multitudes of hydrocarbon derivatives: alcohols and thiols, esters, aldehydes, ketones, acids, amines, etc.; as well as in organometallic compounds, e.g., tetramethyllead; in halogen compounds, such as methyl chloride; in carbohydrates, e.g., methylcellulose; in organic silicon compounds, like methyl chlorosilane, etc. *See also* alkyl; methane; monohydric alcohol.

methyl alcohol.   A liquid monohydric alcohol (CH_3OH), b.p. 64.5°C (148°F), which is both flammable and poisonous, causing blindness if ingested. Its IUPAC name is methanol; it is commonly called wood alcohol, after its original source. It is produced for industrial purposes chiefly by a catalyzed reaction between hydrogen and carbon monoxide or carbon dioxide, or by oxidation of methane and other hydrocarbons. Its major uses are for organic synthesis, especially of formaldehyde, as an automotive antifreeze, as a solvent for cellulosics and natural resins, and as the principal denaturant for ethyl alcohol. It also has some application in the cleaning of metals and equipment parts.

methylamine.   A gas, b.p. −6.79°C (20°F), made

by catalytically reacting methyl alcohol and ammonia; it is commercially available in aqueous solutions up to 40%. Its formula is CH_3NH_2. Both gas and liquid forms are flammable and their fumes are intensely irritating to eyes and the respiratory system. Its major uses are in organic synthesis (dyes, insecticides, accelerators), to retard or prevent premature polymerization, and in photographic developers. *See also* amine.

methyl *tert*-butyl ether. This compound, also called MTBE, is a liquid, b.p. 55.2°C (131.4°F), whose formula is $CH_3OC(CH_3)_3$. Its major use is as an additive to increase the octane number of unleaded gasoline. It is made by the reaction of methanol with the isobutylene in the C_4 hydrocarbon streams from ethylene plant steam crackers and refinery catalytic crackers. The liquid phase reaction operates over a catalyst at temperatures of 38 to 93°C (100 to 200°F) and pressures of 100 to 200 psi. When added to gasoline in concentrations of 7% or less by volume, it has a research octane number (RON) blending value of 115 to 125 and a motor octane number (MON) blending value of 98 to 110. The compound can also be catalytically cracked to yield pure isobutylene for production of *tert*-butylphenol or polybutylene. The methanol co-product from this conversion is recycled to MTBE. *See also* knock; octane number.

methylcellulose. A water-soluble cellulose ether made by reacting cellulose with sodium hydroxide and further reacting the alkali cellulose so formed with a compound containing a methyl group, e.g., methyl chloride, or with propylene oxide. The latter process forms the hydroxypropyl derivative. The product has molecular weights ranging up to 200,000 and is useful as a thickening agent, protective colloid, and general-purpose food additive. It is strongly hydrophilic and forms stiff colloidal dispersions in water at room temperature. *See also* carboxymethylcellulose.

methyl chloride. A toxic compressed gas, b.p. −23.7°C (−10°F), made by reacting methane with chlorine to form CH_3Cl, or by the reaction of HCl and methanol: $HCl + CH_3OH \ 1 \ CH_3Cl + H_2O$. It is flammable and may explode in concentrations between 10 and 17% in air; thus, it should be handled with care. Its primary uses are in the manufacture of butyl rubber, where it acts as a solvent for the catalyst; as a source of methyl groups in organic synthesis; and in the manufacture of silicone polymers. It has minor applications as an herbicide and solvent.

methylene. (1) The divalent hydrocarbon group CH_2—, in which the carbon atom has its normal valence of 4; it is derived from methane by dropping two hydrogen atoms. It occurs in many organic compounds, e.g., methylene chloride (dichloromethane), CH_2Cl_2.

(2) *See* carbene.

methylene blue. A blue dye used in the laboratory as a redox indicator in titrations and as a microscopic stain; on a production scale, it is a colorant for natural textile fibers. The material itself is a green, shiny powder which turns blue in solution. *See also* indicator.

methyl ethyl ketone. Next to acetone, the most important and widely used of the ketones. The formula is $CH_3C\ddot{\ }OC_2H_5$, b.p. 80°C (176°F). It is flammable, and explosions may occur at concentrations of 2 to 10% in air. Its major uses are as a solvent, especially for nitrocellulose, and as an intermediate in the manufacture of synthetic organic chemicals. It is only slightly soluble in water but is quite soluble is most organic solvents. It is also called ethyl methyl ketone. *See also* ketone.

methyl magnesium iodide. *See* Grignard.

methyl orange. *See* indicator.

Mg Symbol for the element magnesium, the name being derived from the Greek city of that name.

mica. A crystalline solid, of which there are two types: ruby (potassium aluminum silicate) and amber (which also contains magnesium). Mica occurs in certain rocky structures, chiefly in India and South America. Its crystals easily split apart into thin leaves or layers. Its high electrical resistivity makes it a useful insulator in electronic, aerospace, and electrical devices. As it is also translucent, it is used as visual inserts or panels in ovens and other high-temperature equipment. In finely divided form, it is a good abherent and solid lubricant, imparting a silvery sheen to surfaces to which it is applied.

micellar flooding. *See* enhanced oil recovery.

micelle. A cluster or aggregate of charged molecules that is large enough to be in the colloidal size range. These characteristically occur in such aqueous suspensions as soaps (fatty acids), milk (casein), etc. In soaps, the individual molecules of fatty acids are made up of a positively charged chain of hydrocarbons which is hydrophobic and a negatively charged carboxyl group which is hydrophilic. When many such molecules are present in a solution, they aggregate or unite to form a particle which has properties similar to those of the component molecules. In milk, the casein combines with calcium to form a calcium caseinate complex, or micelle. *See also* fatty acid; colloid; soap; emulsion.

microanalysis. Analysis of samples weighing from 1/10 to 10 milligrams; semimicroanalysis involves samples weighing from 10 to 100 milligrams.

microcrystalline. A term used to describe materials that can be prepared in a form in which the crystals are much smaller than in the natural product. Paraffin-like waxes derived from lubricating oil wastes and residual oil refining are the most important products in commercial use; they have much higher viscosity and penetration value than normal paraffin wax. They are used in adhesives, paper sizing, carbon paper, cosmetic creams, and many other products. Chlorophyll has a type of structure that is naturally microcrystalline. Some high polymers are now being prepared in this form, including cellulose and nylon. The crystals range in size from 2.5 to 500 millimicrons. *See also* wax.

microencapsulation. *See* encapsulation.

micrometer. *See* micron.

micron. A unit of length in the metric system equivalent to one-millionth meter, or 10,000 angstrom units; a millimicron is one-thousandth micron, or 10 angstrom units. These units are used in physical chemistray and microscopy for measurement of macromolecular and colloidal sizes. The symbol conventionally used for micron is the Greek letter mu (μ), and for millimicron, mμ. 1 micron is generally accepted as the upper limit of the colloidal size range. A micron is also called a micrometer (μm), and a millimicron, a nanometer (nm).

micronutrient. *See* trace element.

microscopy. The magnifying and resolving powers of various kinds of microscopes are of great service to the chemist in observing the structure of solids and the surfaces of particles, in identifying substances, and in visually penetrating the colloidal and upper molecular size ranges. The resolving power of a microscope depends on the wavelength of light used in making the observation. Optical microscopes utilize visible light, and their resolution is thus limited to about the wavelength of violet radiation. The

ultramicroscope introduces a light beam at right angles to the specimen, thus permitting observation of particles beyond the range of an optical microscope, by means of the light which they reflect or scatter. It was invented by Zsigmondy in 1903. Of still greater importance is the electron microscope, which made its appearance in 1940; it utilizes extremely short-wave radiation from clectrons in a vacuum tubc, which focuses radiation in a manner analogous to that of a glass lens. Resolutions as low as 2 angstrom units have been obtained. Its greatest value lies in revealing the fine structure of inanimate objects. *See also* Tyndall effect; resolving power.

microwave spectroscopy. A type of absorption spectroscopy utilizing the portion of the electromagnetic spectrum lying between the far infrared and radiofrequencies, i.e., wavelengths from 1 to 1000 millimeters. Substances analyzed in this manner must be in the gas phase. The klystron tube is used as a source of microwave radiation. The method is especially suitable for study of internal molecular structure. *See also* spectroscopy; absorption.

Midgley, Thomas, Jr. American chemist and inventor (1889–1944). One of the most creative and brilliant chemists of his era, Midgley's early work was in the field of rubber chemistry and technology, especially in the development of synthetic and substitute rubbers, which were being introduced in the late 1930s. He worked with Kettering at General Motors and then became vice-president of Ethyl Corporation, as well as of the Ohio State University Research Foundation. His innovative genius was responsible for the development of organic lead compounds for antiknock gasoline and later for the discovery of fluorocarbon refrigerants, on which he did the basic research. He was recipient of many of chemistry's highest honors, including the Nichols Medal, the Perkin Medal, the Willard Gibbs Medal, and the Priestley Medal.

milk. (1) A natural animal product consisting of a suspension of protein-coated fat globules in an aqueous solution. The percentage composition varies considerably with the animal involved. Cows' milk contains approximately 4.2% fat, 3.6% protein, 5% lactose, and 0.7% minerals. The major proteins are casein, lactalbumin, and lactoglobulin, and the chief minerals are calcium and potassium. Several important vitamins are also present. The solution remaining after removal of the fat and casein is called serum or whey, in which the lactose and certain proteins are dissolved. Coagulation or souring is brought about by bacterial conversion of the lactose to lactic acid, or by addition of acid or the enzyme rennet; the casein is separated in this way. Milk is irreversibly coagulated by mechanical agitation (churning) to form butter. Homogenized milk is made by reducing the fat particles to uniform size by passing the milk under pressure through a colloid mill. This subjects them to a shearing action which reduces their diameter from about 6 microns to 1 micron, thus preventing them from rising to form cream. *See also* pasteurization.

(2) The term *milk* is also used (by analogy) to refer to thick colloidal suspensions of the hydroxide of magnesium, calcium, or bismuth in water (milk of magnesia, milk of lime), prepared for industrial, pharmaceutical, or medical use.

milliliter.   One-thousandth liter, abbreviated ml; the volume occupied by 1 gram of water at 4°C (40°F) and 1 atmosphere pressure; it is equivalent to 1.0000027 cm^3. The volume of one cubic centimeter is usually accepted as the equivalent of a milliliter, though the two are not identical. *See also* gram.

mineral.   (1) A generally used but not precise term referring to the nonliving constituents of the earth's crust, which include naturally occurring elements, compounds, ar d mixtures that have a definite range of chemical composition and properties. Usually inorganic, but sometimes including fossil fuels, e.g., coal, minerals are the raw materials for a wide variety of elements (chiefly metals) and chemical compounds. Minerals can be, and many are, synthesized to achieve purity greater than found in natural products. The term *mineral industry* statistically comprehends the mining and production of metals (ores), fossil fuels, clay, gemstones, cement, glass, rocks, sulfur, sand, etc. The term *mineral* was illogically used by early chemists to describe a variety of substances; many of these uses are obsolescent, but a few persist, e.g.:

mineral acid: many of the major inorganic acids (sulfuric, nitric, hydrochloric).
mineral black: inorganic black pigments.
mineral blue: varieties of blue pigments.
mineral dust: industrial dust; nuisance dust.
mineral green: copper carbonate.
mineral oil: a liquid petroleum derivative.
mineral pitch: asphalt.
mineral red: iron oxide red.
mineral rubber: blown asphalt.
mineral spirits: a grade of naphtha.
mineral water: natural spring water containing sulfur, iron, etc.
mineral wax: a wax found in the earth (ozocerite), or derived from petroleum.
mineral wool: fibers made by blowing air or steam through molten slag.

(2) As used by nutritionists, the term refers to such components of food as iron, copper, phosphorus, calcium, iodine, selenium, fluorine, and other trace micronutrients. *See also* geochemistry; trace element.

mineralogy.   The study and classification of minerals by source, chemical composition, and properties, chiefly physical, such as color, hardness, and crystalline structure. *See also* geochemistry.

mirror-image molecule.   *See* asymmetry; enantiomorph; glyceraldehyde.

misch metal.   A commercially made alloy or mixture of several rare-earth metals, of which cerium constitutes about 50%. It will ignite when exposed to heat or friction. It is used as a component of numerous alloys, as a reducing agent, and as a scavenger (getter) in vacuum devices. A specialized application is as an ingredient of ferrocerium for lighter flints. *See also* getter; scavenger; pyrophoric.

miscible.   A term describing the extent to which liquids or gases can be mixed or blended. Substances in the gaseous phase can blend in any proportion; this may or may not be true of liquids. Some, such as ethyl alcohol and water, will blend uniformly in all proportions; others have limited miscibility, that is, they will blend only until a certain concentration is reached. The term *soluble* is often used in this sense, but it usually applies to solids. *See also* soluble; immiscible.

mixed acid.   A blend of nitric and sulfuric acids used in the chemical industries to introduce nitrogen into a compound; for example, in the manufacture of nitrocellulose, cotton linters are treated with mixed acid of various strenghts to form lacquers and explosives. The process is called nitration. Standard mixed acid is about two-thirds sulfuric and one-third nitric acid. It is also called nitrating acid. *See also* nitration.

mixture.   A mechanical blend of two or more substances in any proportions; the components may or may not be uniformly dispersed. The substances are not united chemically and can be separated by a number of means, as by filtration, distillation, precipitation, etc. Since they are composed of more than one component, they are heterogeneous in nature. Well-known examples of mixtures are air, milk, alloys, petroleum, blood, gasoline, etc. A mixture in which the components are uniformly dispersed is called a solution. *See also* compound; heterogeneous; solution; substance.

ml.   Abbreviation for milliliter.

Mn   Symbol for the element manganese, the name being derived from the Latin word for *magnet*, since manganese ores often occur in association with iron ore. Some authorities give the same derivation as for magnesium.

Mo   Symbol for the element molybdenum, the name being derived from the Greek word meaning "lead-like."

mobile.   A term used to describe the extent to which a liquid moves or flows in response to force. Nonpolar liquids such as hydrocarbon solvents are in general much more mobile than water and somewhat more so than alcohols, because of their lower surface tension. This property is of significance in the handling, transporting, and pouring of such liquids, as well as in fire extinguishment.

modacrylic fiber.   The Federal Trade Commission definition of this term (meaning modified acrylic fiber) states that it is any man-made fiber comprised of a minimum of 35% by weight and a maximum of 85% by weight of acrylonitrile monomer, the balance being made up of various modifying synthetics such as vinyl chloride. *See also* acrylonitrile.

moderator.   A substance used in nuclear reactors to absorb the energy of the fast neutrons produced by fission, to the point where they will be more likely to induce enough new fissions to maintain a chain reaction. Graphite, beryllium, heavy water and ordinary (light) water have been used for this purpose, the last being the most practical. Because the kinetic energy given up by fast neutrons as they collide with the nuclei of the moderating substance appears as heat, the resulting "slow" neutrons are also called "thermal" neutrons. *See also* reactor.

modified cellulose.   *See* carboxymethylcellulose.

modify.   To change the properties of a chemical compound by replacing some or all of its substituent groups with other groups, without impairing its fundamental structure; this is done by various types of chemical reactions (etherification, esterification, etc.). Carbohydrate polymers are especially susceptible to this treatment, yielding modified starches, carboxymethylcellulose, cellulose acetate, and nitrocellulose.

modulus.   *See* elasticity.

moellon degras.   *See* lanolin.

Mohs scale.   An empirical scale of the hardness of mineral or mineral-like materials originally consisting of 10 values, ranging from talc, with a rating of 1, to diamond, with a rating of 10, the rating being based on the ability of each material to scratch the one next below it in the series. The number of materials has been expanded from 10 to 15 with the addition of several synthetically produced substances (e.g., silicon carbide) between the original 9 and 10 positions. The scale is named after the German mineralogist, Friedrich Mohs (1773–1839). *See also* hardness (**1**).

moiety.   A portion of a lot or sample, which has no meaningful mathematical relation to the whole; a unit of indefinite size or weight. *See also* aliquot.

Moissan, Henri   (1852–1907). Native of Paris where he was professor at the School of Pharmacy (1886–1900) and at the Sorbonne (1900–1907). At the former institution, he first isolated and liquefied fluorine (1886) by the electrolysis of potassium acid fluoride in anhydrous hydrogen fluoride. His work with fluorine undoubtedly shortened his life as it did that of many other early experimenters in the field of fluorine chemistry. He won great fame by his development of the electric furnace and pioneered its use in the production of calcium carbide, making acetylene production and use commercially feasible; in the preparation of pure metals, such as manganese, chromium, uranium, tungsten, etc.; and in the production of many new compounds, e.g., silicides and carbides, and refractories. In 1906 he received the Nobel Prize in chemistry.

molal solution.   A solution comprised of 1 mole (gram-molecular weight) of a substance (the solute) dissolved in 1 kilogram (1000 grams) of a solvent. For example, a 1-molal concentration of sodium hydroxide (m.w. = 40) in ethyl alcohol (m.w. = 46.069) would contain 40 grams of sodium hydroxide and 1000/46.069, or 21.75 moles of alcohol. Fractions or multiples of this concentration are expressed as 0.5 molal, 2 molal, etc. The conventional abbreviation is a lowercase italic *m*. It is important to distinguish between the terms molal and molar. *See also* mole; molar solution.

molar solution.   A solution comprised of 1 mole (gram-molecular weight) of a substance (the solute) and enough solvent to make 1 liter of *solution*; for example, a 1-molar concentration of sodium hydroxide (m.w. = 40) in water would contain 40 grams of NaOH dissolved in 1001 grams of water, so that the total volume of the solution is 1 liter. Fractions or multiples of this concentration are expressed as 0.5 molar, 2 molar, etc. The conventional abbreviation is an italic capital *M*. *See also* mole; molal solution.

molasses.   The heavy, viscous liquid remaining after separation of sugar crystals from the mother liquor. In the process of sugar manufacture, this operation is carried out by repeated centrifuging until as much sugar has been recovered as is economically possible. The final molasses (called blackstrap) still contains from 30 to 35% sucrose. Steffen molasses, made from beet sugar, is a source of sodium glutamate. Blackstrap molasses is an important source of ethyl alcohol for industrial use and to some extent for beverage purposes. It can also be used as an animal feed. *See also* blackstrap.

mold.   (1) A type of fungus; it has no chlorophyll and thus must derive its nutrients either from dead vegetation or other organic matter, or by acting as a parasite on living organisms. It is similar to mildew and the so-called rusts or smuts appearing on wheat and other grains. Certain types are cultivated for the manufacture of penicillin, their nutrients being provided in the form of lactose and other sugars. Molds are likely to appear on leather, textiles, and food products in hot, humid climates and can be inhibited by use of benzoic acid (in foods) and formaldehyde (on grain). Metallic naphthenates are used on textiles and similar materials.

(2) A metal or plastic form used to impart a specific shape to a product. The term *molding* is generally used in reference to articles made of rubber or plastics, whereas *casting* applies to the forming of metals.

molding powder.   *See* powder.

mole.   A term that has long been used in the sense of gram-molecular weight, or gram-mole, that is, the molecular weight of a substance expressed in grams. In recent years, its meaning has been extended to include all chemical units (electrons, molecules, atoms, ions, etc.). Specifically, a mole is an amount containing the same number of units as there are atoms in 12 grams of carbon-12, i.e., 6.023×10^{23}. Other weights can be used instead of gram. In general terms, a mole is Avogadro's number of any chemical unit. The spelling *mol* is used in England and some other countries. *See also* Avogadro number; molecular weight.

molecular biology.   That portion of biochemistry devoted largely to the study of cellular biology and genetic mechanisms at the molecular level, including the structure and function of complex nucleotides, DNA, RNA, and related macromolecules. Also involved is the means by which amino acids are arranged in the genes, so that specific characteristics of an organism are passed from one generation to another. Advances of the greatest importance have recently been made in the genetic field by using this approach. *See also* deoxyribonucleic acid; genetic code; recombinant DNA.

molecular sieve.   A term used to describe the function of zeolite materials, which are clay-like in chemical nature (aluminosilicate compounds), and from which all water can be removed without alteration of their molecular structure. As a result of this, the material becomes microporous to such an extent that about half its volume is occupied by very small holes or channels (cages). The material thus readily adsorbs molecules of other substances in much the same manner as activated carbon; however, such molecules must be small enough to enter the pores vacated by the water molecules. Because of the limitation placed by the pores on the size of adsorbed molecules, zeolites act as selective devices which adsorb smaller molecules readily but exclude larger ones. For this reason, they are called molecular sieves. They tend to accept polar molecules as well as unsaturated types. Their more important uses are in water purification by ion exchange and in the cracking of petroleum; they are also active in soil chemistry because of their ion-exchange capability. *See also* zeolite; ion exchange.

molecular weight.   The total obtained by adding together the weights of all the atoms present in a molecule. (To obtain this sum, a reliable table of atomic weights is necessary.) For example, the molecular weight of an oxygen molecule (O_2) is 31.9988; of a molecule of carbon dioxide (CO_2), 44.0103; of a molecule of ethyl alcohol (C_2H_5OH), 46.069. In the last case, the arithmetic is: $2(12.011) + 6(1.0079) + 15.999$. The molecular weight of polymeric substances (proteins, cellulose, plastics) may be several million and is often indeterminate. An important value in chemistry is the gram-molecular weight (now preferably called the mole); it is the molecular weight expressed in grams; thus, for oxygen it is 31.9988 grams and for ethyl alcohol, 46.069 grams. *See also* molecule; mole.

molecule.   A basic unit of matter which may be comprised of one, two, three, or more elements which are present in the form of atoms (or ions); the number of atoms of each element in a molecule ranges from 1 to 150 or more (in some cases), the usual range being from 1 to 30. Molecules that contain only one element are the noble gases and hydrogen, oxygen, ozone, nitrogen, fluorine, chlorine, some gaseous forms of sulfur, etc. A molecule of a gas such as helium has only 1 atom (monatomic); all the others mentioned have 2 atoms (diatomic) except ozone, which has 3 and sulfur, which may have up to 8.

Molecules of *compounds* are made up of at least two elements, in some cases having only 1 atom of each, as in hydrogen chloride (HCl). In this connection, a molecule may be defined as the smallest amount of a *compound* that can exist, for if it were further divided, it would cease to be a compound and would become atoms or ions, i.e., HCl would become H^+ and Cl^-. Even the most complex molecules contain only five or six elements, often linked in chains of various combinations; it is the arrangement or sequence of elements in a molecule, rather than their number, that is of critical importance. Molecules of compounds may be gases (ammonia, carbon dioxide, formaldehyde), liquids (water, alcohol), or solids (calcium carbonate, cellulose). Large, long-chain molecules formed by union of smaller units of similar constitution are called macromolecules or polymers. The atoms within a molecule are held together by forces of electrostatic attraction known as chemical bonds.

See also atom; molecular weight; Avogadro number; mole; bond.

molten salt.   *See* fused salt.

molybdenum.   An element.

Symbol	Mo	Atomic Wt.	95.94
State	Solid	Valence	2,3,4,5,6
Group	VIB	Isotopes	7 stable
Atomic No.	42		

Molybdenum, m.p. 2610°C (4730°F), occurs as the sulfide ore molybdenite, chiefly in western U.S. and Canada and in Chile. It has the highest melting point of any metal except tungsten and tantalum and has strong thermal and electrical conductivity. It also readily forms coordination or complex compounds. In elemental form its chief uses are in steels, to which it imparts toughness for heavy service, and in heat-resistant alloys. Its compounds have applications in grease lubricants and pigments and as catalysts. Molybdenum is necessary in trace amounts in plant nutrition and is also valuable in human metabolism. Neither the metal nor its compounds is considered harmful.

momer.   *See* monomer.

monazite.   A widely occurring earthy material containing metals of the rare-earth series, especially cerium and lanthanum, in the form of phosphates. Though frequently referred to as a "sand," it contains very little silica. It occurs in economically important amounts in Australia, India, South Africa, Brazil, Florida, and several states of the Northwest; minor quantities are found in Egypt and the islands southeast of Asia. *See also* rare.

mono-.   A prefix derived from the Greek word for *one*, appearing in such terms as *monoxide, monohydric, monobasic, monomer, and monomolecular*. In common usage, the prefix in such compounds as monochlorobenzene, monoethanolamine, etc., is considered redundant and is usually omitted. The corresponding prefix *uni-*, with the same meaning, is taken from the Latin *unus*; as a result of semantic convention its use in chemistry is limited to such terms as *univalent, uniaxial,* and *uniform*.

monobasic.   (1) A term used to describe an acid, either organic or inorganic, whose molecule contains only one hydrogen atom that can be replaced by another element to form a salt; examples are hydrogen chloride. HCl, and acetic acid, CH_3COOH. Monobasic inorganic acids liberate one hydrogen ion in solution.

(2) An inorganic compound containing one atom of a basic element, for example, monobasic sodium phosphate, NaH_2PO_4.

monohydric alcohol. An alcohol whose molecule contains only one hydroxyl group. The paraffinic type, often called simple alcohols, have the generic formula $C_nH_{2n+1}OH$. Most of these are mobile, flammable liquids of low boiling point. The simplest is methyl alcohol (methanol), CH_3OH. Ethyl alcohol (ethanol), CH_3CH_2OH, is the most important of all alcohols, with a broad array of industrial uses. Other paraffinic alcohols are amyl, butyl, octyl, and propyl and their branched-chain isomers.

Olefinic alcohols contain at least one double bond, e.g., allyl alcohol, $CH_2{}^{..}CHCH_2OH$; a number of synthetic fatty alcohols ranging from 6 to 22 carbon atoms are derived by reduction of fatty acids. Other types include alicyclic, which are saturated or unsaturated ring structures (cyclohexanol); aromatic, in which one or more benzene rings occur (benzyl alcohol, phenol); and heterocyclic (furfuryl alcohol). There are also a number of triple-bonded alcohols. Sterols are a class of polycyclic alcohols of high molecular weight having one hydroxyl group, e.g., cholesterol.

The terms primary, secondary, and tertiary are used to distinguish monohydric alcohols and their isomers on the basis of the number of alkyl (methyl) groups attached to the methanol carbon atom, a factor which affects their reactivity. Such alcohols can be identified easily by noting the number of hydrogen atoms attached to the carbon atom immediately preceding the hydroxyl group; if this number is two or more, the alcohol is primary; if only one hydrogen is present in this location, the alcohol is secondary; and if no hydrogen is present, it is tertiary. This is shown in the following, where R stands for a methyl group:

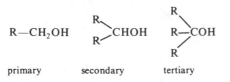

primary   secondary   tertiary

It is also exemplified by the three isomers of butyl alcohol, whose formulas are:

$$CH_3(CH_2)_2CH_2OH \qquad CH_3CH_2CHOHCH_3$$

primary   secondary

$$(CH_3)_3COH$$

tertiary

Methyl alcohol involves no alkyl substitution; the three hydrogens on the methanol carbon comprise a methyl (alkyl) group. *See also* alcohol; ethyl alcohol; methanol group.

monomer. The molecular repeating unit of an addition polymer, either natural or synthetic. For example, the monomer of natural rubber is isoprene $CH_2=C(CH_3)CH=CH_2$; when this product is made synthetically, it is called polyisoprene. Other commercially important monomers are styrene, $C_6H_5CH=CH_2$, from which polystyrene is made; acrylic acid, $H_2C=CHCOOH$, which polymerizes to form acrylate resins; butadiene

$$(H_2C=CHHC=CH_2),$$

the basis for synthetic rubbers; and vinylidene chloride ($H_2C=CCL_2$), the monomer for the plastic saran. The word is sometimes shortened to momer. *See also* polymer.

monomolecular. Having a thickness of one molecule; usually used to describe a film of fatty acid on water. Fatty acids are long-chain molecules; the carboxyl group at one end, being negatively charged, is attracted by the polar water molecule. Thus, when the acid floats on a water surface, its molecules assume a vertical position, forming a monomolecular layer from 150 to 300 angstroms thick. This behavior was investigated in the 1930s by Irving Langmuir (1881–1957), Nobel Prize 1932, and W.D. Harkins of University of Chicago. It is the key factor in emulsification. *See also* film; emulsion.

monoolefin. *See* olefin.

monosaccharide. *See* polysaccharide.

monosodium glutamate. A salt of glutamic acid, a dicarboxylic amino acid occurring in vegetable-derived proteins. It is made by hydrolyzing liquid wastes from sugar beets in an alkaline medium; the formula is $COOH(CH_2)_2CH$-$(NH_2)COONa$. It can also be made synthetically. It is one of the few substances successfully introduced in the food industry as a flavor enhancer; its original application to meats and meat products has been extended to many types of prepared foods. It is effective in concentrations of a few parts per thousand; claims of possible toxicity have not been substantiated. The name is abbreviated to MSG, and the product is also called sodium glutamate. *See also* potentiator; enhancer.

montan wax.   *See* ozocerite.

montmorillonite.   *See* clay.

mordant.   A term used in dyeing technology to indicate a substance that has the property of fixing or binding dye molecules to textile fibers. It is derived from the Latin word for *bite*, which has the associated idea of fastening, as an anchor holds a ship by biting into the mud. Various metal salts act in this way, especially those of chromium, aluminum, and tin; a few organic compounds such as tannic acid are also used. *See also* lake; dye.

morphine.   *See* narcotic.

Moseley, Henry.   (1887–1915). A British chemist, Moseley studied under Rutherford and brilliantly developed the application of x-ray spectra to study of atomic structure; his discoveries resulted in a more accurate positioning of elements in the Periodic Table by closer determination of atomic numbers. Tragically for the development of science, Moseley was killed in action at Gallipoli in 1915.

mother.   (1) A mold containing a type of bacteria that induces formation of acetic acid from ethyl alcohol derived from sugar by enzyme-activated fermentation. Bacteria of this kind are used in the manufacture of vinegar.

(2) Mother of pearl (also called nacre) is secreted by specialized cells of the oyster in response to physical stimuli, e.g., sand grains next to the body. It coats the interior of the shell and may be built up to form pearls. The material is a modified form of calcium carbonate.

(3) Mother liquor is the clarified sugarcane, or beet juice before clarification, from which refined sugar is made after separation of the molasses from the crystals. More generally, it is the solution remaining after a precipitate has been separated from it.

moving boundary.   *See* transference number.

m.p.   Abbreviation for melting point.

MSG.   *See* monosodium glutamate.

MTBE.   *See* methyl *tert*-butyl ether.

mucilage.   A polysaccharide or carbohydrate polymer found in the roots, seeds, or twigs of certain trees and plants, e.g., guar and locust beans, bark of the slippery elm, and some kinds of algae. Mucilages are obtained commercially by extracting the beans, seeds, or roots with water. They do not produce gels, but form thick, slippery dispersions in water; they are used as thickening agents in foods, in paper coatings, and as pharmaceuticals. There is no definite line of demarcation between mucilages and gums. The term should not be used as a synonym for glue or paste, which contain proteins. *See also* gum.

mud.   *See* drilling fluid.

multiple proportions (law).   Formulated in 1804 by John Dalton, an English chemist (1766–1844), this principle states that if two elements can combine to form two or more different products, the weights of one element that unite with a fixed weight of the other are in the ratio of small integers. For example, carbon (atomic weight 12) and oxygen (atomic weight 16) can form two products:

(1) carbon monoxide (CO) 12:16 = 1:1.33
(2) carbon dioxide (CO_2) 12:32 = 1:2.66

Thus, the weights of oxygen that combine with a fixed weight of carbon are expressed by the ratio of 1.33 to 2.66, or 1 to 2. (Approximate atomic weights are used for simplicity.)

municipal waste.   *See* activated sludge.

muriatic acid.   *See* hydrogen chloride.

musk.   *See* butyl alcohol; fixative.

mustard gas.   ß,ß'-dichloroethyl sulfide, $S(CH_2CH_2Cl)_2$, a highly toxic, vesicant war gas, made by the reaction of ethylene and sulfur dichloride; its b.p. is 215°C (419°F). Must be handled with extreme care since both the liquid and the vapor are very poisonous and can be absorbed through the skin. It is used in organic synthesis and has some specialized medical applications. The nitrogen mustard class of compounds, also of special medical use, is related to mustard gas in that the sulfur of the latter is replaced by an amino nitrogen, e.g.,

$$(ClCH_2CH_2)_2NCH_3.$$

The nitrogen mustards are also highly toxic.

mutagen.   Any agent that causes a modification of deoxyribonucleic acid (DNA) in such a way as to affect the hereditary characteristics transmitted by genes or chromosomes. Among these agents are ionizing radiation, heat, and some chemical substances (mustards, diethyl sulfate, and nitrous oxide). The mechanism involved in many cases is not clearly understood.

myelin.   A unique and complex mixture of fatty acids (lipids), proteins, and polysaccharides which forms a laminated sheath around the impulse-transmitting section of nerve cells in the body.

It is particularly rich in long-chain fatty acids containing more than 18 carbon atoms; glycerophosphatides are also present. Development of instability in myelin is thought to be involved in multiple sclerosis and some other degenerative diseases.

myristic acid.   *See* fatty acid; vegetable oil.

N

N Symbol for the element nitrogen; mathematical symbol for Avogadro's number.

N Abbreviation for normal solution.

n Abbreviation for normal, as in *n*-pentane; also symbol for refractive index.

Na Symbol for the element sodium, taken from the Latin word *natrium*.

nacre. *See* mother (2).

name reaction. A chemical reaction, usually organic, that is commonly identified by the name of its discoverers, e.g., Friedel-Crafts, Fischer-Tropsch, Diels-Alder. Many have important industrial applications. For details, see each specific name.

nano-. A prefix meaning one-billionth part; for example, a nanosecond is one-billionth second (often used in expressing the rate of explosions and in advanced computer technology); a nanometer is one-billionth meter, i.e., 1 millimicron or 10 angstrom units.

napalm. A metallic soap made by reacting an aluminum compound with various mixtures of fatty acids; when added to gasoline, it forms a viscous and highly flammable paste or gel that adheres strongly to objects with which it comes into contact. Its chief use has been for military purposes as an incendiary agent. *See also* soap.

naphtha. Any of several liquid mixtures of hydrocarbons of specific boiling and distillation ranges derived from either petroleum or coal tar. As a group, they are highly volatile and flammable; the grades derived from coal tar are toxic. Naphtha obtained by distillation of petroleum has a confusing number of names, many of which are unscientific or chemically erroneous, e.g., petroleum ether, petroleum spirits, mineral spirits, V.M. and P. (Varnish Makers and Painters) naphtha; petroleum naphtha is the most accurate. Coal-tar naphtha is of two grades—heavy (also called high-flash) and solvent; it contains a high proportion of aromatic compounds. Petroleum naphtha can be steam-cracked to form synthesis gas, from which synthetic natural gas can be made. Its widest use is as a solvent and cleaning agent and as a paint thinner. It should never be used near a heat source, and inhalation of fumes should be avoided. Naphtha should not be confused with naphthol ($C_{10}H_7OH$), an alcohol derived from naphthalene.

naphthalene. An aromatic hydrocarbon, m.p. 80°C (176°F), often in the form of flakes or pellets, having a pronounced and distinctive odor suggestive of the coal tar from which it is derived; it can also be made by catalytic conversion of certain petroleum fractions. The $C_{10}H_8$ mol-

ecule is unsaturated and is in the form of a double fused aromatic ring. It is a useful chemical and dye intermediate, e.g., naphthalenesulfonic acids, and is also a fumigant and antimoth agent. The flakes vaporize (sublime) quite readily, especially at temperatures above 39°C (100°F), and the fumes are combustible.

naphthenate. An oil-soluble metallic salt of naphthenic acid, the metals used being copper, cobalt, lead, manganese, and sodium. The products are rather thick, heavy liquids or pastes; they have strong fungicidal properties and are used to impregnate fabrics exposed to tropical insects and in antifouling paints. Some types, notably cobalt and lead naphthenates, are excellent catalysts for resin polymerization in paints and are known as driers. *See also* fungicide; drier.

naphthene.   *See* alicyclic.

naphthenic acid.   One of a group of higher fatty acids containing alkylcyclopentane groups derived from the gas-oil fraction of petroleum by extraction with sodium hydroxide, followed by acidification. Their chief use is in the production of metallic salts (naphthenates). *See also* naphthenate.

naphthol.   An unsaturated alcohol derived from naphthalene, in which one of the hydrogen atoms of naphthalene is replaced by a hydroxyl group ($C_{10}H_7OH$). There are two isomeric forms (alpha and beta), also called 1-naphthol and 2-naphthol, whose respective melting points are 96°C (205°F) and 122°C (251°F). The material in the form of solid flakes or powder is volatile, and the fumes are quite toxic and irritant. The major uses of naphthol are in dye and perfume synthesis and as an insecticide base. Do not confuse with naphtha.

narcotic.   As a noun, this term refers to various derivatives of opium, a plant alkaloid, among the best known of which are codeine, morphine, and their modifications; it also includes synthetic drugs such as methadone. All these can cause addiction and must be administered under medical supervision. Used as an adjective, the term describes the effects of dizziness and drowsiness induced by vapors of benzene and similar compounds regarded as "narcotic in high concentrations." *See also* alkaloid.

nascent.   An obsolescent term referring to the unusually active state of an element in the act of splitting off from a parent compound. Oxygen in the atomic state is generated by hydrogen peroxide when in contact with air; certain sulfur-bearing rubber accelerators split off active sulfur atoms on heating, making addition of sulfur unnecessary for vulcanization. The word is derived from the Latin for *being born,* i.e., generated. It is no longer widely used.

Natta catalyst.   *See* Ziegler catalyst; stereospecific.

natural gas.   A flammable gaseous mixture of straight-chain hydrocarbons, in which methane predominates; it occurs in association with petroleum. Huge volumes are now pipelined as a major energy source for industry and homes. It is also available in liquid form (LNG). In the chemical industries it is an important source of the synthesis gas used in making synthetic organic materials and natural gasoline. Carbon black of the channel or impingement type is manufactured by combustion of natural gas. Natural gas can also be made synthetically (SNG) from petroleum fractions such as naphtha and by gasification of coal. Do not confuse with natural gasoline. See also synthesis gas; carbon black; gasification; synthetic natural gas.

natural gasoline.   A gasoline of low octane number made by extracting and liquefying the C_4 and higher hydrocarbons occurring in natural gas. It is also called casinghead gasoline because the natural gas used as the source is derived from oil wells. It is not suitable for automotive use but is blended with petroleum fractions for use as a source of petrochemicals. *See also* liquefied petroleum gas.

natural product.   As customarily used by chemists, this term refers to organic materials originating in or derived from plants and animals; it includes such basic industrial materials as petroleum, cotton, wool, rubber, coal, leather, sugar, and wood, as well as fatty acids, hormones, enzymes, alkaloids, vitamins, etc. Synthetic or man-made products either duplicate the chemical structure of a natural substance or simulate its physical properties. Rubber is a notable example in which both simulation and chemical synthesis have occurred. In spite of the tremendous accomplishments of synthetic organic chemistry, the depletion of the world's supply of some natural products (fuels) is a matter of concern. *See also* natural gas; plant.

naval stores.   This term has survived from the days of wooden ships, the bottoms of which were coated with the pitch derived from pine trees. It later became extended to include all types of pine-derived products, such as turpentine, resin, tall oil, etc. Most of these products today are obtained from southern slash pine forests. Though obsolescent, the term is still used in industrial literature.

Nb   Symbol for the element niobium, named after the mythological character Niobe, the daughter of Tantalus, because of the close chemical relation of two elements, tantalum and niobium.

Nd   Symbol for the element neodymium, the name meaning the "new twin," as it was one of two rare-earth oxides obtained from didymia.

Ne   Symbol for the element neon, the name being derived from the Greek word for *new*.

neatsfoot oil.   *See* fatliquor.

neodymium.   An element.

Symbol	Nd	Atomic Wt.	144.24
State	Solid	Valence	3
Group	IIIB (Lanthanide Series)	Isotopes	7 stable
Atomic No.	60		

A rare-earth metal, m.p. 1024°C (1875°F), obtained from monazite sands and bastnasite ore. It readily tarnishes and is attacked by atmospheric moisture. It has too little strength for structural use but has some value as an alloying agent. Its scavenging effect makes it useful in refining ferrous alloys. Some of its compounds have been found to have moderate toxicity for experimental animals.

neohexane.   A high-octane aviation fuel (100 or more), made by thermal alkylation of ethylene and isobutane at high temperature and pressure, 480 to 540°C (900 to 1000°F) and 5000 to 8000 psi. It was developed during World War II for military use. Its formula is $C_2H_5C(CH_3)_3$, a branched-chain paraffin. It is extremely flammable and must be kept away from heat sources. *See also* alkylation.

neon.   An element.

Symbol	Ne	Atomic Wt.	20.179
Group	Noble Gas	Valence	0
State	Gas	Isotopes	3 stable
Atomic No.	10		

Neon is a stable, completely inert gas, forming no chemical compounds, though several ions are known. It is a minor component of air (about 0.0001%), from which it can be extracted when in liquid form. It produces colors in electric discharge tubes, one of its largest uses being for this purpose; the color varies with the type of glass used for the tube and the other gases present (argon or helium). Neon is also used in fluorescent lights, but in this case the color produced is due to the phosphors which coat the inner glass lining of the tube. It is also used in lasers and electronic instruments.

neoprene.   A term coined to designate the first successful synthetic rubber-like material (elastomer), introduced in the early 1930s. It is an unsaturated polymer of chloroprene, a molecule similar to isoprene except that the methyl group of the latter is replaced by chlorine:

$$(CH_2{=}\underset{\underset{Cl}{|}}{C}{-}CH{=}CH_2)_n.$$

It can be vulcanized without sulfur. It has sufficient strength and resilience for most mechanical applications and is particularly effective in resisting deterioration due to oil and oxidation. It is not suitable for use in tires but has wide application for use in gasoline and oil transport hose, wire insulation in automotive engines, and other uses involving exposure to petroleum products and electric discharge. *See also* elastomer; isoprene.

neptunium.   An element.

Symbol	Np	Atomic No.	93
State	Solid	Atomic Wt.	237.0482
Group	IIIB (Actinide Series)	Valence	3,4,5,6

An artificial radioactive metal, m.p. 640°C (1184°F), made from uranium by neutron bombardment; it was discovered during experiments with the cyclotron in 1940. The 239 isotope decays quickly to plutonium, but the 237 isotope is comparatively stable. It is produced during the reactor synthesis of plutonium, existing as a reactive metal which behaves as a strong reducing agent. It is now made in small quantities (less than one pound). It forms a large number of compounds. *See also* transuranic.

Nernst, Walther Hermann   (1864–1941). A German chemist and one of the founders of modern physical chemistry, Nernst worked in the fields of electrochemistry, theory of solutions, thermodynamics, solid state reactions, and photochemistry. His famous equation relating the action in a galvanic cell to electromotive force was published in 1889, and his heat theorem or third law of thermodynamics (the maximum work obtainable from a process can be calculated from the heat evolved at absolute zero) was announced in 1906. He was a professor at Gottingen from 1890 to 1905, a professor at Berlin from 1905 on, and director of the Institute for Experimental Physics in Berlin from 1924 until he retired in 1933. In 1920 the Nobel Prize in chemistry was awarded him for his work in thermochemistry. Among his many students for doctoral degrees from around the world was the American chemist, Irving Langmuir. Nernst also contributed to industrial develop/nents of all sorts, such as the conversion of nitrogen to ammonia.

neroli oil.   *See* essential oil.

nerve gas.   One of a group of exceedingly toxic gaseous compounds developed for possible military use. Several types were made in Germany in the 1930s and later in the United States and other countries. Their use in warfare has been prohibited by international agreement. However, many insecticides of the parathion type have been developed and are now in use; though they decompose relatively quickly, they are intensely poisonous and extremely dangerous to handle. Chemically, nerve gases are esters of phosphoric acid derivatives; they immobilize the nervous system by preventing the enzyme cholinesterase from deactivating the acetylcholine formed by nerve impulses. They are often referred to as cholinesterase inhibitors. *See also* insecticide; parathion; acetylcholine; cholinesterase.

neutral.   Derived from the Latin word for neither, this term has two specific meanings in chemistry. **(1)** An aqueous solution in which the concentration of hydrogen ions and hydroxyl ions is exactly equal; such a solution has a pH value of 7.0 at room temperature; it is the pH of pure water.*See also* pH.

(2) The state of an atom in which the positive and negative charges of the protons and electrons offset each other, leaving the atom with no net charge (the normal condition of an atom).

neutralization.   The reaction between equivalent amounts of an acid (acidic compound) and a base (alkaline compound) to form a salt. If either reactant can form hydrogen ion or hydroxyl ion, water is also formed. Usually conducted in an aqueous medium, such as solutions of HCl and NaOH to yield NaCl and H_2O, the reaction may also occur in the absence of water and without

forming water, as between a solid and a gas, e.g., CO_2 and CaO to form $CaCO_3$, or between two gases, e.g., NH_3 and HCl to form NH_4Cl. Neutralization does not mean that the pH of the final water solution is 7, the neutral pH. While true for the reaction of a strong acid and a strong base, such as NaOH and HCl, and of a weak acid and a weak base, such as ammonia and acetic acid, the final pH will be above 7 when a strong base and a weak acid react, such as NaOH and acetic acid, and below 7 when a strong acid reacts with a weak base, e.g., NH_3 and H_2SO_4. In each reaction, a proton is transferred from the acid to the base.

neutrino   *See* fundamental particle.

neutron.   An uncharged nuclear particle, two or more of which are present in the nucleus of every element except hydrogen, which has none. Its mass is 1.008, very slightly greater than that of a proton (1.0) and the same as that of a hydrogen atom. The fact that neutrons have no electric charge enables them to penetrate the nuclei of atoms in the free state, thus causing fission. Free neutrons are generated by bombarding beryllium with alpha particles (helium nuclei); their existence was discovered in this way in 1932 by Sir James Chadwick, an English physicist (b. 1891), Nobel Prize 1935. They are also liberated by nuclear fission, the neutrons thus released serving to initiate a chain reaction. Absorption of neutrons by living organisms is lethal, and nuclear power installations require concrete or lead shielding to protect personnel.

Fast neutrons are those emitted by an atomic nucleus during fission; slow or thermal neutrons are those that have been retarded by contact with a moderating substance in a reactor. The latter are more efficient for power-generating purposes. Neutrons are used in research laboratories to bombard atoms, thus forming radioactive nuclides; the detection and identification of these is the basis of a technique called neutron activation analysis. *See also* fission; nucleus; reactor.

neutron activation analysis.   *See* neutron.
Newtonian liquid.   *See* fluid; liquid.
Ni   Symbol for the element nickel, the origin of the name being Swedish, possibly after Nickelby, Sweden.
niacin.   An alternative name for nicotinic acid (C_5H_4NCOOH) obtained by oxidation of nicotine. It is a member of the B complex group of

vitamins (the so-called antipellagra vitamin) found in meats, milk, and yeast. It is a component of several coenzymes and is essential for the oxidation of carbohydrates and maintenance of body tissues. It is used in fortification of flours, in dietetic foods, and in animal feed supplements. Niacinamide (nicotinamide) is a derivative having a similar function in cellular oxidation. *See also* B complex; nicotine.

nickel.   An element.

Symbol	Ni	Atomic Wt.	58.71
State	Solid	Valence	2,3
Group	VIII	Isotopes	5 stable
Atomic No.	28		

Nickel, m.p. 1455°C (2650°F), is a metallic element of moderate chemical reactivity; it is particularly resistant to corrosion, one of its major uses being as electroplated protective coatings. At high temperatures it is attacked by sulfur compounds, and at normal temperatures by moist air containing sulfur dioxide, ammonia, or chlorine. It is highly resistant to alkalies because a film of nickel oxide forms on its surface in the presence of strongly basic compounds, e.g., sodium hydroxide. Nickel forms a number of coordination compounds; the most important of these, nickel carbonyl, $Ni(CO)_4$, is poisonous, and care should be taken to avoid inhalation of its fumes. In this compound the nickel has a valence of zero; such compounds (called zerovalent) are formed by complexing with ligands that have many available electrons. The chief uses of nickel are as an alloying metal in stainless steel, magnets, and resistance alloys; in cladding and other protective coating; in special types of batteries; and in powder form as a hydrogenation catalyst.

nickel carbonyl.   *See* carbonyl; nickel.

Nicol.   A device used for obtaining ploarized light; it consists of a combination of calcite (Iceland spar) and Canada balsam, forming a translucent prism which allows the passage of a light ray vibrating in only one direction. By using two of these prisms crossed at various angles, this ray is resolved into vertical and horizontal components one of which is reflected away and the other transmitted. The transmitted component is called plane-polarized light. It is named after William Nicol, a Scottish physicist (1768–1851). *See also* birefringent; optical rotation.

nicotine.   A toxic alkaloid having the formula $C_{10}H_{14}N_2$ occurring in both raw and cured tobacco, from which it can be extracted by distillation. Nicotine and related heterocyclic compounds comprise from 5 to 7% of the solids in cigarette smoke. Oxidation of nicotine yields the vitamin nicotinic acid (niacin). Nicotine has few uses other than as an insecticidal dust or spray in the form of compounds, e.g., nicotine sulfate. Nicotine solutions are used as fumigants. Inhalation of fumes should be avoided.

nicotinic acid.   *See* niacin.

nigre.   In soap technology, the brown-to-black liquid remaining after the soap has been separated from the other products of the saponification reaction by the salting-out process. It contains glycerol, sodium chloride, and various impurities. *See also* saponification; soap.

ninhydrin.   An organic hydrate used chiefly in analytical determinations of amino acid content, which is indicated by the amount of carbon dioxide formed during the test reaction. A blue color appears in the product; this is characteristic of the ninhydrin test and is useful also in chromatographic analysis.

niobium.   An element.

Symbol	Nb	Atomic No.	41
State	Solid	Atomic Wt.	92.9064
Group	VB	Valence	2,3,4,5

This element was originally named columbium by its first discoverer, but some years later it was "rediscovered" by another chemist, who named it after the goddess Niobe, the daughter of Tantalus, because of the strong similarity of tantalum and niobium and the close relationship between them. The earlier name is still used by metallurgists, but niobium is approved by the IUPAC. It is a soft, malleable metal, having only one stable form, m.p. 2468°C (4475°F). It is quite stable at room temperature to attack by oxygen and strong acids; it oxidizes readily at high temperature. It is used in stainless steels as a stabilizing agent and is suitable for use in rockets and space vehicles and in nuclear reactor construction.

nitrate.   A compound characterized by the presence of one or more NO_3 groups. They comprise both inorganic and organic substances. Inorganic nitrates are an essential component of soils, where they are formed by bacteria, and also as added fertilizers. Representative compounds of this group are nitric acid (HNO_3), ammonium nitrate, and potassium nitrate. Well-known examples of organic nitrates are nitroglycerin, $C_3H_5(NO_3)_3$, and cellulose nitrate (nitrocellulose). Nitrates are unstable compounds, which tend to evolve oxygen when heated; as a result, they can be hazardous when stored near easily combustible materials. Many of them explode when confined at high temperature or when shocked (particularly nitroglycerin, one of the most powerful and sensitive explosives). *See also* explosive; oxidizing material.

nitrating acid.   *See* mixed acid.

nitration.   A reaction by means of which nitrate nitrogen is added to or incorporated in a compound.This may be done by using nitric acid, as in the manufacture of nitrobenzene: C_6H_6 + $HNO_3 \rightarrow C_6H_5NO_2$ + H_2O. Mixed or nitrating acid is used in making nitrocellulose, nitroglycerin, etc. This term should not be confused with nitriding. *See also* nitric acid; nitriding; nitrogen fixation.

nitric acid.   A strong poisonous and highly corrosive liquid (HNO_3) formerly made by reacting sulfuric acid with sodium nitrate, but now produced by catalytic oxidation of ammonia; it is supplied in aqueous solutions of various strengths. It is a vigorous oxidizing agent and evolves dangerously toxic fumes in the presence of light; it is thus a hazardous material to handle. Its primary uses are in the manufacture of such products as nitrocellulose, ammonium nitrate, and nitroglycerin, and as a component of mixed or nitrating acid (one-third nitric and two-thirds sulfuric). *See also* nitration; fuming.

nitriding.   A term used in metallurgy to describe a process for hardening the surface of alloy steels by heating them in an atmosphere of ammonia gas. The nitrogen released by the decomposition of ammonia ($2NH_3 \rightarrow N_2$ + $3H_2$) reacts with metallic components of the alloys to form nitrides, which are the hardening agents.

nitrile.   A term used to designate an organic compound in which the univalent cyanogen group (CN) is present, the parent substance being hydrocyanic acid, HCN. Among the more important nitriles are acrylonitrile, acetonitrile, and adiponitrile. A synthetic rubber called nitrile rubber is made by copolymerizing acrylonitrile and butadiene. With a few exceptions nitrile compounds are poisonous. *See also* cyanogen.

nitrilotriacetic acid.   A tricarboxylic acid, $(CH_2COOH)_3N$, which has been considered as a promising substitute for phosphates in detergent builders because it forms coordination compounds with metals and is highly biodegradable. This application was not permitted when preliminary (and still debated) reports indicated that it may be a carcinogen. It is useful as an organic intermediate and chelating agent. *See also* builder.

nitrobenzene.   A smooth, viscous liquid at room temperature, becoming a yellowish solid below 5°C (42°F). It is made by reacting benzene with mixed acid, thus introducing nitrogen into the molecule, the formula being $C_6H_5NO_2$. It is also known as oil of mirbane. An important organic intermediate, it is used in making such dye bases as benzidine and aniline, as well as other nitrogenous chemicals. It is very poisonous, and inhalation of its fumes should be avoided. It can be further nitrated to di- and tri-nitrobenzenes.

nitrocellulose.   Any of a series of compounds made by reacting cellulose with nitric acid or mixed acid. It is also called cellulose nitrate. The reactions are closely controlled by varying the ratio of cellulose to acid or other factors (time and temperature) in such a way that the extent of nitration is held to a specified limit. Nitrocellulose in which the nitrogen content is relatively low (10% or less) is a useful lacquer base, especially for automotive coatings, because of its fast-drying properties. Collodion and pyroxylin are plastic forms of low-nitrogen nitrocellulose. If the nitrogen content is allowed to rise to about 14%, the product is a high explosive (guncotton). All types are extremely flammable. Smokeless powder is nitrocellulose having about 13% nitrogen content. *See also* lacquer; explosive.

nitrogen.   An element.

Symbol	N	Atomic Wt.	14.0067
State	Gas	Valence	1,2,3,4,5
Group	VA	Isotopes	2 stable
Atomic No.	7		

Nitrogen is a comparatively unreactive element which comprises 78% by volume of air, though it is unavailable to plants and man from this source. It is as essential to vital processes as are oxygen or carbon, for it is a constituent of amino acids and proteins, the basis of all animal life. These compounds are synthesized by plants from nitrogen "fixed" in the soil by bacteria. The first organic synthesis was of a nitrogenous compound (urea, in 1828); 84 years later the synthesis of ammonia made it possible for the first time for man to utilize atmospheric nitrogen. Nitrogen is the characteristic element of many chemical groups besides those mentioned: amines, nitriles, alkaloids, azides, and purines. It combines readily with fluorine and is the active element in most explosives, rubber accelerators, fertilizers, and many drugs. Nitrous oxide (N_2O) is an anesthetic, but oxides of nitrogen produced by auto exhausts are toxic air contaminants, especially nitrogen dioxide, NO_2.

In elemental form, nitrogen is used as a protective blanket for highly reactive chemicals, as a blowing agent in foamed plastics, as a scavenging and carrier gas, and in special types of refrigeration. It has no active toxicity and is noncombustible. *See also* nitrogen fixation; nitric acid; nitrocellulose.

nitrogen fixation. Utilization of the nitrogen of the air by plants to form critically important compounds; the conversion is performed by certain types of bacteria which live symbiotically on the root hairs of plants, thus enabling them to synthesize nitrogenous compounds (proteins). The term is also applied to various processes for chemically manufacturing ammonia and other nitrogen compounds. *See also* fixed.

nitroglycerin. An extremely sensitive high explosive made by nitration of glycerol; the molecule contains three nitrate groups: $CH_2NO_3CHNO_3CH_2NO_3$. It is a heavy, yellowish liquid and is poisonous as well as a hazardous explosive. It is the original active ingredient of dynamite invented by Nobel (1870). It also has medical application as a vasodilator in coronary occlusion. *See also* dynamite.

nitromethane. *See* nitroparaffin.

nitroparaffin. A group of organic compounds made by nitration of a paraffin (alkane) hydrocarbon, e.g., nitromethane. The generic formula is RNO_2. They are good solvents for cellulose derivatives, many synthetic resins, and a variety of natural products. They are flammable liquids and have a considerable degree of toxicity. Some members of the group, especially nitromethane, may explode when vapors are mixed with air and exposed to open flame.

nitrosamine. An organic compound of the *N*-nitroso type formed by the reaction of amines with nitrogen oxides or nitrite compounds (for example, $N_2O_3 + R_2NH \rightarrow R_2N\text{-}NO + HNO_2$). Also known as nitrosoamines, they have the general formula $R_2N\text{-}NO$, where R is any organic group. They are potent animal carcinogens. Since their precursors are widespread — many in foods (for example, in cured meats and fish) to prevent botulism poisoning—considerable attention is being centered on their formation and their effects on humans. Probably several times as much nitrosamine is formed in the gastrointestinal tract as is ingested in foods. For example, nitrites are produced by reduction of nitrates in saliva and nitrites and amines from the bacterial breakdown of proteins in the intestine. There is no present evidence that nitrites used in curing of meats have any carcinogenic effect on humans. Among the typical nitrosamines that have been studied for carcinogenicity on animals are dimethyl- and diethylnitrosamine, $(CH_3)_2N\text{-}NO$ and $(C_2H_5)_2N\text{-}NO$, liquids of b.p. 152° and 177°, respectively.

nitroso. A prefix indicating the presence of an NO group in an organic compound. It may occur in benzenoid structures, forming compounds that are the basis of dyes (nitrosophenol); it also occurs in fluorocarbon monomers from which so-called nitroso rubbers are made. *See also* nitrosamine.

nitrostarch. A high explosive made by nitration of starch, and having the formula $C_{12}H_{12}(NO_2)_8O_{10}$. It is a finely divided solid of reddish color. For safety in handling, it should be thoroughly wetted with water and protected from heat, vibration, and shock. It is flammable even when water is present to the extent of 20%. Extreme caution should be maintained in storage, shipment, and all aspects of handling. *See also* high explosive.

nitrous oxide. A gaseous compound, b.p. $-88.5°C$ ($-127°F$), used as a general anesthetic (N_2O), chiefly in dentistry and for short operations. It is not toxic but can explode when mixed with air; it should be administered only by trained personnel. It was discovered in 1772 by Joseph Priestley, a British chemist (1753–1804), but its possibilities as an anesthetic were not explored till the 1840s. It was originally called laughing gas.

NMR. *See* nuclear magnetic resonance.

No Symbol for the element nobelium, so named because its dicovery was reported by the Nobel

Institute in Sweden, jointly with Argonne National Laboratory and Harwell Research Institute.

Nobel, Alfred B. (1833–1896). A native of Sweden, Nobel devoted most of his life to a study of explosives and was the inventor of a mixture of nitroglycerin and diatomaceous earth which he called dynamite. He also invented blasting gelatin and smokeless powder. With the fortune he accumulated from his work, Nobel established the foundation that bears his name, which annually recognizes outstanding work in physics, chemistry, medicine, and human relations. The Nobel prize is still the world's most valued scientific award.

nobelium.   An element.

Symbol	No	Atomic No.	102
State	Solid	Atomic Wt.	254(?)
Group	IIIB	Valence	2,3
	(Actinide Series)		

This radioactive element has been isolated by cyclotron experiments, but knowledge of its properties is still fragmentary, as its longest-lived isotope has a half-life of less than 4 minutes.

noble.   A term used to indicate the relative lack of chemical reactivity of an element. For example, of the noble gases, only three (helium, neon, and argon) are actually inert, that is, incapable of forming compounds. Gold is considered more noble than, e.g., silver, because of its comparatively low reactivity. The term should not be confused with "precious" in reference to metals, as it has no economic significance. In the electrode potential series of the elements, an element with the higher potential (Gibbs-Stockholm convention) is referred to as more "noble" than elements of lower potential. *See also* inert; zero group.

nodule.   An aggregate or mixture of manganese and iron (ferromanganese) occurring in huge quantities on the floor of oceans, especially the Pacific. The size averages about 4 cm in diameter, varying from small pellets to masses several feet in diameter. Their origin is unknown. Though they have not yet been recovered on a commercial scale, they may become an important source of manganese in the future.

nomenclature.   The names of chemical substances and the systems used for assigning them, as distinquished from *terminology* which, being a more general term, includes phenomena, processes, and concepts. Many of the names still in wide use originated in the alchemy of the Middle Ages, and others derived from Greek and Latin were introduced in the classically minded nineteenth century. As knowledge of organic structure grew more detailed and explicit, it became essential to systematize these names. A meeting of chemists held at Geneva in 1892 for this purpose introduced a new and logical method of naming organic compounds, which was widely adopted, especially by theoretical chemists. In 1930 another meeting held by the International Union of Pure and Applied Chemistry recommended still further reforms, which were adopted in 1957.

The combination of the older and so-called trivial names with the new and more logical system accounts for the fact that many compounds have two or more synonyms. For example, saturated hydrocarbons are called both paraffins (older name) and alkanes (IUPAC name); unsaturates are called olefins (older name) and alkenes (IUPAC name). Similarly, CH_3OH is known as wood alcohol (trivial), methyl alcohol, and methanol (IUPAC). Many of the older names persist and are preferred by some authorities, e.g., toluene, xylene; but mercaptan has been changed to thiol. *See also* trivial name; IUPAC; numerals.

nondestructive testing.   A type of materials testing made on a finished product rather than on a sample, which involves no damage to the product. Examples are radiography, x-rays, and infrared radiation, which are used to explore large metal structures for flaws, fatigue effects, and similar weaknesses. This technique is also applied to paintings for authentication purposes.

nonferrous.   A term used in metallurgy to refer collectively to all metals other than iron, or to alloys which do not contain iron.

nonmetal.   As used by chemists, this term includes two groups of elements; one group is made up of elements that have little or no similarity to metals, and the other of elements that are somewhat more like metals, especially in their electrical properties. The latter group are semiconductors. They are still sometimes referred to as metalloids, though this term is obsolescent. The first group is made up of gases (noble gases, oxygen, nitrogen, hydrogen, chlorine, fluorine), a liquid (bromine), and several solids (carbon, phosphorus, sulfur, iodine). The second group includes boron, silicon, arsenic, germanium, selenium, tellurium, antimony, and polonium, all of which are solids having semiconducting properites to some degree, especially germanium and silcon. The term is also used in a broad sense by materials engineers for any solid material that is not a metal, e.g., plastics, textiles, etc. *See also* semiconductor; metalloid.

nonnutritive sweetener. A synthetic, low-calorie substitute for sugar, for example, the sodium and calcium salts of saccharin (sulfur-bearing derivative of benzoic acid), which are said to have about 300 times the sweetening power of sucrose. These are used in dietetic foods and soft drinks, and for medical purposes. The corresponding salts of cyclohexane sulfamic acid, known generally as cyclamates, have only about 30 times the strength of sucrose.

nonwoven. A term used in the textile industry to denote a product made by laminating loose fibers placed over each other at random angles and united with a binder of natural or synthetic adhesive. The fibers may be glass, plastic, or natural. The products are used for absorbents, filters, and other disposable purposes.

normal compound. An organic compound whose structure is of the straight- or open- chain type, containing no branching or subsidiary groups. This arrangement is characteristic of many hydrocarbons, alcohols, amines, etc. It is usually indicated by a small italic *n-* preceeding the name, as in *n*-heptane, *n*-butyl alcohol. *See also* chain, molecular; branched chain.

normalize. A term used by metallurgists to refer to a process of tempering steels and other alloys by heating them to a predetermined temperature and cooling at a controlled rate to relieve internal stresses and improve strength and stability. *See also* annealing.

normal solution. The concentration represented by 1 equivalent weight of a solute dissolved in enough water to make a total volume of 1 liter. The dissolved material (solute) may be in the form of ions (sulfuric acid) or organic molecules (alcohol, sugar). The conventional abbreviation is an italic capital N, e.g., $2N$, $0.5N$, etc. *See also* equivalent weight.

novolac. A thermoplastic resin of the phenol-formaldehyde type, which can be cross-linked to a thermosetting (cured) state by addition of an amine, usually hexamethylenetetramine. The novolac molecules are made up of chains of several phenol units joined by methylene (CH_2) linkages. When the nitrogen-containing amine is added, a hard, resinous product is formed that is especially useful for bonding applications (abrasive wheels, brake linings) and for a variety of molded items having high electrical resistivity and good heat resistance. *See also* phenolformaldehyde resin.

noxious gas. Any gas or vapor having a more or less toxic effect when inhaled. The term is restricted to gases resulting from chemicals or chemical reactions that occur either naturally or as by-products of industrial manufacture or use. It does not include so-called poison gases made specifically for military purposes, nor the asphyxiant gases, which have no positive toxicity. Examples of noxious gases are carbon monoxide, chlorine, and hydrogen sulfide, as well as vapors from carbon tetrachloride and other halogenated hydrocarbons. *See also* asphyxiant.

Np Symbol for the element neptunium, named for the planet Neptune, which is next beyond Uranus in the solar system, as neptunium is the next element beyond uranium in the Periodic System.

nuclear energy. Energy of a higher order of magnitude than that of chemical reactions; the latter is derived from the formation and rupture of bonds (electrons) and amounts to about 5 electron volts per bond, whereas the former is evolved from the nucleus of the atom, yielding, for fission, about 200 million electron volts per nucleus. Nuclear energy is released either by splitting (fission) of an unstable, heavy nucleus (uranium 235 or 233 and plutonium 239) or by union (fusion) of the hydrogen isotopes deuterium and tritium to form helium. The latter occurs naturally in the the sun but can be achieved artificially only at the exceedingly high temperature obtainable by a fission reaction, e.g., 10^7 °C. For this reason fusion is called a thermonuclear reaction. The two phenomena are of opposite nature, fission being the dividing of a heavy nucleus to form two lighter ones, and fusion the combination of light nuclei to form a heavier one. The fission reaction is controllable, but thermonuclear reactions are not, though active research is in progress to make them so. *See also* fission; fusion; reactor.

nuclear magnetic resonance. A type of absorption spectroscopy (abbreviated NMR), developed in the 1930's, which has become the most fruitful method of determining the structures of organic compounds; it is particularly effective in identifying isomers, elucidating the associative grouping of hydrogen atoms, and establishing important factors of molecular geometry. It can be used in conjunction with other analytical methods, such as infrared and mass spectroscopy, gas chromatography, etc. Its applications extend to inorganic and physical chemistry, biochemistry, and related areas of chemistry. NMR spectroscopy involves the absorption of energy in the range of radiofrequencies by compounds containing certain types of atomic nuclei which have magnetic properties, for example, hydrogen (protons), carbon-13, oxygen-17, fluorine, phosphorus, and nitrogen. The interaction of the magnetic field generated by the rotation of these nuclei with an imposed magnetic field gives rise to NMR spectra when the energy states of the radiation source and the absorbing nucleus are the same (resonance). By analyzing the shifts in such spectra it is possible to obtain an indication of the internal structure of the molecule.

Closely related types of spectroscopy are microwave, electron paramagnetic resonance (EPR), electron nuclear double resonance (ENDOR) and Mossbauer analysis. *See also* absorption; resonance (2); magnetochemistry.

nuclear reaction. *See* reaction (2); nuclear energy; fission; fusion.

nuclear reactor. *See* reactor.

nucleation. A term used in crystallography and colloid chemistry to refer to the initiation of crystal formation by microscopic particulates, for example, dust in the air or colloidally dispersed solids in a solution. These act as active centers or nuclei on which water droplets may form or crystal structures may develop. Thus, ice particles induced by quick temperature reduction in moist air (by seeding with carbon dioxide or silver iodide) can serve as points of formation for raindrops. Various types of crystals prepared or "grown" synthetically under carefully controlled conditions for scientific purposes require a particulate nucleus for initiation.

nucleic acid. An extremely complex compound of high molecular weight which occurs in the nuclei of biological cells; when associated with a protein, the combined molecule is called a nucleoprotein. Nucleic acids are polymers of nucleotides. They can be partially hydrolyzed to sugar derivatives (ribosides) of purines and pyrimidines (nucleosides), or to nucleotides (phosphate esters of these compounds). The two major types of nucleic acids are ribonucleic acid, which contains ribose, and deoxyribonucleic acid, containing deoxyribose. Both types are of critical importance in cellular biochemistry, as they establish and execute the genetic mechanisms of the organism. The study of nucleic acids and their structure and function comprises the science of molecular biology. *See also* nucleotides; deoxyribonucleic acid; ribonucleic acid; genetic codes; molecular biology.

nuclein. *See* nucleoprotein.

nucleonics. A term used by nuclear engineers to designate the study of the atomic nucleus and its constituents. It is derived from the word *nucleon,* that is, a proton, neutron, and other particles formed as a result of bombardment of a nucleus with high-energy particles, for example, mesons of various types. Nucleonics includes fission, fusion, transmutation, mass defect, binding energy, and the radiation effects that accompany them.

nucleophile. A term used to designate an agent (ammonia, water, alcohol) that contributes or donates a pair of electrons to a chemical structure (nucleus) to form a covalent bond; the element that receives the two electrons is called an electrophile. These occur, for example, in the formation of acids and bases according to the Lewis concept, as well as in covalent carbon bonding of organic compounds. *See also* nucleus (1); Lewis acid.

nucleoprotein. A high molecular weight substance formed by the chemical union of a protein with a nucleic acid in a ratio of about 20 to 1. Such complex molecules are found in cell nuclei of all organisms. They are also called nucleins. *See also* nucleic acid; protein.

nucleotide. A term applied by biochemists to complex molecules which are usually constituents of nucleic acids; they are composed of nitrogen-containing bases (purine or pyrimidine) bonded to either ribose or deoxyribose sugars. phosphoric acid being present to form an ester.

The purine constituents are adenine and guanine, and the pyrimidines are uracil and cytosine. Some nucleotides are physiologically active complexes called coenzymes or prosthetic groups. Nucleotides derived from yeast are used as flavor potentiators in the food industry. A well-known nucleotide is adenosine triphosphate (ATP), which is essential to many biochemical processes, including photosynthesis. *See also* nucleic acid; coenzyme; potentiator.

nucleus. (1) In organic chemistry, a fundamental molecular structure which persists with only peripheral changes through many reactions; examples are the aromatic benzene nucleus, C_6H_6 and the heterocyclic furan nucleus, C_4H_4O.

(2) In physical chemistry, the total number of protons and neutrons, which constitutes virtually the entire mass of an atom. These particles are held together by forces of exceedingly high magnitude known as the binding energy, a portion of which is released by fission.

(3) In biochemistry, the central portion of the cell of an organism which contains the essential reproductive chemicals (nucleic acids).

(4) In crystallography, any microscopic particle that serves as a basis for crystal formation (nucleation).

nuclide. A term used chiefly by nuclear physicists to refer to any isotopic form of any element; unstable types are called radionuclides.

numerals. Arabic numbers are used in organic nomenclature principally to indicate the location or position of subsidiary atoms or groups in aliphatic and aromatic structures, or other active points on the carbon skeleton. For example, in 2-methyl butane (isopentane), the structure is indicated by the name.

$$\overset{1}{C}H_3 - \overset{2}{C}H - \overset{3}{C}H - \overset{4}{C}H_3$$
$$\underset{CH_3}{|}$$

(methyl group of the branched chain is attached to the second carbon).

In the names of alcohols, the numerals designate the position of the hydroxyl group, for example, 1,2,3-propanetriol (glycerol) is ($CH_2OHCHOHCH_2OH$). In benzene rings, the $\overset{1}{}$ $\overset{2}{}$ $\overset{3}{}$ numbers indicate ortho-, meta-, and para-

isomers, as in 1,4-dichlorobenzene (paradichlorobenzene):

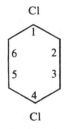

In inorganic compounds, Roman numerals are used to indicate the oxidation state or coordination value in preference to suffixes, e.g., iron (II) chloride is preferred to ferrous chloride for $FeCl_2$, and iron (III) chloride to ferric chloride for $FeCl_3$. *See also* nomenclature.

nutrient. Any substance that is essential to the life and growth of plants, animals, or man either as such or as transformed by chemical or enzymatic reactions. Plant nutrients include numerous mineral elements, as well as nitrogen, carbon dioxide, and water. In animals and man, the primary nutrients are proteins, carbohydrates, and fats that are obtained from plants, either directly or indirectly, supplemented by vitamins and minerals. Water and oxygen are also considered nutrients, of which there are a total of 43. *See also* food.

nutrient culture. A method of growing flowers and vegetables in the absence of soil by supplying the essential elements in a properly balanced aqueous solution. This technique is especially suitable for localities where good soil is scarce, as in arid areas, small islands, and the like. It is called hydroponics by agriculturalists and is popularly known as soilless gowth.

nutrition. The complex of synthesis and decomposition reactions that occurs in the growth and maintenance of living organisms. Plants synthesize carbohydrates from carbon dioxide and water by photosynthesis, proteins from nitrogenous compounds, as well as fats, oils, and vitamins. The essential elements required are carbon, hydrogen, nitrogen, phosphorus, potassium, and a number of trace elements. The nutrients of man are supplied by plants, either as such or in the meats resulting from their ingestion by animals, all the essential factors being obtained from this source. In man, most of the nutritive processes are of the decomposition type, chiefly hydrolysis and digestion, usually catalyzed by enzymes. Thus, the ultimate products are, to a large extent, carbon dioxide and water, which complete the carbon cycle. *See also* food; photosynthesis; trace element.

nylon. A polyamide resin synthesized by a number of different chemical methods; all types contain the characteristic $CONH_2$ (amide) grouping. The two most important grades of nylon are 66 and 6. The former is derived by condensation polymerization of adipic acid and hexamethylenediamine, orginally researched in 1935 by Wallace H. Carothers, an American chemist (1896–1937). The latter, based on caprolactam, was produced in Europe several years later. There are several other types identified by the numbers 610, 4, 9, 11, and 12. Nylon has many advantageous properties, among which are strength, resistance to heat and abrasion, colorability, noncombustibility, and nontoxicity. Its uses are manifold, though its textile applications predominate in the form of fibers for carpets, cordage, tire cord, twine, hosiery, and suitings. As a plastic, it is made in many forms by molding and machining; among these are gears, bearings, and other machine components, piping, brush bristles, insulation, and shoe heels. Its lack of toxicity makes it the material of choice for surgical inserts in the body, prosthetic devices, etc. Its synthesis was one of the great modern achievements of synthetic organic chemistry. *See also* polyamide; adipic acid; caprolactam; Carothers.

O

O   Symbol for the elment oxygen, the name being derived from the Greek words meaning "acid maker," as it was originally thought that oxygen was the characteristic element of acids.

o-.   Abbreviation for the chemical prefix *ortho-.*

ocher.   *See* pigment.

octane.   A paraffinic hydrocarbon (C_8H_{18}) obtained from petroleum; it is a liquid of relatively low toxicity, but its fumes are highly flammable. Its isomer, isooctane, is a branched-chain structure which is the basis of high-octane gasoline. Both have good solvent properties and are used in the synthesis of organic chemicals. *See also* octane number; branched chain.

octane number.   An arbitrary value denoting the antiknock rating of a gasoline; it is the proportion of the branched-chain hydrocarbon isooctane that is contained in a test mixture of heptane and isooctane. The octane number of heptane (a straight-chain hydrocarbon) is rated as zero, while that of isooctane is 100. Mixing these in various proportions and determining the degree of knocking for each blend establishes values which make up the octane rating scale; for example, a mixture of 90 parts isooctane and 10 parts heptane has an octane number of 90. Research octane numbers, (RON) obtained under laboratory conditions, run several points higher than so-called motor octane numbers, (MON) which represent the actual performance of a gasoline in a car engine. Research octane numbers above 100 are possible by using antiknock agents. *See also* knock; methyl *tert*-butyl ether; neohexane.

octanol.   *See* octyl alcohol.

octyl alcohol.   An eight-carbon alcohol having primary, b.p. 195°C (383°F), and secondary b.p. 178°C (352°F), isomeric forms ($C_8H_{17}OH$). It is characterized by a strong and fragrant odor which makes it useful as a solvent in the perfume industry. The straight-chain nature of the molecule gives it good demulsifying and wetting power. Octyl alcohol (or octanol) also finds use in the manufacture of cosmetics and plasticizers. It is often referred to in the trade literature as capryl alcohol, but this term is not accepted by chemists.

odor.   A response of the olfactory nerves to stimulation by the molecules of an inhaled substance. It is an important and useful property for both identification and warning purposes. The intensity varies greatly at a given concentration, from virtually undetectable to highly penetrating and obnoxious, e.g., sulfur-bearing compounds, of which ethanethiol and hydrogen sulfide are examples. Essential oils in general have a pleasing, sweet odor. The effectiveness of many animal repellents is due to their odor, e.g., sulfur dioxide and citronella oil.

oil.   A nonspecific term applied to several groups of organic mixtures having quite different chemical composition and properties. Oils are liquids at room temperature, becoming increasingly viscous to semisolid at lower temperatures. The most important groups are: (1) the mineral type (petroleum, shale oil) which are mixtures of hydrocarbons; (2) the vegetable type (cottonseed, soybean, etc.) which are mixtures of unsaturated fatty acids, many of which are edible; (3) the animal type (sperm and fish oils), also containing fatty acids; and (4) the flower type (essential oils), in which terpenes predominate (though derived from vegetation, they are classified separately from vegetable oils). The terms *oil of mirbane* for nitrobenzene and *oil of vitriol* for sulfuric acid are obsolescent. *See also* petroleum; vegetable oil; edible oil; essential oil; oleum.

oil black.   *See* black.

oiliness.   *See* lubricity.

oil of bitter almond.   *See* benzaldehyde.

oil of mirbane.   *See* nitrobenzene.

oil of vitriol.   *See* oleum; vitreous.

oil of wintergreen.   *See* salicylic acid.

oil sands   (tar sands). Porous sandstone structures occurring on the surface and to depths of 750 meters in certain localities; they contain a high proportion of bitumen composed chiefly of asphaltenes and maltha, together with substantial percentages of sulfur and heavy metals Its viscosity is about midway between that of crude oil and soft asphalt. The largest deposit in North America is in the Athabasca region of Alberta; there are smaller ones in the western U.S. Venezuela and Trinidad have large deposits. The Athabasca area has been producing over 50,000 barrels of oil a day for several years. The yield is from one-half to one barrel per ton of material processed. The solids content is reduced to 70% by addition of hot water (190°F), after which the oil separates by gravity. As its hydrogen-to-carbon ratio is low and its molecular weight high, it is upgraded by thermal cracking (900°F), addition of hydrogen, and removal of sulfur. Plants now, or soon to be in operation, are expected to produce 175,000 barrels a day by 1982 and up to 500,000 barrels by 1990.

oiticica oil.   A mixture of unsaturated glycerides occurring in the seeds of a South American tree. It solidifies (or dries) by oxidative polymerization. Its prinicipal use is in paints and varnishes, though (together with other oils of this type) it has been largely replaced by synthetic resins for this purpose. Although it has no specific toxic properties, it is not edible. *See also* drying oil.

-ol.   A suffix indicating the presence of one or more hydroxyl groups in an organic compound, as in the names *methanol, isopropanol, phenol, sterol,* ets. An exception is thiol, a class of compounds similar to alcohol in which the oxygen of the OH group is replaced by sulfur.

olefin.   A major group of aliphatic hydrocarbons, both both straight- and branched-chain, characterized by the presence of at least one double bond; those having only one are called mono-olefins or alkenes, and those with two are di-olefins, also called alkadienes, or simply dienes. Some olefins have three or more double bonds and are named accordingly (triolefins, etc.) The unsaturated linkages are always between adjacent carbons in monoolefins; in diolefins they may be between successive carbons ($C=C=C$) or between alternate pairs of carbons (conjugated):

$$(C=C-C=C)$$

Ethylene is the simplest monolefin ($CH_2=CH_2$) and is one of the most useful intermediates in synthetic chemistry; butadiene is the most important of the diolefins.

oleic acid.   One of the most abundant fatty acids, m.p. 14°C (57°F), obtained from both animal and vegetable sources. It is an unsaturated compound having one double bond and terminating in the characteristic carboxyl group: $CH_3(CH_2)_7CH=(CH_2)_7COOH$. It is a reddish, viscous liquid at room temperature; the 70% grade is known in the trade as red oil. Oleic acid readily undergoes saponification and hydrogenation. Its high-volume use is in commercial soap manufacture, and is also a component of ointment bases and cosmetic creams. It has no toxic properties. Oleates are derivatives of oleic acid comprising two types of compounds: metallic (heavy-metal) soaps by reaction with lead, cobalt, manganese, etc., and esters by reaction with alcohols, e.g., glycerol. *See also* fatty acid; glycerin.

oleoresin.   A sticky or gummy mixture of essential oils and natural resins occurring in numerous varieties of trees and shrubs, particularly of the evergreen or coniferous type. Turpentine, a well-known example, contains abietic acid, pinene, and dipentene. Most oleoresins have a strong, pleasant aroma; some have value as medicines (cough syrups). *See also* balsam; resin **(1)**; rosin.

oleum.   Alternative name for fuming sulfuric acid, i.e., sulfuric acid mixed with sulfur trioxide, used largely as a sulfonating agent. Oleum is the source of the word *oil* and undoubtedly owes its present meaning to the original name of sulfuric acid (oil of vitriol), derived from the Latin word for glass, because of the refractive appearance of the ferrous sulfate from which it was once made.

oligo-. A prefix derived from the Greek, meaning "several" or "slight;" in chemistry it appears in such terms as *oligosaccharides* (containing from three to ten monosaccharide units) and oligodynamic (slight bactericidal activity).

oligomer. A macromolecule comprised of less then 5 monomer units, i.e., dimers, trimers, and tetramers.

oligosaccharide. *See* sugar.

opacity. A term used in the paint and paper industries to indicate the hiding power of a colorant or the extent ot which a pigment or filler overcomes transparency, for example, in paper intended for letterpress printing. It also may refer to the product itself, as the opacity of a plastic film. *See also* hiding power.

operation. *See* process; unit operation.

opium. *See* alkaloid.

opsin. *See* rhodopsin.

optical activity. *See* optical rotation.

optical brightener. *See* brightener (2).

optical fiber. An extremely fine-drawn glass fiber of exceptional purity that will transmit laser light impulses with high fidelity. Such fibers are made from high-purity quartz coated with germanium-doped silica by vapor deposition. After drawing to a microscopically small diameter, the strands are assembled into cables containing 144 filaments for use in telephone systems as replacements for copper cables. Short experimental lines have been in service for some time, and full-scale use is expected by the early 1980's. Other developing uses are in medicine as a diagnostic aid, in engineering for inspection of airplane engines, nuclear reactors, etc., and in military communications.

optical glass. A type of glass specifically formulated for use in lenses designed to correct astigmatism and other malfunctions of vision, as well as in scientific instruments, cameras, etc. The refractive indexes and dispersion properties of such glass can be precisely controlled by proper selection and proportion of ingredients. Optical glass may be of either the crown or flint type, the former containing lime and the latter lead. Improvements have been achieved by inclusion of many diverse substances in the composition of such glasses, e.g., oxides of boron, tantalum, barium, thorium, etc., as well as fluorides. The properties of the glass are modified according to optometrists' prescriptions by meticulous grinding and polishing.

optical isomer. *See* optical rotation; asymmetry; enantiomorph.

optical microscope. *See* microscopy; resolving power.

optical rotation. A property of some crystalline substances which enables them to turn the plane of polarized light (vibrations transmitted in one direction only) as it passes through Nicol prisms, the rotation being either to the left (levo) or the right (dextro). Many inorganic materials display this property, e.g., quartz and mercury sulfide. It is also characteristic of organic compounds containing one or more asymmetric carbon atoms (turpentine, tartaric acid); most prominent among these are sugars or sugar derivatives, as indicated by the names *dextrose* and *levulose;* some acids, including amino acids; and certain aldehydes and alcohols. The Greek letter alpha is often used as a symbol for optical rotation. *See also* d-; l-; asymmetry; glyceraldehyde; Nicol.

orbital. A mathematical designation of the wave function or energy level of an electron within a shell. Each shell is composed of several orbitals or subshells, the energy state of each being defined by its quantum numbers. No orbital can contain more than two electrons, and these two must have opposite spin (Pauli exclusion principle). The orbital concept, developed by Heisenberg, Schrödinger, and others, reflects the impact of quantum mechanics on chemistry, especially as regards bonding and coordination theory. This concept considers the electron as a light wave as well as a particle; it is described mathematically as a standing wave, the orbital being a function of this wave. The square of the wave function is a real property of the electron, representing the space limits (orbital) within which an electron is most probably located; these limits are established by four quantum numbers, which denote its energy level, angular momentum, magnetic moment, and spin. Thus, an orbital represents the mathematical probability that an electron will be found at any point in a specific volume of space. *See also* shell; electron; uncertainty principle.

order. The regular arrangement or pattern assumed by atoms or other chemical units in a space lattice; absence of such an arrangement is known as disorder, or random distribution. The lattices may be of one, two, or three dimensions, depending on the nature of the material (polymer, colloid, crystal, etc.). Maximum order is characteristic of crystals, while liquids are disordered ; intermediate structures (proteins) may exhibit reversible change from one state to the other, e.g., from helical form to random coils. Statistical study of such changes is helpful in understanding phase transitions in alloys, colloidal systems, and the structure of macromolecules such as DNA and polypeptides.

order of magnitude. A term used in science to indicate a range of values representing numbers, dimensions, distances, etc., which starts at any given value and ends at ten times that value; that is, two units are of the same order of magnitude if one of them is up to ten times as large as the other. Thus, for distance, a foot is the same order of magnitude as any length between 1 foot and 10 feet; for numbers, 100 is the same order of magnitude as any number between 100 and 1000, and a million (10^6) is the same order of magnitude as any number between itself and ten million (10^7). The term is usually applied to very large values such as atomic or molecular populations and astronomical distances.

ore. Any naturally occurring mixture of a metal or metal compound and various silicates, carbonates, etc., found not only in rock strata but also in certain clayey or sandy formations. Bauxite is the principal ore of aluminum, and monazite sands are ores from which titanium and rare-earth metals are obtained. Ores have geological names usually ending in -ite (taconite, carnotite, ilmenite, bastnasite, hematite). Ores must be separated, beneficiated (dressed), and reduced by any of several methods to extract the metal. See also metal; bauxite.

ore dressing. A term used in extractive metallurgy to refer to various nonchemical means of separating metals or their compounds (oxides, sulfides) from the siliceous or calcareous structures in which they occur. Some of the operations performed are beneficiation, elutriation, and flotation.

organic. This term basically means "living" as in the biological word organism; it was thought by early chemists that all organic compounds orginated in plants and animals, until the German chemist, Friedrich Wöhler (1800–1882), discovered in 1828 that urea could be formed from ammonium cyanate by molecular rearrangement. This was the beginning of synthetic organic chemistry, much of which involves the

duplication of natural products in the laboratory. Thus, organic has come to mean any compound containing the element carbon with several exceptions: such binary compounds as the carbon oxides, the carbides, and carbon disulfide; such ternary compounds as the metallic cyanides, metallic carbonyls, phosgene ($COCl_2$), and carbonyl sulfide (COS); and the metallic carbonates, such as calcium carbonate and sodium carbonate. Another view is that organic chemistry deals with the hydrocarbons and their derivatives; however, it also includes compounds containing nitrogen, oxygen, sulfur, etc. in addition to hydrogen and carbon, e.g., carbohydrates. Developments in modern chemistry have tended to make the division between organic and inorganic less and less definite. See also inorganic; natural product; organometallic.

organoleptic. A term used in the food and fragrance industries to refer to trials or experiments carried out by selected groups of people, who taste or smell a series of materials in an attempt to distinguish between them in one or another respect. Such tests have been widely used and statistically evaluated for many years; they often involve subjective or psychological factors, for example, the effect of color on the taste of a food product. The word is apparently derived from organo ("living," i.e., people rather than instruments) and lepto-, meaning "slight" or "narrow," referring to the small differences observed.

organometallic. A compound containing an organic group bonded to a metal atom. An important and versatile class of organometallics is known as Grignard reagents, in which magnesium appears, together with a halogen atom (CH_3MgI); other examples are metallocenes, tetraethyllead, and triethylaluminum. Names of the many types of organometallic compounds are formed by using the prefix organo-, followed by the metal, e.g., organotin, organozinc, organophosphorus, etc. Many of these are highly toxic, and some are fire and explosion hazards, especially triethylaluminum. See also Grignard; tetraethyllead; triethylaluminum.

organophilic. A term descriptive of a solid substance which has an affinity for organic solvents, forming viscous solutions or gel-like dispersions. The thick doughs or pastes resulting from dispersion of crude rubber in naphtha or benzene are instances of organophilic behavior. A substance that repels organic liquids is said to be organophobic.

organosol.   A term used chiefly in the plastics industry to describe a mixture or suspension of a finely divided thermoplastic resin (usually polyvinyl chloride or a cellulose ester) with a plasticizer and an organic solvent. It is a viscous liquid at room temperature, which becomes a rubber-like solid when heated. Organosols are used in adhesive bonding and various types of mechanically applied coatings. Its more general meaning is any dispersion of solids of colloidal dimensions in an organic liquid; if the liquid is water the dispersion is called a hydrosol. *See also* plastisol.

orientation.   A term used in describing the chemical constitution of organic molecules, especially those of the aromatic type, in reference to the relative location of substituent groups or atoms attached to the structure at various points.

Orsat.   An absorption pipette apparatus used in analytical chemistry for determining the composition of gaseous mixtures. After a sample has been collected in the instrument, absorbing media are successively introduced which remove carbon dioxide, olefins, oxygen, and carbon monoxide in that order; the balance is considered as nitrogen.

ortho-.   A prefix used in naming isomers of aromatic compounds; it denotes the position of the second carbon atom in a benzene ring, numbering clockwise from the top of the ring (position 1). Thus, the names of orthoisomers may be prefixed by the numerals 1,2,- indicating that a substituent atom or group is attached to the carbon at these two locations, replacing the hydrogen originally present. The term literally means "straight ahead." It is also used in inorganic nomenclature to indicate the highest degree of hydration which an anhydride can undergo to form an acid or salt, e.g., orthophosphoric acid, H_3PO_4, made from P_2O_5, its anhydride. *See also* meta-; para-; benzene ring.

Os   Symbol for the element osmium, the name being derived from the Greek for *odor*, in reference to the odd smell of the oxide noted by its discoverer, Smithson Tennant, an English scientist (1803).

-ose.   A suffix indicating a carbohydrate compound, usually a simple or complex sugar, for example, sucrose, fructose, glucose, maltose, etc., but often a polymer, e.g., cellulose, cellobiose, amylose (starch).

OSHA.   *See* Toxic Substances Control Act.

osmium.   An element.

Symbol	Os	Atomic Wt.	190.2
State	Solid	Valence	2,3,4,6,8
Group	VIII	Isotopes	7 stable
Atomic No.	76		

Osmium, m.p. 3050°C (5522°F), belongs to the so-called platinum group of metals; it is extremely hard and is one of the heaviest elements, with a density about twice that of lead, i.e., 22.61 g/cc. When heated to 200°C (390°F), it forms osmium tetroxide (OsO_4), which has severely toxic properties. The metal is used chiefly as a component of hard alloys for specialized applications; the tetroxide is used as an oxidation catalyst and for biological staining. *See also* platinum.

osmosis.   Passage of a fluid, either pure or containing a low concentration of solute, through a cell wall or microporous barrier (collodion, parchment) into a more concentrated solution, the openings in the barrier being too small to permit molecules or ions of solute to pass through. The effect of this is to equalize the concentrations and vapor pressures of the two solutions. The barrier or membrane is called semipermeable if it allows the solvent to diffuse but tends to retain the dissolved substance, which in the case of biological osmosis is sucrose. Considerable pressure is exerted by this phenomenon, which is an essential part of the growth process, as it is the means by which nutrients are furnished to living cells.

Reverse osmosis is applied by engineers to reclaiming industrial waste liquors, desalting ocean water, and the like; here the water is forced through a barrier by mechanically exerted pressure, thus effecting separation of the solvent from the dissolved material. *See also* dialysis; membrane.

otto.   An alternative spelling of "attar," the essential oil extracted from roses and similar flowers by steam distillation. The term is used in the perfume industry.

-ous.   A suffix used in naming inorganic compounds to indicate that the multivalent positive element in the compound is in a lower valence state. It is used (a) to designate that the metal in a metallic compound has an oxidation number less than its highest one, e.g., ferrous chloride ($FeCl_2$), ferrous oxide (FeO) or ferrous sulfate ($FeSO_4$), in which the iron valence is 2 rather than the 3 of ferric compounds; and (b) to indicate that the central positive valence element in an acid is in a lower than maximum valence state, e.g., sulfurous acid (H_2SO_3), in which the sulfur has a valence of 4 as compared with its valence of 6 in sulfuric acid (H_2SO_4). In the first case, a more accurate terminology is to use the symbol or name of the metal followed by a Roman numeral in parentheses to show the valence state, e.g., Fe(II) or iron (II) chloride. *See also* -ic.

oxalic acid. A hydrated dicarboxylic acid m.p. 101°C (214°F), which occurs naturally in some plant products in the form of the oxalate; it can be made by oxidation of sugars, starches, and cellulose, as well as of ethylene glycol. The formula is $(COOH)_2 \cdot 2H_2O$. It is used largely as a metal cleaning and bleaching agent, though it also has application in dyeing technology and as a reagent chemical. It is poisonous and should be handled with care.

oxalonitrile. *See* cyanogen.

oxidation. The reverse of reduction. Though oxidation can be regarded as simply the chemical union of oxygen with an atom or compound, as occurs in the combustion of organic materials or the formation of metallic oxides, it is more basically defined as a reaction in which electrons are transferred from one atom to another, either in the uncombined state or within a molecule. The atom that receives the electrons is the oxidizing agent, and the atom that gives up the electrons is the reducing agent. Thus, oxidation is always accompanied by reduction, and such an electron transfer is known as a redox reaction. Regardless of the literal meaning of the term *oxidation,* oxygen is not essential to the reaction, though it is one of the commonest oxidizing agents.

In the reaction $Mg + Cu^{++} \rightarrow Mg^{++} + Cu$, there are two reactants, a magnesium atom in the elemental state, in which the oxidation number is zero, and a cupric ion with oxidation number of $+2$. The magnesium atom gives up two electrons to the cupric ion and by so doing becomes oxidized, while the ion is reduced to el-

emental copper. In cases where a molecule is formed, that is, when oxygen combines with a metal to form an oxide , the electron transfer is limited to a displacement within the molecule; the metal atom is said to be in an oxidized state (in magnesium oxide, the magnesium has an oxidation number of $+2$ and the oxygen an oxidation number of -2).

Electronegative elements such as chlorine and other halogens are excellent electron acceptors and thus are strong oxidizing agents. Removal of hydrogen atoms (each having one electron) from a hydrogen-containing organic compound (dehydrogenation) is a form of oxidation; it is effected by a catalytic reaction with air or oxygen, as in the oxidation of alcohols to aldehydes. Transference of electrons by dissociation of electrolytes in water solution can also be regarded as oxidation. Free radical chains play an important part in the oxidation of such organic compounds as rubber, drying oils, fats, etc. *See also* oxidation number; reduction; free radical; dehydrogenation.

oxidation number. For a given element, the number of electrons it can transfer to another element with which it combines. Conversely, it is the number of electrons it is necessary to add to or subtract from a combined atom to restore it to its orginal uncombined (elemental) state. For example, in the compound magnesium oxide (MgO), the oxidation number of magnesium is $+2$ and of the oxygen -2; since the sum of these is zero, the molecule is electrically neutral. The oxidation number may vary for the same element; and for different elements it may be positive or negative, depending on the number of valence electrons in the outer shell. Thus, there is a close relation between oxidation number (electrons added or subtracted) and the valence, or combining power, of an element. Some metals have as many as six positive oxidation numbers, and many have two or three. An oxidation number of zero indicates the elemental state. *See also* valence.

oxidative coupling. *See* coupling (2).

oxide. An extensive class of compounds formed by chemical combination of oxygen with another element or with an organic compound; the commonest examples of inorganic oxides are water (an oxide of hydrogen) and carbon dioxide, both of which are products of combustion. Virtually all the elements except the noble gases form oxides, especially when heated. They are relatively stable compounds, and in the case of metals, oxygen often acts preferentially, displacing other combined elements. Some aliphatic hydrocarbons are readily oxidized, forming butylene oxide, ethylene oxide, etc. The formation of oxides is called oxidation, though this term has a wider and more sophisticated significance. There is a distinctive group of oxides known as peroxides. *See also* oxidation; peroxide.

oxidizing material.   A term used chiefly in shipping and storage safety regulations to refer to any substance which tends to evolve oxygen when the environmental temperature rises, since this may cause ignition of sensitive organic materials in the same area. It also refers to oxidizing chemicals like chlorates, perchlorates, nitrates, peroxides, etc., which can react spontaneously and vigorously at ambient temperature when in contact with most organic compounds or materials.

oxime.   A compound resulting from a condensation reaction of an aldehyde or ketone, which contain the carbonyl group, $C=O$, with hydroxylamine (NH_2OH), yielding compounds characterized by presence of the group $C=NOH$. These may have two forms which are geometric isomers. Oximes behave as weak acids or bases. Their formation is used as an indication of the carbonyl content of various materials, e.g., essential oils. *See also* carbonyl.

oxirane.   *See* ethylene oxide.

Oxo process.   A method of synthesizing alcohols and aldehydes, developed in Germany in the 1940's. The reaction involves heating high molecular weight olefins, such as heptene, over a catalyst, together with carbon monoxide and hydrogen (obtained from synthesis gas), at about 200 atmospheres pressure. The most effective catalyst is cobalt, which forms a carbonyl complex that is later removed. Several normal and branched-chain alcohols having eight or more carbons are made in this way by hydrogenation of the aldehyde formed in the reaction. *See also* aldehyde.

oxychlorination.   *See* Deacon process.

oxygen.   An element.

Symbol	O	Atomic Wt.	15.9994
State	Gas	Valence	2
Group	VIA	Isotopes	3 stable
Atomic No.	8		

Oxygen comprises 50% by weight of the earth's crust and oceans and thus is the most abundant terrestrial element, equal to the amount of all the other elements combined. Most of this is the oxygen in silicates (silicon is the second most abundant element); oxygen is also present in all oxides, sulfates, carbonates, and nitrates, as well as in carbohydrates, ethers, alcohols, aldehydes, ketones, and most acids and bases. Gaseous oxygen constitutes 21% by volume of the air near the earth's surface, much of which is formed by photosynthesis of plants; water contains 88.8% oxygen by weight. It is vital to all animals and most plants as the supporter of respiration-metabolism processes. Though oxygen is the major agency for support or maintenance of combustion of fuels, it is not itself combustible. Joseph Priestley, an English chemist (1733–1804), is generally credited with the discovery of oxygen in 1774, but his contemporaries, Carl W. Scheele, a Swedish pharmacist (1742–1786), and Antoine-Laurent Lavoisier, a French chemist (1743–1794), also did basic work on the nature of air and the role of oxygen in combustion. Oxygen is a diatomic gas (O_2), b.p. $-183°C$ ($-297°F$), which can be liquefied under pressure. It is manufactured in large tonnages by fractional distillation of liquefied air for use in the metallurgical industries, in making gas for chemical synthesis and medical applications, and as a rocket fuel (as liquid). In compressed and liquid form, oxygen is stored and shipped in steel cylinders.

Oxygen is a reactive, electronegative element, combining readily with almost all other elements. It forms two covalent bonds and also donates electrons for coordination compounds. Its molecules may be ionized by electric discharge, the ions carrying as many as six positive charges. In animal metabolism, oxygen is transported to the tissues by hemoglobin in the blood. *See also* ozone; oxide; combustion.

oxygen acid (oxacid).   *See* per-.

ozocerite.   A mixture of hydrocarbons of high molecular weight, probably petroleum residues, occurring in the earth in the vicinity of oil-bearing strata. It is a solid, low-melting material similar to wax (from which the word is derived). When purified, it is known as ceresin wax. It can be used as a substitute for more expensive waxes for most of their standard applications. Montan wax, a similar material, is derived from lignite and has similar uses. *See also* wax.

ozone.   A triatomic oxygen molecule (O_3) which occurs naturally in low percentages in the air, particularly at high elevations, as a result of radiation; it is formed wherever an electric current is passed through oxygen. It is a powerful oxidizing agent and is also quite toxic. It is prepared commercially for use as a bactericide in water purification and to kill algae and fungi. It also has some application as an oxidizing agent in chemical reactions. Ozone has a characteristic smell which makes identification easy, even in very low concentrations. *See also* oxygen; corona.

P

P Symbol for the element phosphorus, the name being derived form the Greek, meaning "bearer of light," in reference to its luminescent properties.

p- Abbreviation for the the chemical prefix *para-*.

Pa Symbol for the element protactinium, the name being adopted because it forms actinium by radioactive decay.

packing. (1) Any inert material used in absorption, distillation, or rectification columns to baffle the downward flow of countercurrent liquid. It may consist of glass fiber or beads, short pieces of metal tubing, special shapes made of screening or metal, or ceramic forms.

(2) A material used in cylinder heads and pipe joints to prevent escape of steam, water, or oil; it is usually soft enough to conform to the contour of the metals it is intended to seal and thus may be made of a soft metal, plastic, or specially compounded rubber (except for oil lines).

(3) *See* packing fraction.

packing fraction. An energy concept developed in 1927 by Francis W. Aston, a British physicist (1877–1945), Nobel Prize 1922, based on the weights of isotopes as related to mass numbers, which indicates the comparative stabilities of the nuclei of the elements. Positive packing fraction values, which are characteristic of unstable elements, occur with mass numbers less than 25 and more than 175; negative values occur between mass numbers of 25 and 175. The most unstable element is hydrogen, followed by helium, radium, and uranium. The most stable nuclei occur between mass numbers of 50 and 75. The energy released by the fusion of hy-drogen nuclei to form helium can be predicted from this concept.

paint. An organic protective coating in the form of a liquid dispersion comprised of a binder, which hardens to form the protective film, a colorant (pigment), which is usually a dry powder, and a solvent or thinner. The liquid portion of paint (binder plus solvent) is called the vehicle. Heavy-metal soaps such as cobalt naphthenate are often included to accelerate film formation. Paints containing organic solvents may have a drying oil (linseed or tung) or a synthetic resin (alkyd) as binder, plus pigment. Other colorants such as iron oxide, carbon black, titanium dioxide, or organic pigments may be added for decorative effects. Use of lead compounds is no longer permitted in house paints. In emulsion paints, the solvent is water, and the binder an emulsified resin. Acrylic and alkyd resins are widely used in latex emulsion paints for both interior and exterior work. A high-quality binder is essential in exterior paints for weather resistance; drying is brought about by polymerization induced by atmospheric oxygen after evaporation of solvent. *See also* protective coating; antifouling paint.

paired electrons. *See* electron.

palladium. An element.

Symbol	Pd	Atomic Wt.	106.4
State	Solid	Valence	2,3,4
Group	VIII	Isotopes	6 stable
Atomic No.	46		

A member of the so-called platinum group of metals, palladium, m.p. 1552°C (2825°F), is grayish, comparatively soft, and malleable, with an unusual tendency to pick up such gases as hydrogen and oxygen, which may considerably affect its properties. It is fairly reactive chemically and readily forms metallic complexes. It is widely used (where cost permits) in alloys for special electrical systems, as a catalyst in synthetic organic reactions, and to some extent in the fine arts and dentistry. Its chief source in the western hemisphere is Ontario; it occurs in considerable abundance in Siberia. *See also* platinum.

palmerosa oil. *See* geraniol.

palmitic acid. *See* fat; fatty acid.

palm oil. A mixture of triglycerides (esters of unsaturated fatty acids) obtained from a variety of palm native to western Africa; it should more properly be called a fat, as it is a soft solid at 21°C (70°F). It exerts a powerful, though temporary, softening and lubricating action and is used for this purpose in mixing dry ingredients; it has also long been used in the hot-dipping of tin, where it serves to prevent oxidation of the layer of molten tin while it is cooling; the oil retained on the surface after cleaning inhibits oxidation during storage. Palm oil has some application in soap and candle manufacturing and as a tool lubricant

pantothenic acid. A B-complex vitamin associated with a coenzyme and essential in carbohydrate metabolism. It is formed in all living cells and may be obtained from liver, yeast, and vegetable sources; it is also produced synthetically. It is a chemical derivative of the amino acid alanine. *See also* B complex.

papain (pa-pa-in). A heat-stable proteolytic enzyme obtained from the papaya plant. Its ability to hydrolyze or break down proteins makes it useful as a tenderizing agent in meat products and in the bating operation in leather manufacturing. It is also used in the pharmaceutical field as a digestive aid. *See also* pepsin; enzyme.

paper. A cellulosic product made primarily from pulp derived from coniferous trees (softwoods); special types utilize hardwoods (deciduous trees), numerous varieties of plant fibers (esparto, jute, etc.), cotton and linen fibers (rag paper), and synthetic fibers. The largest tonnage produced is kraft paper made from slash pine. After chemical digestion of the woody material, which removes the lignin, the resulting pulp is reduced mechanically by "beating," and filling and coloring substances are added; a dilute suspension of the prepared stock is then fed onto a moving screen (fourdrinier) where the sheet is formed as the water drains out. The sheet then moves to a series of drying rolls which bring the moisture content down to about 5%, after which it is calendered and sometimes coated with a clay-starch-resin mixture. Many types of paper are made for a wide variety of special purposes. Newsprint and other low-grade papers are made from groundwood.

History. Paper owes its name to papyrus, a plant from which the ancient Egyptians made rolls of tissue on which they inscribed historical records. The actual manufacture of paper from plant fibers originated in China about 100 A.D.; the primitive technology traversed Asia and the Middle East over a period of several centuries. By 1400, it had reached Europe. The first use of paper for bookmaking was the Gutenberg Bible (1455). Papermaking began in England about 1500. The first mill in the U.S. was installed in 1690 in Philadelphia. Scheele's discovery of the bleaching power of chlorine (1774) was an important advance; another was the invention of vat sizing with alum and rosin (1800).

On the mechanical side, the invention of the fourdrinier machine was largely responsible for the future development of the industry. Patented in 1807, it was built in England and installed in the U.S. in 1827. Soda pulp was first made in 1854. The principles of sulfite pulping were disclosed by Tilghman in America (1867), and the process was developed in Germany by Mitscherlich in 1874. Groundwood pulp entered commerical use for newsprint about 1870. The sulfate or kraft process, invented by Dahl in Germany in 1884, was first used in the U.S. in 1909. The technique of machine coating was developed in 1875. Continual improvement of the fourdrinier and similar machines by the 1920's resulted in speeds of 1000 feet a minute and continuous operation over a period of several days.

Developments since 1920 include multistage bleaching of kraft; bleaching with peroxides, chlorine dioxide and sodium hypochlorite; use of synthetic resins for wet-strength paper for bags, maps, etc; and use of soluble bases for sulfite pulping, which allowed the use of more species of wood.

The official organization of papermakers in

the U.S. is the Technical Association of the Pulp and Paper Industry (TAPPI), founded in 1915 and located at 155 East 44th Street, New York. Basic research is carried out at the Institute of Paper Chemistry in Appleton, Wisconsin. *See also* pulping; fourdrinier; digestion (1); wood (1); bleach.

para-.   A prefix from the Greek word for opposite and used in naming isomers of aromatic compounds. It denotes location 4 in a benzene ring, numbering clockwise from the top of the ring (location 1); it is thus directly opposite location 1. Names of compounds containing substituent groups which replace hydrogen at these locations are often preceded by the numbers 1,4-. *See also* meta-; ortho-; benzene ring.

paracasein.   *See* casein.

paraffin.   (1) A major group of aliphatic hydrocarbons derived from petroleum; they may be either straight- or branched-chain; the term alkane is also applied to this group. Paraffins are saturated hydrocarbons with the generic formula C_nH_{2n+2}; the simplest is methane, CH_4, followed in homologous series by ethane, C_2H_6, propane, C_3H_8, and butane, C_4H_{10}. The structure of *n*-butane, for example, would appear as:

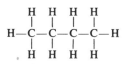

Thus, every paraffin molecule except methane (CH_4) has a CH_3 or methyl group at both ends. In general paraffins are not as reactive as olefins, though substitution of other elements or groups readily occurs along the carbon chain; the term is derived from Latin words meaning "very little affinity." The hydrocarbon chain may be extremely long, with correspondingly high molecular weight; for example, paraffin wax, the distillation end product of aliphatic petroleum, has a chain composed of 36 carbon atoms. Isomeric branched-chain paraffins such as isooctane are used in gasoline; the straight-chain types are unsuitable for gasoline but are used in diesel fuels.

(2) A white wax of high molecular weight ($C_{36}H_{74}$) derived from aliphatic types of petroleum as a distillation end-product. *See* wax; scale (2).

paraformaldehyde.   *See* formaldehyde.

parathion.   The best known of a broad group of organic esters of phosphoric acid originally developed in Germany as nerve gases and later modified and used as insecticides. They are intensely toxic, their action being to inactivate the enzyme cholinesterase, which neutralizes the poisonous acetylcholine produced by nerve impulses in the body. Their application as insecticides has been defended on the ground that their activity is short-lived rather than persistent. Nonetheless, enough contamination may remain several days after use on a field or in an orchard to be critically dangerous. The official permissible concentration is 0.1 milligram in a cubic meter of air. *See also* insecticide; cholinesterase.

partial pressure.   *See* Dalton.

particle.   A term used to refer to individual aggregates of matter having characteristic properties and structure. They include fundamental particles (electrons, protons, and neutrons); atoms and ions (atomic weights from 1 to 259); molecules (molecular weights from 2 to several thousand); macromolecules (weights from about 20,000 to 40 million); colloids, carbon black, and up to palpable sizes such as dust, flour, sand, etc. The activity of electrons, protons, atoms, and ions is primarily chemical and electrical; from colloidal through optically visible sizes, their behavior becomes increasingly physical. Particle size and shape are critical factors in catalysis, reinforcement, coloration, combustion, etc. *See also* fundamental particle; surface area.

particle accelerator.   *See* accelerator; (2).

particle size.   *See* particle; surface area.

partition.   A term which describes a solvent system in which the solute is divided between two components of the liquid phase, the ratio of their concentrations (partition coefficient) being constant for a given substance. Such a system might be comprised of a solute such as a macromolecular substance (dextran), a low percentage of a polyol (polyethylene glycol), and water. Since the liquids are not miscible, they separate into two layers, the solute being distributed between them in concentrations that vary for different materials, e.g., enzymes, proteins, glucose polymers, etc. Separation or fractionation of the macromolecules can be carried out by several procedures, including partition chromatography.

passivity.   A term used in electrochemistry to denote decrease of the reactivity of a metal in an electrochemical system or in a corrosive environment, usually as a result of the formation of an oxide film on its surface. One example of passivity is furnished by immersing iron in nitric acid; the oxidizing action ends abruptly, and the iron is not further attacked. Another instance is the oxide coating of copper anodes during electrolysis. Passivity is inhibited or destroyed in a reducing environment, or in the presence of halogen ions.

pasteurization.   Heat treatment of liquid or semi-liquid food products for the purpose of killing or inactivating disease-causing bacteria. Named for Louis Pasteur, a French bacteriologist (1822–1890), who processed wine by this method, it is legally required for milk in most areas, as tuberculosis bacilli from infected cows at one time were a threat to public health. Milk is held at about 66°C (150°F) for a half hour. The process is also applied to fruit juices. Pasteurization of egg products at 60°C (140°F) for only 3.5 minutes has been mandatory since 1966 in the U.S.

Pasteur, Louis   (1822–1895). French chemist and bacteriologist who made many notable contributions to science. (1) He showed that fermentation results from contamination by infective bacteria (germs), and was the first to identify them as disease-causing agents. (2) He developed the concept of immunizing serums and the antibody-antigen relationship (1880), and was the first to inoculate for rabies and anthrax; he originated the term vaccination (from Latin *vaccus* = cow) in recognition of Jenner's original achievement with smallpox serum (1775). (3) He initiated the practice of heat-treating wine, and later milk and other food products, to kill or inactivate toxic microorganisms, especially the tuberculosis bacillus. (4) He discovered the optical properties of tartaric acid, present in wine residues, which laid the basis for modern knowledge of optical isomers (right and left-handed molecular structure), a phenomenon now often called chirality. *See also* optical isomer; fermentation.

pathway.   A sequence of reactions, usually of a biochemical nature, in which more complex substances are converted to simple end products, as in the degradation of the components of foods to carbon dioxide and water. Its course is largely determined by preferential factors involving coenzymes and other catalysts. An example is the TCA, or Krebs cycle (q.v.).

Pauli exclusion principle.   An important generalization enunciated by Wolfgang Pauli, an Austrian-born physicist (1900–1958), Nobel Prize 1945, relating to atomic structure, which states that it is impossible for two electrons in the same atom to have the same value for all four quantum numbers; two electrons can occupy the same orbital only if they have opposite spin, or direction of rotation, i.e., $+\frac{1}{2}$ and $-\frac{1}{2}$. This principle explains the arrangement of elements in the Periodic Table and accounts for the maximum number of electrons that can occur in the seven shells, i.e., 2 in the first, 8 in the second, 18 in the third, and 32 in the fourth. *See also* orbital; shell; quantum.

Pb   Symbol for the element lead, derived from the Latin *plumbum* for "lead."

Pd   Symbol for the element palladium, the name being adapted from Pallas, an asteroid, which in turn was named for a Greek goddess, no doubt in reference to its whiteness.

peat.   The semicarbonized remains of plants formed in bogs and marshes. It occurs in surface layers from 3 to 10 feet thick. It has a water content of up to 80%, which is reduced to about 35% by drying. The dried product must be protected from autoignition when stored. Peat is easily converted to hydrocarbons and is a good source of gaseous fuel; it can be burned directly after thorough drying. Sources in the U.S. are chiefly in the north central states and Alaska. Besides gas, substantial amounts of oil, ammonia, and sulfur are obtainable as by-products.

pectin.   A colloidal macromolecule, closely related to the polysaccharide group; it is a polymer of galacturonic acid, a carboxylic acid in which the hydrogen of some of the COOH groups is replaced with a methyl group, forming an ester. Pectins are found in plants and especially in fruits. The long chain structure of the molecules enables them to cause stiffening or gelation of 50% aqueous sugar solutions; as the warmed solution cools, the motion of the molecules is reduced, so that they become entangled, thus entrapping the liquid and forming a firm gel structure. Pectins are used almost exclusively in the food industry for jelly and similar products, and as thickeners and stabilizers in mayonnaise and a number of other items. *See also* polysaccharide; gel; agar.

penicillin. The best-known and most widely used antibiotic, penicillin is a product of a particular type of fungus or mold; it is grown in a sugar-containing nutrient medium, and the active substance formed by fermentation is extracted from the solution with solvents or adsorbed on activated carbon. The generalized formula is $RC_9H_{11}N_2O_4S$. Many types of penicillin are made by incorporating chemical agents in the nutrient solution, e.g., organic salts of sodium or potassium. Penicillin is effective against various cocci and spirochetes but somewhat less so against rod-shaped bacteria. It is capable of inducing acute allergic reactions (anaphylactic shock) and should never be taken without the advice of a physician. *See also* antibiotic.

penta-. A prefix meaning five, which may refer to carbon atoms or to atoms of some other element, e.g., pentaborane (boron hydride) contains five boron atoms. In combination with deca- the meaning is 5 + 10, or 15, carbons, as in pentadecane ($C_{15}H_{32}$). The related prefix *pentyl* refers to the hydrocarbon group C_5H_{11} (also called amyl).

pentaerythritol. A polyhydric alcohol derived by a condensation reaction between formaldehyde and acetaldehyde (the so-called aldol reaction); the formula is $C(CH_2OH)_4$. A crystalline solid, it is used in plasticizers, rosin-based varnishes and alkyd resins, and also as the basic component of specialized cross-linking agents and antioxidants. Its most noteworthy derivative is pentaerythritol tetranitrate (PETN), a powerful initiating explosive.

pentane. A liquid hydrocarbon, b.p. 36°C (97°F), derived from petroleum, it is the fifth member of the paraffin series. It is exceedingly flammable, and vapors may explode when present in low concentrations in air (1.5 to 8%). It is used as an industrial solvent, blowing agent, and in special thermometers. The alternative name is amyl hydride. The chemical formula is $CH_3(CH_2)_3CH_3$, or C_5H_{12}. Pentane is the starting point for manufacture of pentanol (amyl alcohol), a solvent used in organic coatings and an intermediate for pharmaceutical products.

pentanol. *See* amyl; pentane.

pepsin. A proteolytic enzyme essential for diges-tion of foods in the animal organism. It is secreted from cells in the stomach glands and is a component of gastric juice. Pepsin is a protein which is capable of decomposing other proteins and thus acts as a catalyst for the digestive process. It can be produced by extraction from the stomach lining of hogs for use in medicine and cheese manufacture. *See also* enzyme; gastric juice.

peptide. *See* polypeptide.

peptization. The stabilization of colloidal solutions (called sols) by means of a so-called electric double layer of adsorbed ions, derived either from a small amount of added electrolyte (peptizing agent) or from water (hydroxyl ions). This results in a liquefying or softening action; the term *peptization* was applied to it by analogy with the digestive action of pepsin on proteins. It is considered to be a form of aggregation.

per-. A prefix used in chemical nomenclature with three specific and different meanings (1) An inorganic or organic compound containing the group—O—O—, in which the oxygen atoms are univalent and are in a negative oxidation state; such so-called peroxy compounds are derived from hydrogen peroxide (H—O—O—H) by substitution of various elements or groups for either or both the hydrogen atoms; examples are peracetic acid (CH_3COOOH) and peroxydisulfuric acid, $H_2S_2O_8$. These are sometimes called oxygen acids or oxyacids.

(2) An inorganic compound comprised of three elements, in which the central element is in its maximum oxidation state; for example, in potassium permanganate ($KMnO_4$) the oxidation states of the three elements are +1 for potassium, +7 for manganese, and −2 for oxygen. Some acids fall into this group, i.e, perchloric, permanganic, and periodic; these should not be confused with the oxygen acids defined in (1).

(3) An organic compound in which complete or maximum chemical substitution has occurred, as in perchloroethane, C_2Cl_6.

peracetic acid. *See* per- (1); bleach.

perchlorate. A salt of perchloric acid, $HClO_4$, in which chlorine has a valence of +7. The most common perchlorate is sodium perchlorate, $NaClO_4$, a colorless crystalline solid salt, m.p. 482°C (900°F), soluble in water, and a powerful oxidizing agent. It is produced by the electrolysis of sodium chlorate in a cell with platinum anodes and steel cathodes, the resulting sodium perchlorate being separated from the residual sodium chlorate by fractional crystallization. It is used in some explosives and is the source of perchloric acid and other perchlorates, such as ammonium perchlorate, NH_4ClO_4, used in solid jet propulsion compositions. While the most stable of the positive chlorine compounds, it is a dangerous fire and explosion risk when in contact with reducing and organic materials and with acids. It is moderately toxic. *See also* chlorate.

perchloroethane.   *See* per- **(3).**

perfect.   This term is used in chemistry in several senses. (1) In reference to gases, it describes a theoretically ideal gas, which behaves in accordance with Boyle's law (*q.v.*); in reality there is no such gas. *See* ideal.

(2) In reference to elasticity, it describes a substance which recovers completely after application of stress, for example, glass; it also is used to denote atomic and molecular collisions in which there is no gain or loss of energy as a result of impact.

(3) As applied to crystals, it means that all lattice locations are occupied by the atoms or ions required by the nature of the molecule, and by no others.

perfume.   Any of a wide range of liquids having an intense, pleasant smell and derived by steam distillation or extraction with solvents from essential oils occurring in flowers, or sometimes from other parts of the plant. The selection and blending of these is an ancient art. *See also* essential oil; fragrance.

period.   **(1)** Any of the seven major horizontal divisions of the Periodic Table.

(2) *See* identity period; periodic (1).

periodic.   A term which has the same spelling for two quite different meanings and pronunciations. (1) When divided pe-ri-od-ic, it describes a quality, condition, or action that repeats itself at regular intervals, for example, the periodic nature of the chemical elements, or periodic temperature increases. (2) When divided per-i-o-dic, it refers to an iodine compound in which the iodine is in its highest oxidation state, i.e., $+7$. *See also* Periodic Table; per-.

Periodic Law.   One of the most important generalizations in the history of chemistry. Essentially it states that the arrangement of electrons in the atoms of the chemical elements and the properties determined by this arrangement (bonding, reactivity, etc.) are closely related to the atomic number, and that the arrangement changes in a regularly repeated (periodic) sequence as the atomic number increases from one element to the next. This law is the basis of the Periodic Table. *See also* element; atomic number.

Periodic Table.   A systematic classification of the chemical elements based on the Periodic Law; it was originally worked out by the Russian chemist, Dmitri Mendeleev (1834–1907), in 1869 by listing the then-known elements in the order of increasing atomic weight. It was later modified by the English physicist, Henry G. J. Moseley (1887–1915), and others, who showed that the periodicity actually depends on the atomic number. Thus, the location of an element in the table is an indication of its electronic structure and chemical properties; elements having a similar electronic pattern occur on a regularly repeating (periodic) basis. There are seven horizontal divisions, called periods, containing, respectively, 2, 8, 8, 18, 18, 32, and 19 elements. (Elements 93-106 are synthetic radioactive elements of the transuranic series.) Each period begins with an element having one outermost electron and ends with one having eight outermost electrons (except helium). The vertical divisions of the table comprise nine major groups; Groups I through VII are divided into two subgroups designated A and B. Thus, Group IA is a vertical series of univalent, reactive elements, while those in the last (noble gas) group are inert, or nearly so, the first three having zero valence. Many variations of the table have been proposed and used; in that now widely approved the groups are as follows:

IA IIA IIIB IVB VB VIB VIIB VIII IB IIB IIIA IVA VA VIA VIIA Noble gas. The generally accepted version of the Periodic Table is shown facing the title page.

Perkin, Sir William Henry   (1838–1907). An English chemist who was the first to make a synthetic dyestuff (1856). He studied under Hofmann at the Royal College of London. Perkin's first dye was called mauveine, but he proceeded to synthesize alizarin and coumarin, the first man-made perfume. In 1907 he was awarded the first Perkin Medal, which has ever since been awarded by the American Division of the Society of Chemical Industry for distinguished work in chemistry. Notwithstanding the fact that Perkin patented and manufactured mauve dye in England, the center of the synthetic organic dye industry shifted to Germany, where it remained till 1914. *See also* Hofmann.

permanent-press fabric. *See* formaldehyde; phenolformaldehyde resin.

peroxide. *See* per-; hydrogen peroxide; initiator; benzoyl peroxide.

pesticide. A broad term that includes all chemical agents used to kill animal and vegetable life which interferes with agricultural productivity; thus, insecticides, herbicides, herpicides, fungicides, and rodenticides are special types of pesticides, regardless of their mode of action. Fumigants, used in orchards and on stored grain, which are toxic to bacteria as well as to animals, also may be considered pesticides. All pesticides are poisonous in varying degrees when ingested directly, and care should be used in their handling and application. The most hazardous are those in the parathion group; next in order are chlorinated hydrocarbons. The chief risks from pesticide intake are exposure during application and accidental ingestion, rather than from intake of residues on agricultural products. *See also* biocide.

PETN. *See* pentaerythritol.

petrochemical. A chemical intermediate obtained directly from petroleum or natural gas by cracking or pyrolysis, or an end-product derived by subsequent chemical processing. The term is so general that its usefulness has been questioned by some authorities. The primary petrochemicals are methane, ethane, propane, ethylene, acetylene, benzene, hydrogen, etc., from which hundreds of chemicals are produced by a variety of reactions. For example, ethylene, obtained largely by heat treatment of hydrocarbon gases, is further reacted to yield such chemicals as ethylene dichloride, ethyl alcohol, ethyl chloride, and polyethylene; some of these are end-products, but others are intermediates for acrylonitrile, ethylene glycol, vinyl chloride, and many other synthetic organics. The same is true of the other primary petrochemicals. The vast number of derivatives are used as solvents, plasticizers, elastomers, fibers, pharmaceuticals, explosives, dyes, fertilizers, etc. Petroleum-derived materials that are used directly as energy sources (gasoline, fuel oil) are not usually considered to be petrochemicals. Some petrochemicals (e.g., benzene) are also derived from sources other than petroleum, such as coal tar. *See also* petroleum; intermediate.

petrolatum. *See* jelly.

petroleum. This term literally means "rock oil." Petroleum is thought to have originated from decomposition of animal life which inhabited the ocean floor in early geologic periods. It is a complex mixture of all types of hydrocarbons, the paraffins predominating in some areas, and aromatics and cycloparaffins in others. It occurs in numerous locations in the world, usually in rock strata, but sometimes close to the surface. The leading producing areas in the U.S. are California and Texas. Deposits have been brought on stream in Alaska and the North Sea, and extensive resources have been found in Mexico; elsewhere it is found in northern Canada, the Mid-east Persian Gulf area, and the Baku peninsula in the Caspian Sea. It is not only the primary source of automotive fuels but of thousands of organic chemicals used in medicines, plastics, dyes, solvents, lubricants, etc. The molecules of some of its components contain up to 50 carbon atoms.

History. Petroleum was found in many locations throughout the world before its discovery in the U.S. The first well was drilled by Drake in Pennsylvania in 1859; its depth was a mere 70 feet, in contrast to the 20,000 feet that can be attained with modern equipment. For many years, production was confined to the eastern states, the oil being used chiefly for lubrication and illumination (kerosine). Around the turn of the century, the great strikes in the southwest and California coincided with the expanding demand for motor fuels. Production figures (in millions of gallons) are 1880, 26; 1900, 64; 1920, 443; 1940, 1,300; 1960, 2,470; and 1970, 3,327. Since 1970, production has declined substantially in spite of new methods of secondary extraction, such as chemical flooding and hydraulic fracturing.

Notable advances in refining techniques occurred at an increasing rate after 1910, when thermal cracking was introduced as a result of research by Burton. This did not yield gasoline of high quality. Its performance was greatly improved by Midgley's discovery in 1921 of the octane-boosting power of tetraethyllead from about 60 to 100 or more. An outstanding achievement was the development of catalytic methods over the next 20 years. Early catalysts were activated clays or silica-alumina. In recent years, these have given way to zeolites (molecular sieves) and to metals of the platinum group. The processing techniques include the fixed-bed method, later replaced by moving-bed and fluidized-bed continuous crackers. The exhaustive study and evaluation of catalysts largely accounted for the present sophisticated techniques that characterize the industry. Many catalytic reforming processes (alkylation, isomerization, hydrocracking, etc.) were researched and developed between 1940 and 1960. Important contributions to this development were made by Ipatieff and Egloff.

The official organization of the industry is the American Petroleum Institute (1919) located at 1271 Avenue of the Americas, New York. *See also* cracking; fluidized bed; catalyst.

petroleum ether   *See* ether; naphtha; ligroin.

petroleum gas.   Any of a number of flammable gases obtained by refining of petroleum and used both for organic synthesis and in liquefied form as fuels. Most important are the saturated compounds butane, isobutane, and propane and the unsaturated butylenes and propylene. Methane is obtained not only from petroleum, but also from other sources such as coal and natural gas. *See also* liquefied petroleum gas.

petroleum jelly.   *See* jelly.

petroleum spirits.   *See* naphtha; spirits.

Ph   Abbreviation of phenyl, the univalent group C_6H_5, characteristic of benzenoid compounds.

pH.   A scale indicating the acidity or alkalinity of aqueous solutions; it was developed in 1909 by S.P.L. Sørensen. (1869–1939) a Danish biochemist. pH is conventionally defined as the negative logarithm of the hydrogen-ion concentration in moles per liter, i.e., $pH = -\log[H^+]$. pH value is a number from 1 to 14, which represents a logarithmic scale indicating the concentration of hydrogen ions. These numbers are actually exponents expressing the molar concentration; water molecules dissociate into H^+ and OH^- and recombine at such a rate that there is at any instant a concentration of these ions of $1/10,000,000$ mole per liter, i.e., $1/10^7$ or 10^{-7}. This is the neutrality point of pure water (pH 7), at which the concentrations of H^+ and OH- ions are equal. When an acid is added to the water, numbers from 1 to 7 are used to indicate the acidity of the solution, i.e., the predominance of hydrogen ions; if alkalies are added to the water, numbers from 7 to 14 indicate the alkalinity, i.e., the predominance of hydroxyl ions in the solution. A pH of 1.0 is represented by $0.1N$ hydrochloric acid, and a pH of 14 is that of $0.1N$ sodium hydroxide. Since the ionic concentration increases logarithmically, the values on the pH scale differ by a factor of 10: a pH of 5 is ten times as acid as a pH of 6 and a pH of 9 is ten times as alkaline as a pH of 8. The pH of solutions is visually shown by indicators and is an important factor in acid base exchange in biochemical systems and in many aspects of industrial technology. Instruments for measuring and recording pH continuously are available. *See also* neutral (1); indicator; titration.

pharmaceutical.   *See* drug.

phase.   (1) Any of the forms or associations in which matter can exist, each depending on the number of atoms or molecules per unit volume of space. These are (a) the gaseous phase, in which the units are widely separated, (b) the liquid phase, in which they are much more highly concentrated, and (c) the solid phase, in which they are tightly packed and strongly bound together. Gases and liquids are amorphous, that is, without structure, though both can be converted into the next denser phase by low temperature and pressure. Solids are crystalline in nature but can be converted into liquids and vapors by heat. Water is a substance that is well-known in all three phases, as solid (ice), liquid, and vapor.

(2) In colloid chemistry, this term is used to designate the portions of a solution or dispersion, namely, the disperse phase and the continuous phase (sometimes called internal and external, respectively). For example, in an oil-in-water emulsion, the oil droplets are the disperse phase and water the continuous phase; in a water-in-oil emulsion, the reverse is the case.

phenanthrene.   A crystalline benzenoid (aromatic) hydrocarbon ($C_{14}H_{10}$), m.p. 100°C (212°F), derived from coal tar. It is closely related to naphthalene and is isomeric with anthracene; its structure is:

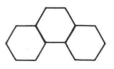

It has been the basis for extensive biochemical study of its possible cancer-inducing effect; prolonged and repeated skin contact should be avoided. *See also* carcinogen.

phenobarbital. *See* barbiturate.

phenol.   The most versatile and industrially useful aromatic alcohol, m.p. 43°C (109°F), derived from benzene by substitution of a hydroxyl group for the hydrogen at the orthoposition (C_6H_5OH);

it is not a typical alcohol in properties, its reactions being strongly acidic, as is evident from its alternative name, carbolic acid. It can be made by a number of processes, oxidation of cumene being the most generally used, with acetone as by-product; it can also be recovered from refinery wastes by solvent extraction. Phenol is a strongly irritant and poisonous material which should be handled with due caution. It is produced in tonnage quantities for use in the manufacture of phenolformaldehyde resins; in making bisphenol A, from which epoxy and polycarbonate resins are derived; in one process for the production of caprolactam for nylon; and in a large number of compounds used as insecticides, germicides, and other chemicals. *See also* phenolic; phenolformaldehyde resin.

phenolformaldehyde resin.   The first truly synthetic resin, developed by Leo Baekeland, a Belgian-American chemist (1863–1944), and trademarked "Bakelite" (1907). It is a condensation polymer of phenol and formaldehyde activated with an alkaline catalyst. Polymerization occurs in three steps or stages. The first (A-stage) is an alcohol-soluble liquid; the second (B-stage) is semisolid and less soluble; the third (C-stage) is a hard, cross-linked solid. The A-stage form is called a resol. The final polymer is a nonconductor of electricity, is not attacked by most chemicals, is nonflammable, and has high strength and durability. Its major limitation is the fact that it cannot be made in any color except black or grayish-brown. Nonetheless it has been one of the most widely used engineering plastics since its introduction in the form of molded components for machinery, electrical equipment housings. and numerous similar applications. As an adhesive, it is used in composites such as plywood, reinforced plastics, and permanent-press fabrics; it also has application as an ion-exchange resin.

Another type of phenolformaldehyde resin (novolac) utilizes an acidic catalyst and is thermoplastic; it is used largely in brake linings and grinding wheels, and for electrical insulation. An additive such as hexamethylenctctramine is necessary to harden novolacs so that they become thermosetting. *See also* novolac.

phenolic.   General name for a class or group of chemicals derived from benzene by substitution of one or more hydroxyl groups, and in some cases methyl groups as well, for hydrogen atoms on the benzene ring. The most important phenolics are phenol (C_6H_5OH); cresol ($CH_3C_6H_4OH$), of which there are three isomeric forms with the methyl group usually shown in the 1 position; and resorcinol, $C_6H_4(OH)_2$. Phenolic compounds are widely used as starting points for synthetic resins, disinfectants, herbicides, pharmaceutical products, and chemical synthesis. In addition, there are a number of special uses for each type. *See also* phenolformaldehyde resin; cresol; resorcinol.

phenolphthalein.   *See* aluminum chloride; indicator; phthalic anhydride.

phenolsulfonic acid.   *See* picric acid.

phenoxy resin.   A hard, transparent polymer made by reacting bisphenol-A with epichlorohydrin, forming a linear, noncrystalline structure which can be cross-linked to a dimensionally stable engineering plastic. It is chemically related to epoxy resins, being composed of the same copolymers, but differs from them because it contains no epoxide groups. Phenoxy resins dissolved in methyl ethyl ketone are useful in the coatings field; they can also be molded by standard techniques for application in piping, machine parts, and similar mechanical products.

phenyl.   The unsaturated aromatic group C_6H_5—, representing a benzene molecule with one open valence where any of numerous atoms, groups, or side chains may be attached. For example, in phenol (C_6H_5OH), the hydrogen originally present in benzene is replaced by a hydroxyl group. In cases where there are two open valences (C_6H_4—), the group is called phenylene, as in phenylenediamine, $C_6H_4(NH_2)_2$, where two amino groups replace the hydrogens.

phenylene.   *See* phenyl.

phenylenediamine. An important intermediate derived from benzene by substitution of two amino groups, $C_6H_4(NH_2)_2$. These may be either ortho-, meta-, or para- to each other in the molecule, forming three isomers, the most widely used being the para- form. All are crystalline solids and have different degrees of toxicity. The ortho- and meta- isomers are not hazardous, but the para- form is toxic and is a tissue irritant. These compounds are used as dye intermediates and in photographic developers; the para- isomer is the basis of rubber accelarators and heat stabilizers.

pheromone. A substance exuded by insects to attract other insects. It may be sex-specific or, more often, species-specific, and can be used artificially as a bait to lead insects to traps. In one notable example, it is being applied in Norway and Sweden in a major program to decrease the population of the destructive bark beetle, which kills spruce trees. A pheromone is a mixture of chemical compounds, in this case methyl butenol, ipsdienol and cis-verbenol, derived from metabolism by the beetle of naturally occurring terpene components of the tree host. Insects respond to pheromones at near-molecular levels and are known to travel miles to reach their source, chiefly for mating.

phosgene. Carbonyl chloride, $COCl_2$, b.p. 8.3°C (47°F), a highly toxic and reactive gas made by the reaction of chlorine and carbon monoxide over an activated carbon catalyst. It is shipped in liquid form under pressure in steel tanks. Phosgene is used in large quantities in the manufacture of isocyanates, polyurethane, and polycarbonate polymers, carbamates, dyes, herbicides, and pesticides. *See also* diisocyanate.

phosphate. Any of a large number of inorganic compounds essentially comprised of an alkali or alkaline-earth metal and a PO_4 group; they form mono-, di-, and tribasic compounds (orthophosphates) depending on the number of metal atoms present and on the hydrogen content. Many of them are hygroscopic or exist in the form of hydrates. Phosphates occur widely as components of rock structures, from which is derived phosphorus for use in elemental form, or in the manufacture of phosphoric acid, fertilizer superphosphates, animal feed supplements, pharmaceuticals, leavening agents, and detergent builders. In the latter case, they contribute to water pollution, as they are a nutrient medium for algae. *See also* algae; phosphoric acid.

phosphatide. Any of a number of complex lipids which contain phosphorus in the molecule; they occur in all living cells and have significant metabolic functions, many of which are not fully elucidated. The alternative name for such compounds is phospholipid. In addition to phosphorus in the form of phosphoric acid, they are comprised of polyhydric alcohols, fatty acids, and a basic group containing nitrogen, all of which can be obtained by hydrolysis. Lecithin and sphingomyelin are among the better-known phosphatides. Uses are limited to the food industry as emulsifiers and antioxidants. Phosphatides are obtained from soybean oil for commercial applications. *See also* lipid; lecithin.

phospholipid. *See* phosphatide.

phosphor. **(1)** A substance which emits radiation in the ultraviolet, visible, or infrared wavelengths during or following excitation by an external energy source; it may be either fluorescent or phosphorescent depending on whether or not it emits light after the exciting source has been removed. Inorganic phosphors such as metallic tungstates, zinc sulfide, calcium phosphate, and other metal compounds are produced in quantity for use in fluorescent lights, color television, scintillation counters, etc.; hundreds of such materials have been prepared for these and similar purposes. For satisfactory results they require the presence of trace quantities of such metals as silver, copper, or manganese, which are called activators. Organic phosphors have also been used for display color effects and textile brighteners. Riboflavin and chlorophyll are examples of natural organic phosphors. *See also* luminescence; phosphorescence; fluorescence.

(2) The term *phosphor* is used adjectivally, as in phosphor bronze, to indicate the presence of a low percentage of phosphorus as a deoxidizing additive.

phosphor bronze. *See* bronze; phosphor (2)

phosphorescence. A kind of luminescence occurring naturally in many minerals and metallic compounds, in some types of organic compounds, and in some living organisms (marine fauna and insects such as the firefly). Phosphorescence is distinguished from fluorescence in two ways: (1) there is a longer time period between excitation and the emission of light, ranging from one-thousandth of a second to several hours, whereas fluorescence occurs almost instantaneously; and (2) phosphorescence may continue for some hours after the exciting source has been removed, the length of emanation depending on the substance, whereas fluorescence ceases when the source is cut off. Thus, phosphorescent materials glow in darkness, but fluorescent materials do not. *See also* phosphor; fluorescence; luminescence.

phosphoric acid. A strong acid which is a solid at 100% strength and liquid at lower concentrations; it is also known as orthophosphoric acid, having the formula H_3PO_4. It is made from phosphate rock either by reaction with sulfuric or hydrochloric acid or by heating in an electric furnace with carbon and silica to evolve phosphorus, which is then burned to P_2O_5 and hydrated. The acid is used to make a number of phosphates and as the basis for a range of fertilizers, soil conditioners, detergents, and metal-cleaning preparations. It is also used in the food industries as an acidulant and in pharmaceutical products. It is irritating to the skin, and strong solutions should by handled with care.

phosphoric acid esters. *See* parathion; nerve gas; cholinesterase.

phosphoric anhydride. Phosphorus pentoxide, P_2O_5, a white solid, m.p. 563°C (1045°F), sublimes at 300°C (572°F); made by the complete combustion of phosphorus in dry air. It reacts exothermally with water to form metaphosphoric acid, HPO_3, or orthophosphoric acid, H_3PO_4, if the water is in excess. It is used as a dehydrating agent and is the source of pure phosphates used in the preparation of many inorganic and organic phosphorus compounds. *See also* flame retardant; phosphate; phosphorus.

phosphorus. An element.

Symbol P		Atomic No. 15
State	Solid	Atomic Wt. 30.9738
Group	VA	Valence 3,5

In some respects phophorus is not unlike carbon; it is a nonmetallic element having several allotropic forms, designated by their color—two white forms differing in their crystal structure, a red-violet, and a black allotrope, the last two being either amorphous or crystalline. The white allotropes are pyrophoric, but the black does not ignite easily. Phosphorus also resembles carbon in its tendency to form covalent bonds, though it does not bond to itself as readily as carbon.

It occurs widely in rock structures, which are its principal source, as well as in the guano deposits found on islands frequented by sea birds. The name is due to its property of emitting absorbed radiation in the dark (afterglow or phosphorescence). It is made by reduction of the calcium phosphate in phosphate rock in an electric furnace with silica and carbon.

Phosphorus is an essential nutrient for plants and animals and has complex metabolic functions (as adenosine triphosphate) in addition to its hardening effect on bones and teeth (as hydroxyapatite). In elemental form it is highly toxic, and some of its organic compounds (nerve gases and insecticides of the parathion type) are lethally poisonous. The chief uses of elemental phosphorus are in the manufacture of numerous compounds (phosphoric acid, combinations with halogens, sulfur, etc., and organophosphorus compounds), as a semiconductor and catalyst, in pyrotechnic devices, and in deoxidizing metals. *See also* phosphoric acid; phosphate; phosphatide.

phosphorus oxychloride. *See* flame retardant.

phosphorus pentoxide. *See* phosphoric anhydride.

phosphorylation. A chemical reaction in which phosphorus is added to or combined with an organic compound, especially in the form of the trivalent phosphoryl group $\equiv P = O$. It is important in cellular metabolism, vitamin activity, and enzyme formation. It is used to produce a modification of cellulose for cation exchange in chromatographic separations.

photochemical smog. *See* sulfur dioxide; photochemistry.

photochemistry. A subdivision of chemistry devoted to the chemical changes induced by various wavelengths of radiation, often brought about through the agency of molecular fragments known as free radicals. Absorption of radiant energy is involved, and the effects of this are permanent rather than temporary, as in spectroscopy, where no chemical changes occur. Photosynthesis is the most important natural photochemical reaction; chemical changes due to radiation are basic to photography, and high-energy radiation is also able to initiate cross-linking of polymers and synthesis of alkyl halides (gamma radiation). Degradation reactions also occur as a result of the ionizing effect of short-wave radiation. Ultraviolet radiation catalyzes the formation of photochemical smog from nitrogen oxides and hydrocarbon air contaminants; it also initiates photolytic decomposition of ketene and accelerates formation of vitamin D in milk. *See also* free radical; radiation; photolysis; ionizing radiation; photosynthesis.

photochromic. Derived from two words meaning "light" and "color" this term denotes a material to which has been added a low percentage of a light-sensitive chemical, the effect of which is to cause the material to darken in the presence of strong light and to resume its original transparency when the light intensity is decreased. This effect is brought about in optical glass by addition of silver bromide, which reduces the light transmission by about 50%. Certain classes of organic dyes exhibit photochromic behavior, changing from colorless to colored by absorption of ultraviolet radiation, which causes a reversible chemical change in the dye molecule.

photography. *See* photochemistry; developing agent; quinone; silver.

photolysis. The decomposition or cleavage of a chemical compound induced by radiation of certain wavelengths; for example, the ketene molecule $H_2C\!=\!C\!=\!O$ can be split into carbon monoxide and methylene (carbene) $(:CH_2)$ upon irradiation with ultraviolet light. Photolysis can be regarded as the opposite of photosynthesis.

Flash photolysis is a method of investigating the mechanism of extremely rapid photochemical reactions involving the formation of free radicals (both inorganic and organic). This is done by irradiating a given reaction mixture with a flash of high-intensity light; this excites the molecules and produces the short-lived radicals which activate photochemical reactions. These products are instantaneously analyzed spectroscopically, which permits identification of the radical species from the spectra obtained. The time lapses involved in this technique are about 1/100,000th second. It has also been applied to study of the exceedingly fast reactions occurring in flames and explosions. *See also* photochemistry; free radical.

photometric analysis. Chemical analysis by means of absorption or emission of radiation, primarily in the near ultraviolet, visible, and infrared portions of the electromagnetic spectrum. It includes such techniques as spectrophotometry, spectrochemical analysis, colorimetry, and fluorescence measurements. *See also* spectroscopy.

photon. The ultimate unit (quantum) of electromagnetic energy which comprises the entire spectrum of radiation, from cosmic and gamma rays to the longest radiofrequencies. An objective description of a photon is a discontinuous "packet of energy" behaving as a wave; an accurate understanding can be gained only from the mathematics involved in quantum mechanics. Photons are unique in that they have no mass independent of their motion, which is the speed of light. An electron gives off one or more photons in dropping from a higher to a lower energy level. Molecules are capable of absorbing the energy of photons, as occurs with chlorophyll, and thus chemical reactions can be activated by this means. Study of such light-activated reactions comprises the science of photochemistry. The nature and behavior of photons was established in the late nineteenth century by James Clerk Maxwell, a British mathematician and physicist (1831–1879); Max Planck, a German physicist (1858–1947), Nobel Prize 1918; and Albert Einstein, a German-American physicist (1879–1955), Nobel Prize 1921. *See also* radiation; photochemistry; quantum.

photosynthesis. The chemical synthesis of carbohydrates (sugars, starches, and cellulose) performed in the cells of plants, the necessary energy being supplied by sunlight. Plant cells absorb water from the soil and carbon dioxide from the air, the reaction being catalyzed by the green pigment chlorophyll. The basic reaction is: $6CO_2 + 6H_2O + e \rightarrow C_6H_{12}O_6 + 6O_2$. This is of the utmost importance, not only for the food substances formed but for the oxygen evolved. Virtually all the oxygen of the air results from photosynthesis, well over half of it being produced by algae and other aquatic plants. Thus, plant life is the primary means of converting radiant energy to chemical energy. The ability of plants to synthesize organic compounds from inorganic constituents is unique, and no life would be possible without it. In the absence of light, oxygen is absorbed and carbon dioxide evolved, with no sugar formation. Combustion can be regarded as the reverse of photosynthesis, yielding carbon dioxide and water, with release of the original energy input. *See also* photochemistry; chlorophyll; carbon cycle; plant.

phthalic acid. *See* phthalic anhydride; polyester.

phthalic anhydride. A crystalline powder, this benzenoid compound is an important industrial chemical derived catalytically from xylene or naphthalene; its formula is $C_6H_4(CO)_2O$, m.p. 131°C (268°F). Addition of water to the molecule gives phthalic acid, $C_6H_4(CO_2H)_2$. The anhydride is a basic component of a broad range of plasticizers and modifiers used in plastics processing, as well as of alkyd and other polyester resins. It is reacted with phenol to produce phenolphthalein, used as an indicator, a dye, and a pharmaceutical. Phthalic anhydride is irritant to tissue and should be handled with care.

phthalocyanine. A porphyrin-derived nitrogenous compound, characterized by four benzenoid groupings, in which the central hydrogen atoms can easily be replaced by metal atoms, forming chelate compounds having strong pigmenting properties. Dozens of such metal phthalocyanines have been made, the most widely used being the copper and iron derivatives. Stable and brilliant colors are obtained in a wide range of shades and intensities. Greens, blues, and reds predominate, the color depending on the metal or halogen atom present. Many such compounds are also used as catalysts in petroleum refining and polymerization reactions. They also initiate photochemical reactions. *See also* porphyrin; chlorophyll.

phycocolloid. Any of several colloidal macromolecular substances occurring in seaweeds; they are chemically classified as polysaccharides, or polymeric forms of certain sugars. They are strongly hydrophilic, forming gelatinous pastes by absorption of water. Because of this, they are used as thickeners and emulsifiers in numerous fat-containing food products, such as ice cream and frozen desserts, and in pharmaceutical products for their laxative action. The most important phycocolloids are obtained from algae (agar) and Irish moss (carrageenan). *See also* polysaccharide.

physical chemistry. Generally regarded as an independent science, physical chemistry includes those areas of chemistry which involve an understanding of the physical principles of mass, motion, electricity, radiation, heat and related phenomena; they include nuclear chemistry, spectroscopy, electrochemistry, atomic structure, chemical kinetics, and colloid chemistry.

physiological salt solution. *See* isotonic.

phytochemistry. That branch of chemistry which deals with the nutrition and metabolism of plants and with the chemical products obtained from them. It thus includes the utilization of mineral nutrients, carbon dioxide, and water, and their transformation into carbohydrates, proteins, oils, vitamins, alkaloids, etc., as well as plant growth regulators, osmosis, soil pH, and other factors affecting growth. *See also* plant.

pickling. (1) Preservation of certain food products by the use of various spices, sodium chloride solution, vinegar (acetic acid), or sodium benzoate in low concentrations. Salt solution is also used in the preliminary treatment of skins for the tanning process in leather technology.

(2) In metal technology this term refers to descaling and cleaning of metallic surfaces with a strong acid such as sulfuric or hydrochloric.

picloram. *See* herbicide; defoliant.

picric acid. A high explosive, trinitrophenol, obtained by treatment of phenolsulfonic acid by nitration; it may be in the form of a yellow powdery solid, a liquid, or a thick paste . Besides its use as an explosive, it has specialized application in dyeing. It reacts strongly with metals, forming salts that are still more sensitive to explosion. It is also poisonous and irritant to tissue. Its formula is

$$C_6H_2(NO_2)_3OH, \text{ m.p. } 122°C (252°F).$$

pigment (1) Any substance, either inorganic or organic, which has positive tinctorial power. Synthetic organic pigments are usually called dyes, though there is no entirely satisfactory line of demarcation between a dye and an organic pigment. A distinction that is more conventional than scientific is that dyes are applied almost exclusively to fibers and fabrics, whereas organic pigments are used for other materials, such as paints, plastics, and ceramics. Inorganic pigments are often insoluble powders, e.g., oxides of metals such as iron, chromium, and cobalt, either made by chemical reaction or occurring naturally as earthy mixtures (ocher). These are less brillant than dyes but are more stable to heat and light. Natural organic pigments are produced by both plants and animals; plant products include flower colors (flavonoids), madder, logwood, carotene, chlorophyll, xanthophyll, indigo and phthalocyanines; animals produce melanin, cochineal, rhodopsin, etc. Synthetic organic types include toners, lakes, lithols, and toluidines. *See also* dye; colorant; rhodopsin.

(2) The term pigment is also applied to a group of particulate materials used in rubber and plastic compositions. Several of these act as both colorant and reinforcing agent. Zinc oxide, a white pigment, was originally used as a reinforcing agent in tires; as more and more carbon black was added to darken the product, its superior abrasion-resistant properties were discovered (an example of serendipity). Thus, the meaning of "pigment" was extended to include clays, whiting, etc., which have neither colorant nor reinforcing power and are more properly classed as fillers.

pile. *See* reactor.

pinene. *See* camphor; oleoresin.

pine oil. *See* fragrance.

pipette (pipet). A slender glass tube open at both ends and having an expanded area at or near the center designed to contain a specific volume of liquid (e.g., 5cc). Liquid is drawn into the tube by oral suction. A pipette is used in transferring measured volumes of liquid from one container to another.

pitch. A carbonaceous residue obtained by distillation of such substances as coal tar, pine tar, rosin, petroleum, and fatty acids. Some types occur naturally, e.g., glance pitch. Most types are of dark brown to black color and are characterized by extreme tackiness, especially when warm. They are used as sealants, in roofing compounds, and in road paving. *See also* asphalt; tar.

pitchblende. *See* radium; uranium.

Planck's constant. The universal constant, h, in the expression $E = hv$, which relates the energy of a quantum of radiation, E, to the frequency, v, of the radiation. The numerical value is 6.6254 $\times$ 10^{-27} erg-seconds. The relation was formulated in 1903 by Max Planck, a German physicist (1858–1947), Nobel Prize 1918. It is interesting to compare it with Einstein's equivalence equation for energy, $E = mc^2$.

plant. This term includes all types of vegetation containing chlorophyll, including algae, trees, bushes, grasses, mosses, etc., but excluding fungi. Their metabolic processes are vital to the maintenance of life on earth resulting in the following products: (1) oxygen (from photosynthesis); (2) carbohydrates (from photosynthesis); (3) amino acids and proteins (from nitrates and nitrogen-fixing bacteria); (4) fats and oils; (5) vitamins; (6) natural fibers; (7) coal; (8) various other substances of value, such as rubber and alkaloid drugs. *See also* photosynthesis; phytochemistry; natural product.

plant growth regulator. A term used in phytochemistry to denote an organic compound present in plants which governs a control mechanism, the action being analogous to that of a hormone in animal metabolism. The functions on which these agents operate include length of dormancy of seeds, the formation of roots, buds, and flowers, and the extension and contraction of cell walls of the stem to allow the plant to bend in phototropic response. Important groups of regulators are the auxins (indoleacetic acid and dichlorophenoxyacetic acid) and the gibberellins. *See also* auxin; gibberellic acid.

plant hormone. *See* plant growth regulator.

plasma. (1) A gas which has been heated to such a degree [upwards of 5500°C (10,000°F)] that electrons are pulled away from the atoms by the force of the greatly intensified collisions, leaving them in an ionized state. The gas then is made up of a mixture of electrons and ionized nuclei which behave quite differently from gases at normal temperatures.

Two kinds of plasma are recognized by physicists, namely, a particle plasma and a reactor plasma. A particle plasma is a neutral mixture of positively and negatively charged particles interacting with an electromagnetic field, which dominates their motion. Temperatures of 10 to 15,000°C can be reached. Such plasma, formed by sudden energy releases, can be utilized as an energy source, as in magnetohydrodynamics. Reactor plasmas, on the other hand, are composed of positively charged ions of hydrogen isotopes (deuterium, tritium); the electric charge is the controlling factor. These are used in nuclear fusion devices called tokamaks, where temperatures of 74 million °C have been attained. These plasmas also respond to electromagnetic forces, which are used to confine them. *See also* fusion; tokamak.

(2) In medicine, this term refers to human blood from which the red and white corpuscles have been removed; it is used primarily to replace fluids lost from the body as a result of.

burns or excessive bleeding. It is often supplemented by so-called volume expanders such as dextran or polyvinylpyrrolidone.

plaster. (1) A mixture of calcium compounds such as lime and gypsum with water; it has a paste-like consistency suitable for spreading or molding, and sets by drying to a hard, infusible mass. Plaster of Paris, a calcium sulfate (gypsum) product which sets with extreme rapidity, is used for making artists' models and molding cores.

(2) Surgeon's plaster is an adhesive applied to a textile or plastic backing for use in treatment of skin abrasions. Some therapeutic plasters contain ingredients which stimulate blood flow to the affected area.

plastic. (1) Descriptive of a material that is easily deformed and retains its shape after deformation. Examples are modelling clays and paraffin wax.

(2) As a noun, a product of the plastics industry.

plastics. Collective term for high polymers (resins) to which various modifying materials have been added, so that the product can be manufactured in whatever degree of quality, hardness, color, and shape is desired. Synthetic resins are the basis of most plastics, but some are also made from modified cellulose, proteins, rubber, etc. The basic material plus added ingredients can by formed by calendering, extrusion, injection molding, foaming. etc. The more important types are derived ultimately from petroleum. Plastics can be broadly classified into two groups: (1) thermoplastic (those that soften when heated) and (2) thermosetting (those that retain their hardness when heated). During the last 50 years, manufacture of plastics has become a dramatically successful industry both in the U.S. and abroad. Not only are plastics used for innumerable utiltarian products, but they have expanded into many fields in which they have been successful substitutes for many natural materials: silk, cotton, animal glues, drying oils, metals, wood, rubber, and leather.

History. The first plastic was a mixture of cellulose nitrate and camphor invented in 1869 by John Wesley Hyatt; it was given the trademark "Celluloid". In 1899, Spitteler developed a method of hardening casein with formaldehyde and thus founded the casein plastics industry (small items such as buttons). The earliest high-

volume plastic, a condensation product of phenol and formaldehyde, was introduced by Leo Baekeland in 1907. Trademarked "Bakelite," it was the first truly synthetic high polymer. Its chief uses were as engineering material, as its dark color limited its application to items in which color was not a factor.

In 1927, cellulose returned to the scene with the development of ccllulose esters in the form of the acetate and butyrate produced as sheet, fibers, and molded items. At this time occurred an important mechanical development: compression molding began to give way to injection molding, a greatly improved technique in which the pelleted plastic is introduced to the mold in semiliquid form under pressure. This was a landmark in the growth of the industry.

During the period from 1920 to 1930, the German chemist Hermann Staudinger was deeply involved in fundamental research on polymerization mechanisms which was to profoundly affect the future of plastics. The importance of his work can hardly be overestimated.

Several notable events took place during the 1930's, when the field was crowded with significant developments that occurred almost simultaneously. One of these was the introduction of amino resins (catalyzed reaction products of urea or melamine with formaldehyde). These had a great advantage over phenol-formaldehyde because they made possible white or light-colored products suitable for scale housings and domestic items such as dinnerware. Two other achievements were due to the genius of Wallace Carothers, from whose research came neoprene (polychloroprene), the first approach to synthetic rubber (1930), and soon after the polyamide nylon, the first wholly synthetic fiber. Two other notable polymers were polystyrene (about 1932) and polyethylene, the latter developed in England (1939).

Other important developments were the technique of emulsion polymerization, in which two or more monomers can be co-polymerized; reinforced plastics (glass fibers impregnated with phenolic and other resins); foamed or ccllular plastics, both flexible and rigid; silicone resins, originally researched by Kipping and developed in the 1940's; the invention of block and graft polymers; and the extremely important stereospecific catalysts devised by Ziegler and Natta and based on the earlier research of Staudinger.

These permit selective control of polymer structure and resulted among other products in the development of polypropylene and polyisoprene (the first actual synthetic rubber).

The industry is represented by the Society of the Plastics Industry (250 Park Avenue, New York); the Society of Plastics Engineers (1942) located in Greenwich, Connecticut, is chiefly concerned with the engineering aspects of plastics technology.

plastic flow. See liquid; plasticity.

plasticity. The tendency of a solid to be deformed as a result of an imposed stress; the amount of force required to begin deformation is called the yield value of the material. If, when the stress is removed, the deformation remains, the material is said to exhibit plastic flow. This property is characteristic of waxes, putty, damp clay, plasters, and some metals, but not of many materials that are generally known as plastics. Special instances of plastic flow are cold flow, which occurs in vulcanized rubber after low stress has been continuously applied for a long time, and creep of metals.

plasticizer. A nonvolatile organic liquid of medium viscosity which is added to rubber and plastic mixtures to act as an internal lubricant and softener. The primary purpose of a plasticizer is to facilitate mechanical processing by increasing plasticity, thus enabling the mixture to flow easily through the gate of injection-molding dies; however, their effect persists to some degree in the finished product by relaxing the bonding between the polymer molecules, thus imparting permanent flexibility. Though hundreds of plasticizers have been developed, those most widely used are ethylene glycol derivatives, tricresyl phosphate, dibutyl phthalate, blown castor oil, etc. Camphor was the original plasticizer used by John W. Hyatt (1837–1920), an American chemist, in 1869 in making the first plastic, "Celluloid" (from cellulose nitrate). See also plasticity.

plastisol. A term used in plastics technology to denote a mixture of vinyl polymer (usually polyvinyl chloride) and from 30 to 50% of a plasticizer; this results in a thick, rubbery mass which is used as the basis for thermoplastic molded products. Various plasticizers are used, among them acetyl tributyl citrate. See also plasticizer.

plate glass. See glass.

platelet. See blood.

platinum. An element.

Symbol	Pt	Atomic Wt.	195.09
State	Solid	Valence	2,3,4
Group	VIII	Isotopes	5 stable
Atomic No.	78		

The most important of a closely related group of six metals (ruthenium, rhodium, palladium, osmium, iridium, and platinum) found in Group VIII of the Periodic Table, all of which are considered to be rare metals.

Their chemical and physical properties are similar, and they occur together in mineral deposits. Platinum is a grayish-white, ductile metal, m.p. 1769°C (3216°F), which is resistant to oxidation and tarnish; it tends to take up large percentages of hydrogen. It is one of the heaviest elements, with a density, 21.45g/cc, almost twice that of lead. It occurs in Ontario and Siberia, with minor deposits in Alaska and South Africa. It is quite reactive chemically, especially in powder form (platinum black), and is one of the most effective catalysts known for the reactions involved in making high-octane motor fuels (platinum sponge) by the reforming process. It has many applications in the electrical and medical fields and is also used in spinneret nozzles, special laboratory vessels, etc., usually in the form of alloys with other members of the group. The extremely high cost of platinum restricts its use to specialized applications.

plutonium. An element.

Symbol	Pu	Atomic No.	94
State	Solid	Atomic Wt.	239.11
Group	IIIB	Valence	3,4,5,6
	Actinide		
	Series		

A highly radioactive and fissionable element of the transuranium series, m.p. 641°C (1186°F), first produced in observable amounts in 1941 by bombardment of uranium with deuterons in a cyclotron. It exists in six allotropic crystalline forms. It is a much more efficient fissioning agent than uranium 235, which it has largely replaced as a nuclear reactor fuel. It is now made in a reactor from natural uranium by neutron bombardment of the 238 isotope to form plutonium 239. Its commercial use is chiefly in electric power reactors; it is extremely dangerous to handle, and operations must be done by remote control and with adequate shielding. *See also* fission.

plywood. *See* adhesive; composite; wood.

Pm Symbol for the chemical element promethium, the name being derived form the mythological hero Prometheus, reputed to have stolen fire from the gods and brought it to earth.

Po Symbol for the element polonium, named after Poland, the birthplace of its codiscoverer, Mme. Marie Curie, a French chemist (1867-1934), Nobel Prizes 1903, 1911.

poise. The standard unit for the viscosity of a fluid; the term is derived from the name of Jean Marie Poiseuille, French physiologist (1797–1869). A centipoise (cp) is one hundredth of a poise. *See also* viscosity.

poison. (1) A term used by organic chemists to denote a substance which contaminates a catalyst to the point where its activity is greatly reduced or destroyed altogether; such poisons are often present as impurities in the reactants. Arsenic and sulfur have this effect when adsorbed on the catalyst, for example, on platinum in the oxidation of sulfur dioxide to sulfur trioxide.

(2) In medicinal chemistry, a poison is a substance which exerts a detrimental effect on the life of an organism, often in minute dosage; the adverse effect may range from severe illness to death. Poisons are chemically active principles, often affecting the central nervous system; among the most severe are arsenic, various cyanides, alkaloids, phosphoric acid esters, carbon monoxide, and mercury compounds. Inorganic substances which cause death by asphyxia (water, carbon dioxide, etc). are not considered as poisons. *See also* toxicity.

polar compound. A molecule which has, or can acquire, electrical charges which enable it to conduct electricity. Many organic compounds are polar in nature, for example, most alcohols and amino acids, and many aldehydes. Among inorganic substances, water is strongly polar; it has a high dielectric constant and conductivity, and a slight tendency to ionize. Metallic salts, hydrochloric acid, and sulfuric acid, which ionize to form electrically conductive solutions, are also polar. Hydrocarbon solvents, carbon tetrachloride, and carbon dioxide are nonpolar, have low dielectric constant, and are poor conductors. *See also* dipole moment; dielectric constant.

polish. (1) Any compound or mixture applied to the surface of a material to give a brillant decorative finish or to remove minor imperfections. Glass is mechanically treated with rouge (red iron oxide) and for special purposes with heat (flame-polishing); metals and ceramic products, leather, wood, and many other materials are finished with a variety of mixtures containing waxes, solvent, and colorant.

(2) The outer coating or husk of cereal grains, which often contains valuable vitamins and is usually removed in processing, for example, rice polish.

pollutant. *See* contaminant.

polonium. An element.

Symbol Po	Atomic No. 84
State Solid	Atomic Wt. 210
Group VIA	Valence 2,4,6

Polonium is a radioactive element; the 210 isotope is the only form produced commercially; it emits alpha particles. It is produced by irradiation of bismuth with neutrons, with subsequent chemical treatment and concentration. It is a silvery crystalline metal, m.p. 254°C (489°F), which resembles lead and bismuth in physical properties more than it does tellurium, the element above it in the Periodic Table. It is used as a neutron source in nuclear reactors and as an alpha particle generator. Investigations as a potential power source have been carried out. Like other radioactive elements, polonium is a hazardous toxic agent and should be handled with proper protective equipment.

poly-. A prefix widely used in chemistry with the general meaning of "many" or "several". In some terms its meaning is more specific: a polymer is usually understood to contain at least five monomer units; a polyhydric alcohol contains three or more hydroxyl groups; a polyunsaturate has two or more double bonds. Mono-, di-, tri-, and tetra- are prefixes meaning, respectively, 1, 2, 3, and 4.

polyacetylene. An electrically conductive polymer of acetylene having alternate single and double bonds; its conductivity can be varied in either direction by doping with either electron acceptors (arsenic pentafluoride) or electron donors (lithium). Thus, it can be either an insulator, a semiconductor, or a conductor. It can be produced in the form of fibers and thin films. *See also* acetylene.

polyacrylonitrile. A polymer of acrylonitrile that is the basic material used in the manufacture of a number of synthetic fibers. When combined with other materials, it produces a hard resin having high solvent resistance and high-temperature stability; from these are made such items as moldings, shoe soling, wall panels, and the like. *See also* acrylonitrile; acrylic resin.

polyallomer. A special type of copolymer of two olefinic compounds, such as propylene and ethylene, in a ratio of about 20 to 1. It is not a uniform mixture, but consists of alternating segments of the two monomers similar to a block polymer. Ziegler stereospecific polymerization catalysts are used, each monomer being exposed to the catalysts separately and blended before the reaction is complete. Polyallomers have a constant crystalline structure, though their chemical composition is not uniform. Their properties are superior to those of dispersed mixtures of conventionally prepared polymers. *See also* block polymer; Ziegler catalyst.

polyamide. A class of high-polymer substances characterized by a $CONH_2$ grouping; it includes such naturally occurring materials as the proteins casein, found in milk, and zein in corn. From these, molded plastics, fibers, adhesives, and coatings can be made, though their use is no longer as extensive as formerly. The two most important synthetic polyamides are the urea-formaldehyde resins and the various forms of nylon. The former are used in white and colored plastic items, and the latter chiefly as a fiber. *See also* amide; nylon; urea-formaldehyde resin.

polyblend. *See* copolymer; homopolymer.

polybutene. *See* butene.

polycarbonate. A type of polyester resin resulting from the catalytic condensation reaction of bisphenol A with a carbonyl-containing compound such as diphenyl carbonate or phosgene (carbonyl chloride). The product is a linear polymer of carbonic acid ester represented by the structure

$$[-OC_6H_5-C(CH_3)_2-C_6H_5-O-C''O]_n.$$

Reaction techniques are by emulsion polymerization at room temperature, by direct reaction under vacuum with heat and catalyst (transesterification), or by polymerization in solution with an organic base as solvent, e.g., ethanolamine. Polycarbonates are thermoplastic and can be colored to any shade desired. They are excellent materials for a wide range of engineering applications. *See also* polyester; carbonate; engineering plastic.

polychloroprene. *See* neoprene.

polycondensation. *See* condensation.

polyelectrolyte. An ionically active, electrically conducting high polymer, either natural or synthetic, used for water conditioning, colloidal clay dispersion, inhibition of calcium carbonate deposition, etc. Polyelectrolytes may exhibit either anionic or cationic properties, and thus their applications depend to some extent on the individual substance. Examples of natural polyelectrolytes are gum arabic, carrageenan, and proteins; sodium polyphosphate is one of a number of synthetic types; ion-exchange resins function in the same manner. *See also* ion exchange.

polyester. Any of a broad class of synthetic resins made by a condensation reaction between a dihydric alcohol such as ethylene glycol and a dicarboxylic acid, e.g., phthalic or maleic. Some types, such as alkyd resins, require the presence of a drying oil. Polyester resins are unsaturated, cross-linked polymers which are made thermosetting by use of a peroxide-type catalyst. They are widely used for foams, fibers, films, and coatings; for laminates and molded parts; for encapsulating electrical wiring and electronic components; and for reinforced plastics for structural work such as boat hulls, building panels, etc. *See also* alkyd resin; polycarbonate; ester.

polyethylene. One of the most versatile and widely used types of plastics; it is also called polythene. It is an addition polymer synthesized by free radical-activated polymerization of ethylene $(C_2H_4)_n$, where the molecular weight may be as low as from 2000 to 4000; in so-called linear polyethylene, it is much greater, the macromolecule containing as many as 10,000 hydrocarbon units arranged in long coiled chains, i.e., amorphous or thermoplastic polyethylene, which may be either high or low density. Polyethylene can be cross-linked, either with a peroxide or by radiation, in which case it becomes thermosetting, with high strength and electrical resistance. A crystalline modification can be made by synthesis with sterospecific catalysts of the Ziegler type. The various forms have a multitude of uses as coatings, packaging films, molded parts, insulation, piping, etc. *See also* cross-linking; stereospecific.

polyethylene glycol. *See* ethylene glycol; antistatic agent.

polyglycol. *See* glycol.

polyhydric alcohol. An alcohol (often called a polyol) which contains three or more hydroxyl groups, each attached to a different carbon atom. They are water-soluble and of sweetish taste, which tends to intensify with increasing hydroxyl content. Examples are glycerol (3 OH), pentaerythritol (4 OH), and sorbitol (6 OH). Polyols may be high-boiling, viscous liquids or (in the case of the higher members) finely divided solids. The synthetic polymer polyvinyl alcohol, though not a typical alcohol, can be considered in this category; made by polymerization of vinyl acetate monomer and subsequent hydrolysis, it has useful properties as a resinous binder and coating agent. *See also* monohydric alcohol.

polyisoprene. A natural high polymer formed in certain species of trees and shrubs as a milky latex containing about 30% of polymerized rubber hydrocarbon $(C_5H_8)_n$. This structure was duplicated by synthetic means in the late 1950s. Polyisoprene has two geometric isomers (*cis* and *trans*); it is the *cis* form that is present in natural rubber, and the problem of reproducing it without contamination by the *trans* isomer, which destroys its rubber-like properties, was not solved until the discovery of the possibilities of stereospecific catalysts. *See also* isoprene; stereospecific.

polymer. A large molecule formed naturally or made synthetically either by (1) the chemical union of five or more molecules of a monomer (addition polymer) or (2) a series of condensation reactions between two or more chemicals, e.g., phenol and formaldehyde (condensation polymer). Proteins are natural polymers made up of amino acids joined by peptide linkages (CONH); cellulose is a natural carbohydrate polymer. Polymers are comprised of a large number of small molecules (monomers) linked together; since this union results in extremely high molecular weights, they are called high polymers. If only a few monomer units combine, the products are low polymers; when only two, three, or four molecules are present, the combinations are called dimers, trimers, and tetramers, respectively. Polymeric macromolecules are commonplace in organic materials, but there are a few cases of inorganic polymers, for example, silica, silane, and black phosphorus. *See also* polymerization; high polymer; monomer; addition polymer; condensation; polypeptide.

polymerization. A chemical reaction in which an indefinite number of small molecules unite or react to form a chain or network structure, which is a single molecule called a high polymer or macromolecule. Polymerization occurs naturally in plants, forming cellulose from sugars and starches, proteins and polypeptides from amino acids, and terpenes from hydrocarbons. The drying of certain vegetable oils is another instance of natural polymerization.

In the chemical industry, polymerization may be performed catalytically at temperatures and pressures ranging from normal to high. Synthetic resins, elastomeric dispersions (latexes), and certain types of gasoline and lubricants are made in this way; initiating agents (peroxides, free radicals, or short-wave radiation) are essential factors in the polymerization mechanism. One type (condensation) involves a multistep catalytic reaction of two chemical compounds, e.g., phenol and formaldehyde; in another (addition), identical molecules (monomers) of an olefinic substance such as styrene combine to form polystyrene. When molecules of two or more such substances are used, the result is a copolymer (styrene-butadiene). In many cases the reactions can be controlled by selective use of catalysts to give "tailor-made" molecular structures called stereospecific polymers. Polymerization techniques are quite varied, including gas phase, solution, batch or block, emulsion, and oxidative coupling. Some substances, notably acrylic acid, polymerize spontaneously at room temperature; inhibitors must be incorporated in handling and storage to prevent this reaction. *See also* polymer; high polymer; emulsion polymerization; Staudinger.

polyol. Contracted form of polyhydric alcohol, i.e., one that contains three or more hydroxyl groups.

polyolefin. A general term referring to any high polymer made by addition polymerization of unsaturated, straight-chain hydrocarbons by catalytic reactions at high temperature. They are thermoplastic resins; some are useful in their linear form, but cross-linking is generally required. Polyethylene is a typical example.

polyoxymethylene resin. *See* formaldehyde.

polypeptide. A sequence of amino acid units

connected by amide groups (CO • NH), known as peptide bonds. These result from the condensation of two amino acids in which water is split off; the remaining portion is called an amino acid residue, represented by —NH • CHR • CO—. The polymers so formed may contain from 2 to x amino residues, the chains being either straight, cross-linked, cyclic, or combinations of these structures. The molecular weight may range as high as 40 million (for bushy stunt virus). Both acidic and basic groups are present, and the products are thus amphoteric. Polypeptides are proteins in all essential respects. They are also called peptides. *See also* amino acid; protein.

polypropylene. A thermoplastic addition polymer of propylene monomer ($CH_3CH^.CH_2$), the reaction being catalyzed with aluminum alkyl, which has stereospecific activity; a crystalline structure characterized by helical-shaped molecules is obtained. Propylene may be copolymerized with other monomers such as styrene and forms a rubber-like terpolymer with ethylene and butadiene called EPDM rubber. Polypropylene is made in a number of forms (powder, film, fiber, etc.) and is readily molded. Its chief uses are in packaging, carpet fibers, athletic surfaces, and a broad range of molded components. Giulio Natta, an Italian chemist, Nobel Prize 1963, pioneered the development of this polymer.

polysaccharide. A broad group of carbohydrate polymers that includes cellulose, starches, and polymers of sugars such as amylose, galactose, mannose, etc., all of which are formed in plants or plant products. Among the subgroups are phycocolloids found in seaweeds, water-soluble gums produced by certain species of tropical trees, and mucilages (guar gum). The monomer units are monosaccharides (simple sugars) which can be derived by hydrolysis of the polymer. The term is derived from the Greek word for *sugar*. *See also* cellulose; gum; phycocolloid; sugar.

polystyrene. The classic example of an addition polymer (homopolymer). Polystyrene is produced from styrene monomer

$$(C_6H_5CH=CH_2)$$

in several degrees of polymerization; it is useful both as such and as a copolymer with butadiene and acrylonitrile. It is as historically important as the formaldehyde condensation polymers that initiated the plastics industry; it was introduced commercially about 1932. Polystyrene is a thermoplastic which has high electrical resistance and is easy to fabricate. It can be made moderately heat-resistant up to 70°C (160°F) and light-stable, and can be colored to any shade desired. It can be manufactured in sheet, bead, and solid forms and can be foamed, injection-molded, and extruded. It is widely used in toys and molded machine housings.

polysulfone resin. *See* bisphenol-A.

polytetrafluoroethylene. *See* fluorocarbon resin.

polythene. *See* polyethylene.

polyunsaturate. This term is usually applied to a type of unsaturated fatty acid found in vegetable fats and oils which has two or more double bonds, e.g., linoleic and linolenic acids. Foods containing these are considered less likely to cause cholesterol deposition than those containing saturated fatty acids (animal-derived). *See also* fatty acid.

polyurethane. A linear condensation polymer made by reaction of a diisocyanate and a dihydric alcohol, e.g., toluene diisocyanate, $CH_3C_6H_3(NCO)_2$, with ethylene glycol, the product being characterized by the urethane group (—NHCOO—). One type is used as elasticated fiber, generally known as spandex; cross-linked forms are used as organic coating bases for wood, masonry, and metals; further reaction with hydroxyl-containing polyesters and cross-linking agents produces excellent elastomeric products, including flexible and rigid foams. These have a broad range of uses both as industrial components and end products. *See also* prepolymer; diisocyanate.

polyvinyl. Any of a broad series of thermoplastic high polymers resulting from the addition polymerization of various monomers containing the vinyl group $CH_2=CH$. Among these are polyvinyl acetate (adhesives, paper and textile coatings, water-based paints); polyvinyl alcohol (water-soluble adhesive and emulsifier) derived from the acetate by reaction with alcohol; polyvinyl chloride (piping), polyvinyl ether (adhesives); and polyvinylidene chloride (food packaging, upholstery). *See also* polyvinyl chloride; polyvinylidene chloride.

polyvinyl acetate. *See* acetate; polyvinyl.

polyvinyl alcohol. *See* polyhydric alcohol; acetate; polyvinyl.

polyvinyl chloride. A thermoplastic addition polymer of vinyl chloride having the formula $(CH_2{=}CHCl)_n$. Readily processed and fabricated by standard methods, it is especially well suited for piping (subterranean, exterior, and interior) and for building components, toys, and containers of various types; a unique application is for surfacing athletic fields. It can be made in a number of forms, including foams and fibers, and is resistant to oxidation and attack by chemicals. It is often abbreviated PVC.

polyvinyl ether. See vinyl chloride.

polyvinylidene chloride. An addition polymer of vinylidene chloride having the formula $(CH_2{=}CCl_2)_n$. The monomer may also be copolymerized with others that are chemically similar to it, such as vinyl chloride. Plastics made from these materials are collectively called saran. One of their specialized uses is for ''shrunk-on'' transparent packaging film for meat products. They are also used for container linings, automobile upholstery, and fiber structures.

polyvinylidene fluoride. See fluorocarbon resin.

polyvinylpyrrolidone. See expander (2); plasma (2).

porcelain. A type of ceramic material which is chemically composed of potassium oxide (from feldspar), silicate (from quartz), and alumina (from clay). A thick, plastic paste is made by mixing these materials with water, followed by high-temperature firing. Porcelain has numerous uses as a refractory in chemical equipment, sparkplugs, electronic devices, etc. In the form of china, it has long been used for household dishes, pottery, and ornaments. When mixed with glass particles and fused on a metal surface, it is called porcelain enamel. See also ceramic; refractory.

poromeric. A term used to describe a material, usually organic in nature, which has a great number of extremely small pores per unit area. This is characteristic of human and animal skins, which transmit moisture from the body; it is also true of leather and some types of synthetic resins used as leather substitutes. These are particularly suitable for shoes, as the pores are so formed as to allow the passage of air and vapor from the inside outward. See also leather.

porphyrin. Any of a number of natural nitrogenous organic compounds characterized by four heterocyclic pyrrole groups, which have high pigmenting power. There are two central hydrogen atoms which can be replaced by metal atoms, forming coordination compounds. The porphyrin molecule has both strong pigmenting properties and catalytic activity. Two of the most important porphyrin derivatives are chlorophyll, where magnesium replaces the central hydrogens, and hemin, in which iron is the central metal atom; chlorophyll is a strong green pigment which catalyzes the photosynthetic reaction, and hemin is the red colorant of hemoglobin in blood cells. Phthalocyanine compounds, which contain benzenoid groupings, are derivatives of porphyrins and are the basis of an important series of synthetic pigments. Porphyrins can be synthesized from ammonia and methane in the presence of water, a fact which may account for the original formation of chlorophyll. See also pyrrole; phthalocyanine.

porpoise oil. See fish by-products.

Portland cement. See cement, hydraulic; concrete.

positron. See fundamental particle.

potash. A generic and commercial name for naturally occurring potassium salts; derived from ''pot ashes'' denoting the ancient (also early American) method of leaching wood ashes for their potassium carbonate content and concentrating the extract in iron pots.

potassium. An element.

Symbol	K	Atomic Wt.	39.098
State	Solid	Valence	1
Group	IA	Isotopes	2 stable
Atomic No.	19		1 radioactive

Potassium, m.p. 63.4°C (146°F), is one of the more reactive of the alkali-metal group. So-named from ''potash'' by Humphry Davy, the English chemist (1778–1829) who isolated the element in 1807 by electrolysis. Its salts, chiefly KCl, occur in many natural deposits; in the U.S., the most important are underground deposits in New Mexico, and in the dry beds of desert lakes in California (Searles Lake) and Utah (Great Salt Lake). The Stassfurt beds in Germany are the chief European source; enormous deposits are located in Saskatchewan, Canada, and the Dead Sea contains great amounts of potassium chloride.

Elemental potassium is dangerous to handle, as it reacts strongly with both air and water, evolving heat and flammable hydrogen to form potassium hydroxide (KOH). It readily combines with halogens and other elements, and its principal uses are in the form of these compounds, as fertilizers and explosives. It is a vital element in plant nutrition and occurs widely in soils; it also has important cellular and metabolic functions in animal organisms. In combination with sodium (NaK), it is useful as a heat-transfer agent. Among the more important compounds are the chloride (fertilizers); the carbonate, also called potash (glass technology); the nitrate and chlorate (explosives); and the permanganate (antiseptic). Many potassium compounds are also used in fireworks and as phosphors.

potassium nitrate. *See* deflagration.

potassium permanganate. *See* antiseptic; per- **(2).**

potentiator. A term used in the food and flavor industries to characterize a substance that intensifies the taste of a food product to a far greater extent than an enhancer. The most important of these are certain enzyme-like nucleotides such as riboflavin 5′-phosphate; the effective concentration is measured in parts per *billion,* whereas that of enhancers such as monosodium glutamate is in parts per thousand. The effect is thought to be due to synergism. Potentiators do not impart any taste of their own, but intensify the taste response to substances already present in the food. *See also* enhancer; flavor; nucleotide.

potting. *See* encapsulation.

pour point. A term used by lubrication engineers to denote the lowest temperature at which a lube oil will flow by gravity from a specified container under standard conditions. Since this point is largely dependent on the wax content of the oil, agents called pour-point depressants are added to reduce the melting point of the wax. Acrylic acid esters and alkylated naphthalenes have been used with some success.

powder. (1) Any finely divided solid material which has been precipitated by a chemical reaction or mechanically reduced by grinding; particle sizes range from less than 1 micron to several millimeters. Typical powders are carbon black, precipitated barium sulfate (blanc fixe), and fine-ground clays, whitings, and talcs. Many metals are comminuted to powders by various methods for use in making compressed products having microscopically small pores (powder metallurgy); they are also used for catalysts (e.g.,

nickel) and, when suspended in solvents, for decorative paints.

(2) In the explosives industry, the term refers to a nondetonating mixture of potassium nitrate and charcoal (black powder or blasting powder).

(3) In the cosmetics field, face and bath powder are specially prepared magnesium silicates (talc).

(4) In plastics technology, molding powders are precompounded resins prepared in powder or pellet form ready for the molding process.

powder metallurgy. *See* sinter; powder **(1)**; tungsten.

Pr (1) Symbol often used in chemical formulas for the propyl group, C_3H_7.

(2) Symbol for the element praseodymium, from the Greek *prasinos* (''green'') and *didymos* (''twin''), denoting the green color of salts of the element.

praseodymium. An element.

Symbol	Pr	Atomic No.	59
State	Solid	Atomic Wt.	140.9077
Group	IIIB	Valence	3,4
	(Lanthanide Series)		

A member of the rare-earth group of elements, praseodymium is a rather soft crystalline solid; m.p. 935°C (1715°F), obtained from monazite sands. The metallic form is similar to iron in color, and many of its compounds have a pronounced greenish shade. It has limited use in ceramics and glass manufacture and as a catalyst and phosphor. It has a number of specialized applications in metallurgy.

precipitate. Solid particles formed as a product of a chemical reaction which can be separated from the reaction medium by gravity settling (including cyclone separators, which intensify the force of gravity), centrifuging, coagulation, and filtration. The formation of precipitates (precipitation) is widely used in the laboratory and in industry to recover desired materials. A precipitate also forms when the solubility limit of a substance is reached by cooling (crystallization) or by removal of the solvent from the reaction system (evaporation). The medium is most commonly liquid, e.g., water, but may be gaseous as when ammonia and hydrogen chloride react to form ammonium chloride, or when zinc vapor reacts with the oxygen in air to produce zinc oxide. The production of most inorganic and many organic chemicals involves the formation and separation of a precipitate at some stage in the process. A precipitate can also be liquid particles formed when a vapor condenses, e.g., rain, which when further cooled forms a precipitate of snow or sleet. *See also* sedimentation; Stokes' Law; suspension.

precipitator. A device used in factory stacks which utilizes an electrostatic field to introduce an electric charge onto solid or liquid suspended particles in a gas stream, so that the particles migrate to one electrode from which they can be collected. This is commonly called the Cottrell precipitator after its inventor, Frederick G. Cottrell, an American chemist (1877–1949), and its efficiency of recovery is in the 90 to 99.9% range. It is not only applied in the prevention of air pollution, but in the recovery of valuable constituents from solid and liquid dusts, fumes, and mists from sulfuric and phosphoric acid plants; roasting, sintering, and calcining operations; blast furnaces; fluid catalyst regenerators; pulp mill chemical recovery furnaces; carbon black plants, etc.

precursor. A term used by biochemists to denote an organic compound occurring naturally in plants or animals which is comparatively inactive until subjected to an external stimulus; when this occurs, the compound assumes an active form which has a definite function, e.g., a vitamin, often as a result of enzymatic activity. Examples are the conversion of prothrombin to thrombin in the blood brought about by release of blood from the circulatory system, which induces clotting; and the activation of ergosterol to calciferol (vitamin D) by ultraviolet light. In these cases, prothrombin and ergosterol are precursors (inactive forms). The term literally means "running before."

preferential. Selectivity of action, either chemical or physicochemical, exhibited by a substance in contact with two other substnaces; it may be due either to chemical affinity or to surface phenomena. An example of a preferential chemical combination is that of hemoglobin for carbon monoxide, with which it unites 200 times as readily as with oxygen when exposed to a mixture of the two; such phenomena as adsorption, corrosion, and the wetting of one material by another are subject to this selectivity.

prepolymer. A liquid, partially polymerized mixture of a polyhydric alcohol and an isocyanate prepared for on-the-spot formation of urethane foams. The mixture is supplied to the user together with a second mixture containing the blowing agent, catalyst, and alcohol; the foam is formed when the two mixtures are blended. The term also applies to other thermosetting resins, such as diallyl phthalate. *See also* polyurethane.

prepreg. A shortened form of the term *preimpregnated,* used by plastics engineers to refer to a material used as the basis for a composite, such as glass fiber or a textile, which has been coated or impregnated with a thermosetting resin prior to assembly and curing.

preservative. Any substance that prolongs the useful life of an organic material, usually in low concentrations. Though generally applied to food additives such as benzoic acid, acetic acid, and ascorbic acid, which have antimicrobial and antioxidant value, the term applies equally well to antioxidants used in petroleum products, rubber, and similar materials. It also includes the various alcohols and aldehydes used in preventing deterioration of biological specimens, e.g., formaldehyde, though in this case a high concentration is necessary, either by injection or immersion.

Priestley, Joseph (1733–1804). Born near Leeds, England, Priestley originally planned to enter the ministry. As a youth, he became interested in both physics and chemistry and his research soon established his position as a scientist. He discovered nitrous oxide (N_2O) in 1772, but his greatest contribution to science was his discovery of oxygen in 1774; in later years he emigrated from England to Pennsylvania, and his research in America resulted in the discovery of carbon monoxide. In 1874, at a meeting at Priestley's old home in Pennsylvania convened to celebrate the centennial of the discovery of oxygen, plans were laid to organize the American Chemical Society, which later established the Priestley Medal, one of the highest awards in chemistry.

prill. A spherical pellet or cylindrical solid form in which some materials are supplied for convenience in handling. Some types of fertilizers and explosives such as ammonium nitrate are made in this form.

primary. (1) A monohydric alcohol in which one of the hydrogen atoms attached to the methanol carbon is replaced by an alkyl group; if the carbon chain is not branched, the alcohol is called normal. Examples are ethyl alcohol (CH_3CH_2OH) and normal butyl alcohol, $CH_3(CH_2)_2CH_2OH$. The term is also applied to amines and numerous other compounds in an analogous manner.

(2) An irreversible electric battery such as a dry cell.

(3) In reference to metals and other mineral products, primary means original production from the source, as opposed to recovery of metals from scrap or tailings, or (for petroleum) by water pressure. *See also* secondary; tertiary.

printing ink. *See* ink.

process. As used in chemical engineering, this term indicates an industrial production method in which a chemical or physicochemical change occurs, as distinguished from an operation, in which there is no chemical change. Among the chemical reactions involved in processes are oxidation, reduction, ionization, polymerization, hydrogenation, halogenation, hydrolysis, and saponification. Thus, petroleum cracking, pulp preparation, rubber vulcanization, leather curing, cooking of food, dyeing of fabrics, and soapmaking are all processes; these and a few other industries of similar type are collectively referred to as the process industries. Operations include comminution, mixing, mechanical coating, and numerous separation methods (distillation, filtration, evaporation, etc). In common usage, the terms tend to overlap, and the distinction is often blurred.

producer gas. A manufactured mixture of gases made by incomplete combustion of coal or by contact of steam with hot coke; carbon monoxide and hydrogen comprise about 40% of the gas, most of the remainder being nitrogen. It is used chiefly as a low-cost fuel, usually on-site. *See also* water gas; synthesis gas.

progesterone. A steroid hormone produced in the body of the female during pregnancy; formula $C_{21}H_{30}O_2$. It is extracted from an ovarian ductless gland of pregnant animals and can also be synthesized. It is also able to inhibit ovulation when injected but has little effect when taken orally. *See also* antifertility agent.

proliferation. A term used in biochemistry to denote production within a plant or animal of specific types of cells or chemical compounds; it implies the idea of abundant multiplication, as occurs in the budding of plants, bacterial growth, and formation of diseased tissue cells.

promethium. An element.

Symbol	Pm	Atomic No.	61
State	Solid	Atomic Wt.	147(?)
Group	IIIB	Valence	3
	(Lan-		
	thanide		
	Series)		

A radioactive rare-earth metal; its longest-lived isotope (145) has a half-life of 18 years. The element has been isolated only in trace quantities. The 147 isotope with half-life of 2.6 years is the only one that has practical use, as in luminescent coatings for instrument dials and as a power source for semiconductor solar batteries for space use. Promethium forms only a few compounds, which are of chiefly theoretical interest.

promoter. An element or compound which increases the surface activity and selective effect of a catalyst; the promoting substance is incorporated in the catalyst in low percentages during commercial preparation. Examples are inclusion of alumina in the iron catalyst for ammonia synthesis and potassium with iron or chromium oxide for dehydrocyclization. The promoter alone has no catalytic action. *See also* catalyst.

proof. An arbitrary term used to designate the volume content of ethyl alcohol in a spiritous liquor or an industrial alcohol. It is twice the percentage of ethyl alcohol present at room temperature; thus, 100-proof alcohol contains 50% of ethyl alcohol and 86-proof alcohol contains 43%, both on a volume basis.

propellant. (1) A term used in aerosol technology to denote a gas used to drive or push various materials through an orifice in the form of a spray or foam. A simple example is a perfume atomizer where air expelled by pressure on a bulb disperses the liquid into droplets as it leaves the nozzle. Propellants of this type are gases such as carbon dioxide, air, nitrous oxide, butane, and propane. They are used to disperse paints, whipped cream, insecticides, and numerous household cleansers and purifiers.

(2) In combustion technology, the term refers to rocket fuels of various types, among which are liquid hydrogen, hydrogen peroxide, hydra-

zine, boron hydride, etc., and solids such as plasticized nitrocellulose, ammonium nitrate, etc.

property. Any of a number of physical and chemical characteristics that establish the identity of an element or compound. Important among these are the following:

state (gas, liquid, solid)
color
odor
taste
boiling point
melting point (solids)
freezing point (liquids)
atomic weight (elements)
molecular weight (compounds)
refractive index
solubility (in water and organic solvents)
combustibility
toxicity
specific gravity (density)
reaction with standard reagents
energy absorption and emission (infrared, ultraviolet, etc.)

See also analysis.

propionic acid. *See* acidulant.

propyl. The univalent group C_3H_7—, the third member of the homologous series of paraffinic hydrocarbons; it is derived from propane by dropping one hydrogen atom and may appear in formulas as $CH_3CH_2CH_2$—. When a second hydrogen is dropped from propane the divalent propylene group is formed (—C_3H_6— or —$CH_2CH_2CH_2$—), which represents the corresponding olefin.

propylene. *See* isopropyl alcohol; isotactic.

propylene oxide. *See* allyl alcohol; methylcellulose.

prostaglandin. One of a number of biologically active derivatives of certain unsaturated fatty acids whose molecules are basically composed of a chain of 20 carbon atoms. Occurring in all tissues of the body, prostaglandins are biosynthesized with the aid of certain enzymes and function as metabolites. They are known to have marked effect on the circulatory system, reducing blood pressure in very low concentrations; they also stimulate smooth muscle contraction and are thought to have critical functions in both male and female reproductive processes. Active

research is continuing on their possible medical and pharmaceutical applications, as well as on their sources; their presence in unusually high concentrations in sea organisms living in coral beds has been discovered. *See also* metabolite.

prosthetic. An active chemical group or metabolite which promotes essential chemical reactions and transformations within the body, for example, oxidation-reduction, decarboxylation, phosphorylation, etc. The term actually means "replacement of a missing part (of the body)," and is used in this sense in medicine. In biochemistry, the meaning is more sophisticated; prosthetic groups function by "fitting in" to active sites afforded by the configurations of molecular complexes. A mechanical analogy would be the teeth on a gear, where each tooth makes an active but temporary connection with a mating gear by locking into an aperture. Some types of prosthetic groups are composed of metal ions bound to porphyrin structures, their effect being produced, at least in part, by chelation, e.g., chlorophyll. Other types are coenzymes, which are organic groups associated with enzymes but easily separated from them. Vitamins of the B complex function by prosthetic mechanisms. *See also* coenzyme; metabolite; Krebs cycle; B complex.

protactinium. An element.

Symbol	Pa	Atomic Wt.	231.0359
State	Solid	Valence	4,5
Group	IIIB	Isotopes	2 natural
	(Acti-		radio-
	nide		active
	Series)		
Atomic No.	91		

A radioactive element whose longest-lived isotope is 231, with a half-life of about 10^6 years; it occurs naturally in trace quantities in uranium ores and is about as abundant as radium. The 233 isotope is made artificially by decay of a thorium isotope. It is a crystalline, whitish metallic solid (density 15.37 g/cc), but there is no present industrial use. Only a few compounds are known.

protective coating. A general term including all types of surface coatings applied primarily to prevent or inhibit deterioration due to oxidation, corrosion, bacterial attack, and the like. These are of two broad groups: (1) organic coatings, which include paints, enamels, varnishes, and lacquers, and (2) inorganic coatings, which are layers of metal of various thickness, the thinner ones being electroplated and the thicker ones mechanically applied. The metals most commonly used for protective purposes are chromium, copper, tin, nickel, and zinc. Glass and porcelain enamels are also widely used. *See also* paint; lacquer; galvanizing; electroplating; cladding.

protective colloid.   A hydrophilic high polymer whose particles (molecules) are of colloidal size, such as a protein or gum; it may be either naturally present in such systems as milk and rubber latex, or intentionally added to mixtures to prevent coagulation or coalescence of the particles of fat or other dispersed material. Protective colloids are also called stabilizing, suspending, or thickening agents; they also act as emulsifiers. Examples are (a) a thin coating of protein on hydrocarbon particles of latex which keeps them from cohering as a result of the impact due to their Brownian motion; (b) gelatin, sodium alginate, or gum arabic which are added to ice cream, confectionery, and other food products to obtain a smooth, creamy mixture. They are readily adsorbed by the suspended particles and reinforce the protective effect of proteins that may be naturally present. *See also* gelatin; gum; emulsion.

protein   (pro-te-in). Any of a large class of natural polymers, often having a molecular weight of a million or more; such macromolecules are made up of amino acid units linked together with amide groups to form a chain-like polypeptide structure of colloidal dimensions. The chains are sometimes crosslinked with disulfides, as in wool. As proteins contain both acidic (COOH) and basic (NH_3) groups, they react in either way and are thus amphoteric. Simple proteins may have only a few amino acid units, but in more complex types there may be several thousand. The sequence in which the amino acids occur in protein molecules is controlled by the nucleoproteins DNA and RNA and is of vital significance in genetic mechanisms. A nucleoprotein is a combination of a protein with a nucleic acid.

Proteins are synthesized by plants from nitrogen fixed by bacteria and from the carbohydrate products of photosynthesis. Though animals can elaborate special types of proteins, they depend on ingestion of plant forms for most of their protein requirements. Proteins are major components of such biologically active complexes as hormones, enzymes, genes, bacteria, and viruses. They are hydrolyzed to their constituent amino acids by proteolytic enzymes of various types. Meat, fish, eggs, milk, and some vegetables are good dietary sources of proteins. Common protein-rich structures are hair and hide (wool, leather), cartilage (glue, gelatin), egg white (albumin), muscle, nails, skin, and blood. *See also* polypeptide; nucleic acid; nucleoprotein, genetic code; amino acid.

proteolytic.   A term used in biochemistry to describe enzymes which have the ability to decompose proteins to their component amino acids, usually by a hydrolytic reaction. This occurs, for example, in the digestion of foods which is catalyzed by the enzyme pepsin, present in gastric juice. Other enzymes of this type are trypsin and papain. *See also* enzyme.

prothrombin.   *See* anticoagulant; precursor.

proton.   The basic unit of mass which is a constituent of the nucleus of all elements, the number present being the atomic number of a given element; the hydrogen nucleus contains but 1 proton (atomic number 1), and the uranium nucleus contains 92 (atomic number 92). The proton has a positive electric charge resulting in a strong repulsion force within the nucleus, which is counterbalanced by the binding energy. A proton may be regarded as a positive hydrogen ion (H^+); it plays an important part in acid-base reactions and chemical bonding. *See also* neutron; electron; binding energy; fundamental particle.

protoplasm.   A complex mixture of components of plant and animal cells. The more diffuse portion lying between the nucleus and the cell wall or membrane is called cytoplasm. The exact composition of protoplasm is uncertain though amino acids, proteins, phosphorus, and ribose sugars are known to occur. *See also* cell (2); nucleotide.

provitamin.   The inactive form of a vitamin, known as a precursor, which is converted to an active form by catalytic mechanisms in the liver (carotene), or by ultraviolet irradiation (ergosterol), forming vitamin A and vitamin D_2 (calciferol), respectively. *See also* precursor.

Prussian Blue.   *See* Iron Blue.

prussic acid.   *See* hydrogen cyanide.

psi. Abbreviation used by chemical and materials engineers for pounds per square inch; lb/in² is also frequently used. In reference to atmospheric and steam pressures, psia (pounds per square inch absolute) and psig (pounds per square inch gauge) are standard abbreviations.

psychotropic agent. *See* drug; tranquilizer; narcotic; hallucinogen.

Pt Symbol for the element platinum; the name is thought to be derived from the Spanish word *plata*, meaning "silvery."

ptomaine (toe-ma-in). Any of a number of extremely poisonous substances derived from decomposition products of proteins formed in animal bodies after death. These toxins are a serious risk in improperly refrigerated meats, especially fish and chicken. A typical ptomaine is cadaverine, $NH_2(CH_2)_5NH_2$. Ptomaine poisoning is a loose term for any type of food poisoning, whether or not it is due to ptomaines.

Pu Symbol for the element plutonium, named after the outermost planet Pluto, by analogy with its location in the solar system. At the time of its discovery, plutonium was thought to be the last of the transuranic elements.

pulping. A term used primarily in the paper industry to refer to any of a number of methods of preparing woody fibers for the papermaking operation. The chief purposes of pulping are the separation and removal of lignin from cellulosic components of the wood, and softening of the fibers. These are accomplished in the following ways: (1) chemical pulping, which involves digestion or cooking of softwood pulp in either an acidic (sulfite pulp) or a basic medium (sulfate or kraft pulp); (2) semichemical pulping, designed chiefly for hardwoods, in which chemical and physical methods are combined; and (3) mechanical or groundwood pulping, mainly for newsprint, in which the pulp is formed by abrading the wood with a rotating stone. The kraft or alkaline process is the most widely used; a relatively new process called holopulping, which does not require use of sulfur compounds, is becoming competitive. *See also* paper.

purine. A fused-ring, basic nitrogenous compound ($C_5N_4H_4$) which can be isolated from uric acid. Many substances containing this structure occur in the metabolic end-products of plants and animals, for example, the alkaloids caffeine and theobromine, guanine, xanthine, etc. Purine

itself is a derivative of pyrimidine, a benzenoid compound having two nitrogen atoms in the ring.

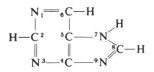

Purine compounds are constituents of nucleic acids, from which they can be obtained by hydrolysis. *See also* uric acid.

purple. (1) Purple of Cassius is the trivial name of a variable mixture of tin oxide and colloidal gold used as a colorant in the manufacture of ruby glass and decoration of ceramic ware.

(2) Visual purple is an alternative name for rhodopsin, a pigment present in the retina which responds to red radiation. *See also* rhodopsin.

putty. *See* linseed oil; sealant; plasticity.

PVA. Abbreviation for polyvinyl alcohol.

PVC. Abbreviation for polyvinyl chloride.

PVT. Abbreviation used by chemical engineers for pressure-volume-temperature, the chief physical factors affecting rate processes in large-scale chemical reactions.

pyridoxine. *See* B complex.

pyrimidine. *See* barbiturate; purine; nucleotide.

pyrites. *See* sulfur dioxide; sulfuric acid.

pyrolysis. A chemical change brought about by heat alone; in the case of inorganic materials (limestone or metal ores) it is synonymous with calcination. It is distinguished from combustion by the fact that oxidation does not occur. It is usually applied to the conversion of organic compounds not only into smaller and simpler compounds but also into substances of higher molecular weight than the original compound. Thus, its meaning is not limited to "thermal decomposition." The temperatures involved may be as high as 1370°C (2500°F), as in the formation of carbon and hydrogen from methane; usually they range from 430 to 820°C (800 to 1500°F), in which hydrocarbon gases are thermally cracked to olefins (ethane to ethylene, propane to propylene, etc.), with evolution of hydrogen. One type of artificial graphite is made by pyrolysis; it is also applied to conversion of organic wastes to fuel gases. *See also* calcination.

pyrophoric. Literally meaning "fire-bearing," this term denotes a substance that ignites in air at or below room temperature without supply of heat, friction, or shock. Typical of such substances are phosphorus and lithium hydride. A number of substances such as sodium, potassium, butyllithium, etc., are spontaneously flammable in contact with humid air, as they react exothermally with water. Pyrophoric materials must be protected from air and moisture by storage in an atmosphere of inert gas or under kerosine. Some alloys (misch metal) are called pyrophoric because they will ignite as a result of friction.

pyrotechnics. The compounding, manufacture, and engineering of fireworks, signal flares, etc.; it involves the use of potassium nitrate, chlorate, dichromate, phosphorus, magnesium, and numerous other flammable and explosive compounds which emit colored light when deflagrated.

pyroxylin. *See* nitrocellulose.

pyrrole. A hereocyclic nitrogenous compound, b.p. 131°C (258°F), having the composition represented by C_4H_5N; it is an unsaturated five-sided ring which is considered to be a resonance hybrid. It is a constituent of indigo and other natural pigments and also is the source of the phthalocyanine group of colors. Pyrrole groups are present in both chlorophyll and hemoglobin. The natural product can be obtained from coal tar and is also synthesized catalytically from ammonia and furan. Pyrroles readily combine with such metals as iron, copper, and nickel, which accounts for their importance in the formation of such biochemically essential complexes as porphyrins. *See also* porphyrin.

pyruvic acid. *See* Krebs cycle.

Q

quad. An energy unit that has come into use in recent years in predicting future energy requirements on a national basis. One quad equals 10^{15} Btu, which is the energy equivalent of 10^{12} cubic feet of natural gas, or 182 million barrels of oil, or 42 million tons of coal, or 293 billion kw-hr of electricity.

qualitative analysis. Examination of a sample of a material to determine the kinds of substances present (that is, whether it is an element, a compound, or a mixture) and to identify each constituent. Often this can be done by observation or testing of physical properties, but chemical separation techniques are usually necessary. In advanced analytical work, instrumental methods such as various types of spectroscopy and chromatography are essential.

quantitative analysis. Examination of a sample of a material to determine the amount or percentage of its constituents. Conventional methods of volumetric and gravimetric determinations are still used for routine work, but instrumental methods are required for precise evaluation. The introduction of such techniques as gas chromatography and mass spectrometry has made continuous analyses possible.

quantum. The smallest possible unit of electromagnetic energy, represented by the photon. Light is transmitted by wave-trains of photons; when these units (quanta) are absorbed or emitted by atoms or molecules exposed to exciting radiation, they yield line or band spectra characteristic of the element or elements present. Quantum or wave mechanics developed around the turn of the century from the theories of electromagnetism advanced by Maxwell, Planck, and Einstein, ultimately proved that matter is essentially a form of energy and can be transformed into it, as occurs in nuclear fission and fusion.

In chemistry, this concept was applied to electrons by Schrödinger, Heisenberg, De-Broglie, and others; by 1927 it had been established that an electron is an electromagnetic wave function rather than a solid particle and that its energy state is defined by four quantum numbers. This discovery gave rise to the orbital concept, which has provided a better understanding of the electronic structure of the elements, chemical bonding, photochemical reactions, catalysis, resonance and other chemical phenomena. *See also* binding energy; photon; radiation; orbital; electron.

quartz. A hard crystalline form of silica (SiO_2) which occurs freely in natural rock formations; it is usually white to grayish but may have a pink coloration due to traces of dispersed iron. Quartz crystals are also produced artificially for special purposes. When melted it is called fused silica, a form of glass used in power tubes, lamps, and electronic devices.

quaternary ammonium compound. A versatile group of ionizing compounds derived from reactions between amines and alkyl- or aryl-containing materials. They comprise a nitrogen atom to which four organic groups (R) are attached (the cation) and a negatively charged group or atom (the anion); thus, the molecule would appear as

The cations formed on ionization are powerful bases which are water-soluble. Their most distinctive property is their bactericidal and disinfecting ability; they are readily adsorbed on surfaces, and their electrical conductivity makes them useful as antistatic agents on fabrics. They are also excellent dyeing auxiliaries, wetting agents, demulsifiers, and accelerators of photographic development. *See also* antistatic agent; disinfectant.

quebracho. *See* tanning.

quick-. An obsolete word meaning "living" or "active," it survives as a prefix in such trivial chemical terms as *quicklime* (calcium oxide) and *quicksilver* (mercury), as well as in *quicksand*.

quicklime. *See* calcium oxide.

quinine (pronounced cwy-nine). An antimalarial medicine obtained from the bark of the cinchona tree (western South America and India). It is an alkaloid product with a repulsive taste and may be toxic unless taken in prescribed dosage. It has largely been replaced by synthetic antimalarial drugs.

quinoline. A bicyclic compound, b.p. 238°C (460°F), occurring in coal tar; its molecule (formula C_9H_7N) consists of a benzene ring fused to a heterocyclic ring containing nitrogen. It is made synthetically from aniline. A strong base, it is quite poisonous and is used as a preservative, in the manufacture of dyes, and to a limited extent as a medicinal.

quinone. Any of several cyclic compounds derived from benzene but having a different arrangement of double bonds. The quinones are strongly colored (red or yellow) compounds as a result of a characteristic chromophore grouping; two atoms of oxygen are double-bonded to the ring in either the para- or the ortho- locations ($C_6H_4O_2$). Quinones are used as dye intermediates and fungicides. They are also the parent compounds of hydroquinones, $C_6H_4(OH)_2$, used in photography as developing agents, as antioxidants for rubber and fatty organic compounds, and as reaction inhibitors. Both quinones and hydroquinones are quite poisonous.

R

R **(1)** Symbol used in generalized formulas of organic compounds to represent a hydrocarbon (alkyl or aryl) group, e.g., CH_3, C_2H_5, C_6H_5, etc. For example, R • OH indicates an alcohol and R • O • R, an ether.

(2) The letter R with a superior dot (R⋅) is used to indicate a free radical, the dot indicating its free valence.

(3) The universal gas constant, R, equal to $p_o v_o/273$. If p_o is one atmosphere and v_o is the volume of one mole, expressed in liters, the numerical value of R is $22.4/273 = 0.08057$ liter-atmosphere per degree C.

Ra Symbol for the element radium, the word being derived from *radius*, a light ray.

racemic. An organic compound in which equal parts of dextro- and levorotatory isomers are blended; as a result, the molecule is not optically active, as the two opposite rotations cancel each other. This is often called external compensation, as opposed to the internal compensation occurring in meso compounds such as tartaric acid, in which 2 asymmetric carbons are naturally present. Racemization can be induced by converting an optically active compound into an optically inactive form by heating or chemical reaction, in which half of the optically active substance becomes its mirror image (enantiomorph).

radiation. The terms *radiant energy* and *light* are essentially synonymous with radiation. Radiation is electromagnetic energy composed of waves of massless units (photons or quanta) emitted as a result of nuclear transformations in astral bodies and moving through space at constant speed (186,000 miles per second). The wavelengths range from a fraction of an angstrom unit (gamma radiation) to several miles for radio and electric waves. The classic mathematical evaluation of radiation is that formulated by Planck in 1903, $E = h\nu$, where ν is frequency and h is a universal constant. The total range of wavelengths comprises the electromagnetic spectrum, which includes the following subdivisions, from short to long: cosmic, gamma, x-rays, ultraviolet, visible, infrared, microwave, and radiofrequencies. Less specifically, the term is also applied to particulate emanations of unstable or split atomic nuclei. These include high-energy electrons (beta), protons or helium nuclei (alpha), and (in the case of fission) neutrons. Gamma radiation is also emitted in the decay of uranium and radium. The many types of spectroscopy utilize radiation of various wavelengths; it is also used in medicine (radioactive isotopes), in industrial measuring devices, and in communications equipment. *See also* photon; ionizing radiation; quantum; beta particle; alpha particle; Planck's constant.

radical. This term strictly denotes any ionized group of atoms characteristic of dissociated inorganic compounds, which behaves as a single unit. Among the more common radicals are the hydroxyl $(OH)^-$, the nitrate $(NO_3)^-$, the ammonium $(NH_4)^+$, the sulfate $(SO_4)^=$, and the carbonate $(CO_3)^=$. It is also (but less correctly) used in the organic field to designate alkyl, aryl, and other groups which are not charged. A specialized meaning includes the active molecular fragments known as free radicals (indicated by R⋅) which initiate reactions by a chain mechanism. *See also* free radical; group; chain reaction (1).

radioactive isotope. An unstable form of an element (also called a radioisotope) whose major isotopes are stable. For example, the 14 isotope of carbon is naturally radioactive, though carbon is a stable element, as less than 1% of its atomic weight is represented by the unstable isotope. Similarly, bismuth has five radioactive isotopes, and potassium, platinum, and lead have one or more. Carbon-14 is used as a tracer and in chemical dating. The nuclei of many stable elements ' are made artificially radioactive by neutron irradiation in a nuclear reactor; these are used for curative purposes in medicine and for specialized industrial uses such as detection of leaks and various types of measuring devices. *See also* radioactivity.

radioactive waste. *See* waste control.

radioactivity. Energy continuously emitted, often over very long periods of time, by an unstable atomic nucleus; it may be natural, as with radium, artificial (caused by bombardment of a stable nucleus with neutrons or deuterons), or induced, as in radioactive carbon. The emanations are in the form of alpha, beta, and gamma rays. The natural radioactive elements are uranium, radium, radon, and thorium (the principal members of the uranium decay series), the ultimate end-products being stable isotopes of lead, and helium. The radiation is useful in cancer therapy, as it quickly destroys diseased tissue; it is also damaging to healthy tissue and must be used with the utmost care. A number of elements, e.g., sodium, iodine, etc., can be made radioactive by bombardment with neutrons, deuterons, or other heavy particles (artificial radioactivity). *See also* radioactive isotope; radiation; decay; half-life.

radiocarbon. Shortened form of radioactive carbon.

radiocarbon dating. A method of determining quite accurately the age of a carbon-bearing material derived from living plants or animals within the last 70,000 years. It is based on the ratio of carbon-14 in the material to that in a modern reference sample by measuring the radioactivity of the carbon-14 in the material. Since the half-life of carbon-14 is 5730 ± 30 years, and the living precursor utilized CO_2 from the atmosphere or some other part of the earth's dynamic carbon reservoir, a process that ceased when the original plant or animal died, the amount of carbon-14 now present gives directly the age of the material. The carbon-14 in the reservoir is constantly being replaced by the sequence: $N^{14} \rightarrow C^{14} + O_2 \rightarrow C^{14}O_2$. This has maintained the ratio of carbon isomers constant during the ages, but burning of fossil fuels since the Industrial Revolution has lowered somewhat the quantity of carbon-14 in the atmosphere during the last few centuries, an effect that does not affect measurements on older objects. The sample to be tested must be carefully prepared to prevent contamination by younger carbon.

The radiocarbon technique was discovered by Willard F. Libby (1908–1980, Nobel Prize 1960) and has been applied with great success in the fields of archeology, geology, geochemistry, and geophysics. Its accuracy has been checked and verified by use of tree-ring counts (dendrochronology) and with the known ages of objects from ancient cultures, such as Egyptian and Chinese. The former shows that, for the 2400–6000 B.P. (before present) age of bristlecone pine tree-rings, 5200 C^{14} years equal 6,000 calendar years. *See also* decay; half-life.

radiochemistry. A subdivision of chemistry devoted to the study of radioactive substances. It includes the nuclear transformations involved, transmutation of one element into another, and the nature and properties of the radiation emitted. It also deals with the use of this radiation in chemical tracer analysis, for geological and archeological research (chemical dating), and for initiation of cross-linking (polymerization). *See also* nucleonics; dating; tracer element; radioactivity.

radioisotope. Shortened from of radioactive isotope.

radium. An element.

Symbol	Ra	Atomic No.	88
State	Solid	Atomic Wt.	226.0254
Group	IIA	Valence	2

A strongly radioactive element, m.p. 700°C (1292°F), occurring in extremely small percentages in uranium ores (pitchblende and carnotite), found in Zaire, northern Canada, Colorado, and USSR. It was first isolated in 1898 by the brilliant analytical research of the Curies, Pierre (1859–1906) and Marie (1867–1934), French chemists and physicists, Nobel Prizes

1903 and 1911. Detection of the nuclear decay of radium, as indicated by the emanation of alpha, beta, and gamma radiation, was in part responsible for the revolution in physics that occurred between 1895 and 1910, for it had previously been thought that atoms were permanent and indestructible entities. Radium is a highly dangerous substance and must be kept in heavy lead containers to block off its ionizing radiation. Its major use is in cancer therapy in the form of salts. *See also* radioactivity.

radon. An element.

Symbol	Rn	Atomic No.	86
State	Gas	Atomic Wt.	222(?)
Group	Noble Gas	Valence	2,4,6

Radon is the first decay product of radium, and its activity is much greater. It is the heaviest of all gases with a density of almost 10 grams per liter (air = 1.3g/1). Radon has limited use in cancer therapy and in research in chemical kinetics. Like radium, it is hazardous to handle and should be used with adequate protective shielding.

rain making. *See* nucleation.

rare. A term applied to several groups of elements to indicate that they occur in very low concentration or in exceedingly small amounts, as follows: (1) the rare earths, the lanthanide series of elements (57–71); (2) the rare gases, i.e., the noble gases in the right-hand group of the Periodic Table; (3) the rare metals, which include the alkaline-earth series as well as many others such as cadmium, beryllium, cobalt, the platinum metals, and uranium.

rare earth. *See* lanthanide series.

rare gas. *See* noble.

rare metal. *See* rare; platinum.

ray. A shortened form of radiation, often used by scientists to refer to beams of light waves emanating from an excited source, such as gamma rays and x-rays, or to beams of high-energy electrons (beta particles) and helium nuclei (alpha particles). The word is conventionally used to denote radiation of extremely short wavelength (an angstrom unit or less for gamma and up to 1000 angstroms for x-rays). *Radiation* and *ray* are virtually synonymous, but the former is preferable, i.e., x-radiation, etc. *See also* radiation; alpha particle; beta particle.

Rayleigh, John W. S. (1842–1919). An English physicist, Lord Rayleigh became head of the Cavendish Laboratory in 1879 and was President of the Royal Society of London from 1905 to 1908. In 1904 he received the Nobel Prize in chemistry for his discovery of the noble gas argon. He also explored a wide range of fields in physics, including sound, electrical phenomena, and elasticity.

rayon. A semisynthetic fiber introduced about 1920, and made by chemical treatment of wood pulp; it is also called regenerated cellulose. It can be made by either the cuprammonium (copper and ammonia) or the viscose process, the latter being the more common. The chemically modified cellulose in liquid form is then extruded through a fine orifice (spinneret) into a hardening bath which causes formation of a smooth, fairly strong fiber. Various modifications with proteins such as casein have been introduced to increase strength and durability. The fibers can be spun to extreme fineness (denier), making them especially suitable for hosiery and other sheer garments. Its applications in recent years have been restricted by the appearance of many synthetic fibers of superior properties.

Cellophane is also regenerated cellulose made by the viscose process and extruded through slits of various widths to form film suitable for packaging. It was the first widely used transparent packaging film, but its use has decreased since the development of synthetic high-polymer films. *See also* viscose process; cellophane.

Rb Symbol for the element rubidium, the name being derived from the Latin term for *dark red,* the spectral area in which it was first identified.

Re Symbol for the element rhenium, the name being derived from the Latin word for the Rhine River (*Rhenus*).

reaction. (1) Any chemical change, regardless of rate, amount of product formed, or whether it occurs naturally (as in plant and animal metabolism) or is purposely induced in a laboratory or in large-scale production. For example, photosynthesis is a relatively slow, natural, energy-absorbing reaction with huge quantities of product; combustion is a rapid, energy-yielding reaction, either natural or induced; an explosion is almost instantaneous, usually artificial, and evolves tremendous amounts of energy compared to the masses of the reactants involved. Heating of a fatty acid and an alkali in a production kettle to make soap is an induced saponification reaction requiring heat and yielding product in commercial quantities. Chemical reactions involve only the electrons of an atom, and the energy absorbed or emitted is small compared with that of nuclear reactions.

Natural reactions usually occur below 36°C (100°F), but the rate of many induced reactions is greatly increased by high temperatures and by use of catalysts; the concentration of reactants and the particle size of the solids involved are also important factors in the efficiency of reactions. Among the more important types are decomposition, replacement, condensation, neutralization, oxidation-reduction, rearrangement, saponification, and polymerization.

(2) Disintegration of the nucleus of an atom either by natural radioactivity or by fission due to bombardment with high-energy particles. A thermonuclear reaction involves the fusion or union of hydrogen nuclei to form helium, with evolution of energy. *See also* change, chemical; chain reaction; fission; fusion.

reactor. A device for maintaining a controlled nuclear reaction, chiefly for electric power production, but also for the manufacture of radioisotopes and military equipment. The essential components of a power reactor are (1) a fissionable element, or fuel, such as uranium-235 or plutonium; (2) a neutron source to initiate the reaction; (3) a moderator to retard fast neutrons for more efficient fission; (4) a coolant (water or liquid sodium); and (5) control rods made of cadmium, which absorb neutrons and thus maintain any desired level of output. Electric power production is now possible with a minimum evolution of waste heat. The term *fuel element*, used by reactor technologists, refers to the "package" of nuclear fuel in a ceramic or metal coating, rather than to the fissionable material itself, the word *element* being used in its electrical rather than its chemical sense. *See also* nuclear energy; reaction (2); breeder.

reagent. Any chemical compound used in laboratory analyses to detect and identify specific constituents of the material being examined. Though reagents may be gases, liquids, or solids, they are usually prepared as solutions (in water or common solvents) of various concentrations, e.g., 1 molar, 0.1 normal, etc. Several thousand chemicals of varying specificity are used as reagents; they are subject to strict specifications, especially as regards purity. Well-known reagents are glacial acetic acid, sulfuric acid, hydrogen sulfide, dimethylglyoxime, and potassium iodide. *See also* analysis.

rearrangement. A type of chemical reaction in which the atoms of a single compound recombine, usually under the influence of a catalyst, to form a new compound having the same molecular weight but different properties. Thus, ammonium cyanate in solution will rearrange to form urea, in which the four hydrogen atoms are equally distributed between the two nitrogen atoms: $NH_4OCN \rightarrow (NH_2)_2CO$. Many instances of such rearrangements have been studied, especially in the petroleum field, where they aid in forming branched chains. Some have been named for their discoverers, e.g., the Fries and Beckmann rearrangements, etc. *See also* Wöhler.

Réaumur (pronounced Ray-a-myoor). A rarely used scale for measuring temperature invented by Rene Réaumur (1683–1757), French physicist. As on the centigrade scale, the freezing point of water is 0°, but the boiling point is 80°. Thus, each °R is 1.25°C, so that 10°R equals 12.5°C, 40°R equals 50°C, and so on. *See also* centigrade.

reclaiming. Any of various processes for recovering the values from scrap or used materials so that they can be incorporated in the manufacture of new products. Waste paper is shredded and deinked by solvents and adsorptive agents; glass and metals are fragmented, and added to fresh batches in limited proportions; scrap rubber is "devulcanized" by heat and solvents or by destructive distillation. The percentage of such reclaimed materials that can be added to new products is restricted, as in many cases their quality has been permanently impaired; often so many impurities and degradation products are present that reclaiming is not economically practicable. The feasibility of reclaiming agricultural wastes for fuels, fodder, and industrial products has been researched; such wastes include corn cobs, bagasse and other cellulosics, as well as manures. *See also* recycling; reprocessing; gasohol.

recombinant DNA (gene-splicing). Genetic material transferred from the genes of one species to those of another by uniting a portion of the DNA of one organism with extranuclear sections of DNA from another organism. This technique has made possible the synthesis not only of certain complex molecules (proteins) but also of entirely new life forms (bacteria). It is being utilized in basic genetic research sponsored by the National Institutes of Health. Experimental work has already produced a number of important results: (1) the production of human insulin from the bacterium *E. coli* by implantation of synthesized genetic material; (2) creation of artificial interferon, used in cancer therapy; (3) a synthetic bacterium that consumes oil spills more efficiently than natural types. Many more achievements of this nature are likely, and the potential for the chemical, agricultural, food, and pharmaceutical industries is enormous.

Note: This is one of the most significant developments in biochemistry in recent years; commercialization of these techniques is already under way. *See also* deoxyribonucleic acid; replication.

reconstitution.   In food technology, replacement at the time of use of the water removed from a dehydrated food product, thus restoring the food to its original edible condition. *See also* dehydration.

rectification.   A term used primarily in the distilled liquor industry to denote the operation whereby an alcoholic product is purified by repeated contact of the liquid distillate with the vapor as it forms in the first evaporative step (countercurrent distillation). *See also* countercurrent flow; distillation.

recycling.   (1) Return of a portion of a reaction product to the beginning of a processing sequence. Separation and recirculating of unreacted components increases the efficiency of conversion and also permits reuse of processing aids that remain unchanged. Recycling is standard procedure in distillation (reflux), the cracking of petroleum, and synthesis of ammonia and other chemicals.

(2) In popular use, the term refers to reclamation of waste paper, glass, metals, etc., a limited proportion of which can be incorporated in new product manufacture. *See also* reclaiming.

red oil.   *See* oleic acid.

redox.   A term coined from the first syllables of "reduction" and "oxidation" and used to describe reactions in which electron transfers of this type occur; it is a convenient short form for "oxidation-reduction." *See also* oxidation.

red oxide.   *See* iron oxide.

reduction.   (1) The reverse of oxidation. The term originally referred to a reaction in which oxygen is removed from a compound and transferred to an element called a *reducing agent*, for example, $NiO + C \rightarrow Ni + CO$. Though reduction can be regarded simply as the removal of oxygen, it also has a broader meaning: the acceptance of one or more electrons (or their equivalent) from another substance. For example, it may be accomplished by transfer of chlorine atoms, as in the Kroll magnesium reduction process, where the metal chloride is reduced to elemental metal while the magnesium is oxidized by combination with chlorine: $2Mg + XCl_4 \rightarrow X + 2MgCl_2$ where X is a metal. Oxygen is not essential to the reaction. Reduction can also be effected by addition of hydrogen to a molecule.

(2) Comminution of larger solid units to smaller ones by crushing, grinding, or similar disintegrating methods, as expressed by the term *size reduction*.

(3) In metallurgy, separation of a metal ore into its wanted and unwanted components by any of several methods, e.g., roasting, flotation, elutriation, etc. *See also* oxidation; comminution; Kroll process.

refining.   (1) In the petroleum industry, this term refers to chemical reactions performed on the various fractions obtained from petroleum by distillation; these include cracking, reforming, polymerization, dehydrogenation, etc., usually with catalysts, for production of gasoline and other derivatives.

(2) In the food industry, refining involves a sequence of operations employed to remove impuriites and to concentrate the desirable portion of a food product to improve its suitability for consumption. Typical is the refining of sugar by such operations as filtration, adsorption on charcoal, centrifuging, crystallization, etc. Flours and other cereals are refined by grinding and sifting. Purification of vegetable oils by adsorption on clays or carbon and the rectification of spiritous liquors are other instances of food product refining. Chemical reactions are not involved in these methods.

(3) In metallurgy, the term also implies removal of impurities by various separation techniques, important among which are electrolysis (used for aluminum, copper, and the precious metals) and a specialized operation called zone refining.

(4) In the rubber reclaim industry, refining involves passing the reclamate repeatedly between close-set metal rollers at high speed, and afterward through a straining device, to remove particles of dirt, metal, and other foreign matter.

reflux. In distillation operations, this term designates vapor that condenses into liquid in a still head or fractionating tower and trickles down through the tower; in doing so it comes into intimate contact with the vapor travelling upward through the tower. This results in more efficient separation of the components. *See also* countercurrent flow; rectification; recycling.

reforming. In petroleum chemistry, any of several specific cracking reactions which materially increase the octane number of alicyclic and paraffinic hydrocarbons in straight-run gasolines by converting them to aromatics or to branched chain structures. Reforming reactions are carried out at from 540 to 800°C (1000 to 1500°F) in the presence of catalysts such as platinum and molybdenum. The reactions most commonly involved are dehydrogenation and isomerization. When platinum is the catalyst, it is necessary to add hydrogen to the reactants to prevent deactivation of the catalyst by contamination with carbon. This type of reforming is called hydroforming. *See also* isomerization; cracking.

refractive. Descriptive of a material, usually a liquid, having a high index of refraction. (Do not confuse with refractory, which has an entirely different meaning.) *See following entries.*

refractive index. A physical constant that defines the extent to which a liquid or solid substance changes the direction of a light ray passing through it; for a given substance, it is the ratio of the sine of the angle at which the light strikes the surface to the sine of the angle at which it enters the substance. Crystals have characteristic refractive indexes, e.g., diamond is 2.41; the refractive power tends to increase with the density. The wavelength used for standard determinations is the sodium D line, and the conventional symbol for refractive index is n_D. For water this value is 1.33; the base value for a vacuum is 1.0. If the incident light is at right angles to the surface of the medium, no refraction occurs. *See also* dispersion (2).

refractory. A term used by materials engineers to designate any substance or mixture which withstands extremely high temperatures without softening or degradation. These are used as fur-

nace linings, crucibles, jet engines, rocket propulsion systems, and similar high-temperature environments. Tungsten, tantalum, and so-called superalloys are examples of refractory metals; silicon carbide, clay, alumina, magnesium oxide, borosilicate glasses, and porcelain are refractory ceramics; temperature of use ranges from 540 to 2750°C (1000 to 5000°F), depending on the material.

refrigerant. Any substance used to preserve food and biological products from deterioration by cooling below 40°F. Many agents have been used, including ice, brines, ammonia, solid carbon dioxide, sulfur dioxide, and ethyl chloride. The first three are still in wide use in food packing and storage plants and refrigerated cars and trucks. In the 1930s, chlorofluorocarbon refrigerants were introduced by T. Midgley, primarily for use in automatic household units; they will produce temperatures well below 0°F which can be continuously maintained. They are neither combustible nor poisonous. A typical compound is dichlorofluoromethane ($CHCl_2F$), but there are many others of similar composition. They are also used in air-conditioning. *See also* fluorocarbon; Midgley.

regenerated cellulose. *See* rayon; cellophane.

regeneration (1) Returning a raw material or natural product to its original state or condition after it has been chemically modified for processing purposes. Examples are regeneration of cellulose to rayon after it has been solubilized with sodium hydroxide and carbon disulfide to permit extrusion into fiber, and regeneration of collagen by neutralization with acid after purification in alkaline solution for use in sausage casings. Regeneration should not be confused with reclaiming, which only partially restores the original properties of a material.

(2) Renewing or reactivating a catalyst, molecular sieve, or ion-exchange medium (zeolite) by any of several methods. Catalysts fouled with coke or other deposits are regenerated by treatment with hot reactive gases such as oxygen, hydrogen, steam, etc. Zeolites are regenerated by replacing the sodium ions exchanged during the removal of calcium compounds from water with a fresh supply of sodium by treatment with a solution of sodium chloride.

reinecke. A hydrated salt resulting from combination of two ammonium compounds, one containing a thiocyanate group (SCN) and the other a diammonochromate group, $Cr(NH_3)_2$. The product is a water- and alcohol-soluble red powder used as a reagent in organic analysis (amines and other organic bases). The formula is $NH_4[Cr(NH_3)_2(SCN)_4] \cdot H_2O$.

reinforced plastic. A composite structure composed of a thermoplastic or a thermosetting resin and a fibrous filler material; the components are pressed together to give an extremely strong product with high impact resistance. Glass fiber is usually used as the reinforcing material, for both lightness and strength. Such composites are called fiber glass-reinforced plastics (FRP); they have many structural uses, especially for bulletproof fuselages, building panels, automobile and boat bodies, switchboards, and filament winding. Their corrosion-resistant properties make them preferable to steel and other metals on an economic basis, for such uses as oil-well drilling equipment, chemical piping, tanks, and similar equipment. Numerous resins are used, epoxy being one of the most important. *See also* composite; laminate; filament winding.

reinforcing agent. A term used in rubber and plastics technology to denote any material that increases the toughness, abrasion resistance, or impact strength of the base material when added to it in substantial quantities. Finely divided powders such as carbon black and zinc oxide greatly enhance these properties in rubber and other elastomers. The impact strength of plastics can be increased by incorporation of asbestos or glass fibers and by fabric insertions. *See also* reinforced plastic.

rennin. An enzyme used in commercial practice for coagulation of milk to obtain the grade of casein used in making buttons and other small plastic items, and also as a food additive. It is one of the digestive enzymes secreted in the stomach and is produced commercially by extraction from the organs of young animals. In commercial terminology it is also called rennet. *See also* enzyme.

repellent. (1) In materials technology, any substance that is not penetrated or wetted by water, i.e., hydrophobic. Chief among these are hydrocarbon solvents, oils, fats, waxes, rubber, and such finely divided materials as carbon black, magnesium carbonate, etc.

(2) In agricultural chemistry, a substance which is so repugnant to an animal or insect as to prevent its approach. Repellents are not neces-

sarily toxic, but are offensive to the subject either because of smell or taste. They are quite specific in their action, no single type being generally effective. They are used on a wide range of animals, including mosquitoes, rabbits, deer, birds, and sharks. *See also* odor; flavor; hydrophobic.

replacement. A type of chemical reaction in which a compound and an element are the reacting substances, the element replacing one of the constituents of the compound, e.g., $2NaBr + Cl_2 \rightarrow 2NaCl + Br_2$, and $C_6H_6 + Cl_2 \rightarrow C_6H_5Cl + HCl$. A reaction between two compounds by which two new compounds are formed is called double replacement, e.g., $NaOH + HCl \rightarrow NaCl + H_2O$, and $BaCl_2 + Na_2SO_4 \rightarrow BaSO_4 + 2NaCl$. Replacement reactions are also called substitution reactions, since one element is substituted for (replaced by) another, the distinction being largely semantic.

replication. Making a reverse image of a surface by means of an impression on or in a receptive material. This technique is used in electron microscopy to obtain plastic replicas of observed objects. In biochemistry, the term refers to the reproduction of the deoxyribonucleic acid molecule, which is composed of two interlocking chains of nucleotide substances, forming a double helix. It reproduces itself by forming two identical daughter molecules, each of which receives one of the two chains of the original molecule, the other chain in each case being synthesized from nucleic acids by the action of enzymes (DNA polymerases). In the over-simplified drawing, the solid lines represent the strands contributed by the original molecule, and the broken lines are the synthesized strands. *See also* deoxyribonucleic acid; genetic code.

parent DNA   daughters

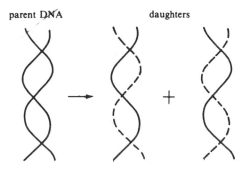

Reppe process. *See* acetylene.

reprocessing.   Treatment of spent nuclear reactor fuel to recover the unconsumed uranium-235 and plutonium; the method comprises a series of complex chemical separations involving solvent extraction, ion-exchange, and precipitation (Purex process). The reclaimed material is sent to fabrication plants for ultimate reuse. Serious radiation hazards are present in reprocessing, which requires shielding and remote-control handling.

Research octane. *See* octane number.

reserpine.   An alkaloid extracted from the root of the rauwolfia plant and used in medicine for blood pressure control and as a tranquilizer. Its formula is $C_{33}H_{40}N_2O_9$. Like most alkaloids, it has toxic properties.

residual.   A descriptive term applied specifically to various lower-grade fuel oils that are heavier than the standard No. 2 oil used for domestic heating. They are so named because they contain a high proportion of the hydrocarbon residues of distillation and cracking. Some types, called bunker fuels, must be heated before use. The term is also applied to ultimate distillation residues, which are viscous asphaltic mixtures used in roofing and road treatment. *See also* fuel oil.

resin.   (1) Naturally occurring water-insoluble mixtures of carboxylic acids, essential oils, and other substances formed in numerous varieties of trees and shrubs. They usually have a strong, pleasant odor, are hard and brittle when cool, and soft and tacky when warm. They are soluble in alcohols and hydrocarbon solvents. Among them are the rosins obtained from pine trees or stumps; the broad group of balsams (Canada, Tolu, copaiba, etc.); elemi, dammar, and similar products of tropical trees; and fossil resins such as amber and copal, which are dug from the ground. Among their uses are special varnishes, electrical insulation, fragrance sources, adhesive formulations, and textile and paper coatings. *See also* rosin; fossil; balsam; gum.

(2) Any synthetic high polymer usually resulting from an addition or a condensation reaction, e.g., polystyrene, phenolformaldehyde. Such substances were originally called resins probably because some early types, such as coumarone-indene, physically resembled natural resins. Thus, the term *synthetic resin,* though still widely used, is strictly inappropriate, since the substances are polymeric compounds rather than mixtures of simple substances. Characteristic properties of such polymers are insolubility in water, resistance to attack by chemicals, high electrical resistivity, and a smooth hard finish. Most types can be readily extruded, machined, and molded, and some can be extruded into filaments and woven into fabrics. There is a definite line of demarcation between thermoplastic (heat-softening) and thermosetting (cross-linked, heat-stable) resins. One major group having rubber-like properties is called elastomers. *See also* high polymer; plastic; elastomer.

resinoid.   A term used by materials engineers to denote a liquid, semisolid, or solid thermosetting resin or plastic, especially those used for bonding agents in such composites as abrasive wheels, plywood, etc. It also includes semipolymerized (heat-bodied) linseed oil and A-stage phenol-formaldehyde resins (also called resols).

resist.   A term used by dyeing technologists to denote a substance used to prevent a dye from adhering to or reacting with certain areas of a fabric substrate. It is in effect a masking medium, application of which to any desired portion of the fabric keeps it from being affected by a given dye. Use of resists thus permits the production of multiple-color textile prints. Neutralizing agents are sometimes used to obtain this result; these may be acidic or alkaline depending on the nature of the dye.

resol. *See* phenolformaldehyde resin; resinoid.

resolution. *See* resolving power.

resolving power.   A term used in optics to denote the degree to which the lens of a microscope can distinguish fine structure, i.e., extremely small distances. This depends on the wavelength of the radiation used to illuminate the sample as well as on the focal depth of the lens. The resolving power of an electron microscope, which uses electron wavelengths only a few angstroms in length, is far greater than that of an optical microscope, which relies on wavelengths of visible light (at least 4000 angstroms). For observation of inanimate objects, the resolving power of a microscope is much more important than its ability to magnify. *See also* microscopy.

resonance.   (1) In organic chemistry, this term denotes a mathematical concept based on quantum mechanical considerations (i.e., the wave functions of electrons), which expresses the true nature of certain compounds which cannot be accurately represented by any one valence-bond structure. It was originally applied to aromatic compounds such as benzene, for which there are many possible approximate structures, none of which is completely satisfactory. The resonance concept indicates that the actual molecular structure lies somewhere between these various approximations but is not capable of objective representation. The same idea has been found useful in regard to inorganic molecules in which an electron pair bond is present. The term *resonance hybrid* denotes molecules to which this concept applies. It is important to indicate that such a molecule does not vibrate back and forth between two or more structures, and that these structures are neither isotopes nor mixtures: the resonance phenomenon is an idealized mathematical expression of the actual molecule of, say, benzene, which cannot be duplicated by any pictorial device. The term *mesomerism* is used synonymously with resonance. *See also* benzene ring.

(2) In spectroscopy, resonance refers to a condition in which a source of radiant energy is in the same energy state as the atom or molecule that absorbs the radiation, as in nuclear magnetic resonance. Thus, the radiation emitted has the same wavelength as the exciting source.

resorcinol.   A benzene derivative, also called metadihydroxybenzene, with formula $C_6H_4(OH)_2$, the two hydroxyl groups being attached to the benzene ring at the 1- and 3-positions. It has many properties similar to phenol, to which it is chemically related. It is used as an antiseptic, with formaldehyde in the manufacture of plastics, and as an adhesive for composites made up of organic materials, such as wood, rubber, and textiles.

retarder.   A benzenoid compound such as salicylic or benzoic acid added to rubber and plastic mixes during processing in low concentrations to prevent cross-linking action (curing) from taking place at the temperatures prevailing during extrusion and other manufacturing steps. They are also used in sponge rubber formulations to withhold vulcanization until the material has had a chance to form a foam. Retarders are special types of inhibitors. *See also* inhibitor; stabilizer.

retting.   *See* jute.

reverse osmosis.   *See* desalination; osmosis.

reversible.   (1) A chemical reaction which can proceed first to the right and then to the left when the conditions are changed; in other words,

the product of the first reaction can decompose to the original components as a result of different conditions of temperature or pressure. Examples are $H_2O + CO_2 \rightleftharpoons H_2CO_3$, in which the carbonic acid reverts to water and carbon dioxide on heating; $NH_4Cl \rightleftharpoons NH_3 + HCl$, in which the ammonium chloride decomposes on heating to ammonia and hydrogen chloride, which recombine on cooling.

(2)   A colloidal system such as a gel or suspension that can be changed back to its original liquid form by heating, addition of water, or other method. For example, evaporated egg white can be restored (reconstituted) by addition of water; pastes of starch or soap are also reversible. *See also* irreversible; gel.

Rh   Symbol for the element rhodium, the name being derived from the Greek word for *rose*, or *rose-colored*, from the deep red color of the chlorine-containing sodium salt from which the element was originally extracted.

rhenium.   An element.

Symbol	Re	Atomic Wt.	186.2
State	Solid	Valence	-1,2,3,4, 5,6,7
Group	VIIB	Isotopes	2 stable
Atomic No.	75		

Rhenium is a silver-gray metal; its density, 21.02 (g/cc,) is almost twice that of lead, and it has a very high melting point, 3170°C (5738°F), which places it in the category of refractory metals. It is quite stable but forms compounds with all the halogens and oxidizes strongly on heating. It is resistant to hydrochloric acid but is attacked by nitric and sulfuric acids. It is molded by powder metallurgy techniques, is a component of thermocouple alloys, and has numerous applications in electronic devices. It can also be electroplated. Its high cost limits its use to specialty items.

rheology.   *See* fluid; dilatant; thixotropy.

Rh factor.   An active substance occurring in the red blood cells of most humans and also of the Rhesus monkey, from which it was named. Humans whose blood contains this factor are said to be Rh-positive, and those who lack it Rh-negative. The transmission of this blood characteristic is hereditary. If Rh-positive blood is introduced into the blood of an Rh-negative individual, as by transfusion or during pregnancy, antibodies are formed just as if the factor were an infectious organism; these cause the Rh-positive cells to clump together (agglutinate), with serious results. An unborn child whose blood is Rh-positive can cause antibody formation in the blood of a mother who is Rh-negative, causing a critical disease to develop in the fetus. *See also* antibody; factor.

rhodium.   An element.

Symbol	Rh	Atomic No.	45
State	Solid	Atomic Wt.	102.9055
Group	VIII	Valence	3,4

Rhodium is one of the members of the so-called platinum group which comprises six of the nine transition metals in Group VIII. It has about the same density as palladium, i.e., it is only about half as heavy as other metals in this group. It has a very high melting point, 1960°C (3560°F), and good electrical conductivity, which make it suitable for high-temperature alloys, electrical devices, furnace windings, and other refractory uses. It can also be electroplated and vacuum-deposited for coatings for specialized purposes. It has only one stable form. *See also* platinum.

rhodopsin.   A visual pigment present in the light-sensitive color receptors located in the retina of the eyes. It is also called visual purple. It is composed of the aldehyde of vitamin A (carotene) and the protein opsin. A reversible color change similar to bleaching occurs when the pigment absorbs light; this reduces the sensitivity of the eye until the rhodopsin is restored, largely by the action of opsin. The word is derived from the Latin for *rose* (red, referring to its color sensitivity) plus *opsin*, the visual protein.

riboflavin.   A yellowish pigment and a member of the Vitamin B complex; also called vitamin B_2. It functions as an enzyme in control of tissue oxidation and is an essential factor in cellular metabolism. Its molecule is made up of three fused benzenoid rings, two of which contain nitrogen, with a hydrocarbon side chain on the central ring; the formula is $C_{17}H_{20}N_4O_6$. It can be synthesized from dextrose. Riboflavin occurs in egg yolk, green vegetables, and liver, and is an essential dietary requirement. *See also* B complex.

ribonucleic acid (RNA).   A nucleotide molecule, similar in composition to deoxyribonucleic acid, which performs a number of important functions in programming the genetic code in cells. There are several types, the most important being messenger RNA, ribosomal RNA, and transfer RNA, the latter including 20 or more variations which correspond to the amino acids of which proteins are composed. RNA molecules impart the amino acid coding information received from DNA to nucleoproteins called ribosomes, the code designating not only the amino acids required for the particular protein under construction, but also (and equally important) the sequence in which they are to be assembled in the chain, the location of disulfide cross-links, etc. *See also* deoxyribunucleic acid; genetic code; messenger.

riboside.   *See* nucleic acid.

ricinoleic acid.   *See* castor oil; fatty acid.

ring.   A cyclic molecular arrangement whose structure is represented by polygons having four to eight sides (cycloöctatetraene). The principal element (the atom at each vertex of the polygon) is usually carbon, as in the hexagonal benzene ring (homocyclic). If an element other than carbon appears in the ring, the compound is called heterocyclic, for example, the pentagonal furane ring. Alicyclic hydrocarbons such as cyclohexane have a ring structure which may have varied configurations, such as boats, chairs, birdcages, etc. Not all rings are completely closed, especially in sterols. Fused rings are those arranged in a pattern of two, three, or more, each pair having one side in common, as in anthracene and phenanthrene. *See also* benzene ring; heterocyclic; cyclic compound.

Rn   Symbol for the element radon.

RNA.   *See* ribonucleic acid.

roast.   A term used in metallurgy to denote high-temperature treatment of sulfide-bearing ores in air or oxygen preparatory to chemical separation of the metal values. The roasting process oxidizes the ores, the metals being obtained by reduction (smelting). *See also* smelt.

Rochelle salt.   *See* antacid; salt.

Roentgen, W. K.   (1845–1923).   German physicist who discovered x-rays in 1895, for which he was awarded the Nobel Prize in 1901. Application of x-rays to a number of important problems during the period 1900 to 1925 was developed by the Braggs in England (crystal structure determination), by Moseley, also in England (exact location of elements in the Periodic Table), by von Laue in Germany and Debye and Sherrer in the U.S. (high-polymer structure). For some years these were called Roentgen rays; the term x-ray, originally used by Roentgen himself because their nature was unknown, is now preferred.

rosin. A natural resin obtained from pine trees (gum rosin) or from distilled pine stumps (wood rosin). It is a mixture of abietic acid and other carboxylic or fatty acids, and is the end-product of turpentine production. It is hard and brittle at room temperature, becoming soft and extremely sticky at higher temperatures. Wood rosin is produced in many color grades ranging from water-white to dark brown. Rosin is used in adhesive formulations, as a binder in printing inks, paper-coating compositions, etc., and in the manufacture of ester gum by reaction with higher alcohols. It is not used in electrical insulating tapes, as it has an adverse effect on the shelf-life of uncured rubber. *See also* oleoresin; resin (1).

rotenone. A cyclic compound ($C_{23}H_{22}O_6$) obtained from the roots of plants of the Derris variety and used as an insecticide especially for houseflies, moths, mites, etc. Its toxicity is highly specific, being lethal to fish but without adverse effects on birds and higher animals. It is insoluble in water. *See also* insecticide.

rouge Fine-ground, purified iron oxide (Fe_2O_3) used as an abrasive for finishing glassware, mirrors, and specialty items, and as a component of cosmetic preparations. *See also* abrasive.

Ru Symbol for the element ruthenium, the name being a transliteration of an old name for Russia where the element was discovered.

rubber. A natural elastomeric polymer of isoprene; it is an unsaturated hydrocarbon, *cis*-1,4-polyisoprene:

$$\left[CH_2 = \overset{\overset{\displaystyle CH_3}{\displaystyle |}}{C}CH = CH_2 \right]_n$$

It is produced chiefly by the tree *Hevea braziliensis,* native to Brazil but commercially cultivated in Malaya. Polyisoprene was synthesized in the 1950s and is called "synthetic natural" rubber. The natural product occurs in the tree as a whitish liquid (latex), from which the rubber hydrocarbon can be separated by coagulation

with a mild acid (acetic or formic). In its original state, rubber is nearly useless for anything but adhesive purposes; it is highly temperature-sensitive (thermoplastic) and has little strength until vulcanized with sulfur, aided by organic accelerators. This has the effect of cross-linking the hydrocarbon chains into a thermosetting structure possessing a unique elastic modulus (recovery after deformation), though there is some permanent loss of strength and resilience on repeated extension. When reinforced with carbon black (30 to 40%) it has remarkable abrasion resistance. Though subject to oxidation and to degradation by fatty acids and petroleum products, considerable protection from these can be provided by antioxidants. The term has been extended to include any synthetic elastomeric substance that has the essential properties of natural rubber. Many of these have superior resistance to solvents and oils, and equivalent abrasion resistance, e.g., styrene-butadiene copolymer, polychloroprene, butyl rubber.

History. While exploring the Amazon in 1735, LaCondamine, a French scientist, noted a resilient material that the natives obtained by tapping certain trees (Hevea) and coagulating the sap (latex) over fires. This they used for making waterproof shoes and large balls with which they played games. He took samples of this to Europe, where it aroused great interest, especially among chemists such as Priestley. Its ability to erase pencil marks by rubbing was the origin of the word *rubber*. The Hevea forests of Brazil remained the primary source of rubber for 150 years. In 1875, an English botanist, Henry Wickham, was requested by his government to obtain a supply of Hevea seeds. He helped himself to several thousand of these; they were planted in Kew Gardens in London, and the seedlings transplanted in Malaya. Thus began the commercial production of plantation rubber.

The rubber industry began in 1844 with the simultaneous discovery of vulcanization with sulfur and inorganic accelerators (metallic oxides) by Goodyear in America and Hancock in England. But it was not until the early years of the 20th century that it became a major industry as a result of three important developments: (1) the discovery in 1906 of the much more efficient organic accelerators (nitrogenous compounds), which were used for some years by one company before being widely accepted (1920); (2) fortuitous discovery of the reinforcing and abrasion-resistant properties of carbon black; and (3) the introduction of age-resisters (antioxidants), which greatly prolonged the life of rubber by inhibiting oxidation. As a result of these advances, the mileage of tires suddenly increased by a factor of 10, with comparable reduction in risk of failure, and thus made possible the vast growth of the automobile industry.

An efficient method of reclaiming vulcanized scrap was developed about 1900, long before the advent of organic accelerators, involving digestion of the scrap with NaOH plus a hydrocarbon softener. This separated the cotton insertions from the rubber and partially reversed the effect of vulcanization. Use of reclaim has long been an important adjunct to the industry, especially for low-cost, high-volume products.

Methods of concentrating latex sufficiently by centrifugation to make long distance transportation economically possible were introduced in the early 1930's, thus simplifying the manufacture of thin and irregularly shaped items, either by dipping or electrodeposition, as well as of cellular rubber (foamed or sponge products).

Significant developments in synthetic rubber began at this time. Outstanding were the introduction of polychloroprene (neoprene) by Carothers and of the oil-resistant polysulfide rubber Thiokol by Patrick. These were soon followed by styrene-butadiene copolymers, nitrile rubber, butyl rubber, and various other types, some of which were rushed into production for the war effort in the early 1940's. The stereospecific catalysts researched by Ziegler and Natta aided this development, including synthesis of true rubber hydrocarbon (polyisoprene). Since 1935, synthetic rubbers have been referred to as elastomers. Natural rubber usage has declined substantially in recent years, though the petroleum shortage may cause its return to favor in the 1980's.

rubber substitute.   *See* factice.

rubidium.   An element.

Symbol	Rb	Atomic No. 37
State	Solid	Atomic Wt. 85.467
Group	IA	Valence 1

A metal, m.p. 39°C (102°F), of the alkali group and hence extremely reactive, especially with air and water; it should be stored in an inert atmosphere or under kerosine. It has only one stable form. It is available in only small amounts, its chief uses being in electronic devices, though research has been carried on in other fields such as heat-transfer and thermionics. It is classed as a flammable material.

ruby   (1)   A natural variety of aluminum oxide (corundum) crystal, which can also be "grown" synthetically. Its color is due to small amounts of iron (III) oxide present in the corundum lattice. Besides being a well-known gemstone, its crystal structure is capable of emitting the coherent light beams produced by lasers when activated by trace amounts of chromic oxide. *See also* laser; aluminum oxide.

(2)   The term is also applied to a reddish glass whose coloration is due to colloidal gold or purple of Cassius.

rust   (1)   The reddish corrosion product formed by electrochemical interaction between iron and atmospheric oxygen; ferric oxide, Fe_2O_3. The reaction occurs most rapidly in moist air, indicating the catalytic activity of water.

(2)   A reddish-yellow fungus which attacks plants, especially cereal grains; it is also called smut. It can be controlled by treatment with formaldehyde.

ruthenium.   An element.

Symbol	Ru	Atomic Wt.	101.07
State	Solid	Valence	3,4,5,6,8
Group	VIII	Isotopes	7 stable
Atomic No. 44			

A hard metal, m.p. 2310°C (4190°F), which is a member of the platinum group; it is present in low concentrations in platinum ores. It is used in the arts in combination with platinum as a means of increasing its hardness; it also has some application as a catalyst in the petroleum industry, in special alloys for electrical devices, and as electrodeposited coatings. It is active in chemical combination and forms a number of coordination compounds.

Rutherford, Sir Ernest   (1871–1937). Born in New Zealand, Rutherford studied under J. J. Thomson at the Cavendish Laboratory in England. His work constituted a notable landmark in the history of atomic research, as he developed Becquerel's discovery of radioactivity into an exact and documented proof that the atoms of the heavier elements, which had been thought to be immutable, actually decay by emitting various forms of radiation. Rutherford was the first to establish the theory of the nuclear atom and to carry out a transmutation reaction (1919) (formation of hydrogen and an oxygen isotope by bombardment of nitrogen with alpha particles). Uranium emanations were shown to consist of three types of rays: alpha (helium nuclei) of low penetrating power; beta (electrons); and gamma, of exceedingly short wavelength and great energy. Rutherford also discovered the half-life of radioactive elements and applied this to studies of age determination of rocks by measuring the decay period of radium to lead-206. He received the Nobel Prize in chemistry in 1908. *See also* Aston; radium; radioactivity.

rutile. *See* titanium.

S

S Symbol for the element sulfur; this spelling is preferred by chemists to the older form *sulphur*. The origin of the name goes back to antiquity but is not definitely known.

saccharose. *See* carbohydrate.

sacrificial. A term used by electrochemists to denote a type of galvanic corrosion in which a coating, most commonly zinc, is attacked in preference to the underlying metal, that is, the coating is "sacrificed" in order to preserve the basis metal. *See also* corrosion; preferential; galvanizing.

safflower oil. *See* fatty acid; vegetable oil.

salicylic acid. A carboxylic acid derived from benzene, having the formula $C_6H_4(OH)(COOH)$. Its major use is in the manufacture of pharmaceutical compounds, especially aspirin, by reaction with acetic anhydride. It has several specialized applications, e.g., as an antiseptic and preservative and as an antiscorching agent in rubber processing. It occurs naturally in wintergreen oil and coumarin. *See also* benzoic acid; acetylsalicylic acid.

sal soda. *See* sodium carbonate.

salt. One of the products resulting from a reaction between an acid and a base, in which the hydrogen atom of the acid replaces the metal atom or NH_4 group of the base, the other reaction product being water. Salts may be either inorganic or organic, depending on the nature of the reactants. An example of an inorganic salt-forming reaction is $2KOH + H_2SO_4 \rightarrow K_2SO_4 + 2H_2O$; an organic salt, e.g., of acetic acid, is formed by the reaction $2CH_3COOH + CuO \rightarrow Cu(C_2H_3O_2)_2 + H_2O$. In the former case a sulfate results, and in the latter case, an acetate; thus, salts of most acids ending in *-ic* have the suffix *-ate*. However, salts of such acids as hydrochloric have the suffix *-ide,* and those of acids ending in *-ous* have the suffix *-ite,* e.g., sodium chloride from hydrochloric acid and sodium sulfite from sulfurous acid, respectively.

Most inorganic salts are water-soluble, dissociating to form electrically conductive solutions; such salts are called electrolytes. Those containing transition metals are used for electroplating; solutions of sodium chloride (NaCl), the most common salt, are a commercial source of chlorine. Double salts are formed from mixtures of ions in solution, e.g., Rochelle salt, which is the potassium-sodium salt of tartaric acid. Soaps may be regarded as salts of fatty acid esters, though in this case the other reaction product is glycerol. *See also* sodium chloride; electrolyte; soap; fused salt; saponification.

salt cake. *See* sodium sulfate.

salting out. An operation used chiefly in the soap industry involving the incorporation of common salt in the product at the end of the mixing cycle to cause aggregation of the desired ingredients and thus the separation of these ingredients from unwanted reaction products. *See also* nigre.

samarium. An element.

Symbol	Sm	Atomic Wt.	150.4
State	Solid	Valence	2,3
Group	IIIB (Lanthanide Series)	Isotopes	7 stable
Atomic No.	62		

Samarium, m.p. 1072°C (1962°F), is a member of the rare-earth series of metals. It is very hard and steel-like in appearance. It is a reducing agent for many metallic oxides, carbon monoxide, and carbon tetrachloride. It is combustible at about 150°C (300°F). It has a high neutron absorption capacity; its compounds have important specialized uses in glass, phosphors, lasers, and thermoelectric devices. Its scarcity restricts wider use.

sandwich compound. *See* ferrocene; metallocene.

saponification. A reaction between an alkali-metal compound and a fatty acid ester, the major product of which is a soap. The reaction of alkali-metal hydroxides with glycerides (esters of fatty acids) yields common soap, with glycerol as byproduct.

$$(C_{17}H_{35}COO)_3C_3H_5 + 3NaOH \rightarrow$$
$$3C_{17}H_{35}COONa + C_3H_5(OH)_3.$$

Alkaline-earth and heavy metals such as lead, cobalt, etc., similarly react with fatty acids to form heavy-metal soaps. Saponification is considered to be a form of hydrolysis, as it is based on the so-called "splitting" reaction in which an ester (fat) reacts with water to form glycerol and a fatty acid. *See also* soap; salt.

sapphire. A form of aluminum oxide or alumina (Al_2O_3); in the natural state, it is identical with corundum or emery. Its blue color is due to a minute amount of a vanadium compound, probably vanadium (IV) oxide. Synthetic sapphire is made by crystal growth techniques for use in such devices as microwave and television tubes and various precision instruments. It is extremely hard and will transmit vibrations in both ultraviolet and infrared; it can be used at temperatures up to 1093°C (2000°F).

saran. *See* polyvinylidene chloride.

saturated. The condition of a substance or material when the limit of its capacity to retain or hold another substance has been reached. It is derived from the Latin word for *enough*. In chemistry, it applies to (1) a solution that contains the highest possible concentration of a solute at a given temperature without becoming unstable; (2) a carbon atom, or a compound containing carbon atoms, each of whose valences forms a single bond with another atom,

which may be carbon or some other element. *See also* supersaturation; unsaturation.

Saybolt viscosity. *See* viscosity.

Sb Symbol for the element antimony, derived from stibnite, the principal ore of antimony.

SBR rubber. *See* butadiene-1,3.

Se Symbol for scandium derived from an old name for Sweden, the country of its discovery (Scandinavia).

scale. (**1**) A deposited layer or incrustation of calcium carbonate or calcium sulfate on the lining of boilers or in water pipes, resulting from prolonged contact with hard water. *See also* boiler scale.

(**2**) A type of paraffin wax which retains several percent of liquid distillate; scale wax has a lower melting point, 52°C (125°F), than fully pressed and refined wax and is the type commonly used for packaging, candles, etc.

(**3**) Any standard set of values or gradations for measuring a property or condition, such as temperature, viscosity, pH, hardness, etc.

(**4**) An extremely accurate type of weighing device used in analytical laboratories, commonly called a balance.

scandium. An element.

Symbol Sc	Atomic No. 21
State Solid	Atomic Wt. 44.9559
Group IIIB	Valence 3

Scandium, m.p. 1539°C (2802°F), is classed as a rare metal and is obtained in very small quantities from the ores of other metals, though there is one scarce ore in which it is present to about 35%. So little is available that the metal has been used chiefly for research and has no significant applications in industry. It is electropositive in nature but forms only a few compounds, e.g., the fluoride and the oxide.

scavenger. (**1**) Any metal or alloy which has the ability to remove traces of gaseous impurities from molten metals by combination or adsorption; these are used in the manufacture of steel to eliminate oxygen, nitrogen, etc., during the melting process. They are also used in vacuum devices to pick up stray gaseous molecules; in this application they are called getters.

(**2**) Gases such as ammonia and nitrous oxide, as well as radioactive carbon and iodine, have been used in research on free radicals and other active entities induced by radiation; by reacting in a particular manner with the latter, they act as indicators. Ethylene containing carbon-14 is useful in free-radical research. The indicator effect is due to the scavenging property of the substance used, which need be present in very low concentration.

Scheele, C. W. (1742–1786). One of the outstanding early chemical thinkers and experimenters. Scheele was a Swedish scientist who discovered a number of previously unknown substances, among which were tartaric acid, chlorine, manganese salts, arsine, and copper arsenite (Scheele's green). He also noted the oxidation states of various metals; observed the nature of oxygen two years before Priestley's discovery; and discovered the chemical action of light on silver compounds, thus laying the foundation of photochemistry and photography. Scheelite, or natural calcium tungstate, is named after him.

scouring. *See* kier-boiling; lanolin.

scrub. A term used by chemical engineers to designate an operation by which impurities are removed from gases and vapors by contact with a countercurrent liquid. It is usually performed in a column or tower through which the gaseous mixture rises through the downflowing liquid. Solid and liquid impurities are removed in this way, and unwanted gases can also be separated. The method is used for elimination of dust particles from air, purification of blast-furnace gases, and numerous other industrial purposes.

Se Symbol for the element selenium, the name being derived from an ancient word for *the moon* because of its close similarity to tellurium, named after the earth (tellus).

sealant. A soft, tacky organic material which will flow into, or be easily inserted in cracks, holes, and other small orifices where it will subsequently harden or polymerize to afford a tight, waterproof joint. Most sealants in use today are synthetic self-curing high polymers (organic cements) typified by silicone resins, urethanes, acrylics, and polyethylene. A few are of natural origin, especially linseed oil (putty) and asphalt, the latter used chiefly in highway construction. *See also* adhesive.

seaweed. *See* phycocolloid.

sebacic acid. *See* castor oil.

secondary. (1) A monohydric alcohol in which two of the hydrogen atoms attached to the methanol carbon are replaced by alkyl groups, e.g.,

$CH_3CH_2CHOHCH_3$ (*sec*-butyl alcohol). The term is also applied to amines and other compounds in an analogous manner. In such cases it is conventionally abbreviated *sec*.

(2) A reversible (rechargeable) electric battery, often called a storage battery.

(3) When referring to metals or other mineral products, this term denotes recovery from scrap or other waste products; when applied to petroleum, it indicates exploitation by means of pressurized water and other special methods of wells that can no longer be pumped. *See also* primary; tertiary.

sedimentation. A separation operation in which particles dispersed in a liquid or semiliquid medium are allowed to fall to the bottom of a container, by gravity. It differs from precipitation in respect to the time required; precipitation is rapid, whereas sedimentation may take hours or even days if the particles approach colloidal dimensions. Sedimentation is applied as a large-scale water purification method, in blood analyses, and in metallurgical operations. *See also* Stokes' Law.

selenium. An element.

Symbol	Se	Atomic Wt.	78.96
State	Solid	Valence	2,4,6
Group	VIA	Isotopes	6 stable
Atomic No.	34		

A semiconducting element, m.p. 217°C (423°F), classed as a nonmetal; it is recovered as a by-product of copper ore refining. It has several allotropic forms and can exist as a solid in both amorphous and crystalline states. Its electrical properties make it particularly suitable for use in rectifiers; its electrical conductivity is increased on exposure to light by the effect of photons, which proliferate electrons and energy deficits (holes) within the atoms. Selenium is thus used in photocells, solar batteries, and similar devices. It has specialized applications in rubber vulcanization and catalysis and as a colorant in glass and ceramic products. Selenium itself is not poisonous, but many of its compounds are (e.g., hydrogen selenide); toxic effects on cattle grazing in locations where selenium is present in the soil have been reported from time to time. *See also* tellurium; hole.

semiconductor. An element or compound whose electrical properties are midway between a conductor and an insulator. Semiconduction was first discovered in the metal germanium in 1948 and subsequently in the nonmetals selenium and silicon and in the compounds silicon carbide, lead sulfide, and a few others. It is caused by two factors: (1) free electrons ejected from low-energy covalent bonds at the site of a lattice defect, leaving an energy deficit called a hole (positively charged); and (2) electrons supplied by atoms of impurity elements. Electrons from both sources pass through the crystal from one energy deficit location to another, so that the deficit itself can be regarded as a flow of positive electricity. If enough electrons are present to satisfy the energy deficit by saturating the covalent bond, the electrons become the conducting medium, and the semiconductor is said to be n-type; but if too few electrons are available, the conductivity is by the positively charged holes, which move through the crystal as the electrons shift location from one bond to another. In this case the semiconductor is p-type. Semiconductors are used in transistors, photocells, computers, and many electrical and electronic devices. *See also* solid state; impurity; hole.

semimicroanalysis. An analytical procedure carried out on a sample weighing not less than 10 or more than 100 milligrams.

semipermeable. *See* membrane; osmosis.

semisynthetic. A substance or material made by chemical modification of an existing natural material, as opposed to a true synthetic resulting from reactions between chemicals. Examples are the many products derived from the natural polymer cellulose (rayon, paper, nitrocellulose, carboxymethyl cellulose, etc.); glass (a mixture of sand, soda, lime, *et al.*); and soap (from natural fats and oils). Though most semisynthetics are made by reaction with one or more chemicals, the base materials occur as such in nature.

separation. The division of a mixture into its component substances by either chemical or physical methods. Chemical methods include precipitation, reduction, acid-base reactions, recrystallization, etc.; physical methods are fractionation, solvent extraction, chromatography, adsorption, sedimentation, flotation, dehydration and drying, filtration, centrifugation, and magnetic techniques. All these methods have several variations; for example, fractionation is performed by distillation, freezing techniques (helium), and gel filtration; chromatographic

separations include gas-phase, liquid-phase, thin-layer, paper, etc.

sequestering agent. A type of chelating or coordination compound (also called a metal-ion deactivator) which immobilizes (sequesters) the ions of such metals as copper, iron, calcium, etc., by binding them into complexes that are both stable and soluble. As a result, the ions are prevented from acting as oxidation catalysts and from forming insoluble precipitates, e.g., boiler scale. Ethylenediaminetetraacetic acid (EDTA) is particularly effective.

Sequestering agents are used in a broad range of industrial products such as antioxidants, preservatives, and purifying agents, for example, to retard formation of precipitates (gumming) in gasoline and fuel oil, to decontaminate materials subjected to radioactivity, to inhibit catalytic breakdown of hydrogen peroxide, and for many similar applications where ion deactivation is required. *See also* chelate; ethylenediaminetetraacetic acid.

serum. (1) The liquid portion of a suspension such as milk after removal of all the solid particles, including those in the colloidal size range. Milk from which the fat particles and casein micelles have been separated is a serum called whey; it contains chiefly water in which sugars, minerals, and vitamins are present in true solution, i.e., in molecular form. Another example is the serum obtained from rubber latex by precipitation of the rubber hydrocarbon micelles.

(2) Blood from which corpuscles, platelets, and some proteins have been separated by centrifugation, especially after it has been treated with an antigen for cure of a disease by inoculation.

sewage treatment. *See* activated sludge; aerobic; biodegradability; BOD.

shale oil. A low-grade crude oil extracted from mountains of sedimentary shale located chiefly in Colorado, Utah, and Wyoming. Two methods are used—surface mining and excavation. In the first, the shale is bulldozed from beds, crushed, and fed into retorts (reaction vessels) in which temperatures of 425 to 535°C (about 800 to 1000°F) are required to separate the oil. In the second, shafts are driven into the mountain and the shale heated by direct combustion in an interior chamber excavated for this purpose. The latter method is the more efficient, and significant amounts of oil are being produced in this way. The oil-bearing component of the shale is called kerogen. Since only 20 to 30 gallons of oil are obtained per ton of shale processed, the operation is extremely inefficient. Though the potential oil content of western shales is estimated to approach the total reserves of the Middle East, only about one-third of it is recoverable. Major obstacles to large-scale production are the vast earth-moving operations necessary, the need for large volumes of cooling water, and disposal of the spent shale. *See also* kerogen.

shell. A term applied by physical chemists to the various energy levels of groups of orbitals occupied by electrons as they revolve around atomic nuclei. There are seven possible shells, designated as K, L, M, N, etc.; the number varies with the element from 1 in the lightest (hydrogen and helium) to 7 in the heaviest (uranium). Each shell can contain only a certain number of electrons, for example, 2 for K, 8 for L, 18 for M, and 32 for N; the shells may or may not be completely filled. The outer shell contains the valence electrons, which are directly involved in chemical bonding. *See also* orbital; electron; Pauli exclusion principle.

shellac. An alcohol solution of a resin obtained from a unique type of tropical insect, especially in India. The unbleached type is called orange shellac due to its color. When dry it forms a hard, transparent coating and thus is used on furniture and other wood products as a finishing and protective coating. Its high electrical resistance makes it useful as an insulating coating. The name is derived from the same base as *lacquer* and is probably of Hindu origin.

sherardizing. *See* cementation.

shielding. Protection of personnel by placing adequate thicknesses of appropriate materials around nuclear reactors, and sources of high-energy x-rays and radioactive emanations. Lead is most frequently used for x- and gamma-ray protection because of its high density. Boron and cadmium are most effective as neutron absorbers. Hydrogen-rich materials such as hydraulic cements, polyethylene, paraffin, and water also afford protection, but only in rather large amounts. *See also* absorption (3).

shortening. *See* cooking.

shortstop. A chemical compound used to stop a reaction at a predetermined point. Such agents are particularly useful in polymerization sequences in which it is not desirable for the reaction to go to completion. The term grew out of experimentation in high-polymer synthesis in the 1940s. Diethylhydroxylamine and sodium dimethyldithiocarbamate have been used.

Si Symbol for the element silicon; the name is said to be derived from the Latin word for *flint.*

silane. A gaseous or liquid compound of silicon and hydrogen, e.g., silicon tetrahydride (SiH_4). *See also* silicon.

silica. The oxide of silicon, SiO_2; sand. *See also* silicon; silica gel.

silica gel. A porous solid material consisting of silica manufactured in pellets of various sizes; it is made by treating sodium silicate (water glass) with sulfuric acid. Its physical structure and uses are not unlike those of activated carbon. It has high adsorptive power for atmospheric moisture and finely divided solids, and thus is largely used as a dehumidifier and clarifying agent and as a carrier for catalysts.

silicate. Any of a broad range of mineral compounds comprised of from one to six silica groups (SiO_2), arranged in either rings or chains. These include many of the more familiar gemstones (zircon, garnet, beryl, emerald), as well as asbestos (magnesium silicate), mica, and various types of clay (aluminum silicate). A well-known synthetic silicate is sodium silicate, a water-soluble glass commonly called water glass. Silicates are used in the manufacture of water-softening agents (zeolites). *See also* silicon; water glass.

silicon. An element.

Symbol	Si	Atomic Wt.	28.086
State	Solid	Valence	4
Group	IVA	Isotopes	3 stable
Atomic No.	14		

Silicon, m.p. 1420°C (2588°F), is the most abundant solid element; it rarely occurs in elemental form, virtually all of it existing as compounds (silicon dioxide, silicates, etc.). Silicon is a nonmetal, like carbon, to which it is chemically similar; it has the same valence and is next below carbon in the Periodic Table. Silicon forms single bonds with itself and with carbon, oxygen, hydrogen, and halogens, but it does not form double or triple bonds nor chains of more than six silicon atoms. This similarity to carbon accounts for its ability to form silanes (with hydrogen), siloxanes (with oxygen), and the industrially important silicone compounds (with oxygen and organic groups). Silicon is one of the few elements that have pronounced semiconducting properties. The most abundant compound is the dioxide (silica, sand) which is the basis of glass and a component of Portland cement; other important compounds are the tetrachloride, used in the preparation of various organosilicon products; the carbide (silica plus carbon); and the broad range of silicone products. In elemental form, silicon is used as an alloying agent in steel manufacture (ferrosilicon), and as a semiconductor, e.g., in solar cells and solid state devices. Though normally crystalline, silicon can be prepared in amorphous form, which has been found especially effective in solar cells. *See also* silicate; silicone; semiconductor.

silicon carbide. A chemical compound of carbon and silicon (SiC), trademarked "Carborundum." It is made by heating silica (sand) with coke in an electric furnace at temperatures of 1900–2600°C (3452–4712°F). It is an extremely hard, heat-resistant material which is a good conductor of heat and electricity. It is made as fine-ground particles, as fibers, and "whiskers" of extraordinary tensile strength. Its major uses are as a refractory, as an abrasive in grinding wheels, etc., and in filament-wound composites. *See also* abrasive; refractory.

silicone. A linear polymeric structure derived from siloxanes by substitution of an organic group for the oxygen atoms above and below the silicon atom; chlorine or other halogen may also be included in the compound:

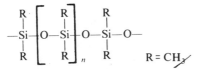

Silicone polymers may be liquids or resinous products; the liquids (sometimes called silicone oils) are used for water-repellency, heat-transfer, and dielectric liquids and foam-inhibiting agents; the resins form excellent adhesive bonding agents, antivibration pads, and electric insulating materials. They have unique elastomeric properties, especially when cross-linked with organic peroxides to give so-called silicone rubber. *See also* siloxane.

silicone oil. *See* silicone; dielectric.
silicone rubber. *See* silicone.
siloxane. A compound of silicon and oxygen in which each atom of silicon is bonded to four oxygen atoms, forming a tetrahedral structure, in a manner analogous to the bonding of carbon to hydrogen in methane, the bonds being of about the same strength in each case:

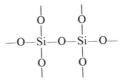

This structure is found in the dioxide and in silicates generally, where the SiO_4 groups occur in chains or rings. *See also* silicone.

silver. An element.

Symbol	Ag	Atomic Wt.	107.868
State	Solid	Valence	1
Group	IB	Isotopes	2 stable
Atomic No.	47		

Silver, m.p. 961°C (1762°F), is still regarded as a precious metal in the arts, but it has many important scientific applications resulting from the unusual properties of the element and its compounds. It is a relatively soft metal which can be readily electrodeposited. It is extremely sensitive to contamination by sulfur and its compounds (tarnish). It has the highest electrical conductivity of any metal and thus is widely used in electrical circuitry. It has antiseptic properties which are utilized in medicine and dentistry, and it plays an essential part in the formation of photographic images. Other applications are in special batteries, alloys, and linings for chemical equipment. Silver forms compounds with the halogen elements, oxygen, and nitrogen. The light-sensitive halides are used in photography; the iodide is effective in atmospheric nucleation for rain-making. Many silver compounds are highly explosive (acetylide, peroxide, nitride, and permanganate), and the greatest caution should be observed in handling them. Sterling silver contains about 7% copper.

silver acetylide. *See* explosive.
silver bromide. *See* developing agent; photochromic.

silver iodide. *See* nucleation.

silvicide. Derived from the Latin *silva* ("forest"), this term refers to nonselective herbicides which kill or defoliate woody plants, bushes, and small trees. Ammonium sulfamate has been used effectively for this purpose. Defoliants such as picloram may also be classed as silvicides if the trees die as a result. *See also* herbicide; defoliant.

sinter. Heating of metal powders, clays and similar earths to a temperature well below the melting point; this treatment results in softening and agglomeration of the particles to form a microporous product which can be pressed or molded to desired shape. The temperatures and pressures used are critical for any given material, since the size and distribution of "voids" can be closely controlled. Sintering is the major operation of powder metallurgy.

sizing. An adherent coating of a soft, tacky nature applied to textile fibers, paper, and various other products to give a smooth finish, to decrease water absorption, and (for paper) to furnish a glossy surface suitable for halftone reproduction. Materials often used are solutions containing starch, rosin, natural gums, casein, etc. Special types of sizing materials called slashing compounds are used in weaving cotton and rayon fibers.

slack. (1) Descriptive of a soft paraffin wax resulting from incomplete pressing of the settlings from the petroleum distillate. Though it has some applications in this form, it is actually an intermediate product between the liquid distillate and the scale wax made by expressing more of the oil.

(2) Specifically, to react calcium oxide (lime) with water to form calcium hydroxide (slacked or hydrated lime); the reaction is $CaO + H_2O \rightarrow Ca(OH)_2 + heat$. The alternative spelling *slake* has the same meaning.

slag. *See* gangue.

slake. *See* slack (2).

slashing. *See* sizing.

slip. A rather thin aqueous dispersion of clays (aluminum silicates) used for coating ceramic ware; a fluxing agent is often added to facilitate melting when the assembly is fired. A glassy finish called a glaze is thus obtained. *See also* flux; glaze; slurry.

slow neutron. *See* neutron; fast (2).

sludge. A thick, viscous mixture of solids and water often resulting from filtration; sewage sludge results from the settling of organic waste products (Imhoff sludge and activated sludge). *See also* activated sludge.

slurry. A dilute suspension of particulate solids in water, for example, thin aqueous dispersions of clay, fiber particles, or metal powders. Slurry explosives are suspensions of ammonium nitrate in water, also containing low percentages of aluminum powder or active explosives.

Sm Symbol for samarium; the name is thought to be derived from the mineral source in which the element was discovered (samarskite).

smelt. A term used by metallurgists to denote the separation of a metal from its ore by heating at high temperature in a confined vessel; the liquefied metal falls to the bottom of the container, the lighter ore components remaining on the surface as slag. A reduction reaction is involved. *See also* roast.

smog. *See* air pollution; photochemistry; sulfur dioxide.

smoke. An aerosol composed of colloidal particles of combustion products dispersed in air combined with the gases resulting from combustion. Smokes from certain woods are used in curing various food products, such as fish, ham, and turkey. Specialized types of smoke include the fumes from heated metal surfaces and so-called military or chemical smoke, which is evolved by reaction of appropriate compounds, e.g., titanium tetrachloride.

smokeless powder. *See* deflagration; ethyl acetate; explosive; nitrocellulose.

smut. *See* mold (1); rust (2).

Sn Symbol for the element tin; derived from the Latin word *stannum* (tin).

SNG. Abbreviation for synthetic natural gas.

soap. A class of chemical compounds formed by a saponification reaction, the soap being a metallic salt of a fatty acid derived either from animal fats and vegetable oils in the form of esters (glycerides) or from petroleum, rosin, etc. Sodium and potassium are the metals used in water-soluble soaps, which are characterized by their emulsifying and detergent action in aqueous solution. They are still widely used notwithstanding the popularity of synthetic detergents. Soaps formed by lead, zinc, and alkaline-earth metals, which are heavier than sodium, are called both heavy-metal soaps and metallic soaps. The former term seems preferable, since sodium soaps are also metallic. The heave-metal soaps are not water-soluble; thus, they are used in lubricating greases, paint driers, and fungicides, e.g., copper naphthenate. *See also* saponification.

Society of Chemical Industry. This society was founded in London in 1881 to advance applied chemistry in all its branches. The American Section was established in 1894. In 1906, on the anniversary of the "birth of the coal-tar industry," it founded the famous Perkin Medal, which has since been awarded annually for outstanding achievement in chemistry in the United States. It has also offered the Chemical Industry Medal since 1932, given for conspicuous service to applied chemistry. Perkin Medalists include L.H. Baekeland, F. G. Cottrell, Irving Langmuir, Herbert Dow, A. D. Little, R. R. Williams, Glenn T. Seaborg, and Roger Adams.

Society of the Plastics Industry. (SPI). An incorporated technical organization serving the needs of the entire plastics industry in the United States. It establishes standards for the properties and selection of materials and for product design and engineering. With it are associated the Plastics Pipe Institute and the Reinforced Plastics/Composites Institute. Its offices are in New York city.

soda. *See* sodium carbonate.

soda ash. *See* sodium carbonate; boiler scale; Solvay process.

sodium. An element.

Symbol	Na	Atomic No.	11
State	Solid	Atomic Wt.	22.98977
Group	IA	Valence	1

Sodium, the sixth most abundant element in the earth's crust and the most common alkali metal (Group IA), is a soft electropositive metal, m.p. 97.83°C (208°F); its compounds are more numerous than those of any other metal. It occurs in many natural compounds in the form of chloride, borate, carbonate, sulfate, and others. The metal is made on a large scale by the electrolysis of molten sodium chloride mixed with calcium chloride. It reacts vigorously with water with formation of hydrogen and sodium hydroxide and the strong evolution of heat, and should be kept out of contact with water, as well as with air and oxygen. It readily forms compounds with the halogens, oxygen, and many other nonmetals, and alloys with a large number of metals, especially lead and mercury. Its chief uses are

in the manufacture of tetraethyl lead and other organic and inorganic chemicals; the production of several metals, e.g., titanium and zirconium; as a heat-transfer agent (liquid) in reactors; and in electric power conduction. Sodium ion is essential to man and other animals whose chief intake of it is in the form of sodium chloride. Sodium is dangerous to handle, as it is corrosive to the skin, and also flammable when exposed to heat.

sodium acetate. *See* buffer; camphor.

sodium alginate. *See* protective colloid.

sodium aluminum hydride. *See* hydride.

sodium amide. *See* amide.

sodium benzoate. *See* pickling (1).

sodium bicarbonate. *See* ammonia; Solvay process; baking powder; blowing agent; antacid.

sodium bisulfite. *See* antichlor; decontamination.

sodium borate. *See* borax.

sodium carbonate. A white powder, the sodium salt of carbonic acid, Na_2CO_3, melts at 851°C (1564°F) and decomposes at higher temperature to yield CO_2 and Na_2O. It is very soluble in water to form alkaline solutions. A tonnage or "heavy" chemical, with a current production of over 8 million tons/year, it is obtained (57%) from natural underground deposits in Wyoming, (11%) from natural salt lakes in California, and (32%) synthetically from the Solvay process, sometimes called the ammonia soda process. The latter formerly was the chief source but has declined in recent years in favor of the natural sources. Soda ash is the common synonym for sodium carbonate. It is essential in the manufacture of soda-lime glass which consumes over half the soda ash used; other important uses are in the production of chemicals, pulp and paper processing, soaps and detergents, water treatment, and the extraction of alumina from bauxite. The decahydrate is called washing soda or sal soda. *See also* Solvay process; alkali.

sodium chlorate. *See* chlorate; chlorine dioxide.

sodium chloride. One of the more abundant salts, NaCl is present in seawater in 2.6% concentration and also in subterranean deposits as brines (Michigan) and solid formations in Louisiana and Texas (salt domes), near Syracuse and Detroit, and in western Utah (Salt Lake). It is obtained either by surface or underground mining or by evaporation of brines or seawater. It is also called rock salt (large crystals) and table salt (ground and refined).

It is readily soluble in water and dissociates to form an electrolytic solution; this also results in raising the boiling point and lowering the freezing point of water to a quite appreciable extent; for example, the 2.6% in seawater lowers its freezing point by 2.8°C (5°F). Its major commercial uses are in the manufacture of chlorine and sodium hydroxide by electrolytic separation, and of sodium carbonate by the Solvay process; there are a multitude of minor uses: in the food industry (seasoning and preservation), leather industry (hide preservation), soap industry (salting out), regeneration of ion-exchange media, etc. Sodium chloride is an essential metabolic factor, the concentration in the body being automatically regulated by the perspiration and thirst mechanisms.

sodium chlorite. *See* chlorite.

sodium fluoride. *See* fluoridation; fluoride.

sodium fluosilicate. *See* fluoride.

sodium formate. *See* formic acid.

sodium hydrosulfite. *See* strip; vat.

sodium hydroxide. A powerful base, often called caustic soda, having the formula NaOH and reaidly dissociating in water solution to form Na$^+$ and OH$^-$ ions. The highest-volume alkaline material used by industry, it is produced in electrolytic cells in which an electric current is passed through a strong aqueous solution of sodium chloride. It is a white or grayish solid when dry, which absorbs moisture from the air. Available in both solid form and water solutions of various strengths, it is highly corrosive to skin and must be handled with care. A noteworthy property of sodium hydroxide is its ability to react with cellulose and to separate the undesirable lignin component from woody structures. This accounts for its use in making kraft paper pulp by high-temperature digestion. It combines chemically with cellulose to form alkali or sodium cellulose, as in the manufacture of rayon and cellophane by the viscose process, and of the water-soluble resin called carboxymethylcellulose. Cotton and other cellulosic fibers are mercerized by room-temperature treatment with a sodium hydroxide solution, followed by washing with water. Another important use is in the manufacture of soap by reaction with fatty acids. *See also* electrolysis; cellulose; mercerizing.

sodium hypochlorite. *See* bleach; hypochlorous acid.

sodium nitrate. *See* deflagration.

sodium perchlorate. *See* perchlorate.

sodium phosphate. *See* dibasic; phosphate.

sodium polyphosphate. *See* polyelectrolyte.

sodium sulfate. An anhydrous salt of sulfuric acid which occurs in nature in the form of salt cake (western U.S.) and is manufactured by the reaction 2NaCl + H$_2$SO$_4$ 1 2HCl + Na$_2$SO$_4$; it is also obtained in the manufacture of phenol. The hydrated from is known as Glauber's salt. It is used in tonnage quantities in the production of paper and glass, and has minor applications in the textile, leather, and pharmaceutical fields. As its heat-storage capacity is several times that of water, it may be useful in solar heating installations.

sodium thiosulfate. *See* antichlor.

soil. *See* nitrate; ion exchange; nitrogen.

soil conditioner. *See* acrylic resin; algae.

soil nutrient. *See* fertilizer, trace element.

sol. A term used by physical chemists to denote a liquid colloidal dispersion of solid particles, for example, gold, in water; the dispersion is stabilized by electric charges, either all negative or all positive, the particles being kept from colliding or flocculating by mutual electrical repulsion. The ''sol'' state is opposed to the ''gel'' state, in which the dispersion is a thick, semisolid mass. The word is a shortened form of *solution*, not of *solid. See also* gel.

solar cell. A device for converting the radiant energy of the sun into electrical or thermal energy for use in power generation and heating and cooling of buildings,. There are two basic types. One type utilizes a p-type semiconductor, such as a ribbon of pure silicon, either crystalline or amorphous, the energy conversion being electrical in nature. Selenium and cadmium sulfide can also be used. The other type is a heat-collecting cell which may be 3 feet wide by 6 or more feet long, utilizing either a circulating gas or glass plates and an aluminum absorber plate. A variation of this is the parabolic concentrator developed at Argonne National Laboratory, consisting of channels made of epoxy resin with vapor-deposited aluminum surface; each wall of the channels is a portion of a parabolic surface.

solder. *See* alloy.

solid state. In physical chemistry, this term refers to matter in the crystalline form, as opposed to the liquid state, which is noncrystalline (amorphous). Some materials that are apparently solid, notably glass and certain high polymers, are classed as liquids because, in spite of their strength and hardness, they lack a crystal structure. The crystals of solid elements and compounds are three-dimensional structures (cubes, rhomboids, tetrahedrons, etc.) held together by chemical bonds of one kind or another in a geometrical pattern called a lattice. Much has been learned about the shape and internal arrangement of crystals by x-ray diffraction.

Lattices may exhibit imperfections of two types: (1) vacancies due to lack of enough atoms to saturate the covalent bonds in the lattice; and (2) the presence of atoms of other elements in trace amounts (impurities). These imperfections have a profound effect on the properties of a crystal, especially as regards electrical conductivity and optical behavior. The study of these defects and of their specific modifying effects led to the development (since 1945) of semiconducting devices (transistors), lasers (coherent light), and numerous applications to computer technology. *See also* crystal; impurity; semiconductor.

solute.   *See* solvent.

solution.   A uniform molecular or ionic mixture of one (or more) substance(s) (solute) in another (solvent). This is called a true solution, in contrast to a colloidal solution, in which the dispersed material is in the form of extremely small particles (1 micron or less); these systems are more properly called dispersions. The molecules or ions of a solution are the solute, and the substance in which they are dissolved is the solvent. Polar compounds, for example, metal hydroxides and halides, form ions in water solutions, but nonpolar compounds such as sugar remain in molecular form. Concentrations of solutions are indicated by such standard definitions as molarity, normality, and molality.

The most familiar type of solution is that of a solid in a liquid. Glass is an undercooled solution of metallic oxides in molten silica, which is technically liquid but physically rigid. There are many instances of solutions of liquids in liquids (alcohol-water). There are also solutions of solids in solids (alloys). Polymerization is often carried out in a solution in which the monomer is the solute. *See also* solvent; concentration; dispersion.

solvation.   A term used by physical chemists to denote the tendency of finely divided solid surfaces to adsorb a diffuse layer or film of a solvent in which they are dispersed as a result of electrostatic forces between the surface of the particle and the solvent molecules. This is most pronounced in water dispersions of colloidal particles. Proteins tend to adsorb or ''bind'' water in this way. *See also* adsorption; hydration (**2**).

Solvay process.   Developed in the 1860s by a Belgian chemist, Ernest Solvay (1838–1922), this process synthesizes sodium bicarbonate ($NaHCO_3$) and sodium carbonate (soda ash) (Na_2CO_3) from sodium chloride, ammonia, and carbon dioxide obtained from calcium carbonate ($CaCO_3$) derived from limestone, oyster shells, etc. It is also known as the ammonia-soda process. Essentially the reactions involve formation of ammonium bicarbonate (NH_4HCO_3) from ammonia, water, and carbon dioxide; this reacts with sodium chloride to yield sodium bicarbonate and ammonium chloride. The sodium bicarbonate breaks down to sodium carbonate (soda ash) on heating. Ammonia is regenerated by reacting the ammonium chloride previously formed with calcium hydroxide, calcium chloride being the by-product. The availability of natural soda ash, from which the bicarbonate can be made, has reduced the economic importance of this process in recent years. *See also* sodium carbonate.

solvent.   Derived from the Latin, meaning ''set free'' or ''loosen,'' this term designates a liquid which can reduce certain solids or liquids to molecular or ionic form by relaxing the intermolecular forces that unite them. This is the exact meaning of *dissolve*, which comes from the same root as *solvent* and *solution*. By far the most common inorganic solvent is water (often erroneously called the universal solvent). It dissolves such solids as sugar, starch, and many ion-forming metal salts, as well as numerous liquids and gases (sulfuric acid, alcohol, formaldehyde, hydrogen chloride). Hydrocarbon liquids such as benzene dissolve organic materials such as fats, waxes, resins, and oils. In all cases the system is a uniform mixture of the solvent and the dissolved substance (the solute) in either molecular or ionic form. *See also* solution; miscible.

solvolysis.   A chemical reaction occurring in a solution between the solvent and the solute in which the solvent molecules lose their identity by decomposition. Examples are hydrolysis, alcoholysis, and ammonolysis, hydrolysis being the most common. *See* hydrolysis.

sorb, sorption.   A term used by chemists in cases when it is not clear whether adsorption or absorption is involved, or when both are occurring simultaneously.

sorbitol.   *See* humectant; polyhydric alcohol.

sour.   As an adjective, this term characterizes (a) one of the fundamental taste sensations, usually evoked by an acid in food products, e.g., milk (lactic acid), vinegar (acetic acid), or grapefruit (citric acid); and (b) a gas or gasoline containing a relatively high proportion of sulfur. As a noun, it refers to several compounds of sodium and fluorine used to neutralize bleaching agents in the washing of textile fabrics (laundry sour).

spandex fiber.   *See* polyurethane.

spar.   (1) A term used by mineralogists to designate a type of crystalline material such as Iceland spar or feldspar, usually containing calcium carbonate or an aluminum silicate; fluorspar is calcium fluoride.

(2) A weather-resistant varnish originally used for coating wooden decks of ships, which may be the reason for its name. It contains a drying oil, rosin, or phenolic resin, and solvent. Some early types contained sulfur.

specific gravity.   *See* density (1); water; Baumé.

spectroscopy.   A branch of analytical chemistry devoted to the determination of the structure of atoms and the chemical composition of molecules by measuring the radiant energy they absorb or emit in any of the wavelengths of the electromagnetic spectrum (gamma and x-radiation, ultraviolet, infrared, visible, microwave, and radiofrequency). When atoms or molecules are excited by energy input from an arc, spark, or flame, they respond in a characteristic manner; their identity and composition are signaled by the wavelengths of incident light they absorb or emit, the emission spectra being in the form of lines (atoms) of distinctive color, such as the yellow sodium D line, or bands (molecules). The intensity of these varies with the amount of an element present and thus affords a basis for quantitative analysis of the sample. The earliest achievement of spectroscopy was analysis of the sun's spectrum in 1814 by Joseph von Fraunhofer (1787–1826), a Bavarian optician; still more important was Moseley's use of x-ray spectro-

scopy to determine precise locations of elements in the Periodic Table. *See also* excited state; infrared spectroscopy; radiation; nuclear magnetic resonance; x-ray.

sp. gr.   Abbreviation for specific gravity.

sphingomyelin.   *See* phosphatide.

spinel.   Any of a group of hard, glassy minerals of various colors of quite wide geographical occurrence. Chemically, spinels are compounds or double oxides of such metals as iron, magnesium, aluminum, chromium, etc., common types being magnesium aluminate ($MgAl_2O_4$) and magnetite ($FeFe_2O_4$). Some types can be made synthetically. They have useful crystallographic properties and are closely related to ferrites. *See also* ferrite; magnetite.

spirits.   This term is an unscientific holdover from the early years of chemistry during which it referred to volatile alcoholic distillates (e.g., spiritus frumenti for grain alcohol); it is still used in pharmacy for alcoholic solutions of ammonia and camphor. In England the term petroleum spirits denotes a petroleum distillate similar to naphtha. It was undoubtedly adapted from the root meaning of "breath," which suggests high volatility.

splitting.   *See* fat; fatty acid.

sponge.   (1) Any material, natural or synthetic, which has or can be made to have a cellular structure, that is, dispersions of gas globules within the solid. They are thus solid foams, exemplified by cake, bread, and cellular plastics, as well as by natural sponges.

(2) A metal converted to a porous particulate form by reduction of its oxide for use as a catalyst or adsorbent; iron and nickel are prepared in this way. Also, the initial porous form of metal obtained by the reduction of a metal compound, e.g., the oxide or chloride, with such agents as carbon, hydrogen, magnesium, sodium, etc., usually at temperatures below the melting point of the metal.

Sr   Symbol for the element strontium, the name being derived from Strontium, a town in Scotland where it was discovered.

stabilizer.   Any substance added to another substance or mixture to prevent or retard a chemical or physicochemical change, at least for a considerable time. As this function can occur in many different ways, the term is a very general one, which includes antioxidants, inhibitors, emulsifiers, protective colloids, etc. It is formed from the term *stable,* meaning "unchanging."

stain. *See* bacteria.

stand oil. *See* linseed oil.

stannic, -ous.   Most inorganic tin compounds are named from the original Latin word for "tin" (*stannum*), e.g., stannic sulfide, stannous fluoride; organic compounds retain the English name, e.g., tributyltin chloride, etc.

starch.   A carbohydrate polymer or polysaccharide which is a mixture of amylose (a linear polymer of the sugar glucose) and amylopectin, another glucose polymer comprised of interlinked chains. The starch molecule is made up of repeating units of these structures joined by oxygen atoms; it contains alcohol units and hydroxyl groups attached to the carbon skeleton. Starch occurs widely in plants and is particularly abundant in corn, potatoes, tapioca, and rice. It is formed by hydrolysis of cellulose and can itself be hydrolyzed to sugars with appropriate catalysts (enzymes). It is hydrophilic with warm water, forming stiff gels which are useful in finishing of fabrics, laundering, etc., and as thickening agents in food preparations (cornstarch). It has a number of specialized uses, for example, in explosives (nitrostarch), as an indicator, as a filler in textile products, and as an abherent. It has important metabolic functions, being stored in the liver as glycogen. A number of water-soluble products (so-called resins) are made by treating starch with acetic anhydride, chlorine, and other chemicals. *See also* carbohydrate; sugar.

Staudinger, Hermann (1881–1965).   A German organic chemist, awarded the Nobel Prize in 1953 for his pioneer work on the structure of macromolecules and polymerization. Modern high-polymer chemistry, especially the selective use of catalysts in the formation of stereospecific polymers, is based on his original discoveries and theoretical observations. *See* stereospecific; Ziegler catalyst.

steam.   Water vapor formed at 100°C (212°F) and atmospheric pressure, having a latent heat of condensation of 540 calories per gram. It has many chemical and industrial uses: production of synthesis gas, cracking of naphtha and gas oil, production of hydrogen by the steam-iron process, in food processing, as a cleaning agent, in rubber vulcanization, in distillation of plants for essential oils and perfumes, and as a source of heat and power. Geothermal steam is being utilized where available for industrial and domestic heating.

stearic acid   (pronounced stee-ar-ic). A saturated fatty acid having the formula $C_{17}H_{35}COOH$; it is a dull whitish solid which softens at about 65°C (150°F). The most widely used material of its kind in industry, it is derived from animal fats such as tallow by hydrolysis, followed by refining and pressing. Commercial grades may contain substantial proportions of other fatty acids. One of the major uses of stearic acid is as a raw material for soap manufacture by reaction of its ester with a metal hydroxide. Another is in rubber mixes, in which it has two distinct functions: (a) it is particularly effective in wetting and dispersing carbon black particles in unvulcanized rubber, and is thus an important ingredient of tire treads and other abrasion-resistant products; (b) it notably increases the vulcanizing efficiency of acidic accelerators. Stearic acid also is used in the manufacture of stearates, candles, and shoe polishes, and as an emulsifying agent. *See also* soap; fatty acid.

steel.   Basically, an alloy of iron and carbon, made from molten pig iron from which all but a controlled percentage of residual carbon has been removed by oxidation. Steels containing from 0.03 to 0.3% carbon are called low-carbon, while those having from 0.75 to 1.7% are known as high-carbon. The presence of some carbon is essential, for it has a significant effect on the physical properties of steel, such as hardness, strength, machinability, etc. Steels are often alloyed with other metals to increase toughness, corrosion resistance, and heat resistance. Stainless steels (noncorrosive) contain chromium and nickel (a well-known type having 18% chromium and 8% nickel); tool steels designed for high-speed metalcutting are alloyed with molybdenum, tungsten, vanadium, etc., often in combination.

ster-, stereo-.   A term derived from the Greek, meaning "three-dimensional" which appears in a number of chemical words both as a prefix and as a basic term. Stereochemistry and stereospecific refer to the spacial configurations of molecular structures; compounds classed as sterols and steroids are so named because of their three-dimensional structure. The same applies to "steric hindrance."

stereoblock.   *See* block polymer.

stereochemistry.   That aspect of chemistry that

is concerned with the three-dimensional structure of organic compounds and the spatial relationships of their constituent atoms or groups. The major subdivisions of this study are in the fields of high-polymer chemistry (stereoregular or stereospecific polymers); isomerism, both optical and geometric; and conformational analysis. *See also* steroid; stereoisomer, stereospecific.

stereoisomer. Either of two isomeric compounds whose only difference is the three-dimensional arrangement of their constituent atoms or groups relative to one another. Such pairs of compounds may be either geometric (*cis-trans*) isomers or optical (D,L) isomers. Among the former are a number of metal coordination complexes, especially of cobalt, where the metal acts as a ligand. In the case of asymmetric (mirror-image) compounds, the enantiomorphs formed are stereoisomers. The four-carbon sugars, for example, have two pairs of enantiomorphs (D and L configurations); in such instances, the two D and the two L forms which are not enantiomorphs are called diastereoisomers. Stereoisomers occur widely in natural products, particularly among the carotenoids, some of which have over 100; for example, beta-carotene has 20. *See also* enantiomorph; asymmetry; isomer.

stereoregular. *See* stereospecific.

stereospecific. This term and its synonym *stereoregular* were introduced into polymer chemistry in the 1940s to designate three-dimensional organic structures whose molecules have an ordered (crystalline) orientation rather than the random (amorphous or atactic) form of most natural high polymers. This regularity of molecular arrangement results either from the nature of the monomer or from the use of unique polymerization catalysts. The original work on developing such catalysts was done by Hermann Staudinger (1881–1965), a German chemist, Nobel Prize 1953, Karl Ziegler (b. 1898), a German chemist, Nobel Prize 1963, and Giulio Natta, an Italian chemist, Nobel Prize 1963. They found complexing or coordination catalysts, such as lithium or aluminum alkyl plus titanium dichloride, to be capable of so controlling the three-dimensional structure (tacticity) of styrene and diene polymers and copolymers as to bring about marked improvement in their strength, heat resistance, and other physical properties. As a result, it became possible to synthesize superior types of elastomeric materials such as polyethylene and polypropylene, as well as polyisoprene and a number of stereo rubbers. *See also* isotactic; syndiotactic; cross-linking; Ziegler catalyst.

steric hindrance. *See* hindrance.

sterilant. *See* antiseptic; disinfectant; chemosterilant.

steroid (pronounced ster-o-id).   Any of a considerable number of organic compounds having a polycyclic 17-carbon phenanthrene nucleus shown below, the numerals indicating location of carbon atoms. Steroids include the monohydric alcohols called sterols, of which cholesterol is a well-known example, hormones such as progesterone, bile acids (cholic acid and its derivatives), and provitamins (ergosterol). Among the hormones one of the most familiar is the adrenal cortical steroid cortisone found effective in alleviating arthritis. Most of the more important steroids have been synthesized. *See also* sterol.

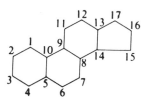

sterol.   A term derived from a Greek word meaning "three-dimensional" and designating a monohydric polycylic alcohol comprised of a series of fused rings plus one hydroxyl group; frequently a side-chain is also present.

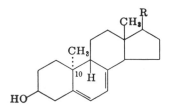

The sterols are closely related chemically to the steroids; their molecules are three-dimensional, each containing 20 or more carbon atoms in the skeleton. They are formed in plants (phytosterols) and in fungi such as yeast, as well as in animals and man. Among the best-known are ergosterol, a precursor of vitamin D$_2$, and cholesterol, an important factor in metabolism. *See also* steroid; cholesterol.

stibine. *See* antimony.

Stoddard solvent. *See* cleaning; solvent.

stoichiometry (pronounced stoy-key-ometry). Study of the mathematics of the material and energy balances (equilibrium) of chemical reactions. It is based on the laws of conservation of mass and energy and on the combining weights of elements. In a given case, it concerns the molecular amounts of the reacting substances and their products at a specific temperature and pressure, which is normally 1 atmosphere and 0°C (32°F), as well as the heat absorbed or evolved (energy balance). Such analysis of a reaction shows that the total material and energy input is equal to the total output, that is, the mass-energy relationship is constant. *See also* material balance; equivalent weight.

Stokes' Law. A mathematical expression of the rate at which particulate solids deposit from a liquid or gas in which they are dispersed, formulated by Sir George G. Stokes, a British mathematician and physicist (1819–1903), famed for his work in hydrodynamics. Factors in the equation are the radius and density of the particles, the density of the liquid or gas, the viscosity, and the gravitational constant g, the velocity of the particle when it reaches the bottom of the container is the distance of fall divided by the time, as computed on the basis of the foregoing factors. Examples of the application of this law are the settling rate of dust particles from air and of solids from water. *See also* sedimentation.

storage battery. *See* battery.

STP. Conventional abbreviation for standard temperature and pressure. The standard temperature is 0°C (32°F) and the pressure is 1 atmosphere (760 mm of mercury).

straight-run gasoline. *See* gasoline.

streptomycin. *See* antibiotic.

strip. (1) In dyeing technology, the removal of a dye from a fabric when this is necessary for any reason, for example, recycling. Chemicals used for this operation are called stripping agents, e.g., sodium hydrosulfite.

(2) In physical chemistry, hydrogen atoms that have lost their electron by the action of intense heat, as in plasmas or on the sun, i.e., stripped hydrogen.

(3) Synonym for desorption. *See* absorption.

strontium. An element.

Symbol	Sr	Atomic Wt.	87.62
State	Solid	Valence	2
Group	IIA	Isotopes	4 stable
Atomic No.	38		

A comparatively soft metal, m.p. 770°C (1418°F), of the alkaline-earth group, similar in many respects to calcium. It is quite reactive and quickly forms an oxide coating when exposed to dry air. It has strong reducing properties, decomposing water to evolve hydrogen; it must be protected from moist air by storage under naphtha. In powder form it catches fire spontaneously at room temperature. It has no important commercial uses; though the bright red color of its flame is utilized in pyrotechnic devices. Two artificial isotopes (89 and 90) formed by nuclear reactions emit highly dangerous ionizing radiation, the 90 isotope having a half-life of 28 years; these are so-called bone-seeking isotopes which contaminate plants and thus appear in milk; progress has been made in reducing this hazard by decontamination with vermiculite. *See also* bone seeker.

structural antagonist. *See* antagonism.

strychnine. *See* alkaloid.

styrene. *See* monomer; polystyrene; addition polymer.

sublime. The change of a substance from the solid to the vapor phase without passing through the intermediate liquid phase. Solid carbon dioxide exhibits this behavior at room temperature; it can also be observed when snow vaporizes on exposure to sunlight on a mild winter day. Other substances that can sublime are sulfur, camphor, and naphthalene. Sublimation can be used as a purification method, analogous to distillation. *See also* heat of sublimation.

subshell. *See* orbital.

substance. Any unit of matter at or above the atomic level that is of identical composition throughout (homogeneous) and that cannot be separated by mechanical means. Thus, all chemical elements and compounds are substances, whereas solutions, dispersions, alloys, and composite materials are mixtures. For example, carbon, water and alcohol are substances; milk and petroleum are mixtures. *See also* mixture; homogeneous; material.

substituent. Any chemical entity that takes the place of another during the course of a replacement reaction (also called a substitution reaction), for example, where the hydrogen of an acid is replaced by a metal. The substituents may be atoms, ions, groups, or radicals. *See also* replacement.

substrate. (1) A term used by biochemists to denote any organic compound that is acted upon by an enzyme, e.g., a protein, carbohydrate, fat, or sugar. The name of the enzyme itself is usually formed by adding the suffix *ase* to the name of the activated substance, for example, protease, lipase, etc. though there are some exceptions.

(2) The term is also loosely used in the pigment and coatings fields to refer to the material being electroplated, painted, etc.

succinic anhydride. *See* acidulant.

sucrose. A disaccharide sugar whose molecule is formed of two glucose units; it contains two alcohol groups ($-CH_2OH$) and six additional hydroxyl groups, the formula being $C_{12}H_{22}O_{11}$. It has a strong positive optical rotation. Sucrose occurs in all fruits and vegetables, being especially abundant in beets and sugarcane. The sugar of commerce is obtained by crushing, extraction, and evaporation of beet and cane juice, the residual product (molasses) being used for manufacture of industrial alcohol by fermentation; it is first hydrolyzed to invert sugar (a levorotatory mixture of fructose and dextrose) by the action of the enzyme invertase; this rearranges to alcohol and carbon dioxide with the aid of another enzyme (zymase). Sucrose is an important energy source in metabolism and is the primary sweetening agent in the food industry. *See also* sugar; dextrose; molasses; fermentation.

sugar. The simplest of the broad carbohydrate series of compounds formed from carbon dioxide and water by the photosynthetic activity of plants. Sugars are subdivided into three groups. (1) The simple sugars are called monosaccharides, as they contain but one characteristic molecular unit having six carbon atoms (hexose sugars). (2) More complex types (oligosaccharides) have from two to ten such units. (3) There are also high-polymer sugars called polysaccharides which occur in natural gums, pectins, seaweeds, as well as in starch and cellulose. Glucose (dextrose) is a typical monosaccharide

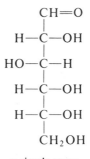

a simple sugar
(monosaccharide)

found in corn and grapes, and sucrose (table sugar) is a disaccharide obtained from beets and sugarcane. These compounds are characterized by optical activity, that is, their molecules rotate plane-polarized light either to the right (dextrose) or to the left (levulose). Invert sugars are slightly levorotatory mixtures of dextrose and levulose obtained by hydrolysis of sucrose. Ethyl alcohol is readily formed from sugars in fermentation reactions catalyzed by enzymes. Sugar also occurs in milk (lactose) and in blood (glucose). *See also* sucrose; phycocolloid, polysaccharide.

sulfanilamide. *See* antagonism.

sulfonation. Addition of an SO_2OH group to an organic compound, the group being bonded directly to a carbon atom. The resulting compounds are called sulfonic acids (e.g., benzenesulfonic acid) and are formed by reaction with sulfuric acid. Since the inclusion of this group provides water-solubility, compounds containing it in combination with amines and various benzene derivatives are widely used as dyes and dye intermediates, for example, 1-naphthylamine-6-sulfonic acid.

sulfur. An element.

Symbol	S	Atomic Wt.	32.06
State	Solid	Valence	2,4,6
Group	VIA	Isotopes	4 stable
Atomic No.	16		

Sulfur is a yellow solid which occurs in elemental form (often in the locality of salt deposits) in southwestern United States, Sicily, and the Middle East. Its low melting point of 115.2°C

(240°F) permits extraction in liquid form by pumping superheated water into the limestone deposit (Frasch process invented in 1890 by Herman Frasch (1852–1914), an American chemist and inventor). It is also extracted from certain iron ores (pyrites) and from gypsum, as well as from smelting wastes and petroleum fractions, e.g., fuel oil. Sulfur has a number of allotropic forms; the vapor molecule may have up to 10 atoms, while the liquid is a mixture of polymer chains and cyclic structures having 6 to 8 atoms. The liquid abruptly increases in viscosity at 160°C (320°F) and acquires unique viscoelastic properties.

Sulfur has chemical properties similar to those of oxygen; it reacts strongly with alkali and alkaline-earth metals, as well as with bromine and chlorine. Many transition metals occur in the form of sulfide ores. Sulfur burns above 260°C (500°F) and sublimes when heated at lower temperatures. A vital raw material of the chemical industry, its major uses are in the manufacture of sulfur chemicals, e.g., sulfuric acid, vulcanization of rubber, processing of paper pulp, certain dyes, animal repellents, and insecticides. Sulfur is a constituent of most proteins. It may be safely ingested and is an ingredient of curative tonics; it also occurs in certain spring waters (Saratoga).

sulfur chloride. A reddish-yellow liquid, b.p. 135.6°C (277°F), which evolves poisonous fumes at room temperature and is extremely irritating to the eyes and skin. Its formula is S_2Cl_2 (sulfur monochloride); it is produced by direct reaction between gaseous chlorine and liquid sulfur. It is used in the manufacture of chemicals, in dilute solution as a vulcanizing agent for dipped rubber products, and in the manufacture of factice and insecticides. It should be kept out of contact with water, with which it reacts strongly. It is classed as a corrosive liquid. Sulfur dichloride (SCl_2) b.p. 59°C (138°F), has similar properties and applications.

sulfur dioxide. A noxious gas (SO_2), b.p. −10°C (14°F), resulting from the combustion of sulfur and sulfur-bearing materials such as iron sulfide (pyrites), petroleum, and other carbonaceous substances. A notorious air contaminant and a component of photochemical smog and acid rain; it has a strongly irritating effect on sensitive body tissues and is also an important factor in the corrosion and tarnishing of metals. Its maximum permissible concentration in air is 5 parts per million. It is noncombustible.

For industrial use, it is produced by burning sulfur or iron pyrites (FeS_2) as the first step in the manufacture of sulfuric acid. Other major uses (including its water solution, sulfurous acid, H_2SO_3) are in making sulfite paper pulp, as an intermediate in the production of inorganic sulfur compounds of commercial importance (sodium sulfate, thiosulfates, hydrosulfites), as a bleaching agent, food preservative, and fumigant. Before the advent of chlorofluorocarbons, it was employed in liquid form as a refrigerant.

sulfuric acid. An inorganic (mineral) acid, probably the most important of all the inorganic chemicals used in industry because of its high reactivity, versatility, and comparatively low cost. It is the highest-volume chemical produced in the U.S. It has the formula H_2SO_4 and is available in many grades and strengths. The pure acid is highly corrosive and will attack most metals. It is a strong electrolyte and forms H^+ and $SO_4^=$ ions in water solution. Sulfuric acid is made by combustion of sulfur or pyrites (iron sulfide); the resulting sulfur dioxide, SO_2, is catalytically oxidized with air to sulfur trioxide SO_3, with vanadium pentoxide (contact process), and is then dissolved in concentrated sulfuric acid to form oleum (fuming sulfuric acid). Direct addition of water to sulfur trioxide to form H_2SO_4 occurs with explosive violence. The oleum is later reduced in strength to form 100% acid. An older process involving use of lead chambers is no longer used in the United States. Among the major uses of sulfuric acid are: as an electrolyte in storage batteries; in the manufacture of fertilizers and many sulfate and sulfide chemicals; to convert cellulose xanthate to rayon fibers in the viscose process; and in photoengraving and etching of metals.

sulfur trioxide. *See* sulfuric acid; anhydride.

superalloy. A term used by metallurgists to designate a group of refractory alloys which resist heat-deterioration up to 1370°C (2500°F). The basic component metals are iron, cobalt, and nickel, together with small proportions of molybdenum, chromium, or tungsten. *See also* refractory.

superconductivity. A phenomenon discovered in 1911 by the Dutch physicist Kamerlingh-Onnes, which causes certain metals, alloys, and compounds near absolute zero to lose both electrical resistance and magnetic permeability, i.e., to have infinite electrical conductivity. Depending upon the substance, the maximum temperature (transition temperature) for the behavior is 0.5-18°K. Superconductivity does not occur in alkali metals, noble metals, ferro- and antiferromagnetic metals. It is well-known in elements having 3, 5 or 7 valence electrons per atom, and is associated with high room-temperature resistivity. A system for transmitting electric current underground by means of superconducting cables has been developed. Superconduction is being used in advanced computer research; it is expected to make an improvement in response time comparable to that obtained when vacuum tubes were replaced by semiconductors. The chief problem in utilizing super-conductivity is to attain and maintain the extremely low temperatures required.

superphosphate. *See* fertilizer.

supersaturation. A condition in which a solute is present in a solution beyond its limit of solubility; this occurs, for example, in the confectionery industry where sugar crystallizes out of solution on cooling, e.g., in fudge and similar candy forms. Supersaturated solutions are unstable and will precipitate on addition of the merest trace of solute; they are often used for demonstration purposes.

surface. As used in physical chemistry, this term refers to a type of interface, or area of contact, that exists between two different phases of matter. Examples are the surfaces between (1) finely divided solid particles and air (solid-gas); (2) liquids and air (liquid-gas or gas-liquid); (3) insoluble particles and a liquid (solid-liquid). Such surfaces are sites of physicochemical activity between the phases, which is responsible for such phenomena as adsorption, evaporation, surface tension, reactivity, and catalysis. *See also* interface; colloid; surface tension.

surface-active agent. This term and its condensed synonym *surfactant* conventionally refer to substances that reduce the interfacial or surface tension of liquids; it could logically be applied to catalysts also, since their action depends directly on surface activity. The terms *detergent, wetting agent,* and *emulsifier* are also used, with only a slight difference in meaning. Detergents aid in cleaning by lowering the surface tension of water, thus enabling the water to form emulsions with fatty and oily materials; fatty acids, long-chain alcohols, and various synthetic products (dodecylbenzene) have this property, and hence they are called emulsifiers. Reduction of surface tension is also the function of wetting agents, permitting water to cover particles of dry powders and other finely divided materials more effectively than would otherwise be possible. *See also* emulsion; detergent; wet; surface tension.

surface area. The total area of exposed surface of a solid material, usually in the form of a powder, including all irregularities, pores, etc. Since activity is greatest at the surface, that is the interface between the particle and its environment, the larger the surface area of a given substance, the more reactive it becomes. Thus, reduction to small particles is a means of increasing the efficiency of both chemical and physical reactions; for example, twigs and wood fragments ignite more easily than large slabs, and potatoes cook faster when cut into small pieces. The coloring effect of pigments is increased by size reduction. Carbon black is notable among solids for its huge surface area (as much as 18 acres a pound for some types); the activity of its surface accounts for its outstanding ability to increase the strength and abrasion resistance of rubber. The ability of activated carbon to adsorb molecules of gases is due to this factor. Surface area is measured most accurately by nitrogen adsorption techniques. *See also* particle.

surface tension. The force (tension) of a liquid which makes the surface act as an elastic enveloping membrane which always tends to contract to the minimum area. Hence, drops of liquid tend to be spherical. The force acts in a direction normal to the surface and into the bulk phase. It is expressed as the work required to increase the surface area by one unit and is usually given as dynes/cm. Values for surface tension are usually measured at a liquid-gas interface, commonly air, and vary with the composition of the liquid and the solutes in it. Mercury has the highest surface tension of any liquid at ordinary temperatures. Polar liquids, of which water is typical, have pronounced surface tension, which limits their ability to "wet" solids and to blend with dissimilar liquids. This accounts for the formation of water droplets and the ability of a water surface to sustain the weight of a razor blade. Nonpolar liquids of low dielectric constant, e.g., hydrocarbons, have much lower surface tension than water; thus, they evaporate faster and are more mobile. Surface tension is reduced by surface-active agents (detergents).

Interfacial tension is the surface tension exhibited by one liquid in contact with another liquid instead of a gas. For example, the surface tension of mercury in air is 487 dynes/cm at 15°C, while the interfacial tension between mercury and water is 375 dynes/cm at 20°C. *See also* surface; interface; surface-active agent.

surfactant. *See* surface-active agent.

suspension. A type of dispersion in which finely divided solid particles are distributed more or less uniformly throughout a liquid medium. If the particles are in the colloidal size range, for example, suspensions of sulfur or gold in water (often loosely called colloidal solutions), they will not settle out by gravity, as they are stabilized by electrical charges. Particles of larger size can be prevented from coalescing or settling by protective colloids, which coat the particles and prevent them from cohering, for example, the protein layer on the fat particles in milk and on the hydrocarbon particles of rubber latex. Suspensions of this type are similar to emulsions. A typical solid-in-liquid suspension of industrial importance is that of wood pulp in water; when fed onto the screen of a fourdrinier machine, the water immediately drains through the meshes, leaving a continuous film of pulp. Polymerization is often carried out with monomers in the form of a suspension. *See also* dispersion; colloid chemistry.

sweetening. *See* doctor test.

symmetrical compound. A compound derived from benzene in which the hydrogen atom on three alternate carbon atoms has been replaced with another element or group, for example, at the 1,3, and 5 positions.

symmetry. Arrangement of the constituents of a molecule in a definite and continuously repeated space pattern or coordinate system. It is described in terms of three parameters called elements of symmetry, namely: (1) The center of symmetry, around which the constituent atoms are located in an ordered arrangement; there is only one such center in a molecule, which may or may not be an atom. (2) Planes of symmetry, which represent division of a molecule into mirror-image segments. (3) Axes of symmetry, represented by lines passing through the center of symmetry; if the molecule is rotated it will have the same position in space more than once in a complete 360° turn; e.g., the benzene molecule, with six axes of symmetry, requires a 60° rotation to return to its identical position. *See also* stereochemistry; asymmetry.

syndet. *See* detergent.

syndiotactic. A type of stereospecific polymer whose molecules are comprised of atoms (hydrogen) and organic groups (R) arranged in a specific and predetermined sequence above and below a chain of carbon atoms. The molecule of a syndiotactic polymer might have the arrangement RHHHRHHH above the carbon chain and HHRHHHRH below it, as indicated:

$$
\begin{array}{cccccccccccc}
R & H & H & H & R & H & H & H & R & H & H & H \\
| & | & | & | & | & | & | & | & | & | & | & | \\
-C & -C & -C & -C & -C & -C & -C & -C & -C & -C & -C & -C- \\
| & | & | & | & | & | & | & | & | & | & | & | \\
H & H & R & H & H & H & R & H & H & H & R & H
\end{array}
$$

Such polymers can be obtained by use of Ziegler and Natta catalysts. *See also* isotactic; atactic; stereospecific.

syneresis. Separation of water or other suspending liquid from a uniformly dispersed colloidal gel, with resulting contraction of the gel. This action is probably caused by a decrease in the capacity of the protein portion of the gel to bind water, induced by change in pH, temperature, surface tension, etc. Examples of syneresis are the formation of whey in cheese-making and the watery exudation from injured tissue. The separated aqueous liquid is actually a solution.

synergism. A phenomenon often encountered in chemistry in which one or more properties of a mixture are affected to a far greater extent than would be indicated by adding the values for the components taken individually. For example, the toxicity to house flies of a mixture of pyrethrum and kerosine is increased by a factor of 10 by the presence of a low percentage of sesame oil, though the latter alone has no effect; another instance is the more than expected increase in strength obtained in a copolymer, compared with the strengths of the two polymers separately. This phenomenon does not contravene any laws of nature, but lack of sufficient knowledge of the many variables involved and the possibilities of interaction between them prevent a complete understanding of this behavior.

synfuel.   Any fuel, either liquid or gaseous, made by chemical treatment of coal, naphtha, shale oil, or agricultural wastes. *See* gasification; synthetic natural gas; fermentation.

synthesis.   The act of combining chemical elements or compounds in such a way as (a) to form new compounds that would not otherwise exist, or (b) to duplicate a natural product. The syntheses essential to life are performed by plants (those of carbohydrates, proteins, hormones, vitamins, etc.) The first synthesis of a natural organic compound by man was that of urea in 1828, by Friedrich Wöhler (1800–1882), a German chemist. Since the mid-1800s, when dyes were synthesized by Sir William Henry Perkin (1838–1907) and others, there have been an increasing number of outstanding achievements in synthetic chemistry which reflect the highest order of chemical ingenuity. Some of these were altogether new materials, others being laboratory duplications of natural products; among them are ammonia, indigo, rubber, insulin, nylon, cortisone and other hormones, chlorophyll, vitamins, and deoxyribonucleic acid. Synthetic organic chemistry has also produced a vast array of new high polymers (plastics), solvents, fibers, and pharmaceuticals. In the inorganic field the production of sulfuric acid, sodium hydroxide and sodium carbonate are noteworthy. *See also* Perkin; Wöhler.

synthesis gas.   A blend of hydrogen and carbon monoxide in varying proportions obtained from coke, natural gas, or other carbonaceous materials, usually by reaction with steam in the presence of air. It contains a low percentage of nitrogen. It is used for a number of important synthesis operations, including production of ammonia, methanol, and a wide range of organic compounds. *See also* Oxo process; ammonia; methanol; Haber-Bosch process.

synthetic natural gas.   Any manufactured fuel gas of approximately the same composition and heating value as gas obtained directly from oil fields. It contains about 85% methane and 15% ethane. It is derived from such carbon-rich materials as coal, naphtha, and agricultural wastes.

synthetic resin.   *See* resin (2).

T

T Symbol for tritium, the name being derived from the Latin for *three*, i.e., H³.

Ta Symbol for the element tantalum, the name being adopted from the mythological character Tantalus because of the difficulties experienced in isolating the element. Tantalus was punished by being placed up to his neck in water, with fruit hanging from a limb just out of his reach. The word *tantalize* has the same origin. *See also* Nb.

tacticity. Derived from a Greek word meaning "to arrange," this term is used by polymer chemists to denote various arrangements or sequences of atoms in the molecules of high polymers. Those in which the sequence of substituent atoms on both sides of the carbon chain is random are called atactic and the polymers are amorphous. Advances in the art of controlled catalysis have made it possible to synthesize crystalline polymeric structures in which the arrangement of substituents is predetermined; these are called stereospecific polymers, which may be isotactic or syndiotactic. *See also* stereospecific; isotactic; syndiotactic.

tagged atom. *See* tracer element.

tailings. *See* gangue.

talc. *See* powder; Mohs scale.

tall oil. A by-product of the alkaline pine-wood-pulping process comprised of about equal parts of rosin acids and fatty acids. Made by processing liquid digester waste (black liquor), it is used in the manufacture of soaps, driers for paints, linoleum, and numerous other products. It is one of the materials included in the term *naval stores*. It is named for the Scandinavian word for *pine tree*.

tankage. A by-product of the meatpacking industry made by boiling slaughterhouse scrap. It contains substantial percentages of ammonia and phosphates extracted from the bones, as well as fats and proteins. Another type containing less ammonia and phosphate is made by high-temperature treatment of garbage. Both kinds are used as fertilizer components.

tannic acid. *See* ink (1); tanning; mordant.

tanning. Conversion of the soft, perishable proteins of the skin or hide to a durable but flexible material (leather) by treatment with chemical compounds; also called curing. There are two major types of tanning; (1) vegetable tanning, in which the active agent is tannic acid derived from the bark of certain rare trees (quebracho and wattle), which functions as a cross-linking agent for the skin proteins; (2) chrome tanning, in which salts of various metals act as complexing agents, e.g., chromium, aluminum, and zirconium; hydrogen bonding is also involved in the conversion. *See also* leather.

tantalum. An element.

Symbol	Ta	Atomic No.	73
State	Solid	Atomic Wt.	180.947
Group	VB	Valence	5

A quite reactive metal, m.p. 2996°C (5425°F), which forms compounds with most other elements; its most common valence is 5, though it has lower oxidation states which are unstable, i.e., 3 and 2. It forms complexes with both organic and inorganic substances. It is not readily attacked by chemicals and forms an oxide film on its surface which increases its electrical resistivity; because of this, it is used in making capacitors and rectifiers. Tantalum was the first metal used for electric light filaments. It also acts as a scavenger, as a component of high-speed tool steels, and in various types of chemical equipment, where corrosion resistance, especially to acids, is a factor. It does not occur in pure form in nature, but only combined with other elements (oxygen, niobium, iron, etc). Like many other metals, tantalum powder is easy to ignite.

tar. The residue obtained by destructive distillation of carbon-rich materials such as coal, wood, or petroleum. Most important to the chemist is coal tar, from which many useful compounds are derived. *See also* pitch; asphalt.

tare. **(1)** The weight of a container, wrapper, or liner that is deducted in determining the net weight of a material.

(2) A weight used in analytical work to offset the weight of a container.

tarnish. A gray-to-black film of sulfide formed on surfaces of silver and gold when exposed over a period of time to sulfur or its compounds. Most tarnishing is due to sulfur dioxide in the air or to sulfur compounds in eggs or other food products in contact with silverware. It does no permanent damage and can be effectively removed with prepared cleaners. A transparent, water-insoluble film is sometimes applied to reduce susceptibility to tarnish.

tar sands. *See* oil sands.

tartaric acid. A dihydroxy compound having the formula $(CHOH)_2(COOH)_2$ and occurring naturally in the juices of many fruits, especially grapes. It has unusual optical properties, as its molecule has two asymmetric carbon atoms and four possible optical isomers. The natural acid is dextrorotatory, but on heating it is converted in part to a meso form which is optically inactive. This was discovered by Pasteur working with wine residues. Tartaric acid is the basis of so-called cream of tartar, used as a leavening agent; it also has use in other aspects of food technology and as a source of certain pharmaceuticals, e.g., tartar emetic. *See also* asymmetry; meso; Pasteur.

taste. *See* flavor.

Tb Symbol for the element terbium, the name being derived from the Swedish town of Ytterby, where its oxide form terbia was discovered.

Tc Symbol for the element technetium, the name being given by its discoverers in 1937.

TCA cycle. *See* Krebs cycle.

Te Symbol for the element tellurium, the name being derived from the Latin *tellus*, meaning "the earth." *See also* Se.

technetium. An element.

Symbol Tc	Atomic No 43
State Solid	Atomic Wt. 98,9062
Group VIIB	Valence 2,4,5,6,7,

This radioactive element, m.p. 2170°C (3938°F), was artificially formed in 1937 from molybdenum by deuteron bombardment; prior to that date, location 43 in the Periodic Table was known to exist but had remained open. Three of its isotopes have half-lives of over 100,000 years, but the others are quite short. One of the latter is used in diagnostic medicine. The element has superconducting properties. Its radioactive nature (gamma) makes it dangerous to handle without adequate protection.

tellurium. An element.

Symbol Te	Atomic Wt. 127.60
State Solid	Valence 2,4,6
Group VIA	Isotopes 8 stable
Atomic No. 52	

A comparatively soft solid, m.p. 450°C (842°F), classed as a nonmetal and recovered as a by-product of copper ore refining; it is similar in many respects to sulfur and, especially, to selenium, which is regarded as its sister element. It has semiconducting properties and forms compounds with many other elements, especially halogens, hydrogen, and oxygen. In elemental state it is not poisonous, but many of its salts are quite toxic, e.g., sodium tellurite, hydrogen telluride. Its major uses are as an alloying agent in steels, copper, etc., as a rubber accelerator, and particularly as a semiconductor. It is also used in electric power generation for specified purposes (cooling, relay signals, and the like). *See also* selenium.

temperature. A value representing a measurement of the degree of hotness or coldness of any matter; the thermal state or condition of a substance or mixture (air). There are several temperature scales, each of which is a measurement of a thermal condition compared with an arbitrary reference point, for example, the freezing and boiling points of water, absolute zero, etc. The best-known are the Fahrenheit, centigrade (Celsius), and Kelvin (Absolute) scales, which are separately defined. The term *ambient temperature* means the temperature of any local environment in which an experiment is conducted or in which an operation is carried out. Room temperature is from 70 to 75°F. Since temperature is a value, it should be described as being high, low, etc., rather than hot, warm, or cold. *See also* heat.

tempering. *See* iron.

tenacity. *See* tensile strength.

tensile strength. The maximum weight that can be supported by a unit cross-sectional area of a material; it is usually expressed as pounds per square inch (psi), or as grams per square centimeter. When under load (stress), a material will stretch (yield); as the stress is increased, the material will eventually break, the tensile strength being the strain at the time of rupture. Some metals and single crystals (whiskers) have tensile strengths running into the millions; plastics, vulcanized rubber, and other elastomeric materials have tensile strengths ranging from about 5000 psi for carbon-reinforced cured rubber to 30,000 psi for nylon. In the textile industry, the analogous term *tenacity* is often used, expressed as grams per denier.

terbium. An element.

Symbol	Tb	Atomic No.	65
State	Solid	Atomic Wt.	158.9254
Group	IIIB	Valence	3,4
	(Lanthanide Series)		

Terbium, m.p. 1356°C (2473°F), is a rare-earth metal occurring in low concentration in monazite sands. The metal is extracted by chemical purification techniques, i.e., by reducing the fluoride. It is extremely reactive when exposed to air or oxygen and must be kept in an inert atmosphere. It has no important industrial uses. *See also* lanthanide series.

terephthalic acid. A dicarboxylic acid whose structure is that of benzene in which two COOH groups replace hydrogen atoms in the 1,4- (para)

position. The formula is $C_6H_4(COOH)_2$; it is made by the oxidation of xylene. It is the basis of an important series of polyester resins when esterified with ethylene glycol; these are widely used in the form of fibers for manufacture of suitings, carpets, and other textile products, as well as films for packaging. The polymers are crystalline and are resistant to moisture, electricity, and chemical degradation. *See also* polyester.

ternary. A uniformly dispersed mixture (solution or alloy) having three components; also, a chemical compound having three constituent elements or groups.

terne plate. *See* tin.

terpene. Any of a broad series of olefinic (unsaturated) hydrocarbons whose molecules are made up of C_5H_8 units; they may be either cyclic or straight-chain structures. Natural rubber is a polyterpene, of which the C_5H_8 unit (isoprene) is the monomer. Terpenes occur widely in plants, especially in the resins and essential oils of many varieties of trees (pinene, turpentine); they are usually removed from citrus fruit oils to reduce spoilage due to oxidation of the terpene. The formula of many simple terpenes is $C_{10}H_{16}$; more complex types are represented by the sesquiterpenes ($C_{15}H_{24}$) and the triterpenes ($C_{30}H_{48}$), the structure in each case being a simple multiple of the $C_{10}H_{16}$ unit. Many are optically active. Well-known alcohols derived from terpenes are borneol, terpineol, and geraniol; other derivatives containing oxygen in the molecule are camphor and menthol. *See also* turpentine; isoprene.

tertiary. In reference to alcohols this term denotes the presence of three alkyl groups attached to the methanol carbon atom; in reference to amines, it denotes the attachment of three alkyl groups to the nitrogen atom. It is customarily abbreviated *tert.*

testosterone. *See* hormone.

tetra-. A prefix indicating the presence of four atoms of an element or four groups in a compound, e.g., carbon tetrachloride (CCl_4), tetraethyllead, $Pb(C_2H_5)_4$.

tetraethyllead. A toxic, organometallic liquid introduced as a gasoline antiknock additive about 1923; its use made possible octane ratings of up to 100 and was largely responsible for the development of more powerful engines with high compression ratios in automobiles and airplanes.

Its use in motor gasolines has been virtually eliminated for environmental reasons. *See* methyl-*tert*-butyl ether; Midgley.

tetrafluoroethylene.   *See* fluorocarbon.

tetrahydrocannabinol.   *See* hallucinogen.

tetrahydrofurfuryl alcohol.   *See* furfuryl alcohol.

tetramer.   A macromolecule comprised of four monomer units.

Th   Symbol for the element thorium, the name being derived from the god Thor, of Scandinavian mythology.

thallium.   An element

Symbol	Tl	Atomic Wt.	204.37
State	Solid	Valence	1,3
Group	IIIA	Isotopes	2 stable
Atomic No.	81		

Thallium is a relatively heavy metal of low melting point, 303°C (577°F), occurring at an average concentration of about 1 gram per ton in the earth's crust; its highest concentrations are in potassium-containing minerals. Thallium compounds can be recovered from by-products of zinc smelting, and the metal obtained by various separation techniques. The metal oxidizes in air to from a protective oxide coating. Its uses in metallic form are chiefly in alloys with bismuth, cadmium, mercury, etc. As it is quite poisonous, its compounds have been used in rodenticides and various insecticides; it also has limited application in photoelectric devices.

theobroma oil.   A fatty semisolid (also called cocoa butter) expressed from the cacao bean, native to central and northern portions of South America, which is the source of chocolate. It is a mixture of fatty acid glycerides and also contains the alkaloid theobromine. It has a bitter taste and a smell similar to chocolate; its major use is in the candy industry and dessert flavoring; it also has some application in pharmaceutical products such as suppositories.

theoretical plate.   A term used in chemical engineering to describe a contacting device that yields the same separation of the components of a liquid mixture in a fractionating column as does one simple distillation. Thus, a column that effects the same degree of separation as ten successive simple distillations is said to have ten theoretical plates. *See also* distillation; HETP.

thermal neutron.   *See* moderator; neutron.

thermite.   A strongly exothermic blend of finely divided aluminum and iron oxide which, when ignited, reacts to produce molten iron and aluminum oxide by the reaction $2Al + Fe_2O_3 \rightarrow 2Fe + Al_2O_3 + 185,000$ cal. The temperatures attained reach about 3000°C (5432°F). When used for welding massive iron parts, the mixture is placed in a magnesia crucible above the site of the joint to be made, is ignited, and the molten iron flows into the space to make the weld. Thermite is also used in incendiary military bombs. Thermite was discovered about 1900 by the German chemist, Hans Goldschmidt. *See also* Goldschmidt process.

thermochemistry.   The branch of chemistry that embraces the determination and study of the heat absorbed or evolved in the formation and dissociation of compounds in chemical reactions and in changes of phase (liquid to vapor, solid to liquid, etc.). This information is essential in chemical engineering, for the heat or energy balance of a reaction is directly related to the design of large-scale equipment, such as heat-exchangers, distillation and fractionation towers, evaporators, and the like. Application of these data involves knowledge of the three laws of thermodynamics. *See also* equilibrium; material balance; chemical engineering.

thermofor.   *See* heat transfer.

thermonuclear reaction.   A reaction in which deuterium and tritium, which are isotopes of hydrogen, are caused to unite to form helium at excessively high temperature (5×10^7°C). For details, *see* fusion (1).

thermoplastic.   Any substance that is solid or semisolid at room temperature but that softens when heated. Such materials can be molded or shaped and will retain their shape until their softening point is reached, for example, waxes, butter, and hard fats. In the plastics industry, the term is applied to high-polymer resins or plastics which can be molded either at room temperature or under low heat but which will soften under high heat, e.g., nylon, polystyrene, and polyethylene. Some thermoplastic materials can be made thermosetting by cross-linking agents or exposure to high-energy radiation; among these are polyethylene and rubber. *See also* thermoset.

thermoset.   Any substance which requires exposure to high temperature to yield a useful article that will be permanently hard, infusible and relatively insoluble. An example among inorganic materials is clay. In the plastics industry, the term is applied to high-polymer resins such as the phenolics, polyesters, and epoxies, which are naturally thermosetting, as well as to those that can be made thermosetting by means of a cross-linking agent. *See also* thermoplastic.

thi-, thio-.   Prefixes used in chemical nomenclature to denote the presence of sulfur in the compound, usually as a substitute or replacement for oxygen. For example, thiols are compounds corresponding to alcohols in which oxygen is replaced by sulfur, with the generic formula RSH. Among inorganic compounds, sodium thiosulfate ($Na_2S_2O_3$) corresponds to sodium sulfate in which one oxygen is replaced by sulfur, and sodium thiocyanate (NaSCN) corresponds to sodium cyanate (NaOCN) in which the oxygen is replaced by sulfur. Organic compounds which contain both sulfur and nitrogen are known as thiazoles. *See also* thiol.

thiamine.   Generally known as vitamin B_1, this compound is one of the most important in the group forming the B complex. Its formula is $C_{12}H_{17}ClN_4OS$. It has an essential catalytic function in carbohydrate metabolism, as do other members of the complex and specifically affects the nervous system; the chief deficiency symptom is development of the disease called beriberi, resulting from malnutrition. Its existence in rice hulls was discovered in 1911 by Casimir Funk (1884–1967), a Polish biochemist, though it was not chemically identified till years later. It is made synthetically and widely used to fortify wheat flour from which the hulls have been removed by processing. *See also* B complex; Funk.

thiazole.   *See* accelerator; thi-.

thin.   In spite of its apparent simplicity, this term has several meanings in materials technology. (1) In electronic metallurgy, a thin film is a vapor-deposited coating having a thickness of only a single atom; such *monatomic* films, e.g., thorium on tungsten, are used in electronic devices such as cathodes. (2) A coating or film of a fatty acid on water which is one molecule thick (about 200 angstroms) is called a *monomolecular* film. (3) In thin-layer chromatography, the term applies to a specially prepared mixture of adsorbents spread on glass slides to a thickness of about 1/100 inch. (4) The term is also used

in the sense of a liquid of low viscosity, as in paint thinner and thin-boiling starch. *See also* film; monomolecular.

thinner.   A low-viscosity liquid such as hydrocarbons (naphtha, benzene, etc.) and turpentine, which are added to oil-base paints to adjust their viscosity to a suitable point for application. Turpentine is most generally used, as it has a low evaporation rate. Water is used as a thinner in emulsion paints.

thiol.   Any of a class of organic compounds, either aliphatic or aromatic, which are structurally similar to alcohols but are characterized by the presence of an —SH (sulfhydrate) group instead of an — OH group. These were formerly called mercaptans, a name which is misleading because it suggests the presence of mercury. Thiols have a pronounced and often offensive odor; ethanethiol is one of the most malodorous of all substances and is a component of the skunk's fluid. Thiols are used as organic intermediates; alcohol derivatives have application as solvents for dyes and in the production of plasticizers, pharmaceuticals, and rubber accelerators. One of the most important of the latter group is mercaptobenzothiazole, a nitrogen-containing thiol derivative which (with its modifications) is generally used in curing tires and mechanical goods.

thiomersal.   *See* tincture; antiseptic.

thiophene.   *See* hydrogen sulfide.

thiourea.   An organic derivative of ammonia, containing a sulfur atom; the formula is $(NH_2)_2CS$. It is widely used in synthesis of organic compounds, especially dyes and hair colorants, as a sensitizing agent in photography, and in amino resin manufacture. It is toxic when taken internally. *See also* urea.

thixotropy.   A term literally meaning a change of form caused by slight pressure. It is applied to certain types of solid-in-liquid dispersions such as clay-water pastes, sand saturated with water, concentrated paints, and colloidal gel structures. Application of slight pressure causes such mixtures to liquefy while the pressure is maintained, after which they resume their original form. Practical applications of thixotropy are in the use of thick paints in paste form, printers' inks, shaving sticks, and the like, which become fluid under service conditions.

Thomson, Sir J. J. (1856–1940).   A native of England, Thomson entered Cambridge University in 1876 and remained there permanently as a professor of physics, especially in the field of electrical phenomena. His observations and calculations of cathode ray experiments led to proof of the existence of the electron as the lightest particle of matter (1896). This proof was announced at the Royal Institution in the following year. This was the keystone of the entire theory of atomic structure and one of the most notable discoveries in the history of science. Thomson was awarded the Nobel Prize in physics in 1906. *See* Rutherford.

thorium.   An element.

Symbol	Th	Atomic No.	90
State	Solid	Atomic Wt.	232.0381
Group	IIIB	Valence	4
	(Actinide		
	Series)		

Thorium is a relatively heavy, radioactive metal, m.p. 1750°C (3182°F), occurring in monazite sands, and is associated with the rare-earth elements of the lanthanide series. It is quite soft and can be easily machined. It decays by both alpha and beta radiation to the 208 lead isotope; it also is converted to fissionable uranium-233 by absorbing a neutron to form thorium-233, which then decays in two steps to this isotope of uranium. Thorium occurs in greater volume in the earth's crust than does uranium and is thus important for future supply of nuclear fuel. It has limited use in alloys with magnesium, in photoelectric cells, x-ray tubes, etc. Cathodes of thoriated tungsten are used in electronic devices. Thorium is toxic when taken internally.

thrombin.   *See* coagulation; anticoagulant.

throwing.   (1) In textile technology, this term refers to an auxiliary called a throwing oil; it is applied to rayon and other fibers as a lubricant in making yarn from filaments, which involves a twisting or "throwing" action. The oil is later removed by scouring.

(2) In electroplating, the term *throwing power* denotes the ability of an electrolytic solution (electroplating bath) to deposit a uniform coating of metal on all parts of the object being plated, including any recessed or intricate surfaces that may be present on hollow or irregular pieces. Throwing power is related to the temperature of operation and the current density of the bath. *See also* electroplating.

thulium.   An element.

Symbol	Tm	Atomic Wt.	168.9342
State	Solid	Valence	3
Group	IIIB	Isotopes	None
	(Lan-		stable
	thanide		2 radio-
	Series)		active
Atomic No. 69			

Thulium, m.p. 1545°C (2813°F), is a minor element of the rare-earth group, occurring in association with yttrium in a number of ores such as xenotime and euxenite found in monazite sands. It is stable in air and reacts slowly with halogen elements at room temperature; it also forms compounds with oxygen, carbon, nitrogen, and hydrogen at high temperatures. It has few important uses; small pellets can be used as a portable source of diagnostic x-rays, and there is also a possibility of application in magnetic ceramic materials (ferrites).

thymol.   *See* antiseptic.

thyroxine.   *See* hormone.

Ti   Symbol for the element titanium, the name being derived from Titan, the mythical giants of antiquity, as also in *titanic*.

tin.   An element.

Symbol	Sn	Atomic Wt.	118.69
State	Solid	Valence	2,4
Group	IVA	Isotopes	10 stable
Atomic No. 50			

Tin is obtained by roasting or smelting the ore cassiterite, which occurs chiefly in Malaya, Bolivia, and central Africa. It is relatively soft and has a comparatively low melting point 232°C (449°F). Its primary uses are in plating steel and other metals and as a component of such alloys as terne plate (lead + tin), low-melting or fusible alloys (solders), bearing alloys (babbitt), and type metal. Tin also has a number of specialized engineering uses. The metal itself is not toxic, but its organic compounds are quite poisonous; they are used as polymerization catalysts and as biocidal agents. Stannous fluoride (SnF_2) and stannous pyrophosphate ($Sn_2P_2O_7$) are used in toothpaste as decay-inhibiting additives; they are nontoxic.

tincture.   Derived from the Latin word meaning "to color slightly" (*tinge*), this term is used in pharmaceutical chemistry to designate a dilute alcoholic solution of a drug or antiseptic, such as iodine or thiomersal. The concentrations vary with the strength of the active substance and may range from as little as 0.1% to 10%; tincture of iodine for medical purposes is from 2 to 3%.

titanium. An element.

Symbol	Ti	Atomic Wt.	47.90
State	Solid	Valence	2,3,4
Group	IVB	Isotopes	5 stable
Atomic No.	22		

Titanium, m.p. 1668°C (3034°F), is derived chiefly from the ores ilmenite and rutile, though it occurs in low concentrations in most of the rocky structures in the earth's crust and was originally discovered in beach sand. It is produced commercially by reduction of titanium tetrachloride with magnesium (Kroll process) or sodium, or by electrolytic means. It is flammable in powder form, and the bulk metal will catch fire at about 1200°C (2200°F). Its outstanding property is its lightness, which, combined with sufficient strength, makes it of unique value as a structural metal in airplanes and aerospace vehicles. It is also useful in manufacture of chemical plant and desalination equipment because of its high resistance to chlorine. It has many applications in alloys. Its most important compound is titanium dioxide, TiO_2, used principally as a white pigment in paints, rubber, plastics, and paper. It readily forms compounds with halogens, such as titanium tetrachloride, which is a source of pure titanium as well as a catalyst and white pigment in skywriting smokes, etc. *See also* Kroll process.

titanium dioxide. *See* titanium.

titanium tetrachloride. *See* titanium; smoke.

titanous sulfate. *See* discharge.

titration. One of the most widely used laboratory methods of routine chemical analysis, titration is a volumetric means of finding the amount of a given substance (e.g., hydrochloric acid) in a solution. This is determined by addition of a standard solution of known concentration (e.g., sodium hydroxide) which will react with the substance in the original solution in an exact and predictable way; this standard solution is called the titrant or the reagent solution. In this case, the volume of titrant (NaOH) required to neutralize the original solution (HCl), that is, bring it to pH 7.0, is called the equivalence point. It is visually shown by a soluble dye, or indicator, which changes color instantly when enough titrant has been added to react quantitatively with the substance being determined. This color change occurs at the end point (usually the same as the equivalence point). The amount of substance in the original solution can be calculated from the amount of titrant added and from the known proportions with which it reacts with this substance.

Titrations are distinguished by the kind of reactions involved, most common of which are acid-base and redox reactions. Besides indicators, end point determinations are made by various instrumental methods, such as amperometric, conductometric, photometric, coulometric, etc. *See also* end point; indicator; neutralization.

Tl Symbol for the element thallium, the name being derived from the Latin word for *green,* since the element produces a green line in its spectrum.

Tm Symbol for the element thulium, named for Thule, a legendary name for the most northerly part of the habitable world.

tobacco. *See* curing.

tocopherol. A group of isomeric organic compounds having the formula $C_{29}H_{49}OOH$; they are designated by the Greek letters alpha, beta, and gamma, of which the alpha form is the most active. It was isolated from wheat germ oil, while the gamma form was found in cottonseed oil. The alpha form is optically active and contains three asymmetric carbon atoms. Tocopherol is commonly known as vitamin E, which was found to be effective in the reproductive processes of laboratory animals. It also inhibits the oxidation of other vitamins. *See also* vitamin.

tokamak. A transliterated Russian word for an assembly for producing nuclear fusion (thermonuclear) reactions experimentally. It consists essentially of a doughnut-shaped evacuated chamber called a torus, 18 inches in cross-section, through which plasma circulates. Surrounding the torus is an electromagnetic field powerful enough to confine the energized plasma sufficiently to achieve the required density of 10^{14} particles per cc per second and a temperature of 44 million degrees C. The vacuum chamber and the magnetic field simulate conditions on the sun; namely, absence of air and immense gravitational forces. Larger and more powerful tokamaks up to 36 inches in cross-section are under development, in which temperatures up to 100 million degrees C may be achieved. *See also* fusion; plasma (**1**).

toluene. An aromatic compound whose molecule is comprised of a benzene ring in which one hydrogen is replaced by a methyl group at any position; its formula is $CH_3C_6H_5$, b.p. 110°C (230°F). It is a flammable and quite poisonous liquid which forms explosive mixtures in air at concentrations of as low as 1.2%. It is obtained either from coal-tar or from petroleum. It has many important applications in the chemical industries; it is a source material for benzene and phenol, a solvent for many resins used in paints, an explosives base (trinitrotoluene), and a source of polyurethane resins (diisocyanates). It is also useful as a diluent in nitrocellulose lacquers. Its sulfonated derivatives are used in dyes, detergents, plasticizers, and organic synthesis.

toner. Any of a group of organic pigments that are free from the inorganic components or substrates which characterize lakes. Among them are various phthalocyanine, diazo, and triphenylmethane pigments. *See also* lake; pigment.

torus. *See* tokamak; JET.

toxicity. The extent to which a substance adversely affects the life process or any of its subcritical aspects as a result of chemical reactions involving tissue or metabolic functions. Many diseases are caused by the toxic products of infectious organisms; other toxic agents act directly on the central nervous system, on the blood, or on the muscular and digestive systems. In all such cases one or more chemical reactions are involved. Toxicity standards are usually based on results of animal experiments, the results of which are presented as LD_{50} ratings (lethal dose 50%), which means that half of the exposed animals die from a given dosage of the substance tested. Toxicity is affected by fineness of subdivision, solubility, and chemical state; many metals that are relatively harmless in elemental form are severely toxic as compounds, for example mercury and beryllium. *See also* poison (**2**).

The following list includes chemical individuals and groups that are generally regarded as having toxic properties. There is considerable variation in the degree of toxicity among these, and the listing is by no means exhaustive.

Individuals	*Groups*
aniline	aldehydes
arsenic and cpds.	alkaloids
asbestos (carcinogen)	allyl cpds.
barium and soluble	barbiturates
cpds.	chlorinated
beryllium and soluble	hydrocarbons
cpds.	corrosive materials
benzidine (carcinogen)	cyanides
cadmium and cpds.	organic phosphate
benzpyrene (carcinogen)	esters
carbon monoxide	radioactive elements
chlorine	
chromium (hexavalent	
cpds)	
coal-tar pitch	
(carcinogen)	
cresol	
fluorine and cpds.	
hydrogen peroxide	
hydrogen sulfide	
hydrogen selenide	
lead compounds	
mercury (soluble cpds.)	
methyl alcohol	
nickel carbonyl	
osmium tetroxide	
oxalic acid	
ozone	
phenol	
pyrene	
pyridine	
phosphine	
selenium cpds.	
stibine	
sulfur dioxide	
thallium (soluble cpds.)	
tin (organic cpds.)	
vinyl chloride	
monomer	
(carcinogen)	

Toxic Substances Control Act. Legislation passed by Congress in 1977 that provides the legal basis for regulations concerning all aspects of the manufacture of such products. Enforcement is carried out by the Occupational Safety and Health Administration (OSHA). Effective control of toxic materials, especially as regards storage and disposal of hazardous wastes, has assumed increasing importance in recent years. *See also* waste control.

trace elements. This term refers to five elements necessary for plant nutrition which are present in the soil in minute concentrations (less than 1000 parts per million). These are generally considered to be boron, copper, manganese, molybdenum, and zinc. Additional elements required in trace concentrations in animal and human nutrition are fluorine, iodine, and selenium. Elements present in the soil in higher concentrations are not regarded as trace elements. The term micronutrient is synonymous with trace element.

NOTE: Trace elements should not be confused with tracer elements.

tracer element. A radioactive isotope, either natural or artificial, introduced into a system to determine its behavior in and reactivity with the other constituents present. This is made possible by the emission of radiation from the tracer which is detected by a sensing device, so that the movement of the isotope can be followed continuously. This method is used in studies of biochemical reactions (carbon-14) as well as in medical research; for example, radioactive iodine is used in treatment of the thyroid gland, which concentrates this element. The terms *labelled* or *tagged* are frequently used to designate these elements or the compounds in which they occur.

NOTE: Tracer elements should not be confused with trace elements.

trademark. The proprietary name of a product or process which is officially registered, and the use of which may be licensed by the owner. It is often erroneously called a trade name. When used in the literature, such names should be identified typographically by capitals or quotation marks, or by means of the letter *R* enclosed in a cirlce. If this is not done, the name may become so commonplace that its owners will lose their proprietary right to it. Chemical literature abounds with trademarks; among them are the well-known products "Celotex", "Vaseline," "Bakelite," "Freon," "Adrenalin," "Celite," "Pyrex," "Lucite," "Mylar," "Celluloid," "Glyptal" resins, "Carborundum," and the "Xerox" photocopying process. On the other hand, the terms nylon, neoprene, and saran are generic names, by courtesy of their manufacturers.

tragacanth. *See* arabic gum; gum.

tranquilizer. More properly called a psychotropic drug, a tranquilizer is a pharmaceutical compound used in treating psychic problems of emotionally unstable persons. Contrary to the meaning of the term in popular usage, some of these agents act as stimulants rather than relaxants, though their therapeutic effect is equally beneficial. The so-called major tranquilizers are derivatives of butyrophenone and phenothiazine (chloropromazine); their clinical action is quite different from that of the minor group, which includes benzodiazepine and glycerol derivatives such as meprobamate. There are three classifications of psychotropic drugs based on their functional effect, namely, antipsychotics, antidepressants, and antianxiety agents. Such drugs should not be taken without the advice of a physician.

trans-. *See cis-.*

transalkylation . A catalytic reaction of the disproportionation type by which benzene and xylene are formed from toluene by hydrogenation without production of methane, that is, by hydrodealkylation. *See also* disproportionation; hydrodealkylation.

transesterification. *See* polycarbonate.

transference number. A term used by electrochemists to designate the fraction of the total current in an electrolytic solution that is carried by any specific type of ion, e.g., either positive or negative. The symbols used are usually t^+ and t^-. The values are worked out by mathematical procedures involving conductance and concentration of the electrolyte. Ions in a solution move with different speeds depending on their charge factors, the more rapidly moving ions carrying more current than the slower ones. This fact permits measurement of the transference numbers by a method developed by the German physicist, Johann W. Hittorf (1824–1914). Another way of measuring them is by the so-called moving boundary method, in which solutions of different colors are used; the boundary between the colors moves in response to an applied electric current.

transfer RNA. *See* ribonucleic acid (RNA).

transformer oil. A dielectric liquid used in transformers as a thermal and electric insulator. Various types include mineral oils, chlorinated hydrocarbons called askarels, and silicone oils. *See also* dielectric.

transistor.   *See* germanium.

transition element.   Any of a large number of metallic elements occurring in the fourth, fifth, sixth, and seventh periods of the Periodic Table. Some authorities prefer the term transitional element. Specifically, they are as follows:

Period 4	Period 5
scandium	yttrium
titanium	zirconium
vanadium	niobium
chromium	molybdenum
manganese	technetium
iron	ruthenium
cobalt	rhodium
nickel	palladium
copper	silver

Period 6	Period 7
lanthanum	actinium
lanthanide series	actinide series
hafnium	
tantalum	
tungsten	
rhenium	
osmium	
iridium	
platinum	
gold	

The transition elements or metals are so named because they represent a gradual shift from the strongly electropositive elements of Groups IA and IIA to the electronegative elements of Groups VIA and VIIA. They are also more versatile in their chemical bonding properties than are other elements, using the two outer shell orbitals instead of only those of the outermost shell. Most are active complexing agents, particularly cobalt, iron, and chromium. *See also* coordination compound.

transmutation.   Conversion of one element into another by introducing one more proton into its nucleus by nuclear bombardment. This was first achieved by the British physicist, Sir Ernest Rutherford (1871–1937), Nobel Prize 1908, who utilized the heavy helium nucleus (alpha particle) to initiate the change in a nitrogen nucleus. One of the protons and the two neutrons of the helium enter the nitrogen nucleus (atomic num-

ber 7), thus increasing its atomic number by 1 and its atomic weight by 3, so that it becomes the 17 isotope of oxygen (atomic number 8). The additional proton is emitted as a hydrogen nucleus. The reaction is represented as $_2He^4 + _7N^{14} \rightarrow _8O^{17} + _1H^1$. Similarly, sodium can be transmuted to magnesium by addition of a proton to its nucleus. By far the most important transmutation is that of U-238 to plutonium, first accomplished in 1940 by deuteron bombardment in a cyclotron. This can now be effected in the breeder reactor; it represents a practical means of extracting maximum energy from uranium for future power supply. *See also* breeder; fission; Rutherford.

transuranic.   A term used by chemists to denote the series of synthetic or artifically created elements which lie beyond uranium in the Periodic Table, beginning with neptunium (atomic number 93) and extending to the most recently discovered element 106. None of these elements was known before 1940, and further discoveries are to be expected.

tribasic.   (1) An acid, either inorganic or organic, whose molecule contains three hydrogen atoms that can be replaced by basic elements to form salts.

(2) An inorganic compound whose molecule contains three atoms of a basic univalent element, e.g., tribasic sodium phosphate (Na_3PO_4).

tricarboxylic acid cycle.   *See* Krebs cycle.

trichloroethylene.   A heavy chlorinated hydrocarbon liquid, b.p. 87°C (189°F), used principally as a solvent for removing greases from metals, for cleaning of precision equipment, and in dry cleaning. It has also been used as an anesthetic. Its formula is $CHClCCl_2$. Its toxic properties are sufficient to cause prohibition of its use in some areas. It is nonflammable.

trichlorofluoromethane.   *See* fluorocarbon.

trichloropicolinic acid.   *See* defoliant.

tricresyl phosphate.   One of the most widely used plasticizers in the processing of plastics such as nitrocellulose, vinyl, and styrene polymers, etc. Made from cresol and phosphorus oxychloride, it is a mixture of isomers whose formula is $(CH_3C_6H_4O)PO$. It is a water-white liquid, b.p. 420°C (788°F), soluble in organic substances but insoluble in water. It is somewhat heavier than water and also has some application as a hydraulic liquid. *See also* plasticizer.

tricyclic.   An organic compound in which three cyclic or ring structures occur, as in anthracene and phenanthrene.

triethanolamine.   A nitrogenous organic base, m.p. 21.2°C (70°F), b.p. dec. at 335°C (634°F), having the formula $(HOCH_2CH_2)_3N$, formed by reacting ammonia with ethylene oxide. It is a transparent, syrupy liquid, somewhat heavier than water. Its major uses are in detergents, emulsifiers, and plasticizers; its ability to absorb water (hygroscopicity) makes it useful as a humectant. It has a number of minor uses as a textile cleaning agent, rubber accelerator, etc.

triethylaluminum.   A toxic organometallic liquid, b.p. 194°C (381°F), having the formula $(C_2H_5)_3Al$, developed for use as a stereospecific catalyst of the Ziegler type. It is dangerous to handle, as it burns on exposure to air and may explode in contact with many common substances such as water and alcohols. It is made by bringing finely divided aluminum into contact with ethylene and hydrogen with heat and pressure. It is usually used in conjunction with titanium dichloride. The catalyst is used primarily for making crystalline polyethylene and polypropylene. See also stereospecific; Ziegler catalyst.

triglyceride.   See glyceride.

trihydric.   An alcohol whose molecule contains three hydroxyl groups (OH) each attached to a different carbon atom, for example, glycerol $(CH_2OHCHOHCH_2OH)$ or $C_3H_5(OH)_3$. Such alcohols are also called polyhydric. See also polyhydric alcohol.

trimer.   A macromolecule comprised of three monomer units.

trimethylamine.   See amine.

triolefin.   See olefin.

triple bond.   A chemical linkage characteristic of acetylene ($HC{\equiv}CH$) and its derivatives, though it also occurs in some other compounds, for example, certain fatty acids. Its presence indicates a highly unsaturated and reactive compound. Triple bonds are more stable to high temperature than single and double bonds and are strong electron receptors. They are active in many types of reactions, e.g., Diels-Alder, vinylation, and polymerization with Ziegler catalysts; they tend to react strongly with water, alcohols, and amines. See also acetylene.

tris-.   A prefix indicating that a certain chemical grouping occurs three times in a molecule, for example, tris (hydroxymethyl) aminomethane $(CH_2OH)_3CNH_2$. See also bis-.

tritium.   The only artificial radioactive isotope of hydrogen (H^3); its nucleus contains two neutrons and one proton, giving an atomic weight of 3.017. It is made by impingement of slow neutrons on lithium nuclei and has a half-life of about 12 years. It is used to induce thermonuclear (fusion) reactions, as well as in biochemical tracer work and specialized research. Its nucleus is called a triton, and its symbol is T. See also hydrogen; deuterium.

triton.   See tritium.

trivial name.   A name applied by early chemists to a number of simple organic compounds, usually based on their sources or properties, e.g., acetone and acetic acid (from Latin acetum, vinegar); urea (from urine); glucose and glycerol from Greek glyc- (sweet). See also nomenclature.

true solution.   See solution.

trypsin.   See bating; proteolytic.

tungsten.   An element.

Symbol W	Atomic Wt.	183.85
State Solid	Valence	2,4,5,6
Group VIB	Isotopes	5 stable
Atomic No. 74		

Tungsten does not occur as such in nature but is obtained in the form of the compound tungstic oxide, WO_3, by reduction of the ores wolframite and scheelite. For this reason, it is often called wolfram, with the symbol W. Most of the supply is located in China, but there are also extensive deposits in the United States, Europe, and South America. The major properties of tungsten which make it an essential natural resource are its extremely high melting point of 3410°C (6170°F) and its strong electrical conductivity. It is one of the heaviest metals, with a specific gravity of 19.3. Its first important application was as a filament in electric light bulbs. It is also an alloying element in tool steels and a component of cathodes in electronic devices and of welding electrodes. Tungsten is also valuable in aerospace technology. One of its most important compounds is tungsten carbide (WC), used in machine tools, metal-cutting dies, etc. Various alloys are used in electric circuitry and as aerospace materials; these and other tungsten products (wire and filament) are made by powder metallurgy techniques.

tungsten carbide.   *See* tungsten.
Turkey red oil.   *See* castor oil.
Turnbull's blue.   *See* blue.
turpentine.   An essential oil belonging to the terpene family of hydrocarbons ($C_{10}H_{16}$). Two varieties are common: one is obtained by distilling the sticky exudation found on the bark of pine trees (so-called gum turpentine); the other is derived by extraction from the stumps of pine trees with an organic solvent (so-called wood turpentine). It is a light, mobile, optically active liquid having a strong and unmistakable odor; its major uses are as a paint thinner and solvent for resins, waxes, and other organic materials. It ignites quite easily and is poisonous if taken internally. Venice turpentine is an oily resinous material obtained from a European larch tree. *See also* terpene; gum (**2**)
Twitchell process.   *See* fatty acid.
Tyndall effect.   The reflection or scattering of light by colloidal particles suspended in a gas or liquid. As the light beam passes through the suspension, the particles (which are too small to be seen in an optical microscope) indicate their presence by reflecting the light at various angles, depending on their position relative to the incident beam. As a result, a visible path or cone of light appears. A similar effect can be seen by passing an intense light beam through a darkened room. The diagram shows that particles at positions 1 and 3 (45° to the beam) give

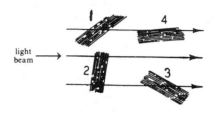

partial reflection; those at position 4 (parallel to beam) give no reflection; while those at position 2 (vertical to beam) give total reflection. This principle was discovered by John Tyndall (1820–1893), a British physicist who was professor of physics at the Royal Institution, London. He made many valuable and original investigations in the fields of magnetism, light, sound, electricity, and heat, and published and lectured extensively on his work. The Tyndall effect is utilized in the ultramicroscope. *See also* Zsigmondy.

U

U   Symbol for the element uranium, the name being adopted from the planet Uranus which was thought to be the outermost planet of the solar system at the time the element was discovered, 1789.

ultracentrifuge.   *See* centrifugation.

ultramarine blue.   *See* blue (1).

ultramicroscope.   *See* microscopy; Zsigmondy.

ultrasonics.   The science of effects of sound vibrations beyond the limit of audible frequencies. Used for dust, smoke and mist precipitation; preparation of colloidal dispersions; cleaning of metal parts, precision machinery, fabrics, etc.; friction welding; formation of catalysts; degassing and solidification of molten metals; extracting flavor oils in brewing; electroplating; drilling hard materials; fluxless soldering; nondestructive testing. Also used for investigation of physical properties, determination of molecular weights of liquid polymers, degree of association of water, and for inducing chemical reaction. Biological effects are also under study.

ultraviolet absorber.   An organic compound which absorbs the ultraviolet radiation present in sunlight; it is added to products which tend to deteriorate from prolonged exposure to light, for example, various plastics and rubber. Various derivatives of benzene, phenol, and acrylonitrile are used. Though not usually classed as a U.V. absorber in the chemical additive sense, glass does not transmit ultraviolet radiation.

uncertainty principle.   A concept based on quantum mechanics which resulted from the investigations of the German physicist Werner K. Heisenberg (1901-76), Nobel Prize in physics 1932, and became generally accepted between 1920 and 1930. It is important to chemistry through its influence on the concept of electron behavior; from it grew the approach to electronic structure of the atom and to chemical bonding that is known as the orbital theory. In the simplest possible terms, this principle may be described as follows: since exact determination of the position and speed of an electron requires measurement, and since the electron has such a tiny mass that it cannot be measured without altering its position to an unknown extent, it must be concluded that there is no way of determining these properties with absolute certainty for any given electron. Thus, the location and speed of electrons can be dealt with only by the mathematics of probability, a conclusion which led to the orbital concept. *See also* orbital; quantum; electron.

uni.   *See* mono-.

uniform dispersion.   A mixture of two or more substances (gas-in-gas, solid-in-solid, liquid-in-liquid, or solid-in-liquid) in which the proportion of components is exactly the same in all parts of the mixture. For example, air at sea level is a uniform mixture of oxygen, nitrogen, carbon dioxide, etc.; a properly blended paint is a uniform dispersion of its solid and liquid components. The word *homogeneous* is frequently (but inexactly) used in this sense. *See also* homogeneous; heterogeneous; mixture.

unit operation.   A term used in chemical engineering to designate an action performed on a material or mixture of materials for a particular purpose which does not involve a chemical change (though it often does involve a physical change). Examples are filtration (separation without physical change), distillation (separation with physical change), and heat exchange (energy transfer with or without physical change). Unit operations are distinguished from unit processes, which involve chemical change, e.g., hydrolysis, polymerization, etc.

unit process. *See* process.

unsaturation. A term used by chemists to indicate that an organic compound contains certain atoms (usually carbon), one or more of whose valences are not satisfied; this results in the formation of double or triple bonds within the molecule. The extra bond is less tightly bound to the carbon atom than is the primary bond and thus readily attaches itself to any available univalent atom or group.

$$
\begin{array}{ccc}
\text{H}\ \ \text{H} & \text{H}\ \ \text{H} & \\
|\ \ \ | & |\ \ \ | & \\
\text{H—C—C—H} & \text{C}=\text{C} & \text{HC}\equiv\text{CH} \\
|\ \ \ | & |\ \ \ | & \\
\text{H}\ \ \text{H} & \text{H}\ \ \text{H} & \\
\end{array}
$$

saturated C atoms	unsaturated C atoms	
(ethane)	(ethylene)	(acetylene)

Compounds containing two or more double bonds are said to be polyunsaturated, e.g., safflower oil, benzene. Paraffins and cycloparaffins are saturated compounds, while olefins, acetylenes, and aromatic and heterocyclic compounds are unsaturated. The negative form of the term is most often used because it is the reactivity of a compound that is of primary concern in chemical reactions. *See also* olefin; double bond; triple bond.

uranium. An element

Symbol	U	Atomic Wt.	238.029
State	Solid	Valence	3,4,6
Group	IIIB	Isotopes	3 natural
	(Actinide		radio-
	Series)		active
Atomic No.	92		

A heavy and strongly radioactive metal, m.p. 1132°C (2070°F), extracted from the ores pitchblende and carnotite occurring chiefly in Zaire, Canada, and Colorado. The principal isotope, U-238, comprises 99% of the metal but is not subject to fission. The 235 isotope, present to only 0.7%, stands at the top of the actinide decay series. It is this isotope that can be split by neutron bombardment to yield nuclear energy. The artificial isotope U-233, derived from thorium, is also fissionable. The third natural isotope, U-234, has little importance. Separation of the very small amount of U-235 from the nonfissionable U-238 was one of the chief achievements of the Manhattan Project (1941) which developed the original "atomic" bomb; separations were effected by gaseous diffusion (uranium hexafluoride, UF_6, through porous barrier sequences) and by electromagnetic means. The energy yield per atom of 235 is about 200 million electron volts, plus two to three neutrons, which in turn initiate new fissions. Plutonium is made from uranium by neutron absorption. *See also* radioactivity; chain reaction (2); fission.

uranium hexafluoride. *See* uranium; fluoride.

urea. A crystalline organic compound, m.p. 133°C (271°F), belonging to the amide group and commercially derived from ammonia and carbon dioxide at high temperature and pressure. It also occurs as a metabolic product in urine. Its formula is $CO(NH_2)_2$. It was the first organic compound to be made synthetically. In 1828 the German chemist Friedrich Wöhler published his observations of its formation by molecular rearrangement of ammonium cyanate; $NH_4OCN \rightarrow (NH_2)_2CO$. Urea is manufactured on a large scale for use in fertilizers and as a component of ureaformaldehyde and urea melamine resins. Its alternative name is carbamide. *See also* amide; rearrangement; organic; Wöhler.

ureaformaldehyde resin. A synthetic high polymer generally classed as an amino resin, though it may also be regarded as a polyamide. It is made by catalytic condensation of urea (an amide) and formaldehyde in a two-stage process. The resins are thermosetting, and may be used in the form of molding powders for a wide variety of plastic items and as adhesives or coatings for textiles, paper, etc. These were the first synthetic resins which could give light-colored, white, or pastel shades; the others available at that time (1930) gave only black products. As a result, the ureas leaped into prominence for a broad range of white molded units for machine housings, scales, dinnerware, and other household articles. They are also used in various laminated structures (plywood). *See also* amino resin; polyamide; urea.

urethane.   *See* polyurethane; diisocyanate; pre-polymer.

Urey, Harold C.   (1894 1981). American chemist who received the Nobel prize in chemistry in 1934 for his discovery of deuterium, the heavy isotope of hydrogen, which became an important factor in the development of nuclear fission and fusion and made possible the production of the first transuranic element plutonium. He was one of the leaders of the Manhattan Project, which constructed the first nuclear reactor at the University of Chicago and eventually produced the first atomic bomb. After obtaining his doctorate at University of California in 1923, he taught at several leading universities, including Columbia University, where he discovered deuterium oxide (heavy water), used as a moderator in early types of nuclear reactors. Later he devoted much study to the origin of the universe and the origin of life on earth. He was author of many scientific treatises and made notable contributions to the cosmological theories of recent years.

uric acid.   A nitrogenous organic compound, $C_5H_4N_4O_3$, formed from nucleic acids by metabolic processes in man and animals, especially in certain species of birds. It is extracted from the excrement of gulls and other sea birds (guano). Its chief uses are in biochemical research and as an intermediate and reagent. It is chemically related to guanine and similar nucleic acid derivatives and is the source of a group of compounds called purines. *See also* purine; urea.

U.V.   Abbreviation for ultraviolet light.

V

V Symbol for the element vanadium, named after a mythological goddess Vanadis.

vacancy. (1) A term used in crystallography to describe a defect in a crystal due to the absence of an atom or ion in the lattice structure. Such defects are commonplace in crystals; they cause energy imbalances within the crystal which often result in marked changes in their electrical and optical properties. *See also* hole; semiconductor.

(2) It also refers to the absence of one or more electrons in the outer shell of an atom.

valence. A whole number indicating for any element its ability to combine with another element. In inorganic compounds, characterized by ionic bonds, this is equivalent to the positive oxidation number, as it represents the number of electrons that the electropositive atom transfers to its partner in a compound, that is, the number of electrons in its outer or valence shell. Negative or polar valences characterize electronegative atoms, which tend to acquire electrons and are thus usually oxidizing agents (oxygen, chlorine, and other halogens). Since a molecule is always electrically neutral, the algebraic sum of the valences of its constituent atoms must equal zero. Many metals have several positive valences. Chlorine (a nonmetal) is exceptional, as under various conditions it can have valences of -1, $+1$, $+3$, $+4$, $+5$, and $+7$. In organic compounds, the valence represents the number of electrons shared between the atoms of a compound as covalent bonds, except in the case of organic nitrogen compounds in which one electron is transferred in addition to those shared. *See also* oxidation number; chlorine; bond.

valerolactone. *See* lactone.

vanadium. An element.

Symbol	V	Atomic Wt.	50.9414
State	Solid	Valence	2,3,4,5
Group	VB	Isotopes	2 stable
Atomic No.	23		

One of the softer metals, vanadium, m.p. 1890°C (3434°F), has unusual resistance to corrosion and has no toxic properties in elemental form. It occurs only in the form of ores, among which are carnotite and phosphate rock, from which it is extracted as V_2O_5; the metal is obtained by the reduction of the oxide with calcium. It is an important transition metal and complexing agent. Its principal use as a metal is as an alloying ingredient (ferrovanadium) for corrosion-resistant steels, especially for those exposed to saline atmospheres and seawater; its resistance to liquid sodium has led to its use in nuclear reactors. Its most important compound is vanadium pentoxide (V_2O_5), which is the chief catalyst used for the manufacture of sulfuric acid by the contact process; it also has other catalytic applications. Both this and other vanadium compounds are quite toxic.

vanadium pentoxide. *See* vanadium; sulfuric acid.

van der Waals forces. Extremely weak forces of interaction between unexcited atoms or molecules of gases; they account for the fact that gases do not behave strictly in accordance with theory, i.e., their behavior varies slightly from that described by the ideal or perfect gas laws. They are weaker than hydrogen bonds and are not involved in chemical bonding. They were discovered and named for a Dutch scientist, Johannes Diderik van der Waals (1837–1923), Nobel Prize 1910, who presented his conclusions in 1873. *See also* perfect; Boyle's Law.

vanillin. A derivative of benzaldehyde occurring naturally in a variety of bean grown chiefly in Madagascar; it is also extracted from the lignin fraction of wood. Its formula is $CH_3O(OH)C_6H_3CHO$, and it melts at 81°C (189°F). It is purified by crystallization; the commercial form is an alcoholic solution. Vanillin is primarily used as a flavoring agent in foods and confectionery and is also the source of a number of pharmaceuticals, notably the amino acid dihydroxyphenylalanine (L-dopa), said to be useful in treating certain degenerative diseases.

vapor. *See* gas.

vapor pressure. A value characteristic of all liquids and some solids. It is the pressure exerted by the vapor when evaporation and condensation exactly balance each other in a closed system, that is, when as many molecules are entering the liquid as are leaving it. This is called the equilibrium state between a liquid and its vapor. It is a function of the substance and of its temperature. This pressure rises with temperature because of the increase in molecular kinetics caused by the added energy. When the pressure of the vapor reaches that of the atmosphere, the liquid boils. Some alcohols and most hydrocarbon liquids have lower vapor pressure than water. It is usually expressed in millimeters of mercury (mm Hg); 760 mm Hg equals 14.7 psi, or 1 atmosphere. *See also* equilibrium; boiling point; Dalton.

varnish. An unpigmented organic coating containing varying proportions of a drying oil or resin, which solidifies on evaporation of solvent due to polymerization of the oil or resin base. Spar varnishes are designed for heavy-duty exterior work and contain a higher proportion of drying components. The solvents used may be any of several alcohols, hydrocarbons, or ketones. Varnishes are used primarily on floors, furniture, and other interior applications. A number of specially compounded types exist for fine work, such as violins, paintings, and other art objects.

vat. A term used in dyeing technology to refer to a group of organic dyes derived from indigo or anthraquinone; the essential process involved in making such dyes is reduction of the insoluble compound to a soluble form to enable it to penetrate the fiber, after which it is oxidized back to the original insoluble state. The reducing agent is sodium hydrosulfite. Indanthrene blue is a well-known dye of this type. Vat dyes are notably stable to water and light and are used for textiles requiring a high degree of light-fastness and color permanence. The term *vat* was presumably adopted from the equipment used in their manufacture.

VCM. Abbreviation for vinyl chloride monomer.

vector. An animal or insect that harbors infectious bacteria or viruses and thus transmits a disease.

vegetable oil. An oil (or when solid, a fat) obtained from a wide variety of vegetables and some fruits and characterized by relatively high percentages of unsaturated fatty acids, though low percentages of saturated acids are often present. These oils are classified as drying, semidrying, or nondrying, based on their tendency to undergo oxygen-catalyzed polymerization on exposure to air, which depends on their content of unsaturated components. The fatty acids occur in the form of esters called glycerides, usually as mixtures of several types. Some of the fatty acids commonly occurring in vegetable oils are: saturated—palmitic, myristic, stearic, lauric, and arachidic; unsaturated—oleic, linoleic, linolenic, eleostearic. Common vegetable fats and oils for the food industry are corn, soybean, coconut, safflower, cottonseed, and peanut; for the paint and plastics industries, linseed, tung, castor, perilla, and oiticica. Most vegetable oils have high nutrient value, and the liquid types are often hydrogenated into hard fats for cooking purposes, or are used as the basis for margarines. *See also* glyceride; hydrogenation.

vegetable tanning. *See* tanning.

vehicle (1) A term used by paint technologists to refer to those components of a paint that are in the form of liquids, solutions or (in the case of emulsion paints) water suspensions. Thus, the vehicle is composed of the binder (drying oil or synthetic resin) plus the solvent or thinner, in which the dry ingredients (pigments, extenders, and driers) are dispersed. It is so-called because the liquid acts as a carrier for the solids. *See also* paint; binder.

(2) *See* vector.

Venice turpentine. *See* turpentine.

vicinal. A term derived from the Latin word for *near* or *neighboring*; in chemical terminology, it refers to locations on a carbon backbone or cyclic structure which are immediately next to each other. Vicinal locations on the molecule shown are occupied by the hydrogen atoms and the hydroxyl groups:

The abbreviation is vic-.

vinegar. *See* acetic acid; pickling (1).

vinyl. The unsaturated organic group $H_2C = CH$ occurring in a large number of reactive compounds which polymerize readily and are thus of great importance in high-polymer chemistry. Examples are styrene, $C_6H_5HC = CH_2$; butadiene,

$$H_2C = CHHC = CH_2;$$

acrylonitrile, $H_2C = CHCN$; and methyl methacrylate, $H_2C = C(CH_3)COOCH_3$. Important vinyl compounds are vinyl acetate, from ethylene and acetic acid,

$$H_2CCOOCH = CH_2;$$

vinyl chloride, from acetylene and hydrochloric acid, $CH_2 = CHCl$; and vinyl ether, $(H_2C = CH)_2O$. From these are derived such polymers as polyvinyl chloride, polyvinyl acetate, polyvinyl alcohol, polymethylmethacrylate, and synthetic elastomers. *See also* polyvinyl; monomer.

vinyl chloride. A gaseous, chlorinated olefin, $CH_2 = CHCl$, b.p. $-13.9°C$ (8.4°F), the most important vinyl compound and the basis for the widely used polyvinyl chloride polymers. Also known as VCM, or vinyl chloride monomer. It is made by (a) dehydrochlorination of dichloroethane, the removal of HCl from $C_2H_4Cl_2$; (b) the addition of HCl to acetylene over a mercury catalyst, $HC = CH + HCl$; and (c) the oxychlorination of ethylene with HCl and oxygen. Its multibillion pounds per year production accounts for the largest single use of chlorine. It is used for the production of polyvinyl chloride and copolymers, in chemical synthesis, and in adhesives for plastics. The gas is highly flammable and a severe fire and explosion risk; the explosive limits in air are 4 to 22%. As a result of indications that prolonged exposure to its fumes is carcinogenic, its concentration in plant atmospheres is limited to one part per million. *See also* polyvinyl chloride; oxychlorination.

vinylidene chloride. A highly reactive liquid $(H_2C = CCl_2)$, b.p. $31.9°C$ (89°F), whose chief use is as a monomer which is copolymerized with vinyl compounds in the manufacture of the thermoplastic polymer called saran (polyvinylidene chloride). The monomer is highly flammable, and its vapors are explosive at from 5 to 10% concentration in air. It requires an inhibitor to prevent unwanted polymerization in storage. *See also* polyvinylidene chloride.

virus. A colloidal complex composed of one or more nucleic acids linked to a protein; most are too small to be retained by the finest filter media and hence are called filterable viruses. Viruses are not true organisms, as are bacteria; they do not absorb nutrients and thus are unable to grow. They do reproduce within cellular tissue and initiate a large number of infectious diseases in man, among which are smallpox, influenza, and poliomyelitis; such animal diseases as rabies, Newcastle disease, and hoof-and-mouth disease are due to viruses. In plants, bushy stunt and tobacco mosaic viruses are well-known; the latter is a nucleoprotein which is about 95% protein and 5% ribonucleic acid. Viruses behave like parasites and have a number of different shapes (fibrous, spherical, etc.), which can be observed clearly in an electron microscope. They are on the borderline between animate and inanimate. *See also* bacteria.

viscose process. The most widely used method for making regenerated cellulose fiber (rayon) and film (cellophane). It involves the following steps: (1) formation of alkali cellulose from delignified wood pulp by treatment with sodium hydroxide (mercerizing); (2) reaction of this product, after purification, with carbon disulfide to form cellulose xanthate, a thick, yellowish liquid called viscose; (3) extrusion of the viscose through extremely small orifices (spinneret) into a solution of sodium sulfate or zinc sulfate and sulfuric acid, which coagulates the viscose and restores the original cellulose by decomposing the cellulose xanthate. For cellophane, the viscose is extruded through a thin slit and passed through a similar solution. *See also* rayon; cellulose; xanthate.

viscosity. That property of a liquid which causes it to resist flow or movement in response to external force applied to it; it is measured in standard units called poises (or for kinematic viscosity, stokes). Viscosity is affected by temperature, especially for liquids comprised of large, complicated molecules such as asphalt or heavy oils. The more complex the molecular organization, the greater is the viscosity of a liquid. The viscosity of water is about 0.01 poise. Pragmatic tests of viscosity involve either dropping a small metal sphere through a container of the liquid being measured and noting its time of fall or placing the test liquid in a glass tube provided with a small outlet and observing the rate of escape (Saybolt method). Kinematic viscosity is the viscosity of a liquid in centipoises divided by its density, expressed in stokes.

visual purple. *See* rhodopsin.

vitamin. A class of biochemically active organic compounds which profoundly affect many metabolic functions, either alone or in conjunction with enzymes or coenzymes; their chemical behavior is essentially catalytic, for example, ascorbic acid (vitamin C) catalyzes cellular oxidation-reduction reactions, and calciferol (vitamin D_2) facilitates deposition of calcium in bony structures. Vitamins include various types of compounds, among which are hydrocarbons, amino complexes, sterols, and acids. All are ultimately derived from plant products, though two (A and D) are present in the body in the form of provitamins and are either elaborated in the liver or activated by ultraviolet radiation. Most important from a nutritional standpoint are the B complex and vitamins A, D and C. The nature and function of vitamins was elucidated beginning about 1912 by such nutritionists as Funk, McCollum, Elvehjem, Szent-Gyorgyi, Sherman, R.J. Williams, and others. The term was coined from the Latin *vita* (''life''), plus *amine*, the terminal *e* later being dropped. All the more important vitamins are made synthetically. *See also* B complex; ergosterol; ascorbic acid; riboflavin; cobalamin; biotin; carotenoid; provitamin.

vitamin A. *See* carotenoid.

vitamin B. *See* biotin; B complex; folic acid; riboflavin; thiamine; pantothenic acid; niacin.

vitamin B_{12}. *See* cobalamin.

vitamin C. *See* ascorbic acid.

vitamin D. *See* ergosterol.

vitreous. A term used to describe matter in the glassy (non-crystalline) state; besides glass itself, it refers to such substances as glazes, fired clays, and porcelain enamels, all of which are of siliceous nature, as well as to metals and alloys that have been cooled so rapidly from the molten condition that crystal formation is prevented. This phenomenon is called vitrification, and its opposite (the unwanted development of a crystalline structure) devitrification. The obsolescent term *vitriol* was formerly applied to a number of metal sulfates and to sulfuric acid (oil of vitriol) because of their glassy appearance. *See also* glass; metal glass.

vitriol. *See* vitreous.

VM&P naphtha. Abbreviation for Varnish Makers and Painters naphtha. *See* naphtha.

volatility. Literally meaning ''to fly away,'' this term refers to the ease with which a liquid or solid passes into its vapor phase (evaporates or sublimes). The rate increases with temperature. Ethyl alcohol and hydrocarbon liquids are more volatile than water; solid carbon dioxide (dry ice) sublimes at $-51°C$ ($-60°F$) and is thus more volatile than ice or snow.

volumetric analysis. *See* titration.

vulcanization. Cross-linking of rubber hydrocarbon and similar elastomers with sulfur, usually with the aid of heat; the time required is notably reduced by use of organic accelerators. Vulcanization without heat, and even without sulfur (with peroxides), is possible under special conditions. Discovered in 1842 by the American inventor, Charles Goodyear (1800–1860), this process transforms rubber from a soft, hydrocarbon-soluble material that is unstable to temperature changes to a strong, resilient, wear-resistant, insoluble product having manifold uses over a wide range of temperatures. High-energy radiation may be used to induce vulcanization. With highly accelerated mixtures, from 2 to 3% sulfur is required, and the time for the process may be as short as 3 minutes. *See also* rubber; cross-linking; accelerator (**1**); Goodyear.

vulcanized oil. *See* factice.

W

W   Symbol for the element tungsten, standing for the word *wolfram* (an alternative name for tungsten). The name *tungsten* is derived from two Scandinavian words meaning "heavy stone," the origin of *wolfram* is not clear.

Waksman, Selman A.,   *See* antibiotic; Fleming.

washing soda.   *See* sodium carbonate; efflorescence.

waste control.   Waste products can be the source of many useful materials, including fuels. They occur in many forms: as gases (industrial smokes, smelter fumes, auto exhaust products); as liquids (chemical plant and paper mill wastes); as solids (asbestos, metals, glass, plastics, rubber, etc.); and as radioactive products from nuclear reactors. The problem of safe disposal of hazardous waste, both chemical and radioactive, has become increasingly severe in recent years. Following is a summary of the major types of waste and methods of treating them.

Type	Treatment
1. Industrial	
(a) chemical	pyrolysis; sorption; oxidation; neutralization; ion exchange; incineration
(b) rubber, plastics	pyrolysis; biodegradation; caustic reclaiming (tires)
(c) metals, glass	remelting and recycling
(d) paper	recycling after deinking and comminution
(e) smoke, fumes, etc.	electrostatic precipitation; SO_2 absorption
2. Agricultural	
(a) corn stalks, cannery waste, off-grade vegetables	fermentation to alcohol; ensilage; incineration
(b) manures	bacterial decomposition to methane
3. Municipal (sewage)	bacterial decomposition by activated sludge; drying to fertilizers
4. Urban (garbage, other solids)	Fermentation; compaction; incineration; landfill
5. Automotive	
(a) exhaust fume	catalytic oxidation
(b) spent oil	conversion to off-grade lubricants
6. Nuclear (radioactive)	dilution and underwater storage; deep burial in salt domes; dispersal in vitreous materials (experimental)

water. An inorganic substance whose formula may be written either H_2O or HOH; it exists in three forms or phases: (a) as a crystalline solid at or below 0°C (32°F) (ice); (b) as a colorless liquid from 0 to 100°C (32 to 212°F); and (c) as a vapor at 100°C (212°F) (steam). Since 1 milliliter (almost exactly 1 cubic centimeter) of water weighs 1 gram at 4°C (39°F), water is the reference standard for specific gravity for liquids and solids. Pure water is a polar liquid with high dielectric constant and heat capacity; it can be decomposed into its constituent gases by passage of an electric current (electrolysis). It is an oxide formed by the combustion of hydrogen: $2H_2 + O_2 \rightarrow 2H_2O$. The water molecule has a slight tendency to ionize, forming hydronium and hydroxyl ions. Its electrical properties have positive significance in electrolytic solutions, as they affect the activity and behavior of ions by lowering the bonding energies of dissolved electrolytes.

Water is the most common solvent (though by no means a "universal" one); it will not dissolve hydrocarbons (oils, fats, waxes, and their derivatives), metals, or transition metal compounds but will dissolve alcohols and other hydroxyl-bearing compounds, as well as sugars, starches, gums, some vitamins, salts of most metals, and many aldehydes, ketones, and acids. Many compounds are soluble in water to only a limited extent.

Though it is not a particularly reactive compound, water enters into chemical reactions in several ways: (1) in electrolysis, where it is the most effective solvent for electroplating and other electrolytic processes due to its high dielectric constant; (2) in hydration, where the molecule acts as a unit, combining with inorganic compounds to form hydrates; (3) in hydrolysis, where the molecule divides, each constituent uniting with another molecule (a reaction frequently occurring with proteins, esters, soaps, and other large molecules); (4) as a catalyst, for example, of corrosion.

Water is essential to all life and comprises about 75% of human tissue. It is used in huge volume in industry, both as a processing ingredient (paper) and as a coolant. *See also* hydronium ion; electrolysis; hydrolysis; hydration; solvent; hardness **(2)**.

water gas. A gaseous mixture derived by reacting steam with hot coke or natural gas in the presence of air; it contains chiefly carbon monoxide and hydrogen admixed with nitrogen and carbon dioxide. The basic reaction is $C + H_2O \rightarrow CO + H_2$. It has a higher Btu value than producer gas and is also called blue gas. It is used in some processes for ammonia synthesis. It can be carburetted by passing it through hydrocarbon vapors to increase combustibility for fuel purposes. *See also* synthesis gas; producer gas; ammonia.

water glass. A water-soluble silicate of sodium (the simplest glass known) made by melting together sand and soda ash. In solid form it is hard and brittle, but when heated with steam, it becomes a colorless liquid with a wide range of viscosities according to water content. Silia gel is formed when the material is neutralized by addition of acid. The major uses of water glass are as a dispersing agent for clay suspensions, as a fire-retardant coating for paper and fiberboard, as a binder for foundry cores, and as a detergent builder. A similar material is made from potassium silicate. *See also* glass.

water of crystallization. *See* hydration.

water pollution. *See* algae; eutrophication.

water softening. *See* ion exchange; ethylenediaminetetraacetic acid.

wattle. *See* tanning.

wave mechanics. *See* electron; orbital; quantum.

wax. A thermoplastic solid of high molecular weight which may be a hydrocarbon derived from distillation of an aliphatic petroleum (paraffin wax) or an ester of unsaturated fatty acids and alcohols which are endproducts of plant and animal metabolism. Of the latter, beeswax, spermaceti, and wool wax (lanolin) from animals and candelilla, carnauba, and bayberry from plants are in most common use. There is also a group of mineral or earth waxes, chief among which is ozocerite or its purified form, ceresin. Certain waxes which have a structure of extremely small crystals are called microcrystalline waxes; examples are paraffin derived from petroleum wastes and also the chlorophyll of plants.

Waxes are physically plastic but are sensitive to temperature. As their chemical structure is not polymeric or resinous, they are not plastics in the accepted sense of the word. Among the major uses of waxes are in candles, shoe and floor polishes, paper coating, light-protective agents for rubber and plastic products, in precision casting of intricate machine parts, and as abherents.

weed killer.   *See* herbicide.

Werner, A.   (1866–1919). A native of Switzerland, Werner was awarded the Nobel Prize for his development of the concept of the coordination theory of valence, which he advanced in 1893. His ideas revolutionized the approach to the structure of inorganic compounds and in recent years have permeated this entire area of chemistry. The term *Werner complex* has largely been replaced by *coordination compound*. *See also* coordination compound.

Werner complex.   *See* coordination compound; Werner.

wet.   When used as a verb, this term refers to the ability of a liquid to spread readily and uniformly over the surface of another liquid or of a solid, especially when the solid is in powder form. A liquid of very high surface tension and electrical conductivity, such as mercury, has virtually no wetting capacity; water, with moderately high surface tension and low conductivity, has reasonably good wetting ability but requires a surfactant (also called a wetting or dispersing agent) for maximum effect. Petroleum solvents and fatty acids, which have low surface tension and conductivity, have excellent wetting ability; for example, a hydrocarbon liquid will quickly spread on a water surface, forming a thin, continuous film. The chemical nature of the substance being wet is also a factor; hydrophobic solids repel water and hydrophilic solids attract it. Wetting ability is of critical importance in dyeing, cleaning, painting, and any operation involving suspensions of dry powders. A notable case is the wetting action exerted by stearic acid on carbon black-rubber mixtures. The wetting phenomenon is closely related to emulsification. *See also* surface tension; emulsion; hydrophilic; hydrophobic.

whiskers.   Fibrous single crystals of metals such as aluminum, cobalt, tungsten, and others made synthetically by crystal-growing techniques. They may be up to 5 cm in length by 10 microns in diameter. They are notable for their fantastically high strength and excellent heat resistance and are used in composites for aerospace construction (ablative agents) and other engineering applications involving high-temperature exposure. *See also* fiber.

white liquor.   *See* liquor.

whiting.   Fine-ground calcium carbonate made from calcite, chalk, or limestone and widely used as a filler or extender in low-cost plastic and rubber mixtures, putty, and similar products. It has no reinforcing power and, in spite of its name, no coloring effect. Though often classed as a pigment, this name is inappropriate. *See also* filler.

Wöhler, Friedrich (1800–1882.). A native of Germany, Wöhler, working with Berzelius placed the qualitative analysis of minerals on a firm foundation. In the early nineteenth century chemists still thought it impossible to synthesize organic compounds. In 1828, Wöhler publicized his laboratory synthesis of urea, which he had obtained four years previously by heating the inorganic substance ammonium cyanate; the synthesis is a result of intramolecular rearrangement:

$$(NH_4)OCN \longrightarrow NH_2-\overset{\overset{\displaystyle O}{\|}}{C}-NH_2$$

This was the beginning of the science of synthetic organic chemistry.

wolfram.   *See* tungsten.

wood.   (1) A closely packed fibrous structure containing about two-thirds cellulose and one-third lignin, which can be separated by hydrolysis, as is done in the manufacture of paper pulp. Cellulose is a carbohydrate polymer; lignin is considered to be a polymer of phenyl propane, which acts as a binder for the cellulose component. Valuable products of wood distillation are methyl alcohol, turpentine, rosin, tall oil, etc. Paper is by far the most important material derived from wood; most of it is made from softwoods (evergreens), though some is made from hardwoods (deciduous trees). Plywood is a composite of layers or plies of wood cemented together with a thermosetting resin, the plies being arranged so that the grain is at right angles on alternate plies to secure maximum strength. Groundwood, used in newsprint, is made by mechanical grinding of wood against a revolving stone.

(2) Used adjectivally, as in *wood rosin* and *wood turpentine*, the word refers to the fact that these products are derived from pine stumps distilled with naphtha, as opposed to "gum rosin"

and "gum turpentine" which are obtained directly from living trees.

wood alcohol. *See* methyl alcohol.

Wood's metal. *See* fusible alloy.

Woodward, Robert B. (1917-79). American chemist born in Quincy, Mass. and widely regarded as one of the world's leading synthetic organic chemists. After receiving his doctorate from M.I.T. (the youngest student in the history of the Institute to do so), he joined the Harvard faculty in 1937 as instructor, and attained full professorship in 1950. He was recipient of the Nobel Prize in 1965 for his brilliant work in synthesizing complex organic compounds, among them quinine, cholesterol, chlorophyll, reserpine and cobalamin (vitamin B_{12}). He was director of the Woodward Research Institute in Basel, Switzerland, and a member of the governing board of the Weizmann Institute in Israel.

wool. A fibrous protein belonging to the keratin group, which also includes hair and feathers. Wool is high in the amino acids cystine and methionine; it is a cross-linked polypeptide with a folded molecular structure, which gives the fiber a spring-like effect. The surface of the fiber is quite rough and scaly. Wool is an amphoteric substance. This factor is important in dyeing, which in this case is essentially an acid-base reaction. The reputation of wool as a "warm" textile material is largely due to the fact that woolen fabrics are often much thicker than those of other fibers. The thermal conductivity of wool *fiber* is 7.3, compared to 11.0 for viscose rayon and 17.5 for cotton (air = 1). *See also* amphoteric; keratin.

wort. *See* malt; brewing.

X

xanthate. A salt of an alkyl dithiocarbonic acid in which the acidic portion of the molecule is represented by OCSSH, the alkyl substituent usually being an ethyl group and the metal, one of the alkali series (sodium or potassium). Xanthates of this type are used in metal separation operations. The most familiar variation is cellulose xanthate, used in the viscose rayon process; it is obtained by reacting alpha-cellulose with carbon disulfide and sodium hydroxide. Xanthates have a characteristic orange-yellow color, from which the name is derived. *See also* viscose process.

xanthophyll. A natural plant pigment of bright yellow shade, which causes the color of autumn foliage as well as of many flowers; it is also formed in certain animals and is present especially in egg yolk. Chemically, it belongs to the carotenoid group of compounds; it is an oxygenated derivative (alcohol) of the hydrocarbon pigment carotene, with two hydroxyl groups, $C_{40}H_{54}(OH)_2$. Xanthophyll is thus entirely different from chlorophyll (a porphyrin), in spite of the similarity in names and its presence in leaves and green vegetation.

Xe Symbol for xenon, a noble gas element, discovered in 1898 by Sir William Ramsay and M.W. Travers, and named for the Greek word meaning "the stranger."

xenon. An element.

Symbol	Xe	Atomic Wt.	131.3
State	Gas	Valence	2,4,6,8
Group	Noble Gas	Isotopes	9 stable
Atomic No. 54			

Xenon is a heavy, unreactive stable gas occurring in air in trace percentages. Until recent years, it was thought to be unable to form compounds, but it is now known to combine with fluorine, oxygen, cesium, and alkali metals to form the corresponding fluorides, oxides, and salts, as well as a number of metallic complexes. All its compounds are toxic, though the gas itself is not. It is obtained from the oxygen portion of liquid air by specialized vaporization and adsorption techniques. It has been used as an anesthetic and also in flash lamps and light bulbs. Some of its compounds have limited application as oxidizing agents in laboratory analysis.

x-ray. Penetrating short-wave electromagnetic radiation emanating from heavy metals subjected to cathode rays; their wavelength ranges from less than one-tenth to about 100 angstrom units, the shorter (hard) x-rays being of greater intensity. They were discovered in 1895 by Wilhelm C. Roentgen (1845–1923), a German physicist (Nobel Prize 1901) and named x-rays because of their then unknown nature; they are still referred to as Roentgen rays in many circles. Max von Laue (1879–1960), a German physicist (Nobel Prize 1914), found that x-rays were diffracted by crystals, and soon thereafter the Braggs, father and son (British physicists, Sir William H (1862–1942) and Sir William L. (1890–1971), joint Nobel Prize 1915) developed the x-ray spectrometer to explore atomic structure. In 1913, another British physicist, Henry G. J. Moseley (1879–1915), used x-ray lines to establish the atomic numbers and thus the correct position of elements in the Periodic Table. X-rays contributed greatly to science by permitting exact determination of the structure of molecules; since crystal lattices act as diffraction gratings, their structure can be ascertained by measuring the angles at which the x-rays are refracted as they pass through the crystal. X-rays are also used as diagnostic aids and in tissue therapy; they kill diseased tissue and can also be harmful to healthy tissue, especially since their effects are retained in the body. Thus, all but medically necessary exposure to x-rays should be avoided. *See also* radiation.

xylene. A derivative of benzene in which two of the hydrogen atoms are replaced by methyl groups, $C_6H_4(CH_3)_2$. It can be distilled from petroleum or coal-tar, or made from straightrun gasoline by catalytic reactions or from toluene by transalkylation. Xylene occurs in three isomeric forms (ortho-, meta-, and para-), the commercial product being a mixture. The ortho- isomer is the source of the plasticizer phthalic anhydride and the para-isomer is oxidized to form terephthalic acid. Xylene is also a useful solvent and intermediate for dyes and pharmaceuticals. While less flammable than benzene, it should be protected from direct ignition sources. The word is adapted from the Greek term for wood (*xylos*), presumably because it was first found there.

Y

Y Symbol for the element yttrium, the name being derived from Ytterby, a town in Sweden. Two other elements, terbium and ytterbium, also were named from this town.

Yb Symbol for the element ytterbium, the name having the same derivation as yttrium.

yeast. A biochemically active material composed of fungus-like, single-celled organisms which are high in protein. It is able to convert carbohydrates into water and carbon dioxide in the presence of air, and into alcohol and carbon dioxide in the absence of air (fermentation). It is an important source of vitamins. Its activity is largely due to the enzyme zymase. Common commercial types used in the food industries are bakers' yeast and brewers' yeast. *See also* fermentation; brewing.

ylang-ylang oil. *See* essential oil.

Young's modulus. *See* elasticity.

ytterbium. An element.

Symbol	Yb	Atomic Wt.	173.04
State	Solid	Valence	2,3
Group	IIIB	Isotopes	7 stable
	(Lan-		
	thanide		
	Series)		
Atomic No. 70			

A rare-earth metal, m.p. 816°C (1501°F), obtained from several ores (monazite, gadolinite) found in Idaho, Florida, Scandinavia, Australia, etc. It is subject to attack by acids and is soluble in ammonia. It forms compounds with oxygen, the halogens, and with carbon and other nonmetals. Its most important use is as a doping agent for garnet crystals in lasers. It has limited use in catalysts composed of mixtures of rare earths. It is not toxic and has several radioactive isotopes, one of which is used in radiography of metal parts and in medical diagnostic work.

yttrium. An element.

Symbol	Y	Atomic Wt.	88.9059
State	Solid	Valence	3
Group	IIIB		
Atomic No. 39			

Though it is not one of the rare-earth metals, yttrium, m.p. 1509°C (2748°F), is very similar to them and is always associated with them in such ores as monazite and gadolinite (Scandinavia, Idaho, Australia, Brazil). It forms compounds with oxygen and the halogen elements, and with carbon. The major use of the metal is as an additive or alloying ingredient with a number of other metals; additions of small percentages increase resistance to oxidation at high temperatures and have other beneficial effects upon the alloy systems to which it is added. Yttrium oxide is used with europium in phosphors for the red color in TV tubes, in yttrium-iron garnets for microwave filters, and in nuclear technology. Yttrium has only one stable form and is nontoxic. It is classed as a rare metal.

Z

zein (pronounced zee-in). The characteristic protein of corn, from which it is extracted on a commercial scale. Its molecule contains the amide group $CONH_2$. It has good nutritional value and thus is used in coatings for capsules and for dietetic and special food preparations. It also has some application in paper coatings and in the textile field, both as a sizing agent and as a fiber for blending with wool.

zeolite. Clays or clay-like structures (aluminosilicates) which have ion-exchange properties and exhibit the behavior of molecular sieves. They are crystalline compounds composed of tetrahedrons of silicon oxide in which aluminum atoms have replaced some of the silicon. The crystalline complex forms channels and cages of varying size which are capable of exerting an ionic selectivity or "sieving" effect when placed in salt solutions. Thus, zeolite ion-exchange units are extensively used in water conditioning to remove calcium and magnesium ions by replacing them with sodium ions. These units can be regenerated by treatment with sodium chloride solutions. Modified zeolites are now widely used in the petroleum refining industry as cracking catalysts, almost to the exclusion of older types. *See also* ion exchange; molecular sieve.

zero group. The gases comprising the righthand column of the Periodic Table (or in older variations, the left-hand column). It is variously designated as the "inert gas," "rare gas," or "noble gas" group. The first three elements of this group do not enter into chemical combination. The zero group was so named at a time when all its members were thought to have zero valence; but it was later found that the last three members (krypton, xenon, and radon) have limited compound-forming ability. The most reasonable name for this group is therefore the noble gas group. *See also* noble; inert.

Ziegler catalyst. A type of organometallic catalyst named after its discoverer, Karl Ziegler (b. 1898), a German chemist, Nobel Prize 1963 shared with Giulio Natta (1903–1979) an Italian chemist. There are many variations in composition, but in general these catalysts are made up of an alkyl metal (the metal being in Groups I, II, or III of the Periodic Table) plus a halide of a transition metal (Groups IV to VIII). An example would be triethylaluminum, $(C_2H_5)_3Al$, plus titanium dichloride ($TiCl_2$) dispersed in a hydrocarbon solvent. These catalysts are of the stereospecific type, permitting close control of the chemical structure of the polymers formed; their discovery was an invaluable aid to the plastics industry. *See also* stereospecific; polypropylene.

zinc. A element.

Symbol	Zn	Atomic Wt.	65.38
State	Solid	Valence	2
Group	IIB	Isotopes	5 stable
Atomic No.	30		

Zinc, m.p. 419°C (786°F), is widely distributed in the earth's crust, usually as a sulfide ore; large deposits are in British Columbia, Ontario, Quebec, Utah, Colorado, and Australia. The metal is obtained by pyrometallurgical and electrochemical techniques. The most notable chemical property of zinc is its pronounced electropositive nature; it is third among the common metals in this respect, exceeded only by aluminum and magnesium. For this reason, it is particularly effective as a protective coating for steel (galvanized coatings), in which the phenomenon called sacrificial protection occurs. Other important applications are in alloys with copper (brass) and with aluminum (die castings). Neither the metal nor its inorganic compounds are toxic. The most important of the latter are zinc oxide and zinc sulfide, used in paint pigment (lithopone), and zinc chloride used in dry cell batteries, textile printing, and as a fireproofing agent for timber and growing trees. Organic compounds of zinc are used as polymerization catalysts and ultra-accelerators of vulcanization; diethylzinc is spontaneously flammable. *See also* galvanizing; sacrificial; brass; zinc oxide.

zinc oxide. A heavy white powder, (ZnO) produced by several methods, chiefly, the burning of zinc vapor in air. It is a strong ultraviolet absorber. It has a number of important industrial uses: as a reinforcing agent and white pigment in rubber products, filled plastics, etc.; as a paint pigment; and as the basis of emollient creams and ointments. It has the unique ability to activate rubber accelerators more efficiently than any other substance; were it not for this property, rubber vulcanizates suitable for most service conditions might never have been possible.

zinc sulfide. *See* lithopone; phosphor.

zirconium. An element.

Symbol	Zr	Atomic Wt.	91.22
State	Solid	Valence	4
Group	IVA	Isotopes	5 stable
Atomic No.	40		

Zirconium, m.p. 1850°C (3362°F), is a strong metal with good mechanical properties. It is found in nature as zircon, zirconium silicate, in beach sands and the beds of streams and lakes, chiefly in Australia and Florida, and is always associated in nature with hafnium. After separation of the hafnium by solvent extraction of the tetrachlorides of the two elements, it is reduced to metal by magnesium (Kroll process). It has a density of 6.5 g/cc and while chemically reactive has excellent resistance to corrosion by most acids and alkalis. It has very low neutron absorption capability and thus an important application is in cladding uranium fuel elements in nuclear reactors, usually in the form of a series of alloys trademarked "Zircaloy." It is also used in the manufacture of corrosion-resistant equipment for the chemical industries. In the form of sponge, powder, or scrap, it will ignite spontaneously at room temperature. Its compounds have many miscellaneous applications in ceramics, catalysts, and special alloys; the sulfate is used in tanning white leather. Neither the metal nor its compounds are toxic. *See also* Kroll process.

Zn Symbol for the element zinc, the name being taken directly from the Greek.

zone refining. *See* refining.

Zr Symbol for the element zirconium, the name being taken from the Greek word *zircon*.

Zsigmondy, Richard (1865–1929). A native of Austria, Zsigmondy received the Nobel Prize in chemistry in 1925 for his work in the field of colloid chemistry, which was initiated by his interest in ruby glass (a colloidal gold suspension). His most important contribution to chemistry was his invention of the ultramicroscope (with Siedentopf) in 1903. By illuminating the sample with a light beam at right angles to the incident light in a compound microscope, the presence of particles in a suspension as small as 3 millimicrons could be detected due to the light-scattering effect of the particles. This phenomenon had been discovered by John Tyndall a few years earlier, but Zsigmondy was the first to apply it to microscopy. The ultramicroscope became of immeasurable value in the study of colloidal suspensions. *See also* Tyndall effect.

zymase. Derived from the Greek word for *ferment*, this term designates a major enzyme occurring in yeast which catalyzes the fermentation of sucrose to alcohol and carbon dioxide.